FOUNDATIONS
OF
NEURAL
DEVELOPMENT

FOUNDATIONS
OF
NEURAL
DEVELOPMENT

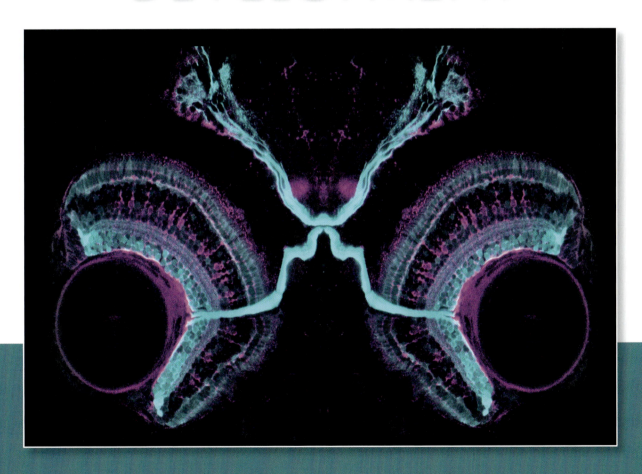

S. MARC BREEDLOVE
MICHIGAN STATE UNIVERSITY

Sinauer Associates, Inc. Publishers
Sunderland, Massachusetts

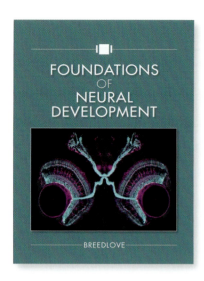

ABOUT THE COVER

Developing zebrafish visual system by Kara Cerveny
This frontal cross-section of an approximately 4.5 day-old zebrafish embryo shows the two eyes sending axons across the midline to the midbrain at the top of the photograph. The fish carried a transgene controlled by the *atoh7* promoter that caused all retinal ganglion cells and most cone photoreceptors to express green fluorescent protein (GFP), which has been immunostained blue. Glial fibrillary acidic protein (GFAP), which is found in glia, was immunostained purple. The photomicrograph was taken by Professor Kara L. Cerveny of Reed College.

FOUNDATIONS OF NEURAL DEVELOPMENT

Copyright © 2017 by Sinauer Associates, Inc.

For information, address
Sinauer Associates, Inc. P.O. Box 407, Sunderland, MA 01375 U.S.A.
Fax: 413-549-1118
E-mail: publish@sinauer.com
Internet: www.sinauer.com

Library of Congress Cataloging-in-Publication Data

Names: Breedlove, S. Marc.
Title: Foundations of neural development / S. Marc Breedlove, Michigan State University.
Description: Sunderland, Massachusetts : Sinauer Associates, Inc., [2017] | Includes bibliographical references and index.
Identifiers: LCCN 2016056386 | ISBN 9781605355795
Subjects: LCSH: Developmental neurobiology. | Neural circuitry. | Neural networks (Neurobiology)
Classification: LCC QP363.5 .B74 2017 | DDC 612.6/4018--dc23
LC record available at https://lccn.loc.gov/2016056386

Printed in U.S.A.
6 5 4 3 2 1

For Cindy

BRIEF CONTENTS

CONTENTS

■ **CHAPTER 2**
Coordinating Fates
DEVELOPMENT OF A BODY PATTERN   51

■ **CHAPTER 3**
Upward Mobility
NEUROGENESIS AND MIGRATION   87

■ CHAPTER 4
Seeking Identity
NEURAL DIFFERENTIATION 127

■ CHAPTER 5
Feeling One's Way
AXONAL PATHFINDING 151

CHAPTER 6
Making Connections
SYNAPSE FORMATION AND MATURATION 185

CHAPTER 7
Accepting Mortality
APOPTOSIS 221

INTERLUDE
THE EMPIRICISTS STRIKE BACK   267

CHAPTER 8
Synaptic Plasticity
ACTIVITY-GUIDED NEURAL DEVELOPMENT   273

CHAPTER 9
Fine-Tuning Sensory Systems
EXPERIENCE-GUIDED NEURAL DEVELOPMENT   305

CHAPTER 10
Maximizing Fitness
SOCIALLY GUIDED NEURAL DEVELOPMENT   333

■ EPILOGUE
IMMANUEL KANT AND THE CRITIQUE OF PURE REASON   373

PREFACE

The growing number of undergraduate neuroscience majors in colleges and universities has prompted Instructors to create a curriculum that covers the breadth of this vibrant science. What neuroscience courses should be offered after the introductory survey course? An argument could be made that developmental neuroscience is the most active and dynamic subfield of the discipline simply because at the annual meeting of the Society for Neuroscience, there are more abstracts submitted under "Development" than any other theme. Perhaps an even stronger argument for the importance of developmental neuroscience is its scope: to understand how a single microscopic cell, a human zygote, divides and grows to form the most complicated machine in the universe, the brain.

At Michigan State University we needed to create a truly comprehensive undergraduate course in Developmental Neuroscience for our neuroscience majors. Having taught the topic to graduate students, both at MSU and at UC Berkeley, for over 20 years, I was drawn to the challenge of creating a new undergraduate course. When team-teaching the course at Berkeley, we used *Principles of Neural Development*, an outstanding textbook by Dale Purves and Jeff Lichtman, published in 1985. Sadly, this wonderful book was never revised. The other neural development textbooks published in the subsequent 30 years seem more appropriate for graduate students (or professors!), so I decided to write a developmental neuroscience text deliberately aimed at undergraduates.

Like virtually all books on developmental neuroscience, the topics are presented in roughly the chronological order in which they appear in brain development. But beyond that I wanted to offer several perspectives that are, as far as I know, unique to this book, reflecting overviews that I've found to be helpful for students. These additional perspectives offer a means of organizing the material into a narrative that makes sense, and should help readers learn and remember the material by showing them the really big picture.

An Evolutionary Perspective

The book emphasizes four watershed events, each required to enable those that came after, that led to the evolution of the human brain:

1. The departure from a mosaic control of cell differentiation to relying upon *cell-cell interactions to direct cell differentiation*. This shift in developmental strategy led to the evolution of extensive modes of cell-cell communication, some relying on direct contact between cell membranes and others relying on diffusible signals to guide development. These interactions between cells not only establish an overall body plan (Chapters 1–2), but also direct cell division and migration (Chapter 3), the differentiation of neurons and glia (Chapter 4), the path of axonal growth cones (Chapter 5), the initiation and maturation of synapses (Chapter 6), and the decision about which cells will be discarded by apoptosis (Chapter 7). These topics make up the bulk of the literature in developmental neuroscience and of the book.

2. The elaboration of cell-cell communication eventually led to mechanisms by which the *electrical activity* of neurons affects other cells, varying the strength or number of synapses between them, or dismantling synapses altogether (Chapter 8). By itself, this activity-guided synaptic plasticity might have been simply one more way for developing cells to affect one another, each relying on information in the genome and spontaneous activity to guide their differentiation. But this mechanism also set the stage for another rich influence on the growing brain.

3. The development of activity-dependent synaptic plasticity enabled *experience-dependent guidance of development* as sensory information began to control synaptic fate. Among other things, this mechanism permitted organisms to learn about their environment and to remember what they'd learned, but it also served to fine-tune developing sensory systems, optimizing them to detect important events in the

physical world (Chapter 9). Thus information from outside the brain, and outside the genome, began shaping development.

4. Experience-dependent neural development in turn enabled an enormous benefit for those species, primarily birds and mammals, that evolved an extended period of parental care and juvenile interactions, so that *social experience* began shaping brain development. Obviously, such socially guided experience is crucial for human intelligence, language, and culture, but social guidance of brain development is crucial for maximizing fitness in many other species, too (Chapter 10).

This progressive expansion of the sphere of influence on the developing brain explains how it's possible to get so much information into a brain using only about 10,000 different genes. The genes alone don't have nearly enough information to specify the wiring of a well-functioning brain, so instead they unfold a program that *pulls in information* from the current environment and even, through inter-generational social interactions, past events. This progression of topics offers an organizing narrative that I think anyone can understand and use as they study neural development. It helps readers to see how all the detailed mechanisms lead to the same goal of a well-formed brain, which is essential for an organism to survive and reproduce.

A Philosophical Perspective

In addition to the evolutionary perspective above, I've included a brief introduction to philosophy as it relates to the study of brain development. The quest of neuroscientists to understand how the zygote divides and grows to form the brain mirrors a much older quest of philosophers. Over two thousand years ago, Plato wrestled with the question of how a newborn can possibly come to learn anything if it starts life knowing *nothing*, and other philosophers fervently discussed this issue for the subsequent millennia. Although they didn't know it, philosophers like Plato, Descartes, Locke, Hume, and Kant, who were wrestling with *epistemology*, the study of knowledge and how we gain it, were in fact also asking how an effective brain can possibly develop before and after birth. They were accidental developmental neuroscientists who knew nothing about neurons, axons, synapses, and the like. Conversely, from the late nineteenth century to today, scientists studying the developing nervous system discovered, in broad strokes, answers to the questions that the philosophers asked. And just as the philosophers didn't know they were asking questions about the nervous system, the developmental neuroscientists, for the most part, didn't seem to realize they were answering questions about epistemology.

To illustrate how these two intellectual histories parallel one another, I relate the history of the philosophical debate in three brief asides: a Prologue, an Interlude (after Chapter 7), and an Epilogue. The developmental neuroscience chapters sandwiched in between illustrate how the philosophers' conclusions were confirmed by scientists' findings. The chapters sometimes refer to the philosophical history, but students need not know that history to understand the neuroscience. Instructors who find the parallels uninteresting or distracting can simply ignore the philosophical installments.

These two perspectives are also complementary—the philosophers used logic to foreshadow the mechanisms that had to arise by natural selection to produce the human mind. As you might have gathered by now, these different perspectives are also meant as a respite from what can otherwise appear to be a textbook crammed with facts. While I've tried to be very, *very* selective about which studies to present, surveying only a few of the most important findings and principles of neural development, there are still nearly a thousand citations. That's a lot of detail! My hope is that by showing how these detailed mechanisms of cell signaling, axonal guidance, and synaptic pruning are part of a larger, coherent story, readers will retain the narrative thread, seeing how the details fit into a larger whole.

Making the Material Accessible

I included several other pedagogical features to make the material accessible to a wide range of readers. Each chapter begins with a **vignette** about a human condition relevant to that topic, which the ensuing material will shed light upon. To further pique readers' interest, I consider several rather applied topics not typically covered in neural development books, such as human sexual orientation, kerfuffles over terminology ("commitment," "innate," "epigenetic"), and the controversy over racial differences in average IQ. Each chapter has several **Researchers at Work**, a feature in which I slow down the pace, deliberately breaking down a classic study to describe the overarching question, the specific hypothesis, the test(s) devised, the results obtained, and the conclusion(s) drawn. I hope this feature helps readers not only understand the experimental process, but also appreciate the underlying logic of science. Most chapters also include one or more **Boxes**, special topics of interest, including episodes in the history of this field. The **How's It Going?** feature appears at the end of each section and offers a few review questions for readers to check their comprehension of the material. If you really were engaged in the reading up to that point, you should be able to answer the questions readily. If you're not sure what the answers are, then you may do well to go back and review the preceding material. Another reader-friendly feature is the boldfacing of **Keywords** and a **Marginal Glossary** (in addition to a glossary and index in the back of the book). Each chapter ends with a **Summary** to help readers self-review. Finally, I've tried to establish a conversational tone throughout, with an active voice, including speculative consideration of various hypotheses ("thinking out loud"), intended to engage the reader. I want to be clear—to accurately convey what I mean—but I feel no need to use long, impressive words when short, widely understood words will do.

I hope these efforts will help you to enjoy reading the pages that follow and to understand this fascinating story, which is far from complete. Maybe reading this book will encourage you to become a neuroscientist and make your own contribution to this ultimate human endeavor, begun thousands of years ago, to understand the origins of our brain.

ACKNOWLEDGMENTS

A lot of people worked long hours to make this book possible. Sinauer Associates Editor Syd Carroll was immensely supportive and encouraging, especially when I needed it most. Production Editor Kathaleen Emerson still amazes me with her ability to keep track of all the figures, boxes, citations, and revisions. Copy Editor Lou Doucette's sharp eye caught many problems that went right over my head (she even tried her best to help me understand the conventions about gene names). Photo researcher David McIntyre worked his usual magic, finding photos and figures and keeping track of all those permissions, with the assistance of Michele Bekta. Joanne Delphia and Chris Small have again assembled a beautiful visage for the book that brings me such pleasure, and Elizabeth Morales did a fabulous job drawing the figures. MSU students Eloise Faust, Eliza Judge, and Alla Kedzierski read a late draft of the book, offering many helpful suggestions. The text also benefited from the selfless efforts of many academic reviewers, including:

Ann Aguanno, *Marymount Manhattan College*
Ethan Bier, *University of California, San Diego*
Sara Clark, *Tulane University*
Gedeon Deák, *University of California, San Diego*
Mark Emerson, *The City College of New York*
Nancy Forger, *Georgia State University*
Alexander Jaworski, *Brown University*
James Jontes, *The Ohio State University*
Ronald Oppenheim, *Wake Forest School of Medicine*
Sarah Pallas, *Georgia State University*
Anita Quintana, *The University of Texas at El Paso*

Finally, my wife Cynthia L. Jordan, to whom this work is dedicated, never stopped believing in the value of this book. I could never have done it without her love for the field and for me.

MEDIA & SUPPLEMENTS
to accompany
FOUNDATIONS
OF
NEURAL DEVELOPMENT

FOR THE STUDENT

Companion Website
(sites.sinauer.com/fond)

The *Foundations of Neural Development* Companion Website contains a range of media and review resources to help students learn the material presented in each chapter of the textbook and to visualize some of the key processes discussed. The site includes the following resources:

- Chapter outlines
- Chapter summaries
- Animations: Detailed animations that cover some of the key processes presented in the textbook.
- Videos: Links to fascinating videos that demonstrate the processes and concepts of neural development.
- Flashcards: An easy way for students to learn and review the key terms introduced in each chapter.
- Glossary
- News Feed: A continuously updated feed of links to science news articles relevant to neural development.

FOR THE INSTRUCTOR

Instructor's Resource Library

The *Foundations of Neural Development* Instructor's Resource Library (available to qualified adopters) includes a variety of resources to aid you in the development of your course and the assessment of your students. The IRL includes the following:

TEXTBOOK FIGURES & TABLES All of the figures (including photos) and tables from the textbook are provided as JPEGs, all optimized for use in presentations. Complex figures are provided in both whole and split versions.

POWERPOINT PRESENTATIONS A PowerPoint presentation containing all figures and tables, with titles and full captions, is provided for each chapter.

VIDEOS A collection of fascinating segments from BBC programs that illustrate important concepts from the textbook.

TEST BANK The Test Bank consists of a range of questions covering key facts and concepts in each chapter. Multiple choice and short answer questions are included, and all questions are ranked according to Bloom's Taxonomy and referenced to specific textbook sections.

COMPUTERIZED TEST BANK The Test Bank is provided in Blackboard's Diploma format (software included). Diploma makes it easy to assemble quizzes and exams from any combination of publisher-provided questions and instructor-created questions.

COURSE MANAGEMENT SYSTEM SUPPORT The Test Bank is also provided in Blackboard format, for easy import into campus Blackboard systems. In addition, using the Computerized Test Bank, instructors can easily create and export quizzes and exams (or the entire test bank) for import into other course management systems, including Moodle, Canvas, and Desire2Learn/Brightspace.

eBOOK

Foundations of Neural Development is available as an eBook, in several different formats, including Vital-Source, RedShelf, Yuzu, and BryteWave. The eBook can be purchased as either a 180-day rental or a permanent (non-expiring) subscription. All major mobile devices are supported. For details on the eBook platforms offered, please visit www.sinauer.com/ebooks.

THE RATIONALIST PHILOSOPHERS

All of humanity's written work is subjected to a ruthless selection process: very few things written today will still be of interest to anyone 5 years from now, including, alas, college textbooks. Even best-selling novels, which account for only a tiny fraction of the novels published (which are themselves a tiny fraction of the novels submitted to publishers), are rarely read 25 years later. Of the many works written in Greece thousands of years ago, the very few available to us are those that were preserved, across hundreds of generations, because they continued to be regarded as valuable.

ALL I KNOW IS THAT I KNOW NOTHING

These ancient Greek writings survived because they reflected an exciting time of new ideas. There arose people who seemed eager to *question everything* about themselves and the world around them. Some practiced **philosophy**, the study of the fundamental nature of knowledge, existence, and reality itself. Among the ancient philosophers, the most famous today are Aristotle (384–322 BCE) and Plato (427–347 BCE), who rejected traditional and religious explanations for the world. Both regarded the Greek gods as metaphors or fairy tales. They also had a healthy skepticism about the way things *seem* to be versus the way things *are*. They knew our senses are not always reliable. For example, Aristotle was interested in visual illusions, such as the so-called waterfall effect: if you stare at falling water for a minute or so and then look away, stationary objects appear to stretch out like drawings on an inflating balloon. You can see this "motion aftereffect" and many other visual illusions at Michael Bach's wonderful website: www.michaelbach.de/ot.

Plato questioned our senses even more deeply than Aristotle. Knowing that under certain circumstances people may hallucinate entire events that never happened, Plato dared to think that *everything* we see may be illusion. I should clarify here that almost all of Plato's writings describe dialogues between his wise teacher, Socrates (469–399 BCE), and other people. This means it is impossible to separate Socrates' views from Plato's, and so, by convention, we attribute them to Plato, since he was the one who wrote them down.

Plato (427–347 BCE) and his great teacher, Socrates (469–399 BCE) (Courtesy of Marie-Lan Nguyen.)

In his famous **parable of the cave**, Plato proposes a "thought experiment." What if people were raised in a cave with a fire constantly blazing in the mouth of the cave? Imagine these people were chained to a rock such that the only thing they ever saw was a wall in front of them on which were cast shadows of people and animals passing in front of the fire behind them (**FIGURE P.1**). The prisoners would believe the shadows represented all there was in the world. If one of them were unchained and allowed to see the fire at the mouth of the cave and the three-dimensional people and animals passing by the fire and casting shadows on the wall, she would be astonished at this deeper reality. Returned to chains, that person would forever have a different concept of the world and would have a hard time convincing any of her neighbors that they were viewing mere shadows, not reality itself (Plato's *Republic*, book VII).

This deep skepticism about our perceptions influences Plato's ideas about **epistemology**, the study of knowledge and how we gain it. He reports on many of Socrates' dialogues with other people that reveal that they knew something without being aware of it. In these *Socratic dialogues*, Socrates almost never provides information, but rather keeps asking the other person questions and using that person's answers to bring up the following questions. In one case Socrates asks a slave boy a question about rectangles that someone who knew the Pythagorean theorem would know, but of course the uneducated boy has never heard of Pythagoras or the theorem. But by replying to a series of questions that Socrates poses, the boy eventually concludes that the square of the diagonal is indeed equal to the sum of the squares of the two sides, and that this must be true of all

philosophy The study of the fundamental nature of knowledge, existence, and reality itself.

parable of the cave Plato's thought experiment about the concept of reality that people would have if they grew up seeing only the shadows of objects in the world.

epistemology The study of knowledge and how we gain it.

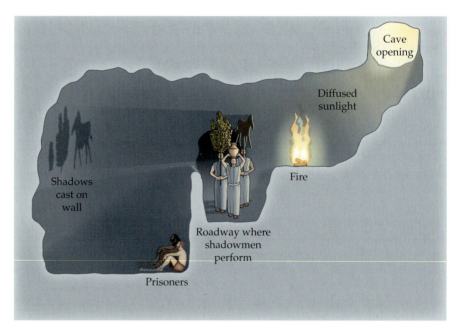

FIGURE P.1 Plato's parable of the cave A revolutionary thinker, Plato thought that anything and everything we learned through the senses might be false.

FIGURE P.2   Is there a platonic Form of dog? Despite their tremendous variation in size, shape, and color, we easily recognize dogs. (Left © iStock.com/ Paul Shlykov; others © iStock.com/Global P.).

rectangles, everywhere (Plato's *Meno*). Socrates concludes that this knowledge about geometry was hidden inside the slave boy's mind all along and the boy was simply *recollecting* what he knew (but didn't know he knew). Furthermore, as the boy had never been instructed on any of the matters Socrates asked him about, he must have *always* known them, and so had always known the Pythagorean theorem. Thus Plato offers the radical idea that knowledge is not something we *acquire*, but something everyone, even uneducated slaves, are *born with*! Likewise, Plato concludes that our understanding of numbers, of equality, of beauty, are not taught to us, but are somewhere inside us before we are born. He quotes Socrates saying, "I cannot teach anybody anything, I can only make them think."

Plato's notion that knowledge is inside us at birth, and not acquired by experience of the world, is reinforced in his writing about *Forms*. We all know what a "bed" is, even though beds come in many different sizes and shapes. In fact, few of us have ever seen two or more beds that are exactly the same in every way, yet we know they are all beds. Likewise, tonight we may see a bed that is unlike any we've seen before, but we will still recognize it as a bed. How did we learn what a bed is? It can't be because this bed is exactly like beds we've seen before, because it's not. Likewise, there are bewildering varieties of tables, horses, and dogs (**FIGURE P.2**), yet we have no trouble recognizing each as a member of that class of things. Plato concluded that every object and every abstract concept, like justice or beauty, is simply reflecting a **Form**, an eternal, perfect "blueprint" of that object or concept that we recognize in its imperfect, highly variable real-life example. Just as the prisoners in the cave must rely on the shadows of real objects to understand their world, we must rely on our perception of imperfect Forms to understand ours. So how do we come to recognize anything, anywhere? Plato's explanation is again that our ideas about the perfect Forms are inside us at birth. We see an object, detect its likeness to the eternal, perfect *Form* of that object in our mind, and so recognize it as another particular, perhaps unique, example of that Form: bed, horse, beauty.

Form   Also called *Platonic Form*. An eternal, perfect "blueprint" of an object or concept that we recognize in its imperfect, highly variable real-life examples.

René Descartes (1596–1650) working at his desk; or is he merely dreaming that he is at his desk? (Image © Mansell/The LIFE Picture Collection/Getty Images.)

rationalist philosophers
Philosophers who believed that the only way to understand the true nature of the world is by use of the intellect and reason, rather than the senses.

In what, for Plato, is an atypically mystic resolution, he concludes that all this knowledge (the Pythagorean theorem, the Forms of numbers, equality, beds, tables, justice, beauty) is stored in a person's *soul*, which existed eternally before the person was born (Plato's *Phaedo*).

Plato's writings were cherished by Europeans through Medieval and Renaissance eras, so the philosophers of later times were all aware of his ideas. For example, René Descartes (1596–1650) explicitly vowed to ignore what previous philosophers had claimed, but of course he had read all of Plato's work, and it's clear that he was influenced by it. Like Plato, Descartes distrusted his own senses, not only because there were times when he knew they misled him, but also because the appearance of objects could change, as when heating wax makes it change from opaque solid to transparent liquid. One night Descartes was working at his desk, writing philosophy, when he suddenly woke up in bed to find that he had been dreaming it all. Yet his experience at the desk had seemed so real that Descartes wondered whether *everything* that he regarded as real life might simply be a very vivid dream.

Descartes' extreme skepticism led him eventually to admit that there was only one thing he was absolutely certain of—his famous phrase "I think, therefore I am" (in Latin, *cogito ergo sum*). Even when he doubted whether this was true, he realized that doubting can only happen if there is a doubter, confirming his existence. From this beginning, Descartes used deduction to reach other conclusions, some of which we might find questionable today, such as that a benevolent God exists. But his doubts still resonate, such as whether his body existed. I can't disprove the idea that our entire world, including everything we perceive, is just a computer simulation. So I may doubt that my body actually exists, but I know that I'm the doubter (even if "I" am merely subroutine in a computer program, or a hallucinating brain in a vat of fluid, kept alive by machines).

Descartes' skeptical ideas about objects in the outside world also call to mind Plato's Forms. Descartes uses an analogy that if you heat up one stone and put it next to others, those other stones can never be any hotter than the original. Just so, whatever we *perceive* of an object can only be less real than the object itself, just as with Plato's cave dwellers living in a world consisting of mere shadows cast by objects themselves. Descartes notes that, like Plato's slave boy who discovered that he already "knew" the Pythagorean theorem, he feels, "I do not so much appear to learn anything new, as to call to remembrance what I before knew" (Descartes' *Meditations* V[4]).

THE BENEFITS OF HAVING AN IMMORTAL SOUL

Also like Plato, Descartes relied on a *soul* that exists before we are born, providing us all this information as a gift from a benevolent God. A little later, Gottfried Leibniz (1646–1716) in Germany would propose that we each possess a *monad*, an irreducible particle reflecting the will of an all-knowing God, which sounds an awful lot like a soul. Like Plato's and Descartes' soul, the monad allows us to recognize objects and theoretical notions like beauty, to know that God exists, and to know right from wrong. Because Plato, Descartes, and Leibniz emphasized that reason, the use of the intellect rather than the use of our senses, was the only pathway to knowledge, they are known as **rationalist philosophers**.

Thus as the Renaissance ended, these rationalist philosophers thought it was obvious that we are born with a storehouse of knowledge, provided by our soul, which was present before we were born. As we grow from baby to toddler to child to adult, we "uncover" more and more of this information that was packed inside us. Thus the rationalists neatly solve the paradox of how we can learn anything if we begin life knowing *absolutely nothing*. How does someone who knows nothing about space ever learn the Pythagorean theorem, or recognize beds? If you know nothing about time, how can you understand how one event can cause another? How do you recognize a face as beautiful when there's no objective definition of what makes a face beautiful or plain? Easy—your soul already knew all that stuff before you were born.

As neuroscientists, we may find this explanation, however tidy, pretty unconvincing. Without any way to detect or measure a soul, we would say that knowledge must be, at least partially, stored in the brain. And the brain does indeed exist before we are born, but only by about 8–9 months, not eternally. By saying this knowledge is inside eternal souls (or monads), rationalists avoid the question of how the souls (or the knowledge) ever came about, but we neuroscientists don't have that luxury. An egg is fertilized, a baby is born, and in a few years the child can, through a series of questions, be shown to "know" the Pythagorean theorem, even if she never takes a geometry course. Where did that knowledge come from?

WHAT DOES ALL THIS PHILOSOPHY STUFF HAVE TO DO WITH THIS BOOK?

As we'll see in the next several chapters, today we know that thousands of genes, each inherited from our parents, guide the initial construction of the brain before birth, a process that is absolutely critical for its proper function. In fact, if those genes don't function properly to produce a reasonably well-formed brain, the individual will never survive to birth. In many ways, this understanding of the role of genes in developmental neuroscience vindicates the views of rationalist philosophers who, knowing very little about inheritance and absolutely nothing about genes, reasoned that we must begin life with *some* knowledge inside us. In English, "knowledge" is something that is possessed by a living individual, while "information" can also be tucked inside an inanimate object, like a book, a computer, or a stretch of DNA. But I don't think the rationalists would have had any problem with saying that information, rather than knowledge *per se*, was tucked inside your soul before you were born. Even the rationalist notion of an *eternal* origin for this information is not too far off the mark: those genes we inherited for constructing the prenatal brain first arose, in one form or another, about half a billion years ago. For us puny humans, who rarely live a hundred years and have existed as a species for just over a million years, 500 million years is pretty darn close to eternity.

In the next seven chapters we'll see how those nearly eternal genes direct the prenatal construction of the human brain, in many ways vindicating the rationalist view of epistemology. What's more, these early processes do not rely on any sensory information, illusory or otherwise, because they take place before any sensory organs exist, or at least before they are functional. Following **Chapter 7,** we'll return to philosophical considerations of epistemology in an Interlude about the empiricist philosophers, British Enlightenment thinkers who emphatically disagreed with the rationalists, insisting that we must rely on our senses and experience to learn about the world.

CHAPTER

The Metazoans' Dilemma
CELL DIFFERENTIATION AND NEURAL INDUCTION

JOINED FOR LIFE Abby and Brittany Hensel are conjoined twins who share a pair of arms and legs but have separate heads, hearts, lungs, and stomachs. From the waist down they share a typical female body, with Brittany controlling the left arm and leg while Abby controls the right. Born in 1990, they have grown up like many other Americans, learning to ride a bike, getting driver's licenses, graduating from high school and college. Along the way they reached out to the world by participating in documentaries and a reality TV show to demonstrate that they are happy, normal people. Many people find Brittany and Abby's story inspiring, the way they work together to do everyday tasks, and the way they talk almost as one person, filling in or finishing each other's sentences. Like many other conjoined twins their joined state seems to be a problem, not for the twins, their family, or friends, but for people who see them in public (Dreger,

2005). Should they have been separated at birth, which would have required that one of them die, to reduce the discomfort of strangers?

Conjoined twins occur either because a single embryo somehow partially splits and begins making duplicates of some body parts or because two separate embryos somehow get merged together and each constructs its own version of some parts, while they work as a unit to form single versions of other structures. Whichever was the case for the Hensel twins, how did these two upper torsos come to be merged to a single pair of legs? Their two spinal cords descend from their brains and then merge before joining a single pelvis. No surgeon could ever accomplish the seamless merging that the twins underwent before birth. How did this happen, and what can conjoined twins tell us about embryonic development?

CHAPTER PREVIEW

A watershed event in evolution was the arrival of multicellular animals, which quickly dwarfed their microscopic single-celled ancestors and eventually diverged from one another with a dazzling array of specialized body parts—digestive systems, fins, eyes, ears, arms, wings, and even sonar systems. These various appendages require that individual cells become specialized in form and function. In this chapter, we'll see how twentieth-century biologists came to understand the role of genes in directing cells to specialize, to become different from one another, in a process called *cell differentiation*. To work properly, the various structures formed by these specialized cells must arise in the right parts of the body, so the next question biologists tackled was how a body plan is organized,

protozoan A single-celled microscopic animal.

metazoan A multicellular animal.

nerve net A system of interconnected neurons found in jellyfish and related animals.

cellular differentiation The process by which individual cells in an organism become progressively more specialized and different from one another.

distributing the various specialized sets of cells to form a body, be it fish, squirrel, or human, that survives and successfully competes to reproduce. We'll see that natural selection solved the puzzle of how to produce specialized structures, each in its proper place, by having *cells extensively communicate in order to coordinate with one another*. In many cases, this *cell-cell interaction* is a competitive process, where one cell "wins out" to take on the form and function to accomplish one task, while its neighbor cell takes on a very different fate. Thus the body, including the nervous system, arises as each of the billions of cells sends instructions to its neighbors, and responds to its neighbors in turn, in deciding whether to form gut, skin, neuron, or eye.

■ 1.1 ■

METAZOANS EVOLVED THE ABILITY TO PRODUCE CELLS WITH VERY DIFFERENT FUNCTIONS

The common ancestor of all the animals is thought to have been a microscopic single-celled organism, a **protozoan** (from Latin *proto*, "earliest," and *zoa*, "animal"), resembling a modern-day species that moves around by wiggling a hairlike structure, a flagellum. This type of protozoan reproduces by simply splitting in two, each offspring of the division swimming off on its own. But at some point natural selection began favoring individuals whose cells stayed together after dividing to form a multicellular body. We presume that in these first multicellular species, every cell was alike in structure and function. But eventually some of the cells in these clumps began taking on specialized functions, as the fossil record indicates the appearance of spongelike species (phylum Porifera) over 600 million years ago, with jellyfish (phylum Cnidaria) appearing about the same time (Laflamme, 2010). These represent the first **metazoans**, multicellular animals with a body composed of more than one type of cell (**FIGURE 1.1**).

A cursory glance might suggest that all the cells in a sponge are alike in structure and function, but closer examination will reveal that only some cells secrete tiny mineralized spines to form a skeleton; other cells migrate, amoeba-like, through the body; and yet other cell types cause parts of the sponge to contract. Still, these microscopic differences between cells don't really form tissues or organs—there is no digestive, circulatory, or nervous system. Neither do sponges have any symmetry or regularity in overall body shape; they simply grow in whichever direction provides the best water flow to feed their constituent cells.

In contrast, jellyfish have obviously different tissues, such as trailing "arms" lined with specialized stinging cells to subdue prey. They have no circulatory system, but they do have a digestive system and a **nerve net**, a diffuse network of neurons distributed throughout the body (**FIGURE 1.2**) (Satterlie, 2011). Originally, the nerve net of jellyfish was regarded as capable only of reflexive responses to encountering prey or other objects, but more recent studies make it clear that the jellyfish nerve net includes specialized sensory cells and connections that coordinate moving toward or away from light, diving in response to rough water, and avoiding rock walls (Albert, 2011).

It would be a bad idea to produce lots of neurons and neglect making a digestive system.

With their array of specialized cells—some detecting stimuli outside the body, others relaying messages from one part of the body to another, and yet others digesting food—one challenge that jellyfish and all other metazoans face is **cellular differentiation**, the process by which individual cells

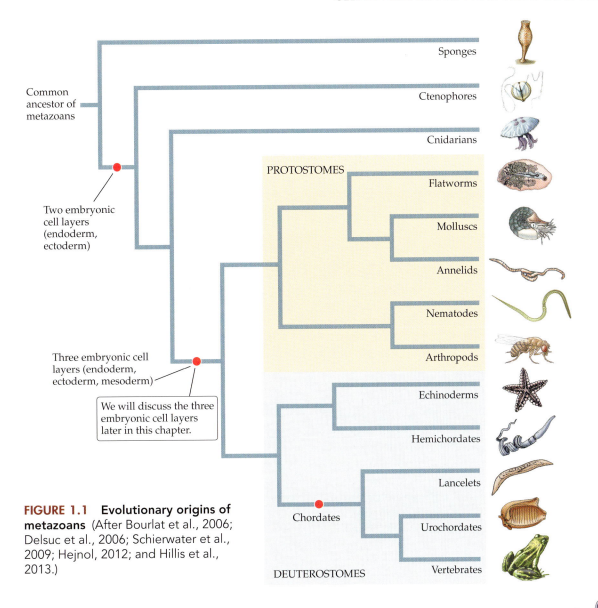

FIGURE 1.1 **Evolutionary origins of metazoans** (After Bourlat et al., 2006; Delsuc et al., 2006; Schierwater et al., 2009; Hejnol, 2012; and Hillis et al., 2013.)

in a developing metazoan become progressively different from one another. Another challenge for metazoans is how to *coordinate* the specialization of cell types. It would be a bad idea to produce lots of neurons and neglect making a digestive system, for example. Sensory neurons must have access to the outside world, while motor neurons must connect to muscles to move the body. In other words, metazoans must establish a *body plan*, an organization of the different cell types to maximize chances to survive and reproduce. How do you arrange to have the right type of cells in the right part of the body?

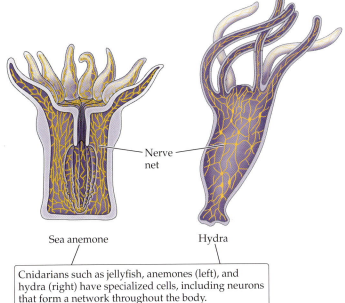

Cnidarians such as jellyfish, anemones (left), and hydra (right) have specialized cells, including neurons that form a network throughout the body.

FIGURE 1.2 **Nerve nets make up the first nervous systems.**

How's It Going?

Why is cellular differentiation required for metazoans?

■ 1.2 ■

PREFORMATIONISM OFFERED AN EASY BUT WRONG SOLUTION, WHILE EPIGENESIS SEEMED INCOMPREHENSIBLE

The ancient philosopher Aristotle (384–322 BCE) addressed the question of how types of cells are arranged in parts of the body by examining hens' eggs at different stages, noting that some structures, like the eyes, head, and heart, appear before others, such as the legs and wings (Aristotle, 350 BCE). Still other structures, like feathers, appear only just before hatching. He proposed that development was a process of **epigenesis**, by which the body gradually changes shape, acquiring new structures and growing more complex, with time. Of course, Aristotle had no rational explanation for *how* the body acquired greater complexity, so he proposed that every person and animal began with a soul that, somehow, directed the orderly growth of the body. Aristotle's vision of development by epigenesis would remain unchallenged in Western circles for centuries.

With the arrival of the Enlightenment in Europe, many thinkers tried to dispose of vague notions of a soul (at least for animals). There was also an optimism that any question could be approached in a scientific manner, including the question of how animals come into being. An important tool used to address this question was the microscope, which made it possible to see details that were invisible to the naked eye. Publication in 1665 of Robert Hooke's *Micrographia*, with his beautiful, detailed drawings of the various parts (including hairs) of fleas, the compound eyes of flies, and so on, alerted humankind to the "invisible world" around us (**FIGURE 1.3**). Hooke also examined thin slices of cork that revealed separate compartments that he called **cells**, which we now know as the basic building blocks of living things. In 1676, Antony van Leeuwenhoek (1632–1723), relying on his innovations in optics, reported seeing tiny organisms in rainwater, which members of the Royal Society in England regarded skeptically until Hooke soon confirmed the observation (C. Wilson, 1995). When Leeuwenhoek trained his microscope on human semen (his own), he spied tiny sperm wriggling like tadpoles.

epigenesis Here, the process in which successively more, and successively more complex, body structures appear in development.

cells The basic building blocks of life.

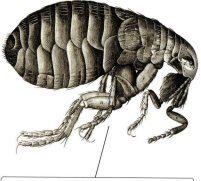

FIGURE 1.3 **The invisible world is made visible.** (From Hooke, 1665.)

Robert Hooke's drawings of a flea, as seen through a microscope, revealed a complexity of body shapes, joints, and hairs that was shocking for many people.

When Hooke trained his microscope on thin slices of cork, he could discern individual compartments, which he called cells.

In a great example of how relying too heavily on new technology can lead us astray, the early microscopists soon began challenging the notion of epigenesis. Jan Swammerdam (1637–1680) used his microscope to find, tucked inside a caterpillar, the wings and body of the butterfly that would have emerged (Cobb, 2000). He concluded that the adult form had been tucked inside the juvenile's body all along. Eventually, this perspective came to be known as **preformationism**, the view that development consists of a simple enlargement of a body plan that was there from the beginning, but simply too small to see. Taking this idea to its logical conclusion, Nicolaas Hartsoeker (1656–1725) published his drawing of a human sperm cell in 1694, showing a "little infant" tucked inside the sperm's head (**FIGURE 1.4**). This perspective indicated that human development consisted of a gradual unfolding of a little man (homunculus) when placed in a uterus to warm it so it could expand, like bread in an oven, to the size of a newborn. Hartsoeker's drawing furnished the iconic image of preformationism, which would supplant Aristotle's epigenetic perspective for over a century. During that time, people argued about whether the homunculus was tucked inside the sperm or the egg, but everyone involved seemed to see what they expected to see.

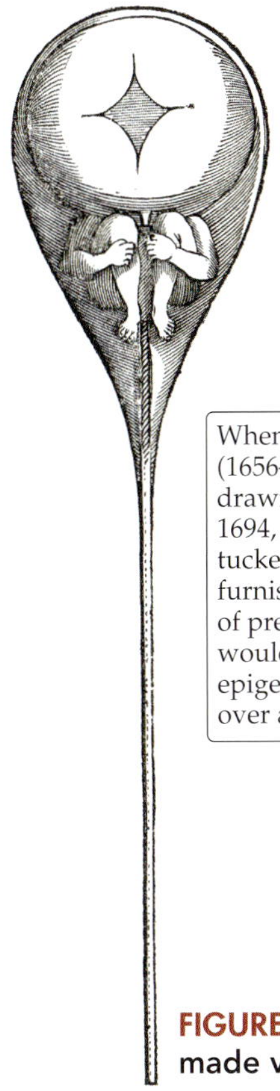

On the one hand, preformationism neatly explains how the developing individual is able to form a head, hands, fingers, legs, and so on—they were already present in tiny form and simply needed to grow. But there is an obvious logical problem with preformationism. If each person begins with a fully formed, microscopic, body tucked into the head of a sperm from Father, then where did Father's body come from? Even from the perspective of Renaissance scholars, who accepted Christian ideas about the creation of humanity, this notion strains credulity. Was every human being that ever was, or ever would be, nested, like a staggering series of Russian dolls, inside Adam's testicles (or Eve's ovaries)?

In addition to this logical problem with preformationism, the steady improvement in microscopes meant that by the early 1800s, scientists began reporting that fertilized eggs from hens and from amphibians such as frogs divided repeatedly to form clusters of cells, each having about the same, roughly spherical shape (**FIGURE 1.5A**). It became clear that the **embryo**, the earliest stage of developing organisms, does not have preformed arms, legs, head, and heart but consists of a clump of spherical cells that all look alike. As the embryo grows, various structures like the head, eyes, and heart begin to take shape as some cells diverge from the plain, simple shape of early embryonic cells and take on shapes that are appropriate for those tissues (**FIGURE 1.5B**). This developmental process of an individual growing up and growing old is called **ontogeny**. Once all the major organs and body parts are in place, we may refer to the developing individual as a **fetus**. As this view clearly refuted the notion of preformationism, scientists abandoned that perspective and returned to Aristotle's concept of epigenesis.

But without Aristotle's easy escape route, relying on an immaterial soul to coordinate and guide the addition of each new body part, scientists could not account for how epigenesis actually *worked*. Clearly the embryo becomes increasingly complex as it develops—that is the core of epigenesis—and so scientists simply accepted this as a fact. But, as Stephen Jay Gould noted, if the fertilized egg "were truly unorganized, homogeneous material without preformed parts, then how could it yield such wondrous complexity without a mysterious directing force?" (Gould, 1977a). How exactly does

When Nicolaas Hartsoeker (1656–1725) published this drawing of a sperm cell in 1694, with a "little infant" tucked inside its head, he furnished the iconic image of preformationism, which would supplant Aristotle's epigenetic perspective for over a century.

FIGURE 1.4 The nonexistent made visible

Was every human being that ever was, or ever would be, nested, like a staggering series of Russian dolls, inside Adam's testicles (or Eve's ovaries)?

preformationism   The notion that all the structures of an individual are present in microscopic form at conception, so development consists of simple growth of structures already present.

embryo   The earliest stage of development of a new individual, typically a spherical collection of cells.

ontogeny   The process of individual development; growing up and growing old.

fetus   A stage of development reached once major organs and body parts are in place.

(A)

(B)

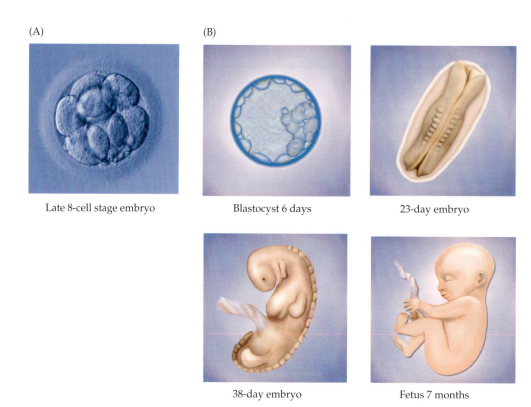

Late 8-cell stage embryo   Blastocyst 6 days   23-day embryo

38-day embryo   Fetus 7 months

FIGURE 1.5   **The process of ontogeny** (A) Early embryos are spherical clumps of cells, which all look alike. (B) As ontogeny proceeds, progressively more and more complex tissues and organs appear. (A © Red Hayabusa/Getty Images; B © Image Zoo/Alamy Stock Photo.)

that increasing complexity come about, and where is the information that guides it? Ultimately, the answer would be genes, so next let's explore how an understanding of genes came about and how that affected the question of cell differentiation.

How's It Going?

Describe and contrast preformationism and epigenesis.

◼ 1.3 ◼

THE REDISCOVERY OF GENES SET THE STAGE FOR UNDERSTANDING DEVELOPMENT

As the twentieth century began, several scientists independently rediscovered Gregor Mendel's 1866 paper demonstrating that pea plants inherit traits as *discrete units*; applied generally, either you inherit the entire gene or you don't. This contrasted with previous beliefs that inheritance consisted of *blending*, such that a parent's prominent trait would be diluted in offspring by the mixing in of the other parent's input.

Among other things, the realization that organisms could inherit in a discrete fashion a trait that would not be diluted as it passed from generation to generation solved a problem that had vexed Charles Darwin: once a new, adaptive trait arose, wouldn't it quickly be lost within just a few generations of breeding with "normal" individuals? Not if the trait, like the color of Mendel's

The fusion of Darwin's theory of evolution by natural selection and Mendel's laws of discrete inheritance became known as the modern synthesis of evolution.

pea blossom, could be inherited in complete form by at least some of the offspring. The fusion of Darwin's theory of evolution by natural selection and Mendel's laws of discrete inheritance became known as the **modern synthesis** of evolution, which underlies evolutionary biology today.

Eventually scientists would propose the word *gene* to describe the discrete hereditary unit that Mendel referred to as a "factor." It would be many years before we realized that genes consist of stretches of DNA. For now, we can define a **gene** as a DNA sequence, usually embedded in a chromosome, that is a functional unit of inheritance, specifying the structure of one or more proteins. Right away scientists began thinking about the relationship between the genes an individual inherits and how that individual turns out in the end. They coined the terms **genotype**, the total genetic makeup an individual inherits, and **phenotype**, the sum total of the physical characteristics an individual actually displays at a particular time. We would eventually learn that our genotype is fixed at the time of fertilization, but of course our phenotype changes year by year, moment by moment, as the outside world affects us. While originally these terms were applied to individual organisms, we will also talk about the phenotype of individual cells within an organism. We'll see many cases where a particular mutation in a gene will change the phenotype of particular cells. Trying to understand the role of genes in development shifted scientists' view of cellular differentiation.

modern synthesis The fusion of Darwin's theory of evolution by natural selection with Mendel's notion that genes represent discrete units of inheritance that are passed on either whole or not at all.

gene A stretch of DNA that represents a functional unit of inheritance, specifying the structure of one or more proteins.

genotype The total genetic makeup of an individual, typically determined at conception.

phenotype The sum total of physical characteristics that an individual displays at a particular time.

■ 1.4 ■
GENE EXPRESSION DIRECTS CELL DIFFERENTIATION

Developmental biologists of the twentieth century immediately appreciated the importance of genes for understanding cellular differentiation, the process by which individual cells in a developing metazoan become different from one another. The progressive divergence of structure across cells would have to be due to a divergence in which genes were being used to produce different traits in different types of cells. Of course, some proteins would be required in every cell for it to function properly, such as those needed for respiration and generating energy. But neurons, for example, would need a different set of proteins than would digestive cells or muscle cells. As cell differentiation continued in growing embryos, different cells would need to use different genes to take on their distinctive structures and therefore distinctive functions.

How would different cells make use of different genes to take on their particular structures and functions? One early hypothesis was that different cells could discard those genes that they did not need (E. B. Wilson, 1896), keeping only the genes needed to be a proper neuron, liver cell, or skin cell, as the case may be. Each cell would zero in on its eventual phenotype as it shed superfluous genes. In some ways, this would be an elegant solution—just jettison the genes you don't need, make use of all the remaining genes, and become specialized for a particular task. But a classic study in 1952 demonstrated that this is not the case. Briggs and King (1952) found that they could extract the nucleus from a cell in a well-developed embryo and implant it into an egg that had been "enucleated" (had its own nucleus removed). In this case, the nucleus from a late embryo could still induce the egg to produce all the cell types of an earlier embryo, showing that all the genes were still present in the older cells. Eventually, scientists would show that even adult cells, such as skin cells, still retained all the genes needed to make a whole organism, as we show next.

RESEARCHERS AT WORK

Do Differentiating Cells Dispose of Unused Genes?

■ **QUESTION**: In the process of cellular differentiation, do individual cells discard unneeded genes and retain only the genes needed for their particular specialized roles?

■ **HYPOTHESIS**: If not, then even a well-differentiated epithelial cell (skin cell) would still have all the genes necessary to make any other type of cell.

■ **TEST**: Extract the nucleus from a skin cell in one frog embryo and implant it in the center of a frog egg that has been enucleated. If the skin cell has retained only the genes needed to make skin, then this egg should not produce any cells except skin cells.

■ **RESULT**: In fact, some of the eggs provided with skin cell nuclei began forming tadpoles that eventually formed adult frogs (Gurdon, Laskey, & Reeves, 1975). Importantly, when epithelial nuclei from albino frogs were implanted into eggs of pigmented frogs, the resulting tadpoles were albino, showing they had indeed followed instructions from the genes in the epithelial nucleus of the donor (**FIGURE 1.6**).

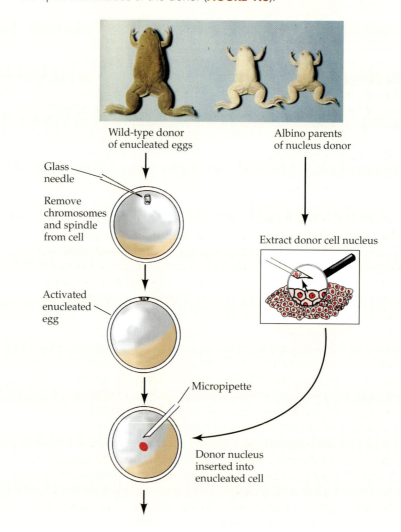

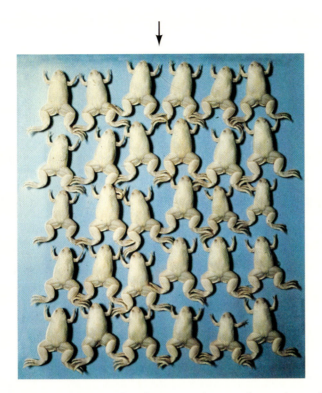

FIGURE 1.6   Bring in the clones Implanting the nucleus of an albino frog's skin cell in an enucleated egg from a pigmented frog can produce a whole frog with the genetic characteristics of the skin cell, including albino coloring. (After King, 1966, photos courtesy of Dr. J. Gurdon.)

■ **CONCLUSION**: As cells become differentiated into specialized cells, they retain all the genes needed to make any type of cell.

This means that cell differentiation *cannot be a matter of discarding unneeded genes* and keeping only those needed to produce a particular type of cell. Rather, every cell retains in its nucleus all of the organism's genes, and cell differentiation is a matter of making use of only a subset of those genes. Eventually this approach would be used to produce "Dolly," the Finn-Dorset strain of lamb that resulted when the nucleus of an udder cell from a Finn-Dorset ewe was implanted in the enucleated egg of a Scottish blackface ewe.

When a cell makes use of a particular gene, we say that it *expresses* the gene. Making use of this terminology, we can offer a more modern definition of *cellular differentiation* as the process by which cells selectively express particular genes and thereby take on a particular structure and perform a particular function within the individual. In other words, to understand cellular differentiation, we must understand the control of gene expression.

> To understand cellular differentiation, we must understand the control of gene expression.

This more modern understanding of cellular differentiation as a matter of gene expression obliges us to update our discussion of genes, as well. A stretch of each gene's DNA encodes for a particular protein, and we can refer to that as the **structural portion of the gene**, a particular sequence of trinucleotides calling for a particular sequence of amino acids and there-

structural portion of the gene
The portion of DNA in a gene that encodes for a particular sequence of amino acids and therefore a particular protein.

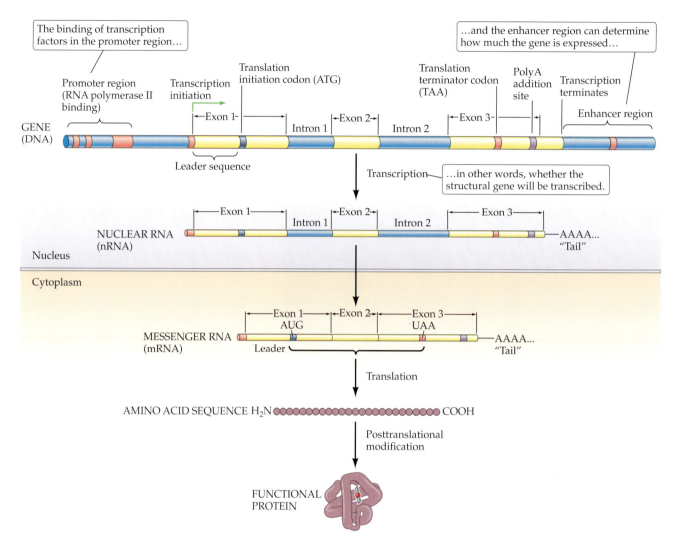

FIGURE 1.7 **The regulatory regions of genes** Recall from introductory biology that exons are portions of the gene that eventually contribute code for the mRNA molecule, while introns are the stretches of DNA that do not contribute information to the final mRNA molecule but may play a role in regulating transcription. An easy mnemonic is that information in *ex*ons *ex*its the nucleus, while *in*trons remain *in*tranuclear.

regulatory portion of the gene The stretches of DNA adjacent to the structural gene, which play a role in regulating transcription of that gene.

promoter region The regulatory region that tends to be upstream from the structural gene.

enhancer region The regulatory region that may be upstream or downstream of the structural gene, which plays a role in controlling transcription of that gene.

fore a particular polypeptide or protein. But there are stretches of DNA before and after the structural portion of the gene that regulate when and in which cells that gene will be expressed (transcribed to produce a molecule of RNA), and we call this DNA the **regulatory portion of the gene**. This regulatory portion is often divided into so-called **promoter regions**, which are upstream from the gene of interest, and **enhancer regions**, which can be found either upstream or downstream from the gene of interest (**FIGURE 1.7**). Although these two regions are named for the ability to increase gene expression ("promote," "enhance"), they can in fact serve to either increase or decrease gene expression.

Within a cell, certain molecules interact with the regulatory portion of genes to make transcription of the adjacent structural portion either more or less likely to occur. These molecules, usually proteins, that regulate gene

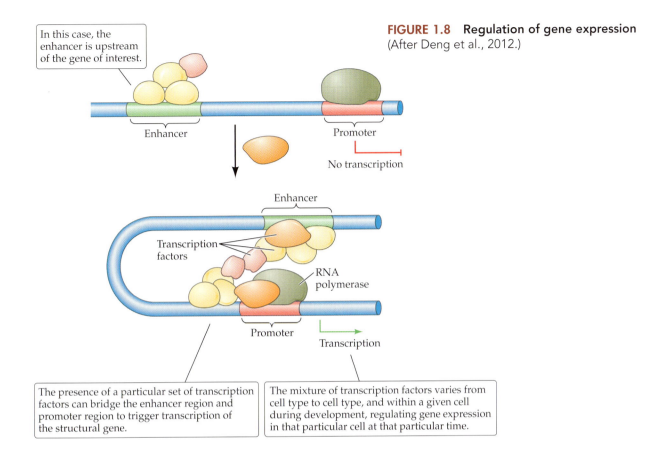

In this case, the enhancer is upstream of the gene of interest.

Enhancer

Promoter

No transcription

Enhancer

Transcription factors

RNA polymerase

Promoter

Transcription

The presence of a particular set of transcription factors can bridge the enhancer region and promoter region to trigger transcription of the structural gene.

The mixture of transcription factors varies from cell type to cell type, and within a given cell during development, regulating gene expression in that particular cell at that particular time.

FIGURE 1.8   **Regulation of gene expression** (After Deng et al., 2012.)

transcription are called, reasonably enough, **transcription factors**. They are sometimes called gene regulatory proteins, to emphasize their role in controlling gene expression (**FIGURE 1.8**).

Why do some cells express a particular gene when others do not? It's because the cells differ in the mix of transcription factors present to interact with the regulatory regions of various genes. The frog's skin cell contained transcription factors that resulted in the expression of only those genes needed by a skin cell. But when its nucleus was inserted into an egg, a different composition of transcription factors, those present in the cytoplasm of that egg, altered gene expression so that an embryo, and eventually a tadpole, was produced. Thus, the egg's cytoplasm was not just empty goo, a blank canvas for the epithelial nucleus to work upon. Rather, the egg's cytoplasm contained the transcription factors needed to trigger the unfolding pattern of gene expression that would result in cell divisions to form an embryo. That was why the nucleus was inserted into an egg rather than into another skin cell or a neuron.

This demonstration is a preview of the rest of this chapter and, in some ways, the rest of this book. One portion of the cell, the cytoplasm, can control gene expression in the nucleus, which then has a major influence on the conditions in the cytoplasm, including what transcription factors will be manufactured next. Those, in turn, will influence gene expression in the nucleus to start the cycle over again. This dialogue between the transcription factors produced in the cytoplasm and the genes they regulate in the nucleus, including those genes producing another round of transcription factors, sculpts the developing cell to take on a highly specialized fate. However simple this may sound in theory, in practice it can be very difficult to track these processes among

transcription factors   Proteins that bind to DNA and regulate the extent to which genes are expressed.

the trillions of cells in a developing chick embryo, for example. To make the task of tracking cellular differentiation less daunting, scientists sought out a simple animal for study, a microscopic worm.

How's It Going?

What is the modern understanding of how cells within a metazoan come to be different from one another?

■ 1.5 ■
SCIENTISTS DOMESTICATED A SIMPLE WORM TO ADDRESS THE QUESTIONS OF CELL DIFFERENTIATION

Inspired by the rapid progress molecular biologists made by studying simple organisms like the bacterium *Escherichia coli*, Sydney Brenner longed to "tame a small metazoan organism to study development directly" (Brenner, 1963). Brenner eventually settled on a microscopic nematode (see Figure 1.1), a common worm found in soil called *Caenorhabditis elegans*. This animal cannot live in human tissue, so there's no danger of a scientist being infected, and no diseases are known that can be transmitted to people by the worm. There are many advantages of studying *C. elegans*, including the fact that its body is fairly transparent (organisms living underground have little use for pigment), meaning you can look inside a living worm to see various internal organs (**FIGURE 1.9A**). Because they are tiny, you can keep many worm subjects in a small space, which means they are inexpensive, too. For example, the worms are typically kept in petri dishes on a layer of agar that is colonized with the worm's favorite food, that same bacterium *E. coli*. Under a microscope, you can use a thin wire (or an eyelash!) to pick up an individual worm and place it on a glass microscope slide coated with agar. Then put on a glass coverslip, and place the slide in a microscope for

Caenorhabditis elegans
A microscopic roundworm that offers a valuable model of cell differentiation.

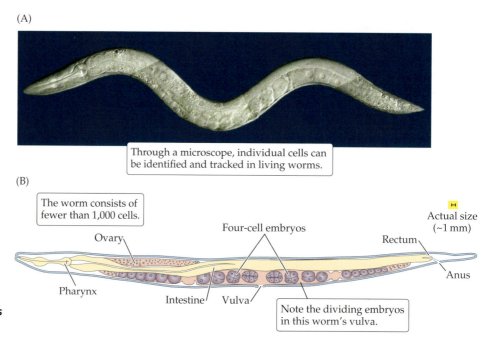

(A)

Through a microscope, individual cells can be identified and tracked in living worms.

(B)

The worm consists of fewer than 1,000 cells.

Ovary

Four-cell embryos

Rectum

Actual size (~1 mm)

Pharynx

Intestine   Vulva

Anus

Note the dividing embryos in this worm's vulva.

FIGURE 1.9   The nematode worm *Caenorhabditis elegans*

a closer look as the worm lives out its life in the layer of agar between the two pieces of glass. The worms are also easy to store—put them in a freezer, where their biological processes stop until you thaw them out again.

C. elegans also offer many advantages for doing genetic studies. As with fruit flies, you can readily induce mutations through exposure to X-rays or chemicals, but *C. elegans* have an even faster generation time than fruit flies, growing from egg to adult in just 3 days (Corsi, 2006). (If that's too fast, you can always "hit the pause button" by putting them in the fridge.) While a very few *C. elegans* are male, the rest are **hermaphrodites** (individuals capable of reproducing as either males or females, in other words that produce both sperm and eggs) that can self-fertilize (**FIGURE 1.9B**). That is very handy for propagating any interesting mutations you find, including those that might impair an animal's ability to mate with another individual.

Another wonderful advantage of studying *C. elegans* is simply one of scale—having so few cells makes brute-force approaches more feasible. For example, scientists were able to section the bodies of several worms for electron microscopy, piecing together all those cross sections to reconstruct the entire interior, including every neuron's synaptic connections. Thus they were able to produce a map of all the synapses in the worm's nervous system, a "connectome" (including 5,000 chemical synapses, 2,000 neuromuscular junctions, and 600 electrical synapses) (White, Southgate, Thomson, & Brenner, 1986), which has not been achieved for any other species (although heroic efforts in other species are underway: www.openconnectomeproject.org). Brenner, Robert Horvitz, and John Sulston were awarded a Nobel Prize in 2002 for their work with *C. elegans*.

Long before the prize was awarded, *C. elegans* proved its value by offering a viable model of cell differentiation, as we see next.

■ 1.6 ■
MITOTIC LINEAGE GUIDES CELL DIFFERENTIATION IN WORMS

The problem of cell differentiation, how to make just the right number of many different types of cells, each in its appropriate place in the body, seems more manageable in *C. elegans* because the entire worm, with a body about 1 mm long, consists of just under 1,000 cells. Indeed, because of its many advantages, we have an excellent idea of how cell differentiation works in this animal.

Researchers could watch worms develop from fertilized egg to full-grown (microscopic) adult and, because the worm's body is mostly transparent, observe the development of internal organs as well as the exterior. The earliest cell divisions in *C. elegans* are asymmetrical, so scientists could name the daughter cells produced by each round of mitosis, keeping track of their order and "lineage," as they descended from the fertilized egg, known as a **zygote**. For example, the first division results in a large cell, designated AB, and a slightly smaller cell, which will end up making the posterior end of the worm, so it's called P1 (**FIGURE 1.10A**). Numerous careful observations revealed that there is a strict order of cell divisions that happens about the same way in every individual worm. In fact, you can make a "map" that shows every cell division that takes place in the transition from zygote to adult, including the final differentiation of every cell (Kimble & Hirsh, 1979; Sulston & Horvitz, 1977). This means that, by keeping track of the **mitotic lineage** of each cell, you can perfectly predict what any particular cell will

hermaphrodite   An individual capable of reproducing as either a male (producing sperm) or female (producing ova).

zygote   A fertilized egg; the single cell that will divide and grow to form a new individual.

mitotic lineage   The sequence of mitosis during ontogeny that gives rise to a particular cell in an individual. In *C. elegans* there is an invariant relationship between mitotic lineage and cell fate.

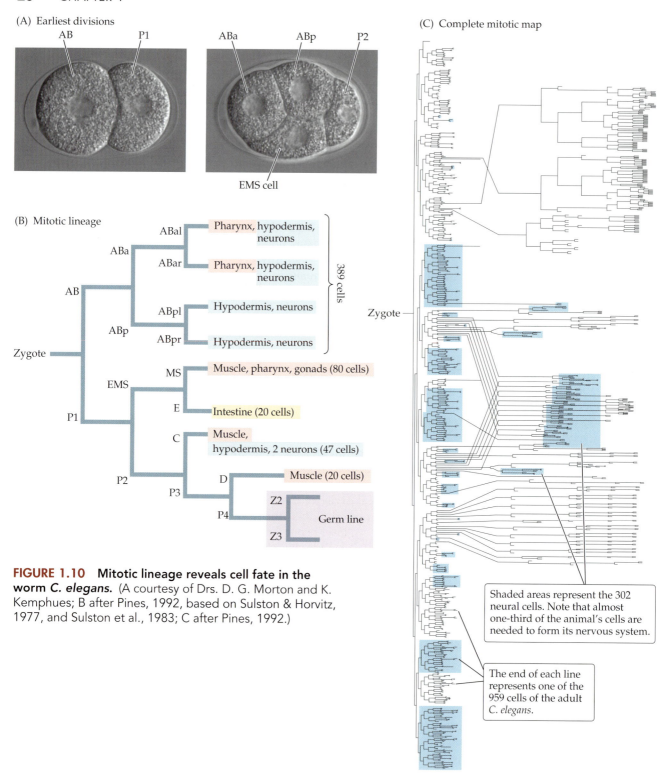

(A) Earliest divisions

AB   P1

ABa   ABp   P2

EMS cell

(B) Mitotic lineage

Zygote

AB

ABa — ABal — Pharynx, hypodermis, neurons
ABa — ABar — Pharynx, hypodermis, neurons

ABp — ABpl — Hypodermis, neurons
ABp — ABpr — Hypodermis, neurons

389 cells

P1

EMS — MS — Muscle, pharynx, gonads (80 cells)
EMS — E — Intestine (20 cells)

P2 — C — Muscle, hypodermis, 2 neurons (47 cells)

P3 — D — Muscle (20 cells)

P4 — Z2 — Germ line
P4 — Z3

(C) Complete mitotic map

Zygote

Shaded areas represent the 302 neural cells. Note that almost one-third of the animal's cells are needed to form its nervous system.

The end of each line represents one of the 959 cells of the adult *C. elegans.*

FIGURE 1.10   Mitotic lineage reveals cell fate in the worm *C. elegans.* (A courtesy of Drs. D. G. Morton and K. Kemphues; B after Pines, 1992, based on Sulston & Horvitz, 1977, and Sulston et al., 1983; C after Pines, 1992.)

fate   Here, the particular structure and function that a given cell takes on in the course of cellular differentiation.

become—neuron, skin cell, or gut or mouth cell (**FIGURE 1.10B**). With enough persistence, you can trace the mitotic lineage of all 959 cells, including the 302 neurons (**FIGURE 1.10C**).

Thus in the normally developing worm there is an invariant pattern of mitosis, and each resulting cell's **fate**, the particular structure the cell adopts and the function it plays within the individual, depends upon where it arises in that mitotic lineage. Is it just a coincidence that the pattern of mitosis perfectly matches every cell's fate, or is there something about the invariant pattern of mitosis that *determines* each cell's fate?

It soon became clear that in many cases mitotic lineage was also *responsible for cells taking on their fates*. For example, if after that first division, the two daughter cells (AB and P1) are separated, the P1 cell goes on dividing as if nothing happened and forms exactly the same set of cells it would have if it had never left its AB sister—basically the back half of a worm (Priess & Thomson, 1987). Without a front half to latch onto, the clump of cells collapses upon itself, but every cell seems to be there and to have taken on its fate (some cells differentiate into muscles, for example). This scheme for development is sometimes called **mosaic specification of cell fate**, where every cell follows its particular destiny no matter what its neighboring cells are up to, like individual tiles in a mosaic floor. In the case of *C. elegans*, it appears that mitotic lineage specifies each cell's fate. Normally, as each descendant of the mitotic lineage follows its own fate, together they make up a functional worm capable of reproducing itself.

If this is an accurate picture of cell differentiation in *C. elegans*, then the death of any one cell in the developing organism should not affect the fate of any remaining cells. To test this idea, several labs used a laser, directed through the microscope, to kill individual cells in developing worms. For the most part, the prediction was fulfilled: killed cells were not replaced, and the surviving cells differentiated just the way they would have anyway (Kimble, 1981; Sulston & White, 1980).

In Chapter 7 we'll see how cell death (apoptosis) is a common, normal part of development in the nervous system (as well as the rest of the body) in all animal species. This is true in *C. elegans*, and in fact we'll also see that studies in worms were crucial in understanding the process. Thus there are times when a cell in developing *C. elegans* undergoes a final mitosis, where normally one daughter cell takes on a particular fate and function while its sister cell simply dies. Immediately after that final division, these two cells must be very similar to one another, in terms of transcription factors present in the cytoplasm. So when experimenters directed a laser beam through the microscope to kill the cell that would have taken on a particular job, did the sister cell pitch in and take over that fate? No, the sister cell stuck with the original program and died, leaving the adult worm one cell shy of the normal 959, just as you'd expect when mosaic specification of cell fate is at work. Such results led people in the field to talk about how cells were committed, or destined to take on a particular fate, no matter what. But scientists soon realized that *commitment* is a relative term, as we discuss in **BOX 1.1**.

Why does the isolated P1 cell undergo these particular patterns of division, and why do the offspring of those divisions take on the particular fates they do? Broadly speaking, the P1 cell must contain a different mix of transcription factors than the AB cell, and these factors must direct P1 to divide in a particular way and produce cells of particular fates. How can P1 have a different mix of transcription factors than AB? After all, just before that first division of the zygote producing AB and P1, they share the same cytoplasm and the same nucleus, right? You don't have to think about it very long to realize that the cytoplasm and/or nucleus of the zygote cannot be completely homogenous, that transcription factors must be distributed unequally such that AB retains one cohort of factors while P1 holds onto another cohort. As further divisions ensue, each cell may, in theory, receive a unique mixture of transcription factors varying in their relative concentrations as they are parceled out among descendants. The particular composition of transcription factors each cell receives may be a result of where, exactly, that cell falls in the mitotic lineage. That particular mixture of transcription factors may direct gene expression in the cell and, thus, differentiation of that cell to take on its fate. It follows that mitotic lineage predicts cell fate in

mosaic specification of fate A strategy for cellular differentiation in which each cell follows a particular fate no matter what neighboring cells might be doing.

In the case of *C. elegans*, it appears that mitotic lineage specifies each cell's fate.

KERFUFFLES IN LANGUAGE:
"CELL FATE" AND "COMMITMENT"

We noted that the experiments using lasers to kill individual cells in developing *C. elegans* resulted, for the most part, in no evidence of self-regulation. In other words, no other cell changed its fate to take the place of the fallen comrade and the animal simply did without whatever the destroyed cell would have become. There was a lot of talk about the cells in developing *C. elegans* being "committed" to whatever fate was dictated by their mitotic lineage. This terminology, of *fate* and *committed*, has a sense of finality, of the last word in what the cell will do in the future, no matter what else happens. I want to urge you to be leery of this interpretation. When we speak of a cell's fate, we certainly do not mean an immutable, unchangeable outcome. For example, what if an experimenter uses a laser to kill a cell? That obviously changes its fate. More broadly, whenever we do any manipulation, short of killing the cell, in an attempt to change its fate, we have to restrict our conclusion to the particular manipulation(s) we tried. Just because our manipulation didn't alter the cell's fate, that doesn't prove that *no* manipulation would. Indeed, if we had enough information and unlimited abilities to manipulate a cell, turning on and off targeted genes, we could shape it into any cell type the animal is capable of.

Likewise, when a cell seems "committed" to a particular fate, we can say that it remained committed to that fate in the face of whatever manipulations we tried. For all we know, there is a simple manipulation that would overcome the cell's commitment and change its fate. In other words, saying a cell is "committed" to a particular fate only has meaning in the context of whatever challenge we presented to the cell. If we don't specify what challenges the cell overcame to reach that particular fate, then saying it is committed conveys no information at all. For a real-life example, my mate seems committed to me, as she has stuck by me in many challenging circumstances (as when I get selfishly absorbed in writing a textbook, for example). So in the face of those challenges, she is committed. But for all I know, a new challenge may present itself tomorrow, a great job offer in Tahiti, that may convince her it's time to change her fate! Thus in developmental neurobiology, **commitment** is the *tendency* for a cell to take on a particular fate even in the face of a particular set of challenges that have been tried so far. *Commitment* no longer carries the implied notion of irreversibility (either in embryology or personal relationships).

---

commitment   Here, the tendency of a cell to take on a particular fate even in the face of particular challenges. Its meaning is restricted to only those challenges that have been tested.

maternal effect   Influences the mother has on ontogeny of offspring apart from the particular genes she contributed.

SKN-1   A transcription factor in *C. elegans* that is more concentrated at one end of an egg or zygote, which will give rise to the posterior half of the individual.

pharynx   The tube lining the mouth that connects to the rest of the digestive system.

unc-86   A gene in *C. elegans* that encodes the protein unc-86, which directs certain late-dividing cells to become touch receptor neurons.

these cases because *mitotic distribution of transcription factors determines gene expression in the cells* and hence their fates.

So if the mother deposits different transcription factors in different parts of the egg, then the invariant divisions of the embryo will parcel those factors into particular cells, directing their fates. We call such influences **maternal effects**, influences the mother has on an offspring's phenotype that are separate from whatever particular genes she contributed to the egg. In the case of *C. elegans* development, one maternal factor has been identified that directs the fate of one "granddaughter" of P1 (**FIGURE 1.11**). The worm mother deposits more of a protein called **SKN-1** in what will be the posterior end of the egg's nucleus, so the first mitosis places most of the SKN-1 in P1. When P1 divides, most of this SKN-1 ends up in the nucleus of the daughter cell called EMS, which then divides, leaving most of the SKN-1 to a cell called MS. At this point, only MS and its "descendants" are able to make the worm's **pharynx** (the tube lining the mouth and connecting to the rest of the digestive system), because only they have enough SKN-1 to direct differentiation down that pathway. If the mother has no functional copies of the *SKN-1* gene and so cannot lay down the protein in the posterior of her eggs, none of her offspring will form a pharynx (Bowerman, Draper, Mello, & Priess, 1993). Instead, those cells normally fated to produce a pharynx will just form extra skin.

There may be relatively few transcription factors that are unevenly distributed in the egg by the mother, but those transcription factors, as they are divvied up in early cell divisions, may in turn regulate the later expression or production of other transcription factors. Then those transcription factors may turn on yet other transcription factor genes and so on. For example, the worms rely on mechanosensory neurons to detect when they encounter something in their agar world. If you use an eyelash to touch them on the rear, they speed away; touch them in the front, and they back up. By producing mutant worms, investigators found several mutants that failed to respond to touch. One touch-insensitive mutant resulted from the disabling of a gene called **unc-86**, which is normally turned on (i.e., begins to be expressed) relatively late in development in cells that normally become the touch receptors (Duggan, Ma, & Chalfie, 1998). Without functional unc-86, the cells never differentiate into touch receptors. We don't know what transcription factor causes expression of *unc-86* in those particular cells, but unc-86 is itself a transcription factor that turns on *yet another* transcription factor called **mec-3**. Then unc-86 and mec-3 proteins bind to each other, forming a **dimer** (because in this case two different proteins are binding together, we say they form a *heterodimer*; when two molecules of the same protein bind together, we say they form a *homodimer*). The unc-86/mec-3 heterodimer then regulates the expression of *over 50 different genes* that turn the cell into a mechanosensory neuron (Zhang et al., 2002) (**FIGURE 1.12**).

So we can see, in theory, how cell differentiation could work in C. elegans. In the early rounds of mitosis, the nucleus and cytoplasm of each resulting cell retain a slightly different mixture of maternally deposited transcription factors, which direct a slightly different overall pattern of gene expression in that cell. As development continues, those maternally derived transcription factors in turn regulate expression of other transcription factors, which may also be unevenly distributed in the next round of mitosis. These later-arising transcription factors then regulate expression of yet other transcription factors until, by the final round of cell division, all the cells have slightly different mixtures of transcription factors, and therefore slightly different patterns of gene expression, so each takes on a unique fate to play its part in the whole animal. Not coincidentally, each cell will also be in the right part of the animal's body to play its particular role in survival and reproduction.

Because there is this remarkable, one-to-one correspondence between mitotic lineage and cell fate in C. elegans, it may have been natural for scientists to wonder whether mitotic lineage relates to cell fate in fruit flies, frogs, chicks, mice, and people. They couldn't apply the methods that worked in C. elegans in these species, where every individual consists of millions or billions of

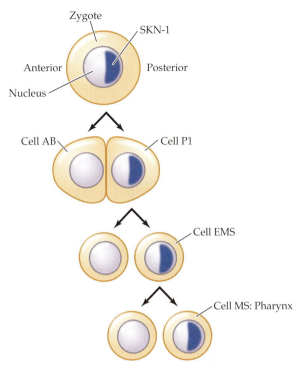

FIGURE 1.11 **Schematic depiction of maternal effects in early divisions of *C. elegans*** The mother constructs the egg such that there is more of the transcription factor SKN-1 in the posterior part of the nucleus than the anterior. The first division results in most of the protein going to the nucleus of cell P1. After several other divisions, a small number of cells have sufficient SKN-1, which triggers a cascade of gene expression to have them differentiate into cells of the pharynx.

mec-3 A gene in *C. elegans* that encodes the protein mec-3, which binds to the protein unc-86. The two proteins together regulate expression of genes important for differentiation into a touch receptor neuron.

dimer A complex of two proteins that bind together to form a functional unit.

Presumably the unknown transcription factors triggering *unc-86* expression were themselves expressed in response to direct or indirect effects of maternal factors, such as uneven distribution of transcription factors in the egg.

FIGURE 1.12 **A cascade of transcription factors leads to the development of touch receptor neurons in *C. elegans*.**

blastula The earliest stage of an embryo, typically a spherical clump of cells.

blastomeres The individual cells that make up a blastula.

blastocoel The hollow, fluid-filled cavity inside a blastula.

gastrula An embryo that has formed a primitive gut, a tube that passes through the embryo.

gastrulation The process by which a blastula becomes a gastrula; the formation of a primitive gut.

blastopore A small dimple on the surface of a blastula that will invaginate to start forming the primitive gut.

cells—how could they keep track of any individual cell in such a crowd? But in fact, however tempting it may have been to think that humans and other vertebrates might use the same strategy as *C. elegans*, relying on the invariant parceling out of maternal factors during mitosis to direct the differentiation of every cell, nearly a century of work had already made it clear that embryos of most species do *not* develop that way. To understand this body of work and how it differs from the development of *C. elegans*, we'll need to back up our story to the nineteenth century, when many talented scientists began training their microscopes on developing embryos.

How's It Going?

Describe how cellular differentiation is controlled in *C. elegans*, and explain why no duplicate cells form.

■ 1.7 ■

EMBRYONIC DEVELOPMENT BEGINS BY FORMING THREE DISTINCT GERM LAYERS

Nineteenth-century improvements in the quality of microscopes allowed more and more scientists to observe developing embryos in a wide variety of species. Most studies focused on aquatic species, such as frogs, fish, and sea urchins, because it was relatively easy to keep the embryos alive (and continuing to develop) in water. One of the few terrestrial embryos studied was the chick, readily available because of commercial demand for eggs, which could be easily opened to gain a peek at the process of development, as Aristotle had (without the benefit of a microscope or even a magnifying glass).

These observations revealed some commonalities in embryonic development. In virtually all animals, the first several divisions of the zygote result in a more or less spherical cluster of cells, leaving a fluid-filled hollow space in the center. At this stage the embryo is called a **blastula** (from Greek *blastos*, meaning "sprout"), and the individual cells are called **blastomeres**. The hollow space within the blastula is called a **blastocoel** ("BLAST-uh-seal") (**FIGURE 1.13**). As the embryo continues dividing and growing, the blastocoel gets smaller and smaller.

The first recognizable adult structure that embryos form is the gastric system, which begins as a hollow tube that pierces the mass of cells. One end of the tube will become the mouth, while the other will become the anus. Once an embryo starts forming this primitive gut, it is no longer called a blastula but is designated a **gastrula**, and the process by which a blastula makes that transformation is called **gastrulation**. The details of gastrulation differ across animal orders, but the classic version is in frogs, where a dimple called the **blastopore** appears in the surface of the blastula and invades the interior of

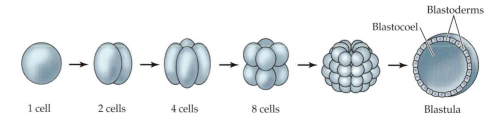

1 cell   2 cells   4 cells   8 cells   Blastula

Blastoderms
Blastocoel

FIGURE 1.13 Earliest embryos form a hollow blastula.

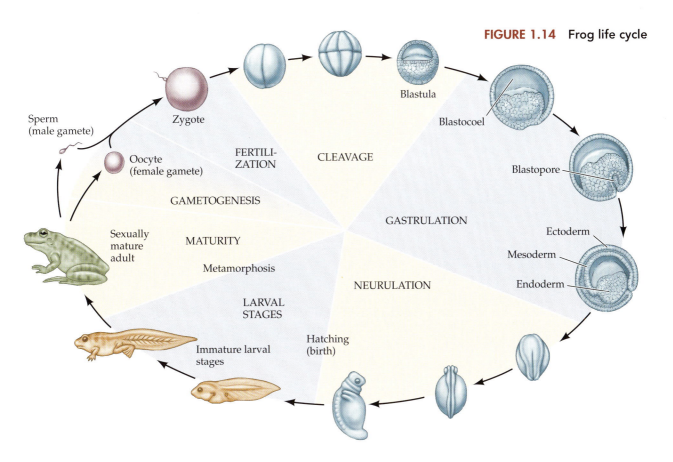

FIGURE 1.14 Frog life cycle

the embryo, eventually reaching the other side. Note that this hollow tube piercing the embryo is distinct from the hollow blastocoel, which is typically distorted by the formation of the primitive gut and eventually filled in with cells (**FIGURE 1.14**).

The formation of the gastrula changes the blastula, which looks, to the untrained eye, like a more or less homogenous sphere, into an embryo with three obvious layers that are called the **germ layers**: the outside layer of cells, called the **ectoderm** ("outer skin"); an internal layer of cells forming the tube of the primitive gut, called the **endoderm** ("inner skin"); and a layer of cells lying between those two layers, the **mesoderm** ("intermediate skin"). The layers are designated as "germ" (sometimes *germinal*) layers because they grow to make so many structures, as the germ in seeds does (the Latin root of *germ* means "bud," "seed," or "sprout"). The three germ layers will continue to divide, and the descendants of those divisions will each make distinctive contributions to the adult body (**FIGURE 1.15**). In this book, we will have little interest in adult cells derived from the endoderm, but the other two germ layers play prominent roles in the developing brain. The nervous system itself will be derived from the outermost germ layer, the ectoderm. But later in this chapter we'll see that formation of the nervous system in the ectoderm critically depends on influences from the mesoderm.

The classic picture of gastrulation we have just described reflects the process in amphibians, such as frogs, which were much studied early in the field. In other vertebrates, such as birds and mammals, gastrulation is different. In mammals, for example, mitosis and cell migration in the blastula result in a clump of cells in the interior of the embryo that forms the **inner cell mass**, which will give rise to all the body, while the exterior of the blastula will form the placenta and other tissues supporting prenatal growth. Gastrulation

germ layers Here, the three layers of cells formed in the course of gastrulation.

ectoderm The outermost of the three germ layers.

endoderm The innermost of the three germ layers.

mesoderm The germ layer that forms between the ectoderm and endoderm.

inner cell mass In mammals, the clump of cells found inside the blastula, which will give rise to the individual's body. The remainder of the blastula will contribute to the placenta and related tissues.

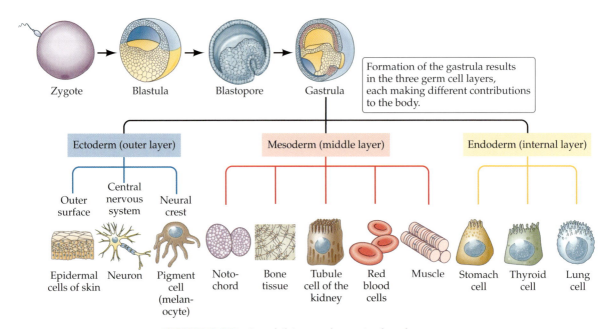

Zygote   Blastula   Blastopore   Gastrula

Formation of the gastrula results in the three germ cell layers, each making different contributions to the body.

Ectoderm (outer layer)   Mesoderm (middle layer)   Endoderm (internal layer)

Outer surface   Central nervous system   Neural crest

Epidermal cells of skin   Neuron   Pigment cell (melan-ocyte)   Noto-chord   Bone tissue   Tubule cell of the kidney   Red blood cells   Muscle   Stomach cell   Thyroid cell   Lung cell

FIGURE 1.15 Amphibian embryonic development

primitive streak The beginnings of the nervous system in the vertebrate embryo, marking the midline of the developing individual.

node Here, the anterior-most portion of the primitive streak, which will give rise to the brain.

neural plate The earliest stage in the development of the vertebrate nervous system from the ectoderm.

begins when the inner cell mass forms a disc two cell layers thick inside the embryo (**FIGURE 1.16A**). Cells on the upper surface of the disc (away from the blastocoel) migrate toward the center and plunge underneath the surface, forming a crease called the **primitive streak**, which is aligned with what will be the midline of the animal (the axis of bilateral symmetry) on the dorsal surface (back) (**FIGURE 1.16B**). The first cohort of cells to dive beneath will form the endodermal layer; then later-arriving cells will fill the space between the outer layer and the endoderm, forming mesoderm (**FIGURE 1.16C**). The endodermal cells will continue to divide and spread out within the blastocoel until they meet, forming the primitive gut, eventually forming a mouth at one end, an anus at the other, and several digestive organs such as the liver and pancreas along the way.

Following gastrulation, the second adult structure formed by the embryo is the nervous system arising from ectoderm, as we'll see next.

How's It Going?
What are the three germ layers, and how do they come about?

■ 1.8 ■

THE VERTEBRATE NERVOUS SYSTEM BEGINS AS A SIMPLE TUBE

In vertebrate embryos, the primitive streak that forms as the germ layers are created in gastrulation marks the midline position, where the nervous system will develop in the ectoderm. The crease formed by the primitive streak is more pronounced at one end, called the **node** (see Figure 1.16B; in birds, it is called Hensen's node), and it marks where the animal's brain and head will eventually form.

The nervous system begins as an elongated layer of ectodermal cells on the surface, called the **neural plate** (**FIGURE 1.17A**). As cell divisions and

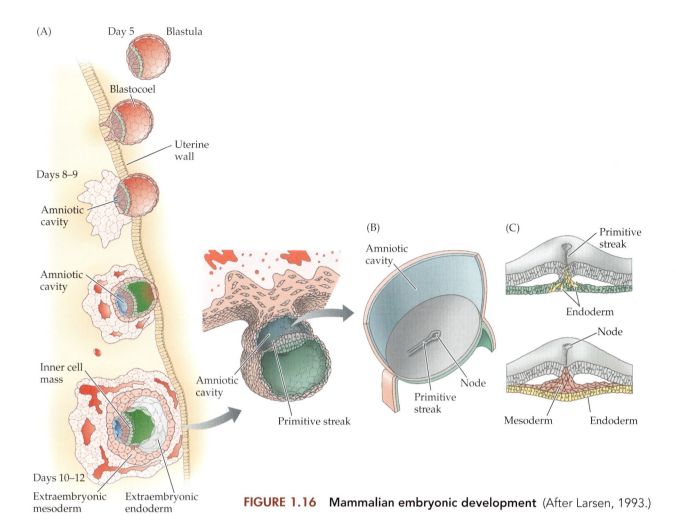

(A)

Day 5   Blastula

Blastocoel

Uterine wall

Days 8–9

Amniotic cavity

Amniotic cavity

Inner cell mass

Amniotic cavity

Primitive streak

Days 10–12

Extraembryonic mesoderm   Extraembryonic endoderm

(B)

Amniotic cavity

Node

Primitive streak

(C)

Primitive streak

Endoderm

Node

Mesoderm   Endoderm

FIGURE 1.16   Mammalian embryonic development (After Larsen, 1993.)

migrations continue, the sides of the neural plate rise up to form the walls of a *neural groove* (**FIGURE 1.17B**), and these walls eventually meet and fuse at the midline to form a **neural tube** (**FIGURE 1.17C**). The end of the neural tube arising from the node will form the brain, while the rest of the neural tube will produce the spinal cord (**FIGURE 1.17D**). As the neural plate grows and folds to form the neural tube in the ectoderm, mesodermal tissue beneath migrates to the midline and forms a rodlike structure, the **notochord**, which will eventually contribute to production of the vertebral column (spine). It is the embryonic presence of the notochord that defines the chordates (see Figure 1.1), including the vertebrates (Stemple, 2005). In Chapters 2 and 4, we'll see that the notochord also plays an important role in directing development of the vertebrate nervous system.

The rest of the nervous system will arise from a group of ectodermal cells that were at the peaks of the two sides of the neural groove. This formation resembles the crest of a wave (see Figure 1.17B, right), and so it is referred to as the **neural crest**. Once the neural tube forms, these neural crest cells separate from both the tube and the overlying ectoderm that forms over the neural tube (see Figure 1.17C, right).

Thus the ectoderm will make three contributions of interest to us:

1. The outermost layer of ectoderm will give rise to the skin, the **epidermis**, which is the largest organ of the body.

neural tube   The early, tube-shaped stage of vertebrate nervous system development.

notochord   An embryonic rod-shaped structure that is derived from mesodermal tissue in all vertebrates.

neural crest   A collection of ectodermal cells that break away from the developing neural tube to lie sandwiched between the tube and overlying ectoderm.

epidermis   Skin tissue, derived from ectodermal cells that do not become neural tissue.

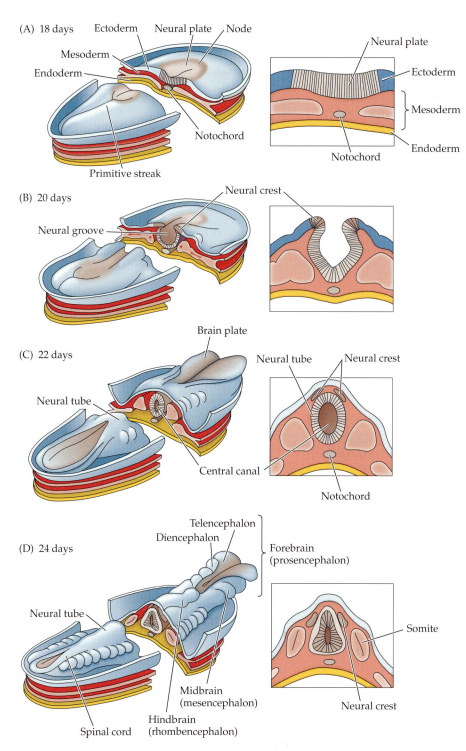

FIGURE 1.17 The neural plate and neural tube

central nervous system (CNS)
The brain and spinal cord in vertebrate species, derived from the neural tube.

peripheral nervous system (PNS)
The entire nervous system other than the central nervous system; it includes the enteric and autonomic nervous system. It is derived from neural crest cells.

2. The neural tube will give rise to the brain and spinal cord, constituting the **central nervous system** (**CNS**).

3. The neural crest cells sandwiched between those two structures will wander all over the body, giving rise to, among other things, the **peripheral nervous system** (**PNS**), the neurons that will send sensory information into the CNS and relay commands from the CNS to the rest of the body (**FIGURE 1.18**).

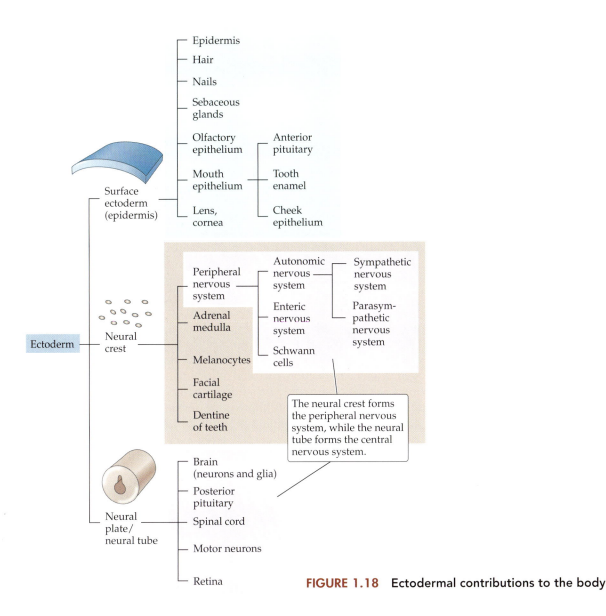

FIGURE 1.18 Ectodermal contributions to the body

If you think about it, your skin represents the interface between you and the outside world, meaning it serves as a two-way conduit of information: the world to you, you to the world. So perhaps ectodermal cells were "preadapted" to take on the role of the nervous system, detecting events in the world and determining what the organism will do in that world.

Once the nervous system appears, we can call the embryo a **neurula** ("NOOR-you-la"). The process by which a gastrula transforms into a neurula is sometimes called **neurulation**. We'll return to the developing nervous system shortly, but to understand the mechanisms at work in that process, we need to return to the work of early embryologists, who discovered the remarkable plasticity of development.

neurula An embryo that has begun forming a nervous system, typically after the completion of gastrulation.

neurulation The process in which an embryo transitions from gastrula to neurula.

How's It Going?

What are the three major contributions of ectoderm, and which two parts will make up the nervous system?

■ **1.9** ■

MANY EMBRYOS, INCLUDING ALL VERTEBRATES, DISPLAY "SELF-REGULATION"

Depending on how much neuroscience you've been exposed to, you may have learned that children have much more plastic brains than adults. Here we're using the term *plastic* to mean "very flexible, capable of a lot of stretching and changing." For example, if surgeons were to remove my left cortical hemisphere, I would lose the power of speech and never gain it back. But there have been several cases of children who, in order to control epilepsy, had their left cortical hemisphere removed, yet they recovered language (Borgstein & Grootendorst, 2002; Smith & Sugar, 1975). This happens because neurons that would have served one purpose can, in children, adapt to serve another purpose, such as language. This ability of young cells to change fates was first discovered by nineteenth-century scientists studying embryos.

Because embryos are small, it's not easy to manipulate individual cells in a blastula. In the first attempts, scientists heated up a fine needle and, guided by a microscope, tried to kill one cell while leaving the other(s) alive. In one study, killing one cell in a two-cell frog embryo seemed to result in formation of only half a neurula (**FIGURE 1.19A**). This result suggested that frog embryos developed in a mosaic fashion, each cell following its predetermined fate, like the worm *C. elegans* that would be studied nearly a century later. But it was always possible that the dead cell affected the development of her neighboring sister. To get around this potential confound, in 1891 Hans Driesch found a way to separate all four cells in an early embryo of a freshwater microscopic invertebrate known as a hydra. He found that each of the four cells went on to produce a full-bodied, if slightly smaller, hydra (**FIGURE 1.19B**). This outcome indicated that every cell in the embryo was capable of producing any type of cell. So, was the very different outcome in frogs due to vertebrates using different developmental mechanisms than hydras, or did the dead frog cell affect its sister? A few years later, Thomas Hunt Morgan was able to gently remove one of the two blastomeres in a frog embryo and found that the remaining blastomere made a whole frog, not half a frog (Morgan, 1895). So neither hydra nor frogs seemed to rely on mosaic specification of cell fate.

In 1903, Hans Spemann (1869–1941) found a way to separate an early newt embryo into two parts, keeping *both* halves alive. His technological innovation was to tie a loose knot in a strand of human hair, position the knot around the blastula, then use forceps to pull the ends of the hair to constrict the blastula. Spemann's hairs were too thick, so he used his baby daughter's fine hair instead (Spemann, 1938). The results matched those of Driesch and Morgan: as long as Spemann made sure there was at least one nucleus in each half of the embryo, two small but complete newt embryos developed (**FIGURE 1.19C**). Eventually, scientists were able to split two-celled embryos in mammals and again found that either cell could form a whole animal (Seidel, 1952). In other experiments, scientists found they could scoop out a clump of cells from a blastula, yet the embryo would go on to form a whole body (**FIGURE 1.19D**). In other words, there was no part of the resulting animal that was missing as a result of the removal of those embryonic cells. For example, when a chick embryo first sprouted recognizable buds where the four limbs would grow, scientists could surgically remove one of the limb buds, and the remaining cells near the site would multiply and fill in for the bud, so the chick would have all four limbs at hatching. Limb bud regeneration would later be demonstrated in rat embryos, too (Deuchar, 1976).

(A) Wilhelm Roux's experiment

Fertilized egg → Cleavage → 2-cell stage

Hot needle

Dead tissue   Living tissue

Blastula

Half-embryo

Destroyed half   Neurula stage

(B) Hans Driesch's experiment

Remove fertilization envelope

Normally, the four cells would make one larva.

Normal larva

Separate into four cells

Note that the four larva are smaller and all differ in size.

Whole individuals developed from each cell of a four-cell embryo.

(C) Hans Spemann's experiment

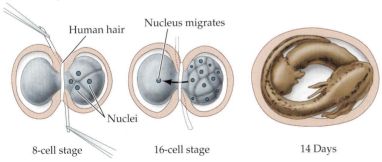

Human hair

Nucleus migrates

Nuclei

8-cell stage   16-cell stage   14 Days

(D) August Weismann's experiment

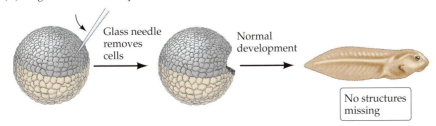

Glass needle removes cells

Normal development

No structures missing

FIGURE 1.19 Demonstrating self-regulation in embryos (A) Early results suggested that amphibians display mosaic determination of cell fate, each half of the two-cell embryo making one half of the body. (B) Driesch's results were very different, suggesting that any of the blastomeres at the four-cell stage could make a complete organism. (C) Using a hair to split an early blastula, Spemann found that, as long as he waited to be sure both halves got at least one nucleus, they would each form a whole salamander. (D) Later in embryonic development, removing a batch of cells does not seem to disrupt the body plan—no particular body structure is missing.

self-regulation Also known simply as *regulation*. Here, the process by which embryos manage to compensate for missing or damaged cells and nevertheless produce an entire individual.

totipotency Total potency; the ability of early embryonic cells to differentiate into any type of cell in the individual.

embryonic stem cells Cells found in embryos that display totipotency.

Embryologists described this ability to change fates in order to compensate for the loss of other cells as **self-regulation**, or simply *regulation*. Self-regulation was seen in the embryos of many different species, invertebrates and vertebrates alike. Just as the brain exhibits less plasticity as we grow older, the ability of embryos to self-regulate becomes more limited as the embryo grows. If you remove a limb bud in a chick embryo late in development, it will not regenerate. And obviously adult vertebrates cannot regenerate amputated limbs (with the fascinating exception of some salamanders [Brockes & Gates, 2014]—how have their limbs retained embryonic self-regulation?). Thus, the later in development you look, the less self-regulation you see. It is as if each cell narrows in on its eventual fate, differentiating in a particular way, and becomes unable to broaden the range of possible cell fates again.

The ability of an embryonic cell to take on any fate is referred to as **totipotency**. In the early blastula, for example, we've seen that any cell can produce an entire newt, as each cell is totipotent. This ability to take on any fate is obviously powerful and is the reason there is great excitement over the possibility that such cells, called **embryonic stem cells** (or simply *stem cells*), may be able to regenerate functioning circuits in the damaged brain or spinal cord (Castillo-Melendez, Yawno, Jenkin, & Miller, 2013) or, for that matter, damaged heart, liver, or lungs (Evans, 2011). Lesion and transplantation studies have made it clear that, as development proceeds, the range of potential fates that cells can take on progressively narrows, so cells are sometimes described as being *pluripotent*, able to take on several fates, but not totipotent. Eventually, each cell differentiates into its adult role and is unable to take on new roles

> As development proceeds, the range of potential fates that cells can take on progressively narrows.

unless it is extensively manipulated, as was the frog epithelial cell that had its nucleus transplanted into an enucleated egg, discussed earlier. Just as that demonstration disproved the idea that cells discard the genes they don't need as they narrow in on their adult differentiation, the demonstrations of self-regulation in embryos also cast doubt on that idea. If any cell can eventually give rise to any structure, then all the cells must retain all the genes needed to differentiate into any type of cell.

The only explanation for self-regulation, this ability of remaining cells to "pitch in" and compensate for missing cells, is that the surviving cells somehow are able to "sense" a loss of comrades and then change their fates. In other words, the cells must be communicating with one another to know what structure each should form.

■ 1.10 ■

SELF-REGULATION SEEMS INCOMPATIBLE WITH MITOTIC LINEAGE–DIRECTED DIFFERENTIATION

Note that these nineteenth-century and early twentieth-century results are the opposite of those later reports we described for the worm *C. elegans*. We described worm development as *mosaic specification of cell fate*, each cell going on to follow the fate predestined by its mitotic lineage, no matter what neighboring cells do. The development seen in these experiments with hydras and amphibians and mammals, in contrast, can be described as **conditional specification of cell fate**, in which cell fate depends on environmental conditions. Cells take on a fate that is appropriate for their location in the body, which is determined by the type of cells that surround them. In other words, cellular differentiation is guided by **cell-cell interactions**, the communication and influence between developing cells.

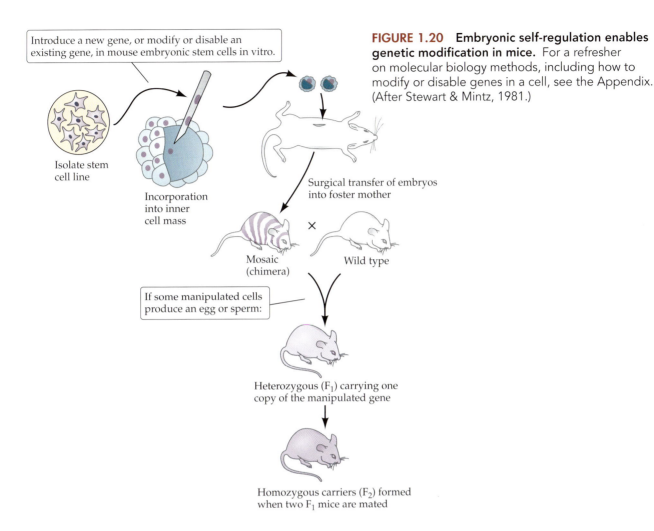

Introduce a new gene, or modify or disable an existing gene, in mouse embryonic stem cells in vitro.

Isolate stem cell line

Incorporation into inner cell mass

Surgical transfer of embryos into foster mother

Mosaic (chimera) × Wild type

If some manipulated cells produce an egg or sperm:

Heterozygous (F_1) carrying one copy of the manipulated gene

Homozygous carriers (F_2) formed when two F_1 mice are mated

FIGURE 1.20 **Embryonic self-regulation enables genetic modification in mice.** For a refresher on molecular biology methods, including how to modify or disable genes in a cell, see the Appendix. (After Stewart & Mintz, 1981.)

Sydney Brenner, who pioneered work on *C. elegans*, has characterized the difference between the worm strategy of mosaic cell differentiation and the more common, cell-cell interaction strategy for determining cell fate as the "European plan" versus the "American plan." The European plan emphasizes "Who is your ancestor?" where each cell uses that information (who was my mother, grandmother, great-grandmother, and so on?) to decide whether to become a neuron or a skin cell, to live or to die. He was likening this to the traditional European reliance on ancestry to decide who would be peasant or lord, commoner or queen. The American plan emphasizes "Who is your neighbor?" where each cell uses that information (what sort of cells are around me?) to decide which cell fate to take on. He was likening this process to the idea, commonly admired in American culture (although one might argue how accurately it describes American society in reality), that a person from any station in life can work to take on any role in society, from trash collector to tycoon, peon to president. It's a matter of being in the right place at the right time.

It is self-regulation that allows cells from two embryos to merge together to make a single body, as is sometimes deliberately done by molecular biologists to produce mice with genetic manipulations. For example, you might introduce a new gene, called a **transgene**, in embryonic stem cells in a dish. If you inject those cells into a blastula with enough skill and luck (or simple perseverance), the manipulated stem cells will be incorporated into the inner cell mass (**FIGURE 1.20**). Now, thanks to self-regulation and the ability of cells to take on whatever fate is appropriate given their neighbors, the stem cells will divide to make their contributions to the animal's body. At this point, the

conditional specification of cell fate   The developmental strategy in which each cell's fate depends on environmental conditions, primarily the fate of neighboring cells.

cell-cell interactions   Here, the process by which developing cells communicate with one another and direct each other's fate.

transgene   A gene that has been artificially introduced into a model organism.

chimera   Here, an individual made up of cells displaying more than one genotype, formed from the combination of cells from two separate zygotes.

transgenic   Referring to an organism in which foreign DNA has been deliberately inserted.

CRISPR   *c*lustered *r*egularly *i*nter-spaced *s*hort *p*alindromic *r*epeats. A system of gene manipulation that evolved in single-celled organisms and is exploited by scientists for gene editing. For more detail about CRISPR and other molecular biology methods, see the Appendix.

knockout   Here, an animal in which a particular gene has been deliberately removed or disabled.

mouse is a mosaic of both the cells of the original blastula, which have one genotype, and the manipulated embryonic stem cells, which have another. Such individuals with cells of two very different genotypes are sometimes called **chimeras** (from the Chimera of Greek mythology, composed of parts of several different animals; note that a modern-day mouse chimera may have four distinct parents). If you're really lucky, some of those embryonic stem cells will contribute to the production of eggs or sperm, in which case when the chimera reproduces, it may produce offspring carrying the gene you wanted. Now you can simply breed those offspring to get more animals, called **transgenics**, carrying that introduced gene. Scientists can also delete an existing gene, which these days is usually done by using **CRISPR** technology described in the Appendix. In this way, one can make a **knockout** animal, in which the gene of interest has been removed or disabled.

Self-regulation also allowed the embryos of Brittany and Abby to merge, seamlessly joining a single pelvis and lower spinal cord to the upper spinal cords and brains of the two sisters. At the points where their cells met, each participating cell simply communicated with the others to produce whatever structures were appropriate for where they happened to end up in the body. There were no gaps between cells from the two different embryos, as local cell-cell communication merged the bottom half of one body to the top halves of two girls.

These early twentieth-century studies of cell-cell interactions directing development generated the concept of induction, as we'll see next.

How's It Going?

1. What is self-regulation, and why does it argue against mitotic lineage–directed cell differentiation?
2. Relate the "European plan" and "American plan" to two different forms of embryonic development.

■ 1.11 ■
EXPERIMENTAL EMBRYOLOGY REVEALED INDUCTIVE PROCESSES UNDERLYING SELF-REGULATION

As the twentieth century began, Hans Spemann spent many hours perfecting his skills in microsurgery, heating glass rods and pulling them to form almost invisibly thin, sharp tips to cut into tiny embryos. He and others had noted that the vertebrate eye develops when the neural tube extends two cup-shaped structures, called the optic cups, out to the epithelium. The optic cup will develop the retina, while the overlying epithelium will produce the transparent lens and cornea that protect the retina while admitting light (**FIGURE 1.21**). Thus the retinas in the backs of your eyes, where a lot of neural processing of visual information takes place, can be legitimately considered extensions of the brain. There were two possible explanations for how the epithelium in front of the optic cup differentiated to form the cornea and lens. One possibility was that this particular stretch of epithelium was always fated to produce cornea and lens, and the optic cup always managed to encounter the epithelium in that particular spot. An alternative possibility, suggested by the many instances of embryonic self-regulation, was that the optic cup instructed whatever epithelium it encountered to produce the tissue needed to complete the front half of the eye. Spemann favored this latter idea, theorizing that the optic cup might release signals to cause the epithelium to change its

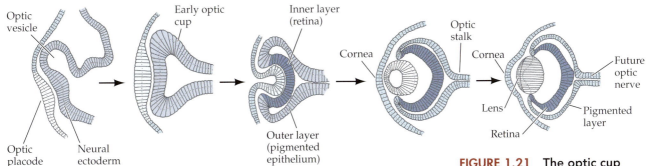

Optic vesicle

Early optic cup

Inner layer (retina)

Cornea

Optic stalk

Cornea

Future optic nerve

Optic placode

Neural ectoderm

Outer layer (pigmented epithelium)

Lens

Retina

Pigmented layer

FIGURE 1.21 **The optic cup meets the epithelium.** (After Cvekl & Piatigorsky, 1996.)

fate, differentiating into cornea and lens rather than skin. This process, when one tissue directs the differentiation of some other tissue, is called **induction**.

To test the idea that the optic cup might induce the overlying epithelium to differentiate into lens and cornea, Spemann first removed the optic cup on one side of a newt and found that the remaining epithelium did not form a lens. This was consistent with his hypothesis, suggesting that this patch of epithelium was not preordained to form an outer eye. But it was always possible that the surgery had damaged the epithelium and kept it from forming a cornea and lens. A better test would be to move an optic cup over so that it lay under a *different* patch of epithelium, to see if the cup could induce a lens and cornea in a *new* location. Spemann had some modest success in such demonstrations, but they were almost immediately mired in controversy as other researchers had mixed success in replicating them. To some extent their poor results could be explained as failure to complete the surgery properly, as it was technically very difficult, and these mixed outcomes were probably also due to differences in the species that various workers used and in which patch of epithelium, exactly, one tried to induce to form a lens. Eventually it became clear that not all epithelium is competent to respond to the optic cup. The optic cup cannot induce a lens in abdominal epithelium, for example, but can do so in epithelium of the head (Grainger, 1992), even if it's not the portion of head epithelium that would normally form a lens (**FIGURE 1.22**).

The conclusion that the optic cup indeed induces lens formation was consistent with the very well established principle of self-regulation. So everyone was looking for examples where having one sort of tissue nearby could alter the differentiation of cells. In Spemann's lab, one study in particular galvanized the field of biology, as we'll see.

induction   Here, the process by which one group of cells directs the differentiation of other, nearby cells.

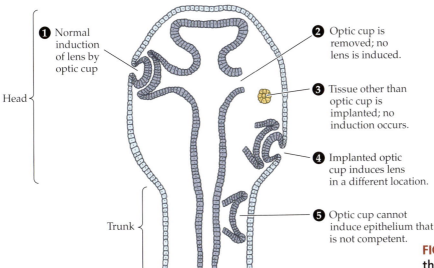

❶ Normal induction of lens by optic cup

Head

Trunk

❷ Optic cup is removed; no lens is induced.

❸ Tissue other than optic cup is implanted; no induction occurs.

❹ Implanted optic cup induces lens in a different location.

❺ Optic cup cannot induce epithelium that is not competent.

FIGURE 1.22 **The optic cup induces the ectoderm to form a lens.**

■ **1.12** ■

A REGION OF THE VERTEBRATE EMBRYO SEEMS TO "ORGANIZE" DEVELOPMENT

By the 1920s, the work on embryonic self-regulation, which indicated that developing cells must be in communication with one another, combined with demonstrations that this communication could allow one set of cells to induce a particular fate in neighboring cells, convinced scientists that conditional specification of cell fate could account for epigenesis, the unfolding complexity from a single zygote that produced various body structures, including the brain. Given this perspective, you could understand, in principle, how the various cells, each communicating with its neighbors, could form an entire individual, with each limb, muscle, and organ where it belonged. But if every blastomere was indeed totipotent, there was still the problem of how the *original* decision was made about which cell, exactly, would get which role. In other words, there must be some *anchor point*, some early event that resulted in the *first* cells getting their instructions, so they could differentiate and induce other cells to take on other, complementary fates until a whole individual was made. A classic study in 1924 seemed to unveil that anchor.

Working at the University of Freiburg, Germany, in the lab of Hans Spemann, who was already famous for work on optic cup induction of epithelium, graduate student Hilde Mangold (née Proescholdt) began transplanting portions of one newt blastula into other blastulas. In order to keep track of which cells came from the donor blastula and which belonged in the recipient blastula, Mangold used two newt species with different pigmentation. Let's see what she found.

dorsal lip of the blastopore
An embryonic region of the blastopore that can induce the development of a second nervous system, and therefore a second individual from an embryo.

organizer   Here, a hypothesized signal from the dorsal lip of the blastopore that induces formation of a nervous system.

─── ■ ───

RESEARCHERS AT WORK

The Dorsal Lip of the Blastopore Can Organize a New Individual

■ **QUESTION**: What is the starting point of embryonic specification of fate, which allows an entire body to form?

■ **HYPOTHESIS**: One of the first recognizable structures at the start of gastrulation is the **dorsal lip of the blastopore**, which will be enfolded into the blastula, forming the three germ layers. Perhaps this first structure directs the fate of other cells.

■ **TEST**: Transplant the dorsal lip of the blastopore from a pigmented newt into the blastula of an unpigmented newt (**FIGURE 1.23A**). Observe subsequent development to see whether the transplanted tissue induces development of any other structures in the host (**FIGURE 1.23B**).

■ **RESULT**: The transplanted dorsal lip not only induced the formation of some structures, it seemed to induce the formation of an *entirely new individual*, with a head and central nervous system (**FIGURE 1.23C**). Importantly, the pigmented donor cells constituted only a minority

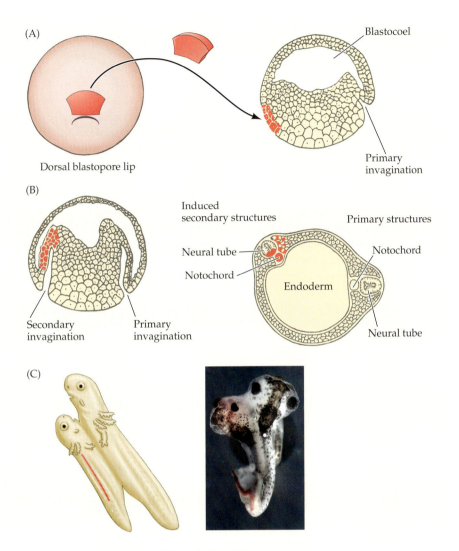

(A)

Blastocoel

Dorsal blastopore lip

Primary invagination

(B)

Induced secondary structures

Primary structures

Neural tube

Notochord

Notochord

Endoderm

Secondary invagination

Primary invagination

Neural tube

(C)

FIGURE 1.23   The dorsal lip of the blastopore can organize a new individual. (A,B after Hamburger, 1988; C photo by Dr. A. Wills, courtesy of Dr. R. Harland.)

of the cells in this second individual, forming the notochord and a few spinal cord cells. The rest of the cells in the supernumerary newt were derived from the host blastula (Spemann & Mangold, 1924). So the transplanted tissue had induced host cells to change their fate, forming a head.

■ **CONCLUSION**: The transplanted tissue seemed to organize the formation of an entire newt, with differentiation of various cells, each in the proper place for the body plan of a newt. Clearly this was an inductive process, like the optic cup inducing ectodermal cells to form a cornea and lens. But unlike the optic cup, which had a fairly limited effect, the dorsal lip of the blastopore appeared to produce an entire animal. Thus Spemann hypothesized that the cells in this region serve as an **organizer** (Spemann, 1921), a "super-inducer" that triggers a cascade of inductive processes, starting with development of the nervous system, to produce an individual. It is like a first falling domino that initiates a long series of events. As gastrulation proceeds, the cells of the dorsal lip are enfolded and form the mesoderm. These mesodermal cells then induce the overlying ectoderm to produce the nervous system.

Tragically, Hilde Mangold never got to see her groundbreaking paper in print, as she died at age 26.

Using the term *organizer* really caught the imagination of developmental biologists, having a massive influence on biology that continues today. Tragically, Hilde Mangold never got to see her groundbreaking paper in print, as she died at age 26 when a gasoline heater in her kitchen exploded. She left behind an infant son, who later died in World War II (Hamburger, 1984). Spemann was awarded a Nobel Prize in 1935, primarily for the organizer, Mangold's dissertation work. But Nobel Prizes are not given posthumously, so it was awarded to Spemann alone.

It would be over 50 years before any techniques would be available to even attempt to isolate and identify the particular molecule(s) from the dorsal lip of the blastopore that organizes a new individual. But in fact, by the 1940s there were doubts about whether the organizer really existed—several people, including another student of Spemann, found that inert implants, or even injections of saline, into a host embryo could sometimes trigger the formation of a second nervous system, just as Mangold's transplants had (Holtfreter, 1944). These results suggested that the Spemann/Mangold dorsal lip transplants held no "organizer" but simply irritated the embryo in some nonspecific way, triggering a birth defect that happened to be another nervous system. Thus people would remark that the organizer had "set back developmental biology for decades." But in the 1990s multiple labs would bring powerful new molecular biological techniques to bear in the search for a signal from the organizer and find not just one, but several!

How's It Going?

Describe the experiments that indicated that there is an "organizer."

■ 1.13 ■

LONG ABANDONED, THE ORGANIZER WAS UNCOVERED THROUGH MOLECULAR BIOLOGICAL TECHNIQUES

Xenopus laevis The African clawed frog, a valuable vertebrate model species.

noggin A gene that encodes the protein noggin, which exerts an organizer-like effect on ectoderm, shifting it from an epidermal to a neural fate.

The advent of molecular biological tools encouraged several labs to look for a substance in the dorsal lip of the blastopore that might be Spemann's fabled "organizer." One lab exploited two manipulations that were known to distort developing embryos in the African clawed frog, ***Xenopus laevis***: exposure to ultraviolet (UV) radiation causes them to develop without any head; on the other hand, exposing embryos to lithium chloride (Li^+Cl^-) causes them to grow enormous heads and brains. Reasoning that the Li^+Cl^- may somehow increase expression of the organizer while UV may eliminate it, they sought a gene that was sensitive to these manipulations, as we'll see next.

RESEARCHERS AT WORK

A Gene Is Discovered that Acts as an Organizer

■ **QUESTION**: Is there a gene expressed in the dorsal lip of the blastopore that produces the effects of the Mangold/Spemann organizer?

■ **HYPOTHESIS**: If so, then manipulations that affect the extent of brain development in *Xenopus* may affect expression of this organizing signal, offering a way to isolate the molecule.

- **TEST 1**: Expose one set of *Xenopus* embryos to Li⁺Cl⁻ to increase neural development, then extract mRNA from the dorsal lip of the blastopore of those embryos, which should be enriched for the organizer signal (if it exists). Inject this mRNA into embryos that have been exposed to UV and therefore normally would not develop a head. Can the mRNA cause a head to form?

- **RESULT 1**: Injecting the mRNA extracts into UV-irradiated embryos "rescued" the animals—they developed normal heads (**FIGURE 1.24A**).

FIGURE 1.24 Isolating *noggin*

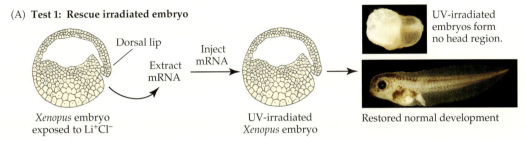

(A) Test 1: Rescue irradiated embryo

Dorsal lip

Extract mRNA

Inject mRNA

Xenopus embryo exposed to Li⁺Cl⁻

UV-irradiated *Xenopus* embryo

UV-irradiated embryos form no head region.

Restored normal development

- **TEST 2**: Extract the RNAs expressed in the dorsal lip of the blastopore of the embryos treated with Li⁺Cl⁻ and make a cDNA library. Inject various subsets of mRNA from this library into UV-treated embryos to see which, if any, can rescue head development.

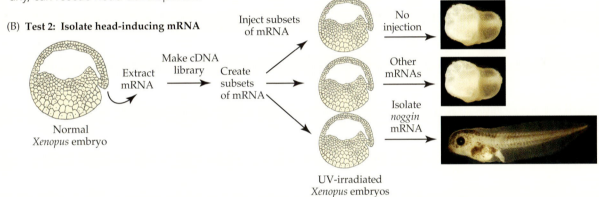

(B) Test 2: Isolate head-inducing mRNA

Extract mRNA

Make cDNA library

Create subsets of mRNA

Inject subsets of mRNA

No injection

Other mRNAs

Isolate *noggin* mRNA

Normal *Xenopus* embryo

UV-irradiated *Xenopus* embryos

- **RESULT 2**: Eventually the scientists isolated a single mRNA, for a gene they named **noggin**, that rescued UV-irradiated embryos (**FIGURE 1.24B**). Furthermore, exposing irradiated embryos to increasing amounts of noggin had a dose-dependent effect on how much of a head developed (**FIGURE 1.25A**). In situ hybridization (see the Appendix) confirmed that *noggin* mRNA is concentrated in the dorsal lip of the blastopore in normal embryos (**FIGURE 1.25B**).

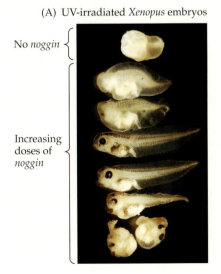

(A) UV-irradiated *Xenopus* embryos

No *noggin*

Increasing doses of *noggin*

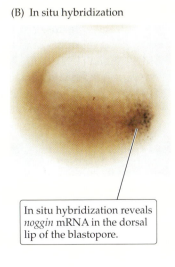

(B) In situ hybridization

In situ hybridization reveals *noggin* mRNA in the dorsal lip of the blastopore.

FIGURE 1.25 Testing the organizing ability of *noggin* (A) UV-irradiated *Xenopus* embryos fail to develop a head, but injecting mRNA for noggin can rescue their development. However, too much noggin results in embryos that are all head (bottom). (B) In situ hybridization reveals *noggin* mRNA in the dorsal lip of the blastopore of normal embryos. (Courtesy of Dr. R. M. Harland.)

- **CONCLUSION**: The signal noggin is found in the organizer region (the dorsal lip of the blastopore) that ends up in the mesodermal tissue, which then induces overlying ectoderm to produce neural tissue.

FIGURE 1.26 Look after your noggin.
(A) The black arrow points out the ear that is present in the wild-type mouse but missing from the *noggin* knockout mouse in (B). (C) The white arrow points to rhombic lip, where the cerebellum would normally form. The rest of the brain is missing from the double knockouts. (From Bachiller et al., 2000.)

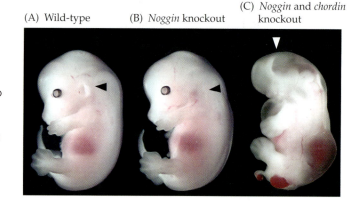

(A) Wild-type   (B) *Noggin* knockout   (C) *Noggin* and *chordin* knockout

chordin A gene that encodes the protein chordin, which exerts an organizer-like effect on ectoderm, shifting it from an epidermal to a neural fate.

follistatin A gene that encodes the protein follistatin, which exerts an organizer-like effect on ectoderm, shifting it from an epidermal to a neural fate.

bone morphogenetic protein (BMP) A class of growth factors that act to encourage ectodermal cells to take on an epidermal, rather than a neural, fate. It is part of the transforming growth factor beta (TGFβ) family.

TGFβ receptors A class of receptors that bind transforming growth factors, including BMP.

With other labs taking other approaches to search for signals from the organizer, the field soon had an embarrassment of riches—not just one signal, but at least three. In addition to *noggin*, genes called **chordin** (Oelgeschlager, Kuroda, Reversade, & De Robertis, 2003) and **follistatin** (Hemmati-Brivanlou, Kelly, & Melton, 1994) were found to be expressed in the dorsal lip of the blastopore, and their protein products could also induce ectodermal cells to differentiate into nervous system. Homologues of these genes were promptly found in mammals, where their role in nervous system development was soon made clear. For example, *noggin* knockouts, mice in which the gene is disabled so that it makes no functional protein, show abnormalities in brain development, and animals with *both noggin* and *chordin* knocked out develop almost no head at all (Bachiller et al., 2000) (**FIGURE 1.26**).

The molecular structure of these proteins indicated that they were indeed secreted from the cells that made them, and it turned out all used the same basic mechanism to induce ectoderm to form the nervous system. Noggin, chordin, and follistatin all act as antagonists for a class of secreted proteins called **bone morphogenetic protein** (**BMP**), a family of growth factors (Chen, Zhao, & Mundy, 2004), one of which (BMP4) is indeed secreted by ectodermal cells. The BMP proteins are part of a larger family called transforming growth factor beta (TGFβ), so the secreted BMP molecules act by binding with **TGFβ receptors**. Thus ectodermal cells both secrete BMP and possess the TGFβ receptor to respond to that signal. In this way they mutually inhibit each other from a neural fate. When the mesoderm slides beneath the ectoderm (**FIGURE 1.27A**), it releases organizers to disrupt that signaling. By blocking TGF-β receptors, the mesoderm induces the overlying ectoderm to become nervous system instead of skin (**FIGURE 1.27B**). Without blockade of BMP signaling, the ectodermal cells will become epithelial cells instead of nervous system and the embryo will develop without a brain (or a head).

> Ectodermal cells are constantly inducing each other to become epithelial cells, but organizers block that signal to induce neural development.

So the organizer seems to work like this: Ectodermal cells secrete BMP, stimulating their ectodermal neighbors through the TGFβ receptor. As long as they all keep sending and receiving this signal, these cells will differentiate into epidermis (skin). If one of these cells is isolated, deprived of BMP signals from neighbors, it will become a neural cell, which we can consider a "default" fate. So the ectodermal cells are suppressing each other from that fate, constantly inducing each other to become epidermis. But when the underlying mesoderm secretes the organizer proteins, they interfere with that BMP signaling, so the ectodermal cells switch fates, becoming neural cells instead.

First this signaling mechanism induces the formation of the neural plate. Then the mesodermal cells secreting the organizers compress laterally until

(A)

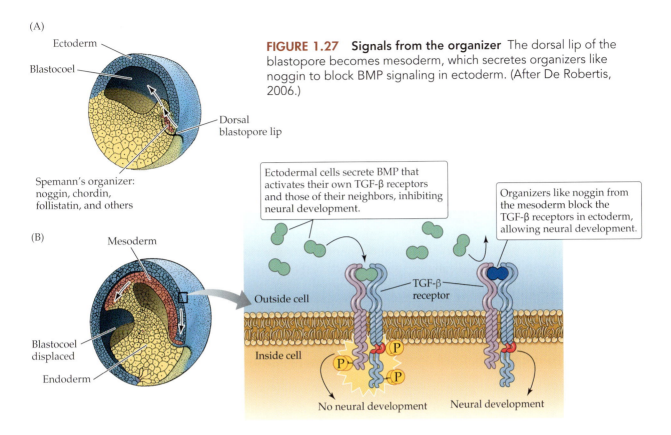

FIGURE 1.27 Signals from the organizer The dorsal lip of the blastopore becomes mesoderm, which secretes organizers like noggin to block BMP signaling in ectoderm. (After De Robertis, 2006.)

Ectoderm

Blastocoel

Dorsal blastopore lip

Spemann's organizer: noggin, chordin, follistatin, and others

(B)

Mesoderm

Blastocoel displaced

Endoderm

Ectodermal cells secrete BMP that activates their own TGF-β receptors and those of their neighbors, inhibiting neural development.

Organizers like noggin from the mesoderm block the TGF-β receptors in ectoderm, allowing neural development.

Outside cell

TGF-β receptor

Inside cell

No neural development

Neural development

they form that rodlike structure on the midline, the notochord (see Figure 1.17). As the neural groove forms, these signals from the notochord continue shifting ectodermal cells from an epidermal (skin) fate to a neural fate. Then the ectodermal cells meet at the top of the neural groove, forming the neural tube, which detaches from the rest of the ectoderm. The cells of that ectodermal layer above the neural tube are exposed to few signals from the organizer, as the neural tube is between them and the notochord. That means the overlying ectodermal cells, which have been secreting BMP all along, continue to activate their own and their neighbors' TGFβ receptors, and they collectively steer away from a neural fate and start differentiating into epidermis.

In fact, there is a dose-response relationship between the BMP blockade from the notochord and the differentiation of overlying tissues. The neural tube gets the highest dose of organizers, and therefore the strongest BMP blockade, and so forms the central nervous system (**FIGURE 1.28**). The overlying ectodermal cells experience the least BMP blockade (i.e., the greatest BMP signaling) and so become epidermis. Those ectodermal cells that are stranded between the neural tube and the overlying epidermis experience only a moderate blockade, which directs their fate to become the neural crest, which will form the peripheral nervous system (and several other structures; see Figure 1.18).

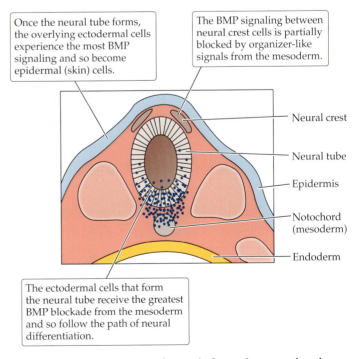

Once the neural tube forms, the overlying ectodermal cells experience the most BMP signaling and so become epidermal (skin) cells.

The BMP signaling between neural crest cells is partially blocked by organizer-like signals from the mesoderm.

Neural crest

Neural tube

Epidermis

Notochord (mesoderm)

Endoderm

The ectodermal cells that form the neural tube receive the greatest BMP blockade from the mesoderm and so follow the path of neural differentiation.

FIGURE 1.28 Gradient of signals from the notochord

◼ **1.14** ◼

WHAT ORGANIZES THE ORGANIZER?

The higher expression of several signaling proteins, including noggin, chordin, and follistatin, from the organizer region in the dorsal lip of the blastopore, cells that end up as the mesodermal layer underlying the ectoderm, induces those cells to differentiate into the nervous system. But wait a minute—how did the dorsal lip of the blastopore come to express the organizer signals in the first place? From our discussion of *C. elegans*, you may have already anticipated that to understand how those signals came to be in one part of the embryo, we will eventually need to consider maternal factors. Let's begin by confining ourselves to how the organizer comes to be in the dorsal lip of the blastopore (which will wind up in the mesoderm and, eventually, the notochord).

> How did the dorsal lip of the blastopore come to express the organizer signals in the first place?

The first part of the answer to that question is that cells in the endoderm induce the cells in the dorsal lip of the blastopore to start expressing organizer signal genes so that, when those cells get enfolded during gastrulation to form mesoderm, they will be secreting organizer signals to induce the overlying ectoderm to make neural tissue. You may be getting a headache by this point—the endoderm induces the dorsal lip of the blastopore, which becomes the mesoderm and then induces the ectoderm to make the brain? It may sound overcomplicated, but it is just this sort of chain of inductive events that you would need to make a really complicated metazoan like a frog (or a human). And one advantage of having many different inductive steps is that if a chunk of embryo is removed or damaged, the embryo will self-regulate in compensation, and you'll still end up with a complete individual.

β-catenin A transcription factor that plays a role in several stages of neural development.

The protein that endodermal cells secrete to induce the dorsal lip of the blastopore to start making organizers is called β-**catenin** (**FIGURE 1.29**). This protein is a transcription factor that plays a role in many steps in embryonic development. In early blastulas, maternally deposited mRNA for β-catenin is expressed pretty much everywhere. But as development proceeds, β-catenin mRNA and protein become concentrated in a quadrant just ventral to where the dorsal lip of the blastopore will form, which by the way is directly opposite where the sperm penetrated the egg at fertilization. How does the β-catenin come to be concentrated in that location? Many players are involved, including enzymes that break down β-catenin and factors that inhibit those enzymes (Gilbert & Barresi, 2016). Those various proteins were unequally distributed in the egg cytoplasm, such that the early divisions of the embryo resulted in the concentration

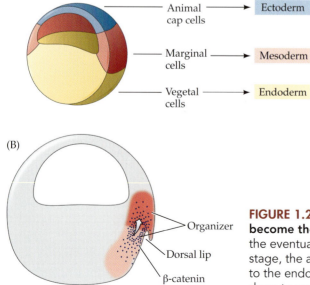

(A)

Animal cap cells ──→ Ectoderm

Marginal cells ──→ Mesoderm

Vegetal cells ──→ Endoderm

(B)

Organizer

Dorsal lip

β-catenin

FIGURE 1.29 Endodermal β-catenin induces mesoderm to become the organizer. (A) Portions of the blastula are named for the eventual germ layer they will make up in the gastrula. (B) At this stage, the accumulation of β-catenin in the region that will give rise to the endoderm induces the nearby cells that will give rise to mesoderm to express the organizers.

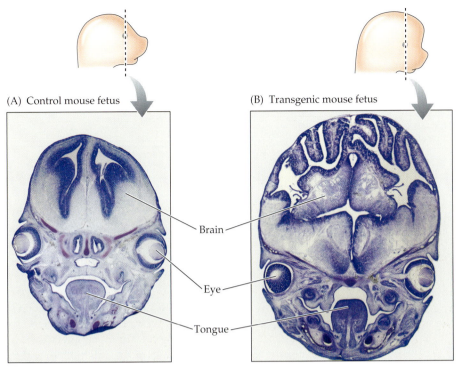

(A) Control mouse fetus

(B) Transgenic mouse fetus

Brain

Eye

Tongue

FIGURE 1.30 Too much orga-nizer of the organizer (A) This cross section shows the extent of brain development in a late fetal wild-type mouse. (B) In transgenic mice that overex-press β-catenin, too much neural development occurs and they die at birth. (From Chenn & Walsh, 2002, photos courtesy of Dr. Anjen Chenn.)

of β-catenin in that one region next to the blastopore. So the mother strategically placed mRNAs and proteins in the egg cytoplasm so that β-catenin would be concentrated in the right place in the endoderm to induce the nearby dorsal lip of the blastopore that would form the mesoderm to become the organizer, which then would induce overlying ectoderm to become the nervous system, and so on until an individual was formed. Whew!

While most of these mechanisms were worked out in studies of *Xenopus*, there have been several demonstrations of the importance of β-catenin for forming the nervous system in mammals. For example, in transgenic mice that overexpress β-catenin, prenatal development of the brain is so amplified that it cannot fit in the animal's normal skull (**FIGURE 1.30**).

While the details differ considerably from what we've seen in vertebrates, most invertebrate embryos display self-regulation and many examples of conditional regulation of cell fate, including in those cells that will form the nervous system. In fact, the same mesodermal induction of ectoderm to become neural tissue is at work in insects, too, as we'll see next.

How's It Going?

How do maternal effects lead to induction of the organizer?

■ 1.15 ■

IN INSECTS, EPIDERMAL CELLS COMPETE TO BECOME NEUROBLASTS

In comparing development of the nervous systems in vertebrates and inver-tebrates such as insects, the first difference we have to deal with is relatively superficial—the vertebrate nervous system is close to the dorsal surface of the body, our back, while the nervous system typical of insects and other invertebrates is close to the ventral surface, the belly. Nearly 200 years ago, Étienne Geoffroy Saint-Hilaire (1772–1844) proposed an explanation for this difference that remains controversial today (St.-Hilaire, 1822). He suggested

FIGURE 1.31   Did a vertebrate ancestor start swimming upside down?

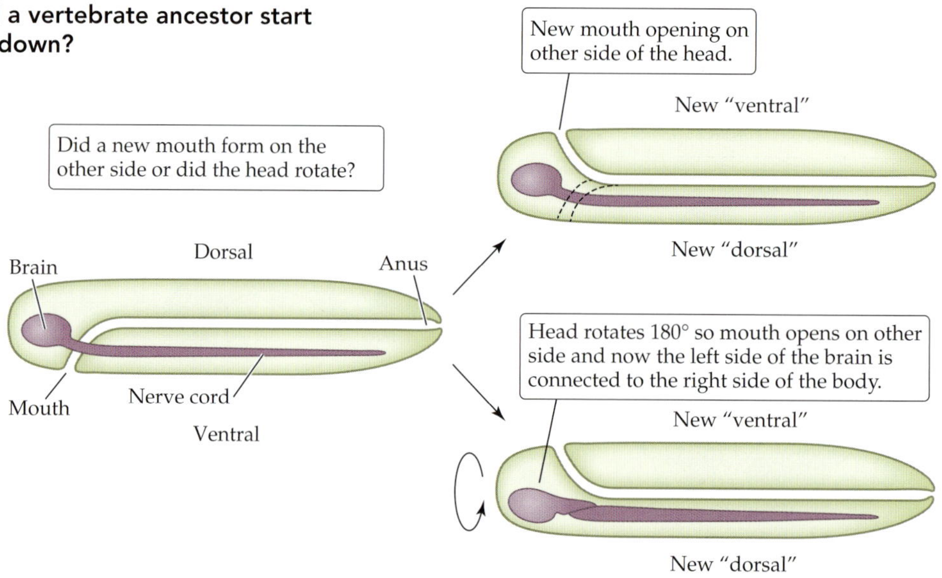

New mouth opening on other side of the head.

New "ventral"

New "dorsal"

Did a new mouth form on the other side or did the head rotate?

Brain

Dorsal

Anus

Mouth

Nerve cord

Ventral

Head rotates 180° so mouth opens on other side and now the left side of the brain is connected to the right side of the body.

New "ventral"

New "dorsal"

that an aquatic ancestor of the vertebrates had its nervous system in the belly, as invertebrates do today, that is, on the same side of the body as the mouth. But what if at some point this species evolved a new mouth opening on the opposite side of the body? Maybe there was some advantage to swimming upside down, and in that case natural selection favored having the mouth move to facilitate feeding. Then the mouth would be on the opposite side of the body from the nervous system (**FIGURE 1.31**). Thus our arbitrary terminology, that the ventral side is the one on which the mouth opens, would mislead us into thinking the nervous system switched sides in vertebrates when, in fact, it may be the *mouth* that switched sides (Mizutani & Bier, 2008).

> **What if at some point this species evolved a new mouth opening on the opposite side of the body?**

Whether it was the mouth or the nervous system that switched sides, the many parallels we'll see between vertebrates and invertebrates indicate that the common ancestor of the invertebrates and vertebrates produced a nervous system much as we do today—by having mesoderm direct ectoderm to produce either neural tissue or skin.

Formation of neural tissue by insect ectodermal cells turned out to involve many genes that were first identified by geneticists studying the fruit fly ***Drosophila melanogaster***. Mutations in those genes altered the fate of epidermal tissues and, so, often affected the external appearance of the flies, making their skin more or less hairy, smoother or rougher. These genes would prove to be part of a signaling pathway determining whether ectodermal cells become epidermis (skin) or neural tissue.

Four of the genes, including those called *achaete* and *scute*, were identified as essential for the development of the nervous system in *Drosophila* so were termed **proneural genes** (the prefix *pro* here means that their expression is a prerequisite for neuronal differentiation). The protein products of these genes bind together to form an **achaete-scute complex (AS-C)**, which serves as a transcription factor (Alonso & Cabrera, 1988). The expression of AS-C is the first step ectodermal cells must take on the path to becoming neural cells.

What determines whether ectodermal cells will express AS-C? In insects, as in vertebrates, mesodermal factors induce ectodermal cells to take on neural fates rather than form epithelium (skin). In vertebrates, the mesoderm secretes organizing signals like noggin and chordin, which interfere with BMP signaling among ectodermal cells to shift their fate from epidermal to neural. In insects, the mesoderm releases a chordin homologue called **sog** (short gastrulation), which

---

Drosophila melanogaster   The common fruit fly, a valuable model organism for studies of genetics and cellular differentiation.

proneural genes   A collection of genes that tend to be expressed in cells that will go on to differentiate into neurons and glia.

achaete-scute complex (AS-C)   The complex of proneural gene products that bind together and serve as transcription factors to direct early differentiation into neurons.

sog   A gene that encodes the protein sog, an insect homologue of chordin, which blocks BMP signaling to direct ectodermal cells to differentiate into a neural, rather than epidermal, fate.

(A) Vertebrate

Vertebrate organizers from the mesoderm block BMP signaling in the dorsal ectoderm to induce those cells to form neural tissue rather than skin.

(B) *Drosophila*

In insects, the chordin homologue sog interferes with signaling of the BMP homologue dpp in ventral ectoderm to induce those cells to form neural tissue.

FIGURE 1.32 **Mesoderm induces ectodermal neuralization in vertebrates and insects.** (After Ball et al., 2004.)

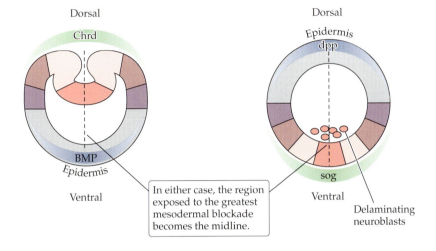

Dorsal

Chrd

BMP

Epidermis

Ventral

In either case, the region exposed to the greatest mesodermal blockade becomes the midline.

Dorsal

Epidermis

dpp

Ventral

sog

Delaminating neuroblasts

interferes with signals from a BMP homologue called **dpp** (decapentaplegic) that would normally direct ectoderm to become skin (**FIGURE 1.32**). When sog interferes with that signal, the ectoderm starts differentiating into neural tissue, and the first step on that pathway is expression of the proneural AS-C genes.

Mapping the expression of the proneural genes in early embryos revealed they were expressed not in the entire ectodermal sheet, but in *rosettes*, circular collections of cells along the ventral ectoderm, which at that stage is a single cell layer (**FIGURE 1.33**). These rosettes are not scattered willy-nilly in the ectoderm, but are spaced out in a very regular, grid-like fashion, on the left and right sides of each body segment, corresponding to the positions of the ganglia that will serve each segment (Doe, 1992).

In vertebrates, the mesodermal induction of ectodermal cells to form the nervous system causes those ectodermal cells, as a sheet, to bend and grow to form the neural tube that ends up inside the body cavity. In insects, the mesodermal influence doesn't induce an entire sheet of cells to become neural, but rather causes many individual cells to move, one by one, into the body interior and start dividing to form individual clusters of neural cells

dpp The gene *decapentaplegic*, which encodes the protein dpp, an insect homologue of bone morphogenetic protein (BMP), which induces ectodermal cells to an epidermal, rather than a neural, fate.

(A) Low power view

(B) Rosette of AS-C-expressing cells

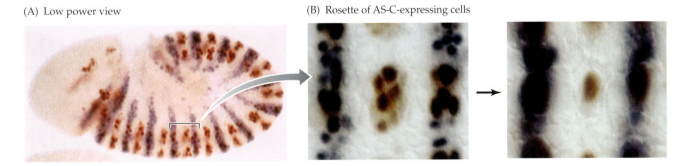

FIGURE 1.33 **Many are called, but few are chosen.** (A) At low power, rosettes of cells expressing AS-C, the proneural genes (brown), are seen in each segment. (B) Early on a cluster of cells express proneural genes, but eventually only one cell continues to do so. (A from Skeath et al., 1992; B courtesy of Dr. James Skeath.)

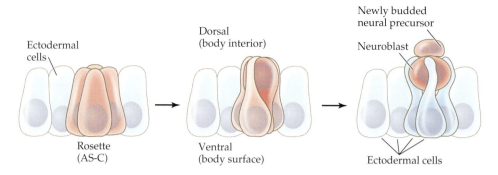

Ectodermal cells

Dorsal (body interior)

Newly budded neural precursor

Neuroblast

Rosette (AS-C)

Ventral (body surface)

Ectodermal cells

FIGURE 1.34 A neuroblast is chosen. Within each rosette of AS-C–expressing ectodermal cells, one cell, typically near the center, intensifies AS-C expression while its neighbors cease expression. This cell delaminates, rising above its fellows into the body interior, and begins dividing, producing cells that will provide the neurons and glia for the segmental ganglion in that part of the insect. (After Doe, 1992.)

delamination Here, the process by which one cell in a cluster of proneural gene–expressing cells detaches from the sheet of neighboring cells to enter the insect body interior. The cell then differentiates into a neuroblast.

neuroblast A cell that will divide to produce neural cells.

(see Figure 1.32B). That decision about which particular cells will migrate to the body interior to produce the nervous system relies on a second signaling system, this time between individual ectodermal cells.

We see evidence of this second signaling system at work when, as development proceeds, AS-C expression intensifies in one cell from each rosette, typically near the center, while AS-C expression dims in its neighbors (see Figure 1.33B). Soon the single cell strongly expressing AS-C within each rosette breaks out of the layer in a process called **delamination**, rises into the body cavity, and continues differentiation. Each delaminated cell then becomes a **neuroblast**, a cell that divides to produce neural progeny, in this case the neurons and glia that will populate that particular segmental ganglion (**FIGURE 1.34**). The ectodermal cells left behind in the single layer then divide and differentiate to epidermal cells forming the ventral skin.

The fact that AS-C expression changes in each rosette, from being expressed in all the cells in the rosette to being expressed in only the one that delaminates, suggests that there is some competition going on among the cells in the rosette, to determine which will take on the neuroblasts' fate. That suggestion was confirmed when scientists used a laser to kill a cell in the process of delaminating. Soon another cell among the rosette began delaminating to take the place of the destroyed cell (**FIGURE 1.35**). Zapping this second neuroblast candidate resulted in the delamination of yet another ectodermal cell (Doe & Goodman, 1985). These results make it clear that the cells within each rosette are communicating with each other, such that if one cell has begun delamination, the others stay put. But if the delaminating cell

> There is some competition going on among the cells in the rosette, to determine which will take on the neuroblasts' fate.

FIGURE 1.35 Ectodermal cells compete for the role of neuroblast. Zapping a delaminating ectodermal cell causes other cells to delaminate to take its place.

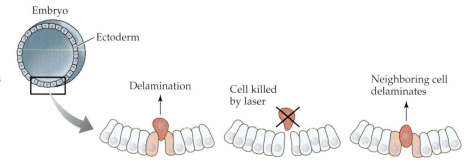

Embryo

Ectoderm

Delamination

Cell killed by laser

Neighboring cell delaminates

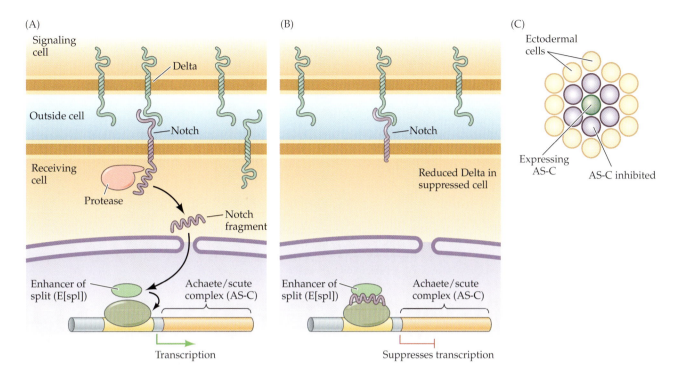

FIGURE 1.36 Delta activates Notch to suppress proneural gene expression. (After Dr. K. Koziol-Dube, Pers. Comm.)

is eliminated, the other cells in the rosette detect this loss and start competing to be the neuroblast. Here we are seeing embryonic self-regulation at the level of individual cells, rather than whole structures like limb buds.

How do the cells within a rosette communicate with each other, and how is it that one, and only one, cell within a rosette delaminates to become a neuroblast? The answer was first hinted at in the 1930s, when a mutation of a gene called *Notch* in fruit flies resulted in an overabundance of developing neuroblasts (Poulson, 1937). It turned out that **Notch** was a receptor protein (Wharton, Johansen, Xu, & Artavanis-Tsakonas, 1985) that is normally stimulated by another protein, **Delta** (Fehon et al., 1990). The proneural cells within a rosette use these proteins, Notch and Delta, to communicate with each other so that eventually one and only one cell within each cluster differentiates into a neuroblast.

How does this work, exactly? First, one of the actions of the proneural AS-C complex is to promote expression of both *Notch* and *Delta* genes, thus all the cells within a rosette are making these proteins, which are inserted into their cytoplasmic membranes, allowing the cells to communicate with each other. The extracellular domain of the Delta protein in one ectodermal cell binds to the extracellular domain of Notch in an adjoining cell. The binding of Delta to the extracellular domain of Notch causes the intracellular portion of the Notch protein to detach (**FIGURE 1.36A**). This Notch fragment binds to yet another protein, called **enhancer of split (E[spl])**, and the two form a heterodimer that acts as a transcription factor to inhibit any further expression of the proneural AS-C genes (**FIGURE 1.36B**).

Thus within each cluster, one AS-C expressing cell comes to suppress its neighbors (**FIGURE 1.36C**). Using Delta to activate Notch in the cell next door causes that cell to stop producing AS-C, which means it will start making less Delta to compete in the contest. Eventually, one cell makes more Delta than

Notch   A membrane-bound protein that binds to Delta found on the surface of adjoining cells.

Delta   A membrane-bound protein that binds to Notch found on the surface of adjoining cells. Delta directs that target cell away from a neural fate and toward an epidermal fate.

Enhancer of split, or E(spl) A protein that dimerizes to a Notch fragment. Together they suppress proneural gene expression.

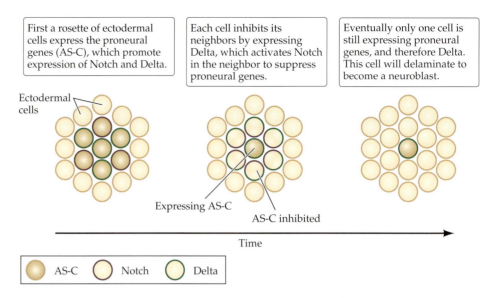

First a rosette of ectodermal cells express the proneural genes (AS-C), which promote expression of Notch and Delta.

Each cell inhibits its neighbors by expressing Delta, which activates Notch in the neighbor to suppress proneural genes.

Eventually only one cell is still expressing proneural genes, and therefore Delta. This cell will delaminate to become a neuroblast.

Ectodermal cells

Expressing AS-C

AS-C inhibited

Time

AS-C   Notch   Delta

FIGURE 1.37 Lateral inhibition in insect ectodermal cells

lateral inhibition   The process by which neighboring cells in a tissue layer inhibit one another, as in the competition between ectodermal cells for a neural fate that is mediated by the Delta-Notch system.

its neighbors and thereby suppresses those cells from a neural fate, and they remain in the layer to become epidermal cells instead. This process, of every cell trying to suppress its neighbors until eventually one cell (or, in other cases, a few cells) gets the upper hand, is called **lateral inhibition** because each cell is trying to inhibit cells on either side (**FIGURE 1.37**).

If the first victor in the Delta-Notch competition dies (or is killed by a laser from a researcher), the remaining cells are released from that suppression and renew the competition among themselves until there is another winner, which then delaminates to become a neuroblast. Why does the eventual neuroblast tend to come from the center of each rosette? Probably because, being in the center of the mesodermal signaling for proneural gene expression, it got slightly more signal and therefore began with a slight edge in Delta. Given the rules for lateral inhibition, slightly more Delta early on will lead to an even greater difference as development proceeds. In other words, the rules are such that those rich in Delta are likely to get richer.

Why does the eventual neuroblast tend to come from the center of each rosette?

Earlier we asked, What organizes the organizer? Here we can ask, How do these clusters of cells expressing proneural genes come about in the first place? So far all I've said is that the signals come from mesoderm. How the mesoderm comes to release this neutralizing signal hasn't been answered completely, but scientists do know that proneural gene expression is eventually decided by the action of a hierarchy of genes that set down the entire body patterns of fruit flies, mice, and humans, which is the subject of the next chapter.

How's It Going?

Describe how Delta-Notch signaling mediates competition among ectodermal cells for selection of a neuroblast.

SUMMARY

■ *Epigenesis* is the progressive development of increasingly complex tissues and structures as a result of ongoing differentiation. **See Figure 1.5**

■ *Cellular differentiation* in *metazoans* represents the selective expression of a particular subset of genes, as regulated by *transcription factors*. **See Figure 1.6**

■ In *C. elegans*, cell division allocates the transcription factors that were unevenly distributed in the egg (a *maternal effect*), and consequently each resulting cell has a unique mix of transcription factors, therefore a unique complement of gene expression, and therefore a particular *fate* depending on *mitotic lineage*. **See Figure 1.10**

■ In other species, this *mosaic specification of cell fate* through mitotic lineage has been supplanted by *conditional specification of cell fate* through *cell-cell interactions*. This means that embryos of most species display extensive *self-regulation*, with cells directing each other's fates, as when the optic cup *induces* formation of a cornea and lens. **See Figure 1.22**

■ In vertebrates, the *organizer*, a group of cells of the *dorsal lip of the blastopore*, releases at least three signals (*noggin, chordin, follistatin*) that block BMP signaling to induce neural formation. The organizer cells migrate to become mesoderm and, eventually, the *notochord*. **See Figure 1.23**

■ The nervous system arises when *mesodermal* tissue blocks *BMP* stimulation of *TGFβ receptors* in overlying *ectoderm* so the tissue follows a neural fate rather than becoming *epidermis*. **See Figure 1.27**

■ The organizer induces the dorsal ectoderm to form the *neural tube*, the future *central nervous system*, and the *neural crest*, the future *peripheral nervous system*. **See Figure 1.28**

■ The induction of an organizer in the dorsal lip of the blastopore is itself a result of β-catenin expression in the underlying *endodermal* tissue, which is an indirect result of maternal effects. **See Figure 1.29**

■ In insects, mesoderm secretes the chordin homologue *sog* to block BMP-like signaling in ventral ectoderm, inducing expression of the *proneural genes AS-C*, which are necessary, but not sufficient, for the cells to take a neural fate. **See Figure 1.32**

■ Within a rosette of AS-C–expressing cells, each exerts *lateral inhibition* to suppress proneural gene expression of its neighbors through a system of *Notch-Delta* signals. **See Figure 1.33**

■ The cell that produces the most Delta activates the Notch receptors of its neighbors to suppress their expression of AS-C, and then it *delaminates* to become a *neuroblast*, which will give rise to the segmental ganglion in that region. **See Figure 1.36**

Go to the Companion Website
sites.sinauer.com/fond
for animations, flashcards, and other review tools.

CHAPTER

2

Coordinating Fates
DEVELOPMENT OF A BODY PATTERN

TOO MANY BODY PARTS, OR TOO FEW Young Inigo Montoya watches his father forge a special sword, perfectly balanced for a mysterious man who has six fingers on his right hand. But when the sword is finished, rather than pay for the masterwork, the six-fingered man kills Inigo's father and slashes the young boy's cheeks. In William Goldman's novel and screenplay *The Princess Bride*, Inigo grows up to be a master swordsman, scouring the earth to identify the six-fingered man so he can avenge his father's murder. Meanwhile, the six-fingered man, a sadist named Count Rugen, conspires to kidnap and murder a beautiful farm girl, setting events in motion that will bring Inigo Montoya and his nemesis together.

Count Rugen is just one of a long line of fictional villains who have too many or too few body parts. Other six-fingered villains appear in the Beatles' movie *Yellow Submarine*, the TV series *Monk*, and the novel (but not the movie) *The Silence of the Lambs*. But perhaps the oldest villain with an abnormal number of body parts is Polyphemus (meaning "many legends"), the ogre-like giant with only one eye, whom Odysseus encounters in the ancient Greek epic poem, *The Odyssey*. Polyphemus is better known for his single eye as Cyclops (but in fact that was the name of the island he lived on, today called Sicily). There is evidence that myths of evil one-eyed monsters may be far older than the *Odyssey* (d'Huy, 2013), stemming from prehistoric times, an example of the vilification of people with physiques that differ from the norm (Chin, 2004).

CHAPTER PREVIEW

In this chapter we'll learn about the mechanisms that typically organize a body plan with the usual number of parts, including brain regions, in the usual places, and how interference with the function of particular genes can result in too many or too few structures. To understand how developing individuals manage to make all the right cells in all the right places, we'll begin by asking why various species that have such different bodies as adults often look so much alike as embryos. That leads to work showing how the unfolding of a body plan begins when the mother provisions signals, either proteins or mRNAs, in particular portions of the egg. Our best understanding of subsequent events comes from work in fruit flies, where a series of changes in gene expression lays down a basic distinction between the head and tail ends of the animal. Then we'll follow the cascade of gene regulation, all mediated by interactions between cells, that demarcates the rest of the body such that cells in different regions are instructed about which structures they are to build. In some cases, the mutation of a single gene in flies will be enough to induce cells that should have made one body part, such as an antenna, to make a very different body part, such as a leg.

This remarkable unfolding of body structure in flies was at first viewed as important only for specifying the fate of each of the body segments that are so prominent in insects. But the genes that proved crucial for establishing a body plan in flies were found to have homologues in vertebrates, in fact usually several different homologous versions of each fly gene. Not only were these same genetic players at work in establishing the vertebrate body plan, but they interacted with each other in much the same way, including the same signaling systems between cells. It soon became obvious that vertebrate bodies and nervous systems are segmented, too. By the chapter's end, we'll see that these signaling systems play a crucial role in specifying the fate of the first neurons to differentiate in vertebrates, the motor neurons of the spinal cord. We'll also understand how the basic layout of the nervous system is established, which will set the stage for Chapter 3, where we'll watch newly differentiated neurons migrate to their final destinations.

■ 2.1 ■

DARWIN NOTED THAT VERTEBRATE EMBRYOS START OFF LOOKING ALIKE

In his landmark book *On the Origin of Species*, Darwin emphasized the importance of embryology as a source of support for his theory of natural selection. He cited many embryologists who noted that the earlier one looks in development, the more similar the body plan appears across species. For example, the American naturalist Agassiz tells Darwin that "having forgotten to [label] the embryo of some vertebrate animal, he cannot now tell whether it be that of a mammal, bird, or reptile" (Darwin, 1859) (**FIGURE 2.1**).

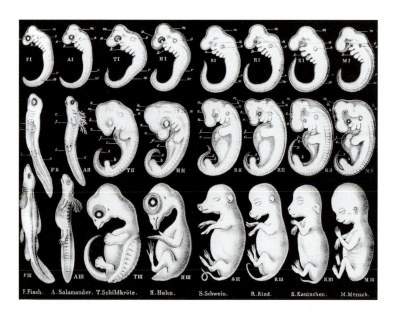

FIGURE 2.1 **Ernst Haeckel's illustration of the similarity of early embryos** Haeckel's illustration of early embryos emphasized similarities, but they are not as similar as Haeckel claimed.

For Darwin, the explanation for the similarity among all vertebrate embryos is that the common ancestor of all the vertebrates also looked like that as an embryo. It was another clue to his idea that new species arose, and changed across time, as a result of "descent with modification." The commonality in embryonic structure across species was evidence of a common ancestor. Why do the bodies of various species differ more from each other as adults than they do as embryos? Darwin reasoned that if a new trait is adaptive for an adult of the species, to compete for food and mates or to evade predators, then there is no necessity, in terms of natural selection, to have that trait appear early in life. If the trait is only adaptive for individuals once they are adults (which would certainly be true of traits that help in competition for mates), then there is no selective advantage for the trait to appear earlier in life.

> If a new trait is adaptive for an adult of the species, there is no necessity, in terms of natural selection, to have that trait appear early in life.

To support his reasoning, Darwin noted that the *adults* of various strains of domesticated pigeons looked very different from one another, in terms of the width and shape of the beak, the proportions of the wings and feet, and so on, because humans selectively bred adults that displayed the characteristics the humans favored. But when he measured the various pigeons as *hatchlings*, he found they were nearly identical. The artificial selection that humans used to create the diverse breeds of pigeons, choosing which individuals to mate and which to discard, focused on their *adult* qualities, not their qualities as hatchlings. In this case, very diverse physical traits were made possible by altering late development, after the embryonic stage.

Darwin concluded that very diverse physical traits, including evolutionarily newly arisen traits that are very adaptive, can arise rather late in development. Thus embryos of related species, such as the vertebrates, are similar to one another because natural selection tends to favor modifications that happen late in development. Indeed, we talk of the fierce competition among individuals in wild populations as "nature red in tooth and claw," but it's significant that most mammals are born without either teeth or claws. They provide a selective advantage only later in life, and so that's when they tend to grow out. An exception is in spotted hyenas, where newborn siblings fight and even kill one another to establish dominance. Not only are they born with teeth, they also arrive with very strong neck muscles to effectively attack their rivals (Frank, Glickman, & Licht, 1991).

So first of all, modifications that appear late in life can be very effective at providing new traits to provide a selective advantage. But there may be another reason why embryos of different vertebrate species look so similar—most modifications of early embryonic events are lethal. For example, the early processes of gastrulation that we described in Chapter 1 involve a very precise sequence of events, with one set of tissues inducing others to take on particular fates, and those tissues in turn altering the fates of yet other, later-arising cells. One little mistake early in development may be magnified as it affects later events. We've seen that embryos can self-regulate, to some degree, if some cells are lost or damaged. But in those experiments, the resulting individuals are usually somewhat deformed, such as Driesch's four stunted hydras that resulted from dividing the four-cell embryo (see Figure 1.19). So despite the ability of embryos to self-regulate, early perturbations in development usually have severe, and often catastrophic, effects.

To illustrate why alterations late in development can produce such varied species, and why alterations early in development are likely to be lethal, I like to make an analogy to donut shops. If you enter a donut shop, you may be struck by how many different shapes and colors of raised donuts are on display—regular glazed donuts, but also bear claws, twists, and jelly-filled

BOX 2.1

A STEP TOO FAR

One of Darwin's most prominent early supporters, the German biologist Ernst Haeckel (1834–1919), went a step further in seeing the similarity of vertebrate embryos as a proof of evolution. Haeckel declared that not only were the embryos of various vertebrates similar to each other, but they re-created simulations of the *adult* features of their ancestors. For example, all early vertebrate embryos have structures called the pharyngeal arches that in fishes will develop into gills. In mammals, they will form the mouth, jaw, and throat. But Haeckel declared that the arches were not just the *precursors* of gills, but actually gills, and for him, the fact that mammalian embryos had these gills was proof that one of their ancient ancestors was a fish. For Haeckel, there was a mystical aspect to his idea that "ontogeny recapitulates phylogeny"—a German Romanticist idea of a "great chain of being" signifying progress for life on Earth. Even in his day, many biologists rejected this idea, and modern-day biology thoroughly rejects Haeckel's recapitulation idea.

Yes, early human embryos have structures that *resemble* the gills of adult fishes, but in fact these structures (called *pharyngeal arches* or *branchial arches*, to be discussed later in this chapter) are only superficially like gills. These arches in embryonic fishes will develop into gills for the adult, but they never form gills (and never function as gills) in *mammalian* embryos. In mammals those branchial arches take a very different developmental path than that taken in any fishlike ancestor to the mammals.

Unfortunately, creationists sometimes confuse Haeckel's muddled, obsolete thinking as an important pillar of evolutionary theory. You can find creationist websites that first declare that "ontogeny recapitulates phylogeny" is an important cornerstone for evolution, then disprove the outmoded idea of recapitulation and declare that they have disproved evolution! Worse yet, in one of Haeckel's early books, some of the illustrations were mislabeled (a mistake Haeckel corrected in later editions), and these websites use those errors to "show" that the whole idea of recapitulation was a fraud from the beginning, a lie to bolster Darwin's false teachings.

So let's be clear—embryonic forms of closely related species are very, very similar in appearance (although not nearly as similar as Haeckel claimed (see figure) (Richardson et al., 1997), and most of the differences between species result from differences in later developmental processes. This is indeed what you'd expect if they descended from a common ancestor, as natural selection predicts. But there is no mystical force that causes embryos to recapitulate *adult* forms of *any* ancestor (much less *all* of them), and there is nothing in Darwin's theory of evolution by natural selection that would predict such a recapitulation. You can (and should) reject Haeckel without rejecting evolution.

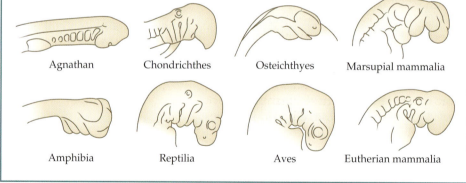

Agnathan Chondrichthes Osteichthyes Marsupial mammalia

Amphibia Reptilia Aves Eutherian mammalia

Haeckel's overzealousness revealed Drawings of actual embryos show they are similar, yes; they all have pharyngeal arches, for example. But clearly there are also differences in appearance. (After Richardson et al., 1997.)

donuts that may be coated with glazed icing, powdered sugar, chocolate icing, or sprinkles. But almost all of those donuts began with the very same dough, with the same proportions of flour, milk, sugar, salt, and yeast. They appear so different from one another because of differences in the *late stages* of production—how the dough was shaped before baking and what was added afterward. Messing with the *early stages*, the proportions of ingredients, is more likely to have a big, unpleasing result—dough that doesn't rise, or doesn't taste good. Put another way, if you are going to introduce a random change in procedure, changes early in the process (leaving out yeast or sugar) are more likely to have a bad effect, while changes late in the process (sprinkling something new on top) are more likely to be benign.

Unfortunately, one of Darwin's disciples took the idea of similarities in embryonic stages to an absurd length, as we discuss in **BOX 2.1**.

How's It Going?

Why does natural selection tend to favor mutations that affect later developmental processes rather than earlier processes?

■ 2.2 ■
MOTHER KNOWS BEST: MATERNAL FACTORS ESTABLISH A BASIC POLARITY OF THE BODY

Early twentieth-century biologists developed **vital dyes**, so called because they were relatively nontoxic and so would not kill cells. The scientists could daub such vital dyes onto a living embryo to stain a few cells and then follow their fate—would the dyed cells end up in the legs, brain, abdomen? From such work they knew that there was an orderly map of body parts that would arise from various parts of the embryo's surface, which they called a **fate map** (**FIGURE 2.2**). These results made it clear that the fate of any particular cell was determined by where it ended up on the blastula. But only once molecular biology tools became available late in the century were scientists able to understand *how* these fate maps were established and to show that the cells were communicating with one another to coordinate their fates.

In Chapter 1 we saw that the "anchor" for unfolding the overall body plans was provided by *maternal factors*, the uneven distribution of *transcription factors* (mostly proteins that regulate the expression of genes) in the egg. Recall that in *C. elegans*, maternal provisioning of SKN-1 protein on one side of the egg's nucleus means that the first cell division sequesters most of that protein in the nucleus of one cell, which we call P1 because SKN-1 is a transcription factor that causes that cell and its progeny to become the posterior half of the worm. If the mother does not supply SKN-1 in one half of the egg nucleus, the body plan will not unfold.

We also saw that in vertebrates, maternal factors trigger a chain of events to form the nervous system. Transcripts and proteins deposited by the mother in particular places in the egg cause an eventual accumulation of β-catenin in blastomeres that will become the endoderm. The β-catenin causes those cells above it, the cells that will form the dorsal lip of the blastopore (which upon involution will form mesoderm in the gastrula), to secrete organizing proteins such as noggin and chordin. These mesodermal organizers block BMP signaling in the overlying ectoderm to induce those cells to switch fates

vital dyes Relatively nontoxic synthetic dyes that can be used to label living cells.

fate map A representation of which parts of an embryo will give rise to various parts of the adult body.

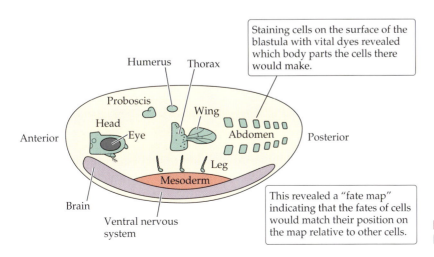

Staining cells on the surface of the blastula with vital dyes revealed which body parts the cells there would make.

This revealed a "fate map" indicating that the fates of cells would match their position on the map relative to other cells.

FIGURE 2.2 **Fate map of *Drosophila* blastula** (After Benzer, 1973.)

FIGURE 2.3   Establishing polarity in the egg (A after Palacios, 2007.)

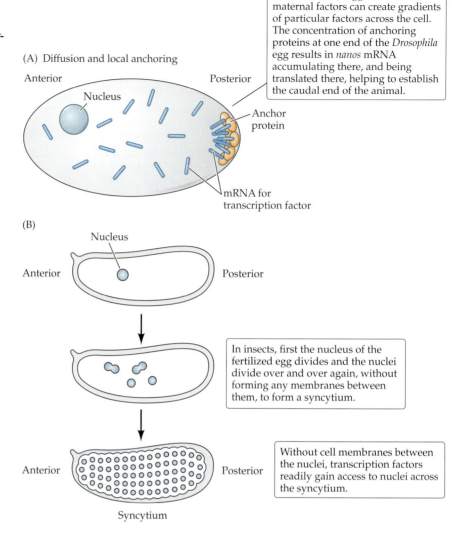

(A) Diffusion and local anchoring

Anterior

Posterior

Nucleus

Anchor protein

mRNA for transcription factor

Even before the egg is fertilized, maternal factors can create gradients of particular factors across the cell. The concentration of anchoring proteins at one end of the *Drosophila* egg results in *nanos* mRNA accumulating there, and being translated there, helping to establish the caudal end of the animal.

(B)

Nucleus

Anterior

Posterior

In insects, first the nucleus of the fertilized egg divides and the nuclei divide over and over again, without forming any membranes between them, to form a syncytium.

Anterior

Posterior

Without cell membranes between the nuclei, transcription factors readily gain access to nuclei across the syncytium.

Syncytium

from epidermal cells to neural cells, forming the neural tube. This formation of the neural tube seems to organize development of the head and the rest of the body. So a cascade of gene regulation sorts various cells into groups, each with its marching orders to form one structure or another.

Despite all these insights from *C. elegans* and vertebrates, our understanding of the forces that map out a body plan advanced most fully with studies of the fruit fly *Drosophila melanogaster*. As we'll see, the powerful tools of genetics—being able to induce many mutations, to quickly screen the many resultant mutants for interesting phenotypes, and then to target particular genes for particular changes—revealed a detailed hierarchy of transcription factors that interact with one another to map out every detail of cellular differentiation, meaning every detail of the fly's body. Happily, this understanding of the genetic establishment of the body plan in flies turned out to be completely relevant to the unfolding body plan in vertebrates like us, too.

Before we discuss those experiments in fruit flies, we need to take stock of some ways in which fruit fly embryogenesis differs from that of vertebrates. In insects, the early embryo does not divide into separate cells; rather, the nucleus alone divides repeatedly until the egg consists of a single cell, which we call a **syncytium**, with one continuous cytoplasm containing many nuclei. This cytoplasm is continuous, without any cell membranes breaking the syncytium into compartments, but it is *not* homogenous. Rather, there are transcription factors distributed in a particular fashion, and this differentially affects these thousands of nuclei within that cell (**FIGURE 2.3A**). You can see how the absence of cytoplasmic membranes

syncytium   A single cell containing several nuclei.

between the nuclei can help set up gradients of transcription factors unimpeded by cell membranes. Thus the position of each nucleus within the syncytium determines what mix of transcription factors it is exposed to and, hence, exactly what genes are expressed in that nucleus (**FIGURE 2.3B**). Only later do the individual nuclei become separated from each other by cellular membranes, a process called **cellularization**. By then, the cells are well targeted to a particular fate, and then it will be signaling mechanisms between the cell membranes that fine-tune each cell's fate.

Among the maternal factors that are unevenly distributed in the *Drosophila* egg are the mRNAs for a transcription factor called **bicoid**, which accumulates in one end of the embryo (Seeger & Kaufman, 1990; Spirov et al., 2009) and so the protein mostly acts upon nuclei there (**FIGURE 2.4**). The mRNAs for another transcription factor, called **nanos**, accumulate at the other end (Nusslein-Volhard, 1991). The end of the embryo containing *nanos* mRNA will become the tail (see Figure 2.3A). At the other end, the concentration of *bicoid* mRNA will cause the cells there to develop as the head of the animal. The first clue that bicoid was responsible for developing head structures came from *Drosophila* mutants in which the *bicoid* gene was disabled. These *bicoid* knockout embryos began developing tails at both ends (Driever & Nusslein-Volhard, 1988); without a head, the embryo stalled in its development. While this finding is consistent with bicoid playing a causal role in the formation of the head, you always have to worry that disabling a gene has had spreading effects on many other genes and that one of those might actually be responsible for the effect. Put another way, there are probably an infinite number of ways you can disrupt development and produce very abnormal embryos. So you can't be sure whether your manipulation worked in a *specific* way or simply disrupted a delicate process to produce a catastrophe.

> The position of each nucleus within the syncytium determines what mix of transcription factors it is exposed to.

cellularization The process in which cell walls are formed between the nuclei in a syncytium.

bicoid A transcription factor that is concentrated in the presumptive anterior end of the embryo and directs formation of the head.

nanos A transcription factor that is concentrated in the presumptive posterior end of the embryo.

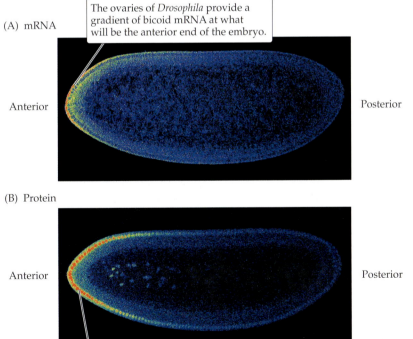

(A) mRNA

> The ovaries of *Drosophila* provide a gradient of bicoid mRNA at what will be the anterior end of the embryo.

Anterior

Posterior

(B) Protein

Anterior

Posterior

> Bicoid protein is therefore concentrated at the anterior end and, because bicoid is a transcription factor, it binds to DNA in the nuclei of the syncytium.

FIGURE 2.4 *Bicoid* mRNA and protein are concentrated at one end of *Drosophila* eggs. (After Spirov et al., 2009; courtesy of Dr. S. Baumgartner.)

Thus a much more convincing demonstration of the role of bicoid in developing the anterior-posterior axis involved injecting *bicoid* mRNA, as we'll see next.

RESEARCHERS AT WORK

Two Heads Are Not Better Than One

■ **QUESTION**: How is the anterior-posterior axis of the body plan established in *Drosophila*?

■ **HYPOTHESIS**: The transcription factor bicoid, which is normally concentrated at the end of the *Drosophila* egg that will become anterior, actually directs the differentiation of the head at that end.

■ **TEST**: If bicoid plays a causal role in differentiating that end of the egg as anterior, then injecting *bicoid* mRNA into other regions of the syncytium should affect the placement of head structures. Inject *bicoid* mRNA into the middle of the egg or the end that normally forms the tail, and look for characteristics normally found only in the head.

■ **RESULT**: As noted above, embryos with disabled bicoid developed tails at both ends (**FIGURE 2.5A**). When *bicoid* mRNA was injected into the middle of a syncytium, head structures began developing there, with tail structures developing at both ends, each developing in response to the "head" in the middle. Injecting *bicoid* mRNA into the posterior end of a syncytium resulted in heads developing at both ends (**FIGURE 2.5B**) (Driever, Siegel, & Nusslein-Volhard, 1990).

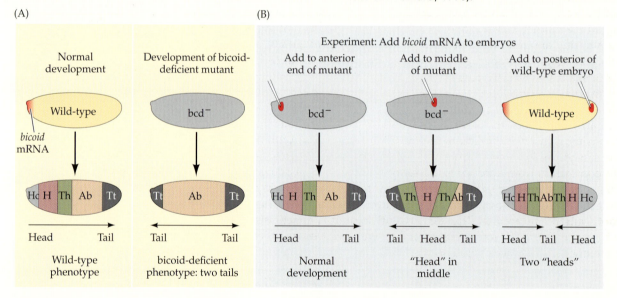

FIGURE 2.5 Testing a maternal polarity gene (After Driever et al., 1990.)

| Hc Head cap | H Head | Th Thorax | Ab Abdomen | Tt Tail tip |

■ **CONCLUSION**: Bicoid normally plays a causal role in the development of the head at one end of the *Drosophila* egg. In other words, bicoid alone is sufficient to induce formation of the head. Later experiments would confirm this role and demonstrate that bicoid and other transcription factors that are unevenly distributed in the egg establish the anterior-posterior axis in fruit flies, directing the differentiation of body structures appropriate to each point of that axis.

The first major axis of the body plan, the anterior-posterior axis, is already foreshadowed in the egg as the result of uneven distribution of the protein products of *bicoid, nanos,* and several other genes (*caudal, swallow, oskar*), which are collectively called **maternal polarity genes**. If you're the sort of person

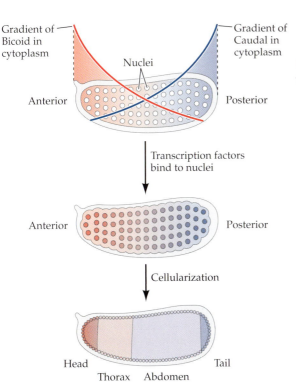

FIGURE 2.6 The syncytium has gradients of transcription factors.

who wonders, Which came first, the chicken or the egg? you may be wondering how nanos came to be concentrated at one end of the egg while bicoid was concentrated at the other. The brief answer is that gradients of these different transcription factors are laid down in the process of manufacturing the egg in the mother's ovary (and, of course, in her mother's ovary before her). Numbering about a dozen or so, the maternal polarity genes include those that play roles in arranging the gradients of bicoid and nanos in the oocyte (Becalska & Gavis, 2009; Johnstone & Lasko, 2001).

In what will develop as a theme in this chapter, *bicoid* and *nanos* have antagonistic effects—genes that are turned on by one of the pair are inhibited by the other. Thus having a gradient of just two opposing transcription factors means that each nucleus within the syncytium is being instructed about exactly where it falls in the continuum from anterior to posterior (**FIGURE 2.6**). Another theme of this chapter—the remarkable contributions these tiny little flies have made to biology—echoes throughout the book. This contribution is no accident. The early twentieth century geneticists who chose *Drosophila* for an animal model were well aware of its advantages, which we discuss in **BOX 2.2**.

maternal polarity genes A class of genes provided by the mother such that their products are unevenly distributed in the zygote and thereby specify the anterior-posterior axis.

The anterior-posterior axis is foreshadowed in the egg as the result of uneven distribution of the protein products of maternal polarity genes.

MEET *DROSOPHILA MELANOGASTER,* THE WELL-SEGMENTED ORGANISM

It would be impossible to overstate the contributions that work in fruit flies has made to our understanding of genetic mechanisms in any domain, including development of the nervous system. Initially flies were studied for practical reasons: they are tiny, easy to feed (they don't actually eat fruit; they eat yeast cells that grow on the fruit), and have a short life cycle of about 2 weeks, so they are very inexpensive

to maintain. That early work revealed additional advantages—fruit fly salivary glands contain hugely inflated chromosomes, visible with light microscopes, which proved crucial for demonstrating a link between chromosomes and heredity (Morgan, 1911), and mutations could be readily induced through the use of X-rays (Muller, 1928). It was the groundbreaking

continued

BOX 2.2
continued

MEET *DROSOPHILA MELANOGASTER,* THE WELL-SEGMENTED ORGANISM

work of Ed Lewis in cataloging homeotic mutations (Lewis, 1978) that led to our understanding the cascade of gene regulatory proteins that lays down a body plan of cell differentiation in flies, which turned out to be entirely relevant to human development.

The rapid life cycle of flies (see figure) is another reason why research with these animals is still going strong (Jennings, 2011). The relatively small size of the fly genome (about 14,000 genes) means it was quickly sequenced (Adams et al., 2000), and it turns out that about 75% of the genes that have been

related to disease in humans have homologues in the fly genome (Reiter, Potocki, Chien, Gribskov, & Bier, 2001). Thus many researchers interested in how alleles of a gene may cause disease in humans have begun asking what the homologue of the gene does in flies and are trying to see whether mutations of the gene in flies can re-create the disease (Lloyd & Taylor, 2010). In future chapters we'll see many examples of work in fruit flies that aided our understanding at every stage of neural development.

Drosophila life cycle

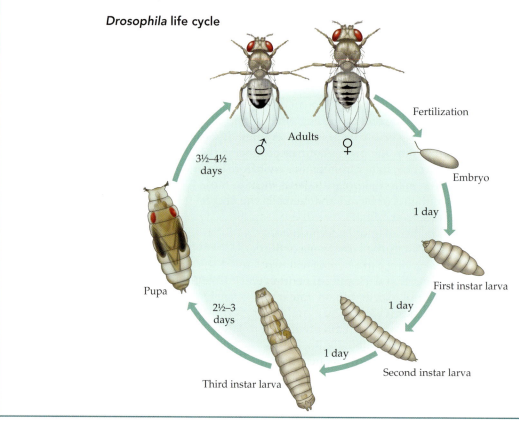

How's It Going?

What conditions provide the initial information to set up the anterior-posterior axis of the body?

■ 2.3 ■

A CASCADE OF GENE REGULATORY PROTEINS ORGANIZES A BODY PLAN

The maternal polarity genes represent the beginning of a hierarchy of transcription factors that will continue to demarcate the *Drosophila* larva along the anterior-posterior axis so that nuclei (and, once membrane walls form

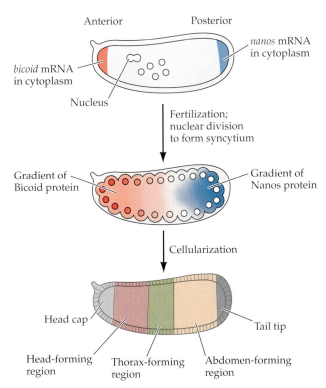

FIGURE 2.7 **Mapping out the body** A cascade of transcription factors demarcates the fruit fly embryo into compartments that will form various parts of the body, from head to tail.

between those nuclei, cells) in different regions will express different sets of genes, all in preparation for those cells to take different developmental paths in order to form different structures. The genes governing this coordination of cell fate were first discovered in flies, but they proved to be at work in vertebrates, too. Later in this chapter we'll discuss signals that tell cells where they fall on the dorsal-ventral axis, but for now let's see how these cascading sets of transcription factors further demarcate the anterior-posterior axis in flies.

As development ensues, the maternal polarity genes will regulate the expression of other transcription factors, which in turn will regulate the expression of yet other transcription factors, in a cascade of gene regulation that eventually will direct cell differentiation and hence cell fate of the entire fly. Working in a coordinated sequence, these factors will parcel the embryo out into various successively smaller subdivisions and direct each one to produce the body part that belongs in that position (**FIGURE 2.7**).

We noted that the maternal polarity genes *bicoid* and *nanos* are mutually antagonistic—genes that are turned on by one tend to be inhibited by the other. For example, nanos protein generated from the posterior end of the zygote inhibits expression of a transcription factor called hunchback, while bicoid protein from the anterior end inhibits *nanos* mRNA translation (Singh, Morlock, & Hanes, 2011), which disinhibits hunchback expression. Thus more hunchback protein is found in the anterior end of the embryo. Importantly, the antagonistic action of the two maternal polarity proteins on hunchback expression results in a distribution of hunchback protein that is more sharply demarcated, less graded, than that of either of the maternal polarity proteins (**FIGURE 2.8**).

Hunchback is just one of a class of genes called **gap genes**, all of which are regulated, directly or indirectly, by the maternal polarity genes. As with hunchback, the antagonistic interaction of diffuse gradients of maternal polarity proteins results in more discrete, tighter boundaries of expression of the other gap genes. Expression of *hunchback* and other gap genes, like *Krüppel* and *knirps*, produces transcription factor proteins that

hunchback A transcription factor that is one of the class of gap genes that is inhibited by the nanos protein.

gap genes Genes that encode a class of transcription factors, the expression of which is regulated by maternal polarity genes.

The antagonistic interaction of diffuse gradients of maternal polarity proteins results in more discrete, tighter boundaries of expression of the gap genes.

FIGURE 2.8   Bringing transcription factors into focus The antagonistic action of maternal polarity genes like *bicoid*, *nanos*, and *caudal* result in sharper boundaries of expression of gap genes like *hunchback*.

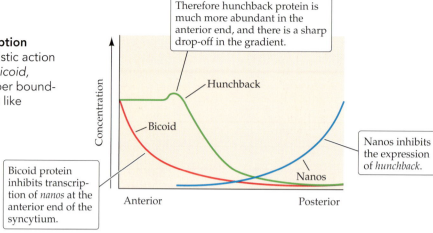

Therefore hunchback protein is much more abundant in the anterior end, and there is a sharp drop-off in the gradient.

Nanos inhibits the expression of *hunchback*.

Bicoid protein inhibits transcription of *nanos* at the anterior end of the syncytium.

pair-rule genes   Genes that encode a class of transcription factors, the expression of which is regulated by gap genes.

segment polarity genes   Genes that encode a class of signaling factors, the expression of which is regulated by pair-rule genes.

will regulate the expression of *another* class of transcription factors, called **pair-rule genes**, such as *fushi tarazu* and *even-skipped*, which are expressed in alternating stripes across the long axis of the embryo (**FIGURE 2.9**). Note that regions expressing one of these genes do not express the other. This exposure to different transcription factors will guide these cells to express different genes at this stage in development and in the future. The pair-rule genes regulate yet another class of transcription factors, called **segment polarity genes**, so named because each tends to be expressed in a band corresponding to one segment of the insect, either an odd-numbered segment (1, 3, 5, etc.) or an even-numbered segment. Thus eventually each segment expresses a unique pattern of these genes, which will cause it to take on a unique identity (**FIGURE 2.10**). Now those unique sets of transcription factors in each segment will activate yet another set of transcription factors, the *homeotic selector genes* that we will discuss in detail shortly.

It is the sequential action of these five classes of gene regulatory factors that maps out the anterior-posterior positions of cells in the embryo and thereby coordinates their fates for each position. Here is the hierarchy of regulatory factors that are expressed in a particular sequence in the embryo:

1. Maternal polarity genes (*bicoid, nanos, caudal*)
2. Gap genes (*hunchback, Krüppel, knirps*)
3. Pair-rule genes (*even-skipped* [expressed in even-numbered segments], *fushi tarazu* [Japanese for "not enough segments"; expressed in odd-numbered segments], *hairy, runt*)

Transcripts for the pair-rule gene *even-skipped* (red) are expressed in the parts of the embryo that will become the even-numbered segments.

Another pair-rule gene, *fushi tarazu* (black), is expressed in the regions that will become the odd-numbered segments.

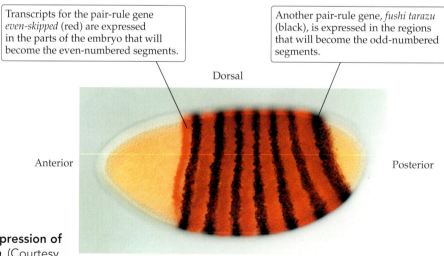

FIGURE 2.9   Complementary expression of two pair-rule genes in *Drosophila* (Courtesy of Dr. S. Small.)

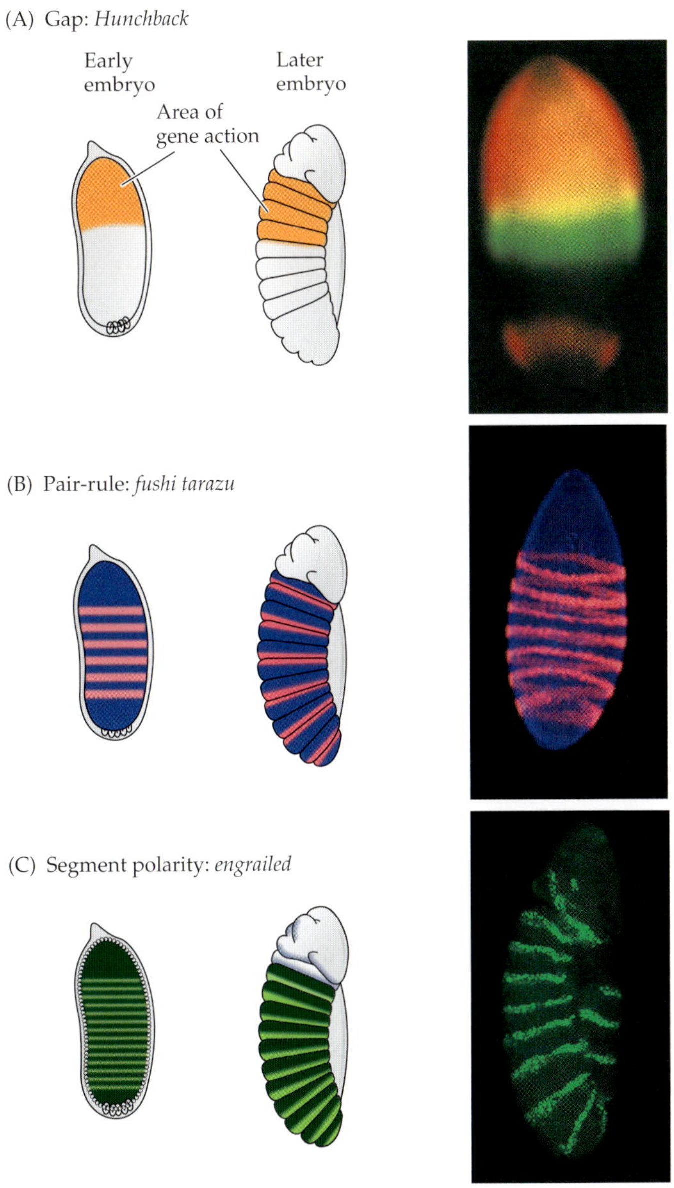

(A) Gap: *Hunchback*

Early embryo

Later embryo

Area of gene action

(B) Pair-rule: *fushi tarazu*

(C) Segment polarity: *engrailed*

FIGURE 2.10 Divide and conquer The homeotic hierarchy gets progressively more discrete in expression as you go down the hierarchy (and through development). (A) Gap genes, such as *hunchback* (orange) and *Krüppel* (green), may sometimes overlap (yellow). (B) The activity of the gap genes directs the pattern of expression of pair-rule genes, such as *fushi tarazu* shown here. Note that pair-rule genes are expressed in several different domains, sandwiched between areas that do not express the gene. (C) The pair-rule genes direct the pattern of expression of segment polarity genes, such as *engrailed* shown here. Note that such segment polarity genes are expressed in more compartments than the pair-rule genes and seem to have sharper boundaries of expression. At this point, cellularization takes place, so the segment polarity genes use cell-cell signaling across the membranes to direct gene expression. (A photo courtesy of Drs. C. Rushlow and M. Levine; B photo courtesy of Dr. D. W. Knowles and the Berkeley Drosophila Transcription Network Project, http://bdtnp.lbl.gov/Fly-Net/; C photo courtesy of Drs. S. Carroll and S. Paddock.)

Up to this point, the protein products of these genes act as transcription factors, diffusing across nuclei in the syncytium to regulate their gene expression. At about this point in development, the embryo ceases being a syncytium as cell membranes form between nuclei. Thus the next tier of factors, the segment polarity genes, use cell-cell signaling across those membranes to regulate gene expression in neighbors.

4. Segment polarity genes (*engrailed, wingless, hedgehog, armadillo*)

5. Homeotic selector genes (*bithorax, Ultrabithorax, Antennapedia*)

We'll discuss homeotic selector genes in detail shortly. But for now, you should know that they are regulated not just by the segment polarity genes immediately above them in the hierarchy, but also directly by the previous action of gap and pair-rule genes. Thus the *history* of various gap, pair-rule, and segment polarity influences on a particular nucleus will determine which homeotic selector gene(s) it will express as an individual cell. The homeotic selector products will then regulate gene expression in that cell and its nearest neighbors such that they will take on a particular fate, including what body part they will form.

If you'd like a mnemonic to help you remember the order of these transcription factors, you can use "**m**any **g**ood **p**eople **s**eek **h**appiness" (**m**aternal polarity, **g**ap, **p**air-rule, **s**egment polarity, **h**omeotic selector), or you can make up one of

FIGURE 2.11 The cascade of gene regulation in fruit flies

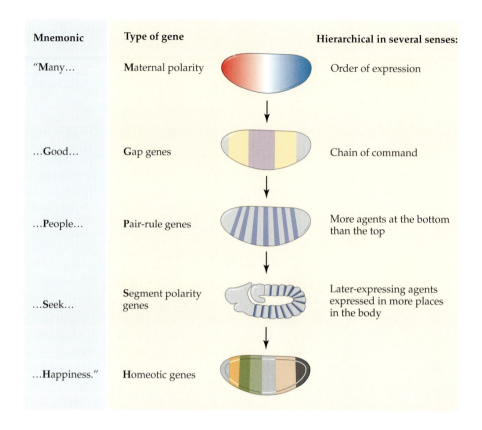

Mnemonic	Type of gene		Hierarchical in several senses:
"Many...	Maternal polarity		Order of expression
...Good...	Gap genes		Chain of command
...People...	Pair-rule genes		More agents at the bottom than the top
...Seek...	Segment polarity genes		Later-expressing agents expressed in more places in the body
...Happiness."	Homeotic genes		

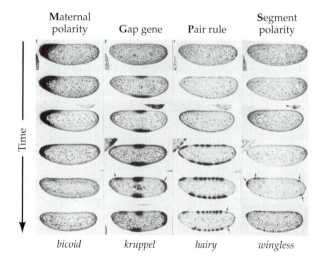

Maternal polarity | Gap gene | Pair rule | Segment polarity

Time

bicoid | *kruppel* | *hairy* | *wingless*

FIGURE 2.12 **The hierarchical nature of gene regulatory mechanisms establishing the anterior-posterior axis in *Drosophila*** The cascade of gene regulatory factors acts in a particular sequence in embryonic development, the earlier factors direct expression of the later factors, and the later factors are expressed in more numerous and smaller domains of the embryo. (From Ingham, 1988.)

your own (nothing naughty please). In the one given, the only verb, *seek*, marks the first tier where gene expression is regulated via cell-cell signaling rather than via transcription factors diffusing across the syncytium.

The transcription factors delineating the anterior-posterior axis in *Drosophila* are hierarchical in several senses (**FIGURE 2.11**). First, they are expressed at different times in embryonic development. The maternal polarity genes are expressed first, then gap genes, then pair-rule genes, etc., in the order listed above. Second, the genes higher up in the hierarchy actually dictate where in the embryo the genes in the tier below them will be expressed. In other words, there is a chain of command: the earlier transcription factors control expression of the next tier of factors. Third, the genes that are higher in the hierarchy (i.e., expressed earlier in development) are expressed over larger areas of the embryo (**FIGURE 2.12**). For example, the maternal polarity gene *bicoid* is expressed in a single zone comprising about half the embryo, while the gap gene *hunchback* is expressed in a band comprising about a third of the embryo. Fourth, the lower in the hierarchy a gene resides (i.e., the later in development it will be expressed), the more separate territories will express that gene. In this way, as the hierarchy of expression is played out, smaller and smaller regions of the embryo are expressing a particular combination of transcription factors, each directing the cells of that region to a separate fate.

The hierarchy of gene regulatory factors from maternal polarity genes to segment polarity genes divides the embryo into about 14 different compartments, each the width of a segment, from presumptive head to tail. Thus each embryonic cell at this stage has a particular history of gene expression that differs depending on which of those 14 compartment the cell happens to reside in. At this stage the final-tier transcription factors direct cells to begin constructing specific organs and other structures, as we'll see next.

How's It Going?

1. What are the five tiers of gene regulatory factors that demarcate the fly anterior-posterior axis?
2. In what senses are these tiers hierarchical?

■ 2.4 ■
SOME MUTATIONS IN *DROSOPHILA* TRANSFORM BODY PARTS WHOLE

While we are describing the homeotic selector genes last, because they come into play later in development, in fact these genes were studied first because mutations in homeotic genes can cause such obvious, even bizarre, body modifications in flies. The **homeotic selector genes**, also called **Hox genes**, are those in which mutations seem to swap out one body part for another, as when a leg replaces an antenna in flies mutant for the Hox gene ***Antennapedia*** (**FIGURE 2.13**). The root *homeo*, meaning "similar," refers to the fact that the body parts that are swapped are similar to each other. In another example, of the three segments making up the fly thorax, the second segment normally makes a pair of wings while the third segment produces a pair of small structures called **halteres**, which are thought to help maintain balance during flight. Mutations in the Hox gene ***bithorax*** can cause the third thoracic segment to develop like the second segment, transforming the halteres into wings and effectively producing a second thorax (Lewis, 1978). There is in fact a complex of *bithorax* genes, including ***Ultrabithorax***, that affect fate for cells in the thorax (**FIGURE 2.14**).

homeotic selector genes or Hox genes Also called simply *homeotic genes*. A class of genes in which mutations tend to result in swapping out one body part for another, as when mutation of *Antennapedia* results in the formation of legs where antennae normally form.

Antennapedia A Hox gene in which mutations result in the formation of a leg where an antenna normally forms.

halteres Paired structures that serve as counterweights to maintain balance in the flight of some flies.

bithorax A Hox gene in which mutations can result in the doubling of the thorax in *Drosophila*.

Ultrabithorax A Hox gene complex that affects the fate of cells in the thorax of *Drosophila*.

Mutations in homeotic genes can cause obvious, even bizarre, body modifications in flies.

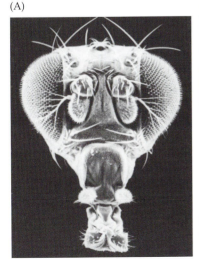
(A)

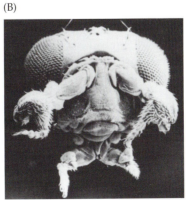
(B)

FIGURE 2.13 **Homeotic mutants** Two antennae in the wild type (A) have been transformed to legs in an *Antennapedia* mutant (B). (From Kaufman et al., 1990, courtesy of Dr. T. C. Kaufman.)

FIGURE 2.14 **Four wings are not better than two.** Mutations in the *Ultrabithorax* complex of Hox genes can result in a fly with two complete thoraxes, including a second pair of wings. (Courtesy of Dr. Nipam Patel.)

homeobox   A nucleotide sequence that produces a DNA-binding domain in many transcription factor proteins. It is found in Hox genes and many other transcription factors.

realizator genes   A class of genes, the expression of which is controlled by Hox genes, that direct the actual construction of particular body parts.

The advent of molecular biological tools to sequence genes soon revealed that homeotic selector genes had more in common than dramatic effects on body structures. A highly conserved stretch of about 180 nucleotides, encoding a 60-amino-acid chain that binds to DNA, was discovered in the various homeotic selector genes and dubbed the **homeobox** (McGinnis, Levine, Hafen, Kuroiwa, & Gehring, 1984; Scott & Weiner, 1984). Eventually other genes, including some that do not obviously transform body parts, were found to contain a homeobox as well. These genes are evolutionarily ancient, as they are found in nearly all eukaryotes (Holland, 2013). What all homeobox genes have in common is the ability of their proteins to regulate gene expression, using the protein's homeobox domain (sometimes shortened to *homeodomain*) to bind DNA in the process. They play crucial roles as transcription factors well beyond our interest in neural development. There are many homeobox genes, but only a subset of those cause body transformations like those caused by *bithorax*, and these particular homeobox genes are the ones called Hox genes or homeotic selector genes (Sivanantharajah & Percival-Smith, 2015). The terminology is a bit confusing: every Hox gene is a homeobox gene, but not every homeobox gene is a Hox gene (sorry). Both Hox genes and homeobox genes are subsets of the many transcription factors at work directing cellular differentiation (**FIGURE 2.15**). The Hox genes (which make up a small subset of the many homeobox genes) like *Antennapedia* play an important role in mapping out the anterior-posterior organization of the body and nervous system.

Note that the whole point of transcription factors directing cell fate is that eventually they must regulate expression of genes that are not themselves transcription factors, but rather genes that produce proteins that actually affect the final structure and function of the cell so it can play its role in building a body part. Some researchers have referred to these genes, which help to *realize* a developmental program to build some structure, as **realizator genes** (Garcia-Bellido, 1975; Maeda & Karch, 2010; Pradel & White, 1998; Redline, Neish, Holmes, & Collins, 1992). It is a rather ugly-sounding phrase, which may be why it is not widely used (searching PubMed for "realizator genes" retrieves only a few articles). An alternative phrase is "terminal differentiation genes" (Hobert, 2011), which may be even uglier. Despite the fact that these genes do all the actual work of making legs, antennae, spiracles (Lovegrove et al., 2006), and the like, research has focused on the transcription factors

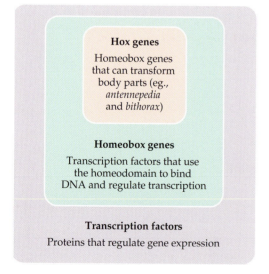

Hox genes

Homeobox genes that can transform body parts (eg., *antennepedia* and *bithorax*)

Homeobox genes

Transcription factors that use the homeodomain to bind DNA and regulate transcription

Transcription factors

Proteins that regulate gene expression

FIGURE 2.15 **Not all homeobox genes are Hox genes.**

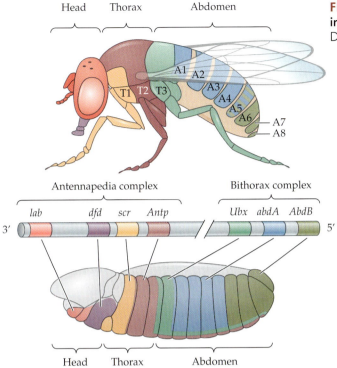

FIGURE 2.16 Colinearity of Hox gene expression in the developing fly (After Kaufman et al., 1990, and Dessain et al., 1992.)

that direct those genes. (If you want to include realizator genes in your mnemonic, try "**m**any **g**ood **p**eople **s**eek **h**appy **r**elationships.")

There is a curious characteristic of Hox genes that is not understood. In flies, they are all found in two clusters, the *bithorax* cluster and the *Antennapedia* cluster, on chromosome 3. Fascinatingly, the sequence of Hox genes in the chromosome aligns with the anterior-posterior sequence of the fly body portions that express those particular genes (**FIGURE 2.16**). For example, the homeotic selector gene *proboscipedia* is located near the 3' end of the chromosome and expressed in the head (null mutants for *proboscipedia* produce feet where the proboscis of the mouth normally appears) (Pultz, Diederich, Cribbs, & Kaufman, 1988). The *abdominal* genes are at the other end of the chromosome (the 5' end) and are normally expressed in the abdomen (as the name suggests). The *bithorax* complex, which is normally expressed in the thorax, lies in between *proboscipedia* and *abdominal* on the chromosome. This correlation between the position of the Hox genes on the chromosome and the anterior-posterior region of the body where they are normally expressed is called **colinearity**. While we still don't understand why this colinearity of gene position and bodily expression occurs in *Drosophila*, it is likely to be important, because when scientists searched for Hox genes in mammals the colinearity of gene position and body expression was found there, too, as we'll see next.

colinearity The property of Hox genes in which their order on the chromosome matches the order in which they are expressed along the anterior-posterior axis of the body.

> The sequence of Hox genes in the chromosome aligns with the anterior-posterior sequence of the fly body portions that express those particular genes.

How's It Going?

1. What is distinctive about mutations of Hox genes in flies?
2. Are all Hox genes homeobox genes?
3. Are all homeobox genes Hox genes?

■ 2.5 ■

HOX GENES ARE CRUCIAL FOR VERTEBRATE DEVELOPMENT, TOO

gene duplication and divergence The evolutionary process by which a gene duplication is followed by successive divergence in the sequence and function of the two copies.

Looking for genes with sequences similar to the homeobox revealed that vertebrates do indeed have an abundance of such genes. As in flies, a subset of those many homeobox genes are homologues of the Hox genes. In fact, unlike *Drosophila*, which carry two Hox complexes on a single chromosome, mammals have Hox complexes, each homologous to either the *bithorax* or *Antennapedia* complex of flies, distributed on four different chromosomes, and these have been designated Hox A, B, C, and D. Individual genes within each complex are often designated by the letter designating the chromosome and a number within that Hox complex, such as *HoxA3, HoxB4*, or *HoxD8*.

It appears that at some point since our divergence from the ancestor we share with flies, duplications of the Hox complexes occurred, presumably as a mistake in cell division, a "stutter" in chromosome replication. Once a duplication happens, the initially redundant second copy of the gene can take on a new role while the other copy continues to take care of whatever the original function was. As natural selection relies upon one of the duplicates to take on another role, the nucleotide sequences of the two copies diverge, producing different proteins with different functions. The differences will be slight at first but over evolutionary epochs may diverge considerably. This evolutionary process of genes first being copied and then taking on different functions is known as **gene duplication and divergence** (Taylor & Raes, 2004). Thus every *Drosophila* Hox gene has at least one homologue in mammals, although not every fly gene has a homologue on each of the four chromosomes (Holland, 2013; Sivanantharajah & Percival-Smith, 2015). In other words, after duplication of a Hox complex of genes onto a particular chromosome, some individual genes within the complex were apparently lost over time.

As natural selection relies upon one of the duplicates to take on another role, the nucleotide sequences of the two copies diverge, producing different proteins with different functions.

Just as *Drosophila* Hox genes show colinearity, expressed across the anterior-posterior axis in the same order as the genes appear on the chromosome, vertebrate Hox genes were found to be expressed in a colinear order, for all 4 chromosomes (**FIGURE 2.17**). This colinear expression was most obvious in the spinal column, where the Hox genes are expressed in, and direct the fate of, cells forming the bony vertebrae protecting the spinal cord (Mallo, Wellik, & Deschamps, 2010). The duplication of the Hox genes in vertebrates seems to have

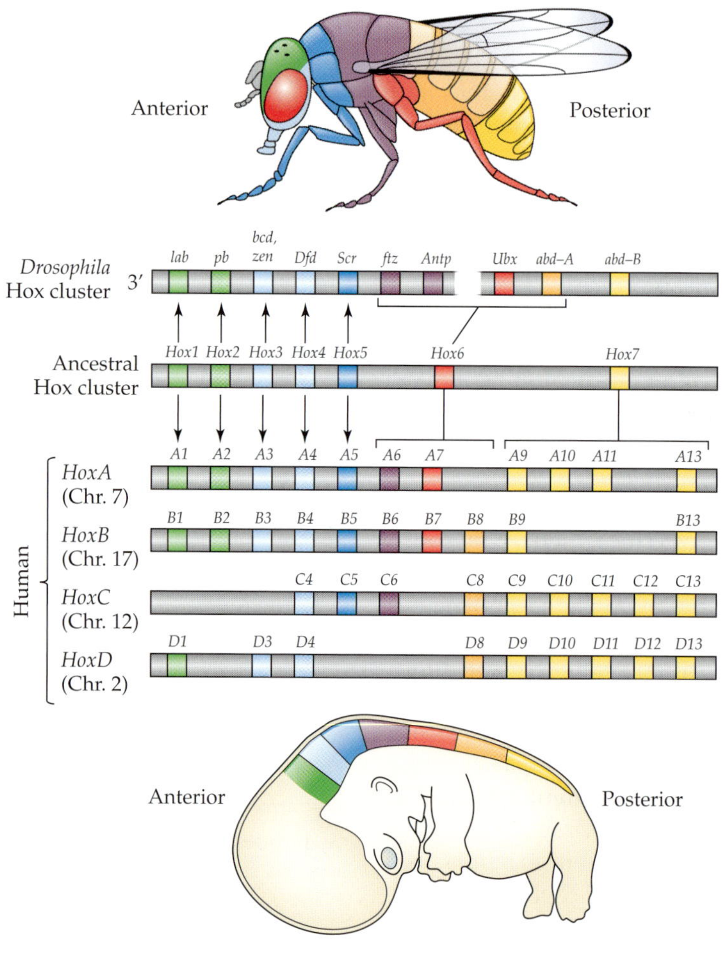

FIGURE 2.17 **Mammals also show colinearity of Hox gene expression.** (After Veraksa & McGinnis, 2000.)

(A) AA ; DD   (B) aa ; DD   (C) AA ; dd   (D) aa ; dd

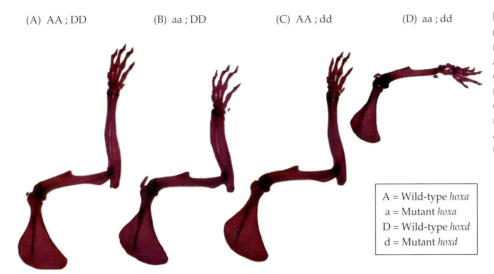

A = Wild-type *hoxa*
a = Mutant *hoxa*
D = Wild-type *hoxd*
d = Mutant *hoxd*

FIGURE 2.18 **Hox gene mutations in mammals** There may be some redundancy among the four clusters of Hox genes in mammals, as single mutations have relatively subtle effects. However, deletions of more than one Hox gene have a greater effect on phenotype. (From Davis et al., 1995.)

retained some redundancy in the system. Typically, a single mutation in a Hox gene in mammals may have a relatively subtle effect, so it may take deletion of more than one Hox gene to have a noticeable effect (**FIGURE 2.18**). Of course, we are more interested in Hox influences on neural fate, so we'll turn to their effects on brain development next.

How's It Going?

What is meant by the colinearity of Hox gene expression in flies and mammals?

■ 2.6 ■
HOX GENES DIRECT "SEGMENTATION" IN THE MAMMALIAN BRAIN

You may not have thought of the vertebrate central nervous system as a segmented structure, but that segmented nature is rather apparent when you examine the fetal brain (**FIGURE 2.19A**). As the rostral end of the neural tube develops, three prominent swellings develop one behind the other: the **prosencephalon** (forebrain), the **mesencephalon** (midbrain), and the **rhombencephalon** (hindbrain). Of these three divisions, the mesencephalon will grow the least and form the adult midbrain, including the *tectum* (roof) on the dorsal surface, and the *tegmentum* on the ventral side. The other two divisions grow much more than the mesencephalon. The prosencepalon will later form the **telencephalon**, which includes the cerebral cortex and several subcortical structures such as the hippocampus and basal ganglia, and the **diencephalon** (thalamus and hypothalamus). The rhombencephalon will form the **pons** and **cerebellum** (sometimes together called the **metencephalon**) and the **myelencephalon** (medulla) (**FIGURE 2.19B**).

These divisions of the brain are directed toward their different fates by the action of Hox genes and other homeobox genes during development. As when they are specifying body parts in flies, the genes direct different cell fates in different anterior-posterior compartments of the vertebrate brain. In addition to providing a distinct fate to midbrain versus hindbrain, for example, the genes also set up "inducing zones" that direct the fates of their neighbors, much the way the dorsal lip of the blastopore induces formation of the neural

prosencephalon Also called *forebrain*. The most anterior aspect of the embryonic vertebrate brain. It will develop into the telencephalon and diencephalon.

mesencephalon Also called *midbrain*. The middle segment of the embryonic vertebrate brain. It will develop into the adult midbrain.

rhombencephalon Also called *hindbrain*. The caudal-most segment of the embryonic vertebrate brain. It will develop into the metencephalon (pons and cerebellum) and myelencephalon (medulla).

telencephalon The anteriormost portion of the vertebrate brain, consisting of the cerebral cortex and related subcortical structures such as the basal ganglia and hippocampus.

diencephalon The portion of the vertebrate brain that consists of the thalamus and hypothalamus.

pons The portion of the brainstem caudal to the midbrain, to which the cerebellum is attached.

cerebellum A brain region attached to the pons that plays an important role in coordination of movement.

metencephalon A subdivision of the hindbrain that includes the cerebellum and the pons.

myelencephalon Also called *medulla*. The caudal-most portion of the vertebrate brainstem, which blends into the rostral spinal cord.

(A) Development of the human brain

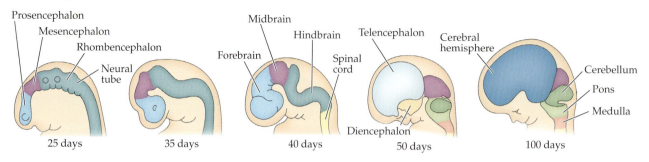

FIGURE 2.19   **Vertebrate nervous systems are segmented, too.** (B after Moore & Persaud, 1993.)

plate. Also as in flies, some Hox genes are expressed over large stretches of the anterior-posterior axis, while others are expressed over smaller domains. Let's consider one of each.

Homeobox gene Otx2 specifies the vertebrate forebrain and midbrain

We saw that in flies the transcription factors interact to determine their pattern of expression, and that pattern of expression then directs differentiation of cells in a way that is appropriate for each segment. In vertebrates, too, the homeobox genes interact to produce a very particular pattern of expression in the developing nervous system, and that pattern of expression directs cells to differentiate in a way that is appropriate for the nervous system in that particular segment of the anterior-posterior axis.

For example, the homeobox (but not Hox) gene **Otx2** is required for the forebrain and midbrain to develop. Thus *Otx2* knockout animals fail to develop any head at all (Matsuo, Kuratani, Kimura, Takeda, & Aizawa, 1995). Normally *Otx2* is expressed throughout the forebrain and midbrain, but not in the hindbrain (Millet, Bloch-Gallego, Simeone, & Alvarado-Mallart, 1996), and this pattern of expression is itself dependent on the earlier action of another

Otx2   A homeobox gene required for development of the vertebrate midbrain and forebrain.

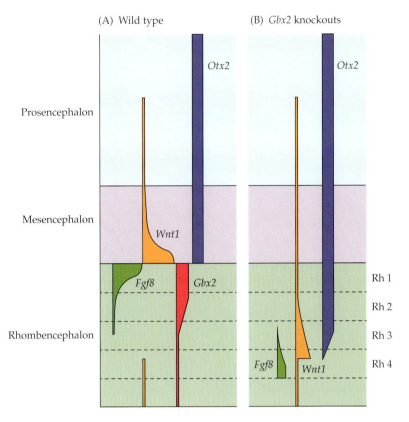

FIGURE 2.20   Homeobox gene *Otx2* is expressed in the anterior nervous system. (A) Normal distribution of gene regulatory proteins Otx2, Gbx2, Wnt, and FGF in the vertebrate nervous system. (B) The normal pattern of expression of these genes depends upon their interaction with each other. Here, animals with *Gbx2* knocked out show an abnormal expression pattern of the other three. (After Millet et al., 1999.)

homeobox gene, *Gbx2*, interacting with yet other homeobox genes (**FIGURE 2.20**) (Millet et al., 1999).

By the way, the *Drosophila* homologue of *Otx2*, called *orthodenticle* (*otd*), is important for directing differentiation of the anterior brain in flies (Reichert, 2002), so this mechanism for forming the anterior nervous system was probably at work in the common ancestor of insects and vertebrates—that's a long time.

Homeobox gene engrailed *marks the boundary of midbrain and hindbrain*

While *Otx2* is expressed throughout the forebrain and midbrain, another homeobox gene, *engrailed* (*en*), is expressed solely in the posterior portion of the midbrain, seeming to mark the boundary between the midbrain and hindbrain. Once this pattern of gene expression has been in place for a while, this transition zone itself takes on the role of inducing the fate of nearby structures, becoming a "local organizer" or a "morphogenetic field."

The presence of a midbrain/hindbrain local organizer was first suggested by experiments that transplanted tissue between embryonic chickens and quail. These species are closely enough related that tissue from one will develop in the other, but at the microscopic level their cells can be distinguished (quail cells have larger, darker nucleoli in standard histological stains [Le Douarin, 1973]). This difference allowed the researchers to tell whether developing brain

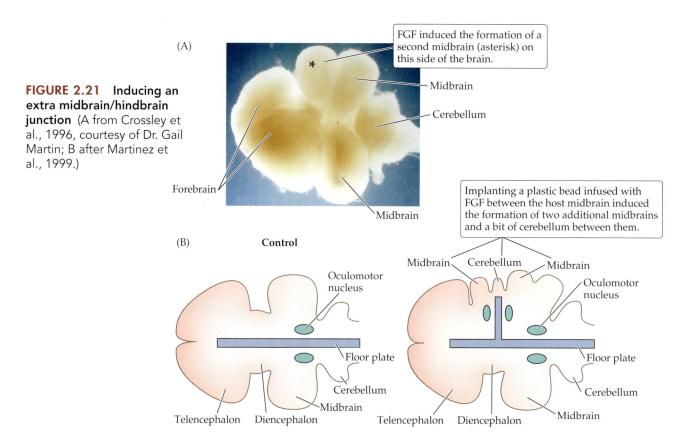

(A)

FGF induced the formation of a second midbrain (asterisk) on this side of the brain.

Midbrain

Cerebellum

Forebrain

Midbrain

FIGURE 2.21 Inducing an extra midbrain/hindbrain junction (A from Crossley et al., 1996, courtesy of Dr. Gail Martin; B after Martinez et al., 1999.)

(B)   **Control**

Implanting a plastic bead infused with FGF between the host midbrain induced the formation of two additional midbrains and a bit of cerebellum between them.

Oculomotor nucleus

Midbrain   Cerebellum   Midbrain

Oculomotor nucleus

Floor plate

Floor plate

Cerebellum

Cerebellum

Midbrain

Midbrain

Telencephalon   Diencephalon

Telencephalon   Diencephalon   Midbrain

regions were derived from the grafted tissue or the host (just as Mangold and Spemann used differently pigmented amphibians to show that the grafted dorsal lip of the blastopore recruited albino host cells to form a second nervous system). We'll return to this technique in Chapter 3 when we discuss neural crest cell migration and differentiation.

Transplanting a bit of tissue from the midbrain/hindbrain border into the forebrain induced the tissues there to form a tiny midbrain and cerebellum in the middle of the cortex (Alvarado-Mallart, 2005; Hidalgo-Sanchez, Millet, Bloch-Gallego, & Alvarado-Mallart, 2005)! The midbrain/hindbrain region secretes several different factors that might be responsible for such induction of a midbrain and cerebellum, and probably all play a role in normal development, but one of the **fibroblast growth factor (FGF)** proteins has been shown to be sufficient. Placing an acrylic bead that had been infused with FGF into an embryo just behind the forebrain could induce neighboring tissue to form an additional midbrain (**FIGURE 2.21**) (Martinez, Crossley, Cobos, Rubenstein, & Martin, 1999).

Homeobox genes influence cortical differentiation as well

Later in development, when the cerebral cortex begins developing in the telencephalon, two homeobox genes are expressed in anterior-posterior gradients. **Emx2** is concentrated in posterior cortex, while **Pax6** is concentrated at the anterior end (**FIGURE 2.22A**). These genes appear to play a role in regulating how much cortex is devoted to various functions, because animals with null mutations of *Emx2* (*Emx2-/-*) have an expanded motor region in frontal cortex, while *Pax6* knockouts (*Pax6-/-*) show expanded visual regions in the occipital cortex (Muzio & Mallamaci, 2003) (**FIGURE 2.22B**).

These examples are enough to give you an idea of how successive waves of transcription factors, one after the other, with differing factors and different *sequences* of factors, induce each part of the neural tube to produce the particular brain region appropriate for that position. Natural selection has arranged it so that, in most cases, these successive inductions happen at the right times and the right places to make a whole brain, organized in such a way that the individual can fit its ecological niche.

fibroblast growth factors (FGFs)
A family of proteins that act upon a family of receptor tyrosine kinase (RTK) receptors. They are concentrated in the posterior portions of the developing vertebrate nervous system and demarcate the midbrain/hindbrain junction.

Emx2 A homeobox gene highly expressed in the posterior portion of the developing vertebrate cortex.

Pax6 A homeobox gene highly expressed in the anterior portion of the developing vertebrate cortex.

(A)

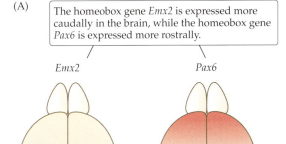

The homeobox gene *Emx2* is expressed more caudally in the brain, while the homeobox gene *Pax6* is expressed more rostrally.

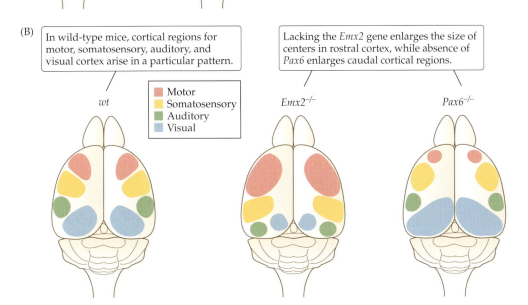

Emx2

Pax6

FIGURE 2.22 **Role of homeobox genes in cortical specification** (After Muzio & Mallamaci, 2003.)

(B)

In wild-type mice, cortical regions for motor, somatosensory, auditory, and visual cortex arise in a particular pattern.

Lacking the *Emx2* gene enlarges the size of centers in rostral cortex, while absence of *Pax6* enlarges caudal cortical regions.

■ Motor
■ Somatosensory
■ Auditory
■ Visual

wt

Emx2⁻/⁻

Pax6⁻/⁻

Although the research on homeotic genes in *Drosophila* began with a focus on how the identity of various segments is established, the abdominal segments versus the thorax, for example, it became clear that the issue was not really segmentation per se. Rather, the issue was a much larger one—how the overall body plan is established. From that perspective, we vertebrates are segmented, too, as we discuss in **BOX 2.3**.

BOX 2.3

KERFUFFLES IN LANGUAGE: "SEGMENTATION"

As scientists working with *Drosophila* were dramatically uncovering the gene mechanisms regulating cell fate in flies, coming to understand how each segment of the fly develops a unique set of structures, there was of course interest, and perhaps a little envy, among scientists working with mammals. For a time, there were people questioning whether understanding the rules of segmentation in flies

would have any relevance to mammals because, as everyone "knew," mammals are not segmented.

But after the homeobox was found to be present in the Hox genes, researchers looked for similar sequences in vertebrates. Lo and behold, homeobox genes were found in frogs (Wright, Cho, Fritz, Burglin, & De Robertis, 1987) and mammals

continued

continued

KERFUFFLES IN LANGUAGE: "SEGMENTATION"

(Hauser et al., 1985), too. Then came the revelation that the Hox genes in mammals, like those in flies, show colinearity: the positions of Hox genes on chromosomes align with the body regions where they are expressed, especially in the nervous system (see Figure 2.17). In fact, every neuroscientist knows that the central nervous system is segmented: the major divisions of the brain are prominently segmented in embryonic development (prosencephalon, mesencephalon, rhombencephalon, etc.), and the spinal cord is *explicitly* segmented, with pairs of dorsal and ventral roots coursing between vertebrae, even in adulthood (see figure).

So the work on genetic specification of the anterior-posterior segmentation in *Drosophila* turned out to be relevant to mammals after all. In fact, as the text explores, mammals have *four sets* of Hox genes, suggesting they are very important, and mutations of these genes can have devastating effects on brain development. If you conceive of Hox genes as directing segmentation in flies, then it turns out mammals are segmented, too. Was the reluctance to embrace Hox genes as relevant to humans a failure of imagining vertebrate bodies as segmented, a failure to appreciate the shared lineage of flies and mammals, a reluctance to appreciate lessons from a pesky little bug, or just one set of researchers reluctant to give credit to another?

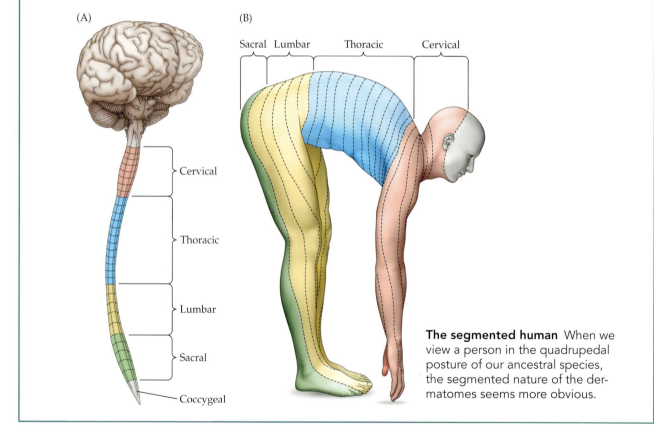

(A)

(B)

Sacral Lumbar   Thoracic   Cervical

Cervical

Thoracic

Lumbar

Sacral

Coccygeal

The segmented human When we view a person in the quadrupedal posture of our ancestral species, the segmented nature of the dermatomes seems more obvious.

How's It Going?

In what sense is the vertebrate brain segmented, and how do homeobox genes play a role in that segmentation?

■ **2.7** ■

HINDBRAIN RHOMBOMERE FATES ARE DIRECTED BY HOMEOBOX GENES

As development proceeds, the rhombencephalon becomes quite prominently segmented, as eight swellings arise, forming structures called **rhombomeres**. Each of these rhombomeres will grow to form a distinct part of the nervous system (Cooke & Moens, 2002). Some cells in the first rhombomere (r1) will form the cerebellum, cells from r1–5 will form the pons (metencephalon), and r6–8 will form the medulla (myelencephalon). The neural crest cells associated with each rhombomere, basically sitting along either side of the dorsal surface, will also make distinct contributions to the six pharyngeal arches we mentioned earlier (that Haeckel erroneously identified as "gills"; **FIGURE 2.23**), and they in turn will form cranial nerve ganglia and construct the face, as we'll discuss in Chapter 3.

The rhombomeres are not only structurally distinct swellings, they are also distinct in their expression of various transcription factors, including homeobox genes. The rhombomeres also differ in the expression of **ephrins**, a family of membrane-bound proteins that bind to a family of ephrin receptors, membrane-bound tyrosine kinase receptors (RTKs). We'll encounter other members of the RTK superfamily in Chapter 4. Note that ephrins do not themselves bind to DNA to regulate gene expression, so they are neither transcription factors nor homeobox genes; they affect gene expression indirectly by altering signaling pathways inside the target cell. As both the ligand and receptor are membrane bound, using the ephrin system requires physical contact between cells to trigger second messengers in the receptor that can, among other things, lead to changes in gene expression in the target cell (although often both cells express both the ligand and the receptor to respond to the ligand from the neighbor—in other words, the ephrin signaling is often reciprocal). In Chapter 5 we'll see that ephrin signaling plays a role in axon guidance, but at this stage of development it regulates cell migration to sharpen the boundaries between rhombomeres (Cooke, Kemp, & Moens, 2005; Cooke & Moens, 2002). During ephrin signaling, cells that are "out of place," expressing a homeobox gene when surrounded by neighbors

rhombomeres A group of prominently segmented portions of the embryonic rhombencephalon.

ephrins A family of membrane-bound signaling molecules that bind to ephrin receptors, which are part of the receptor tyrosine kinase (RTK) superfamily.

Ephrins affect gene expression indirectly by altering signaling pathways inside the target cell.

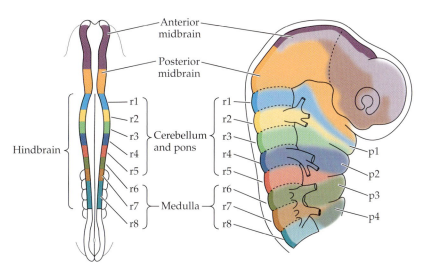

FIGURE 2.23 **The eight vertebrate rhombomeres** (After Le Dourarin, 2004.)

(A)

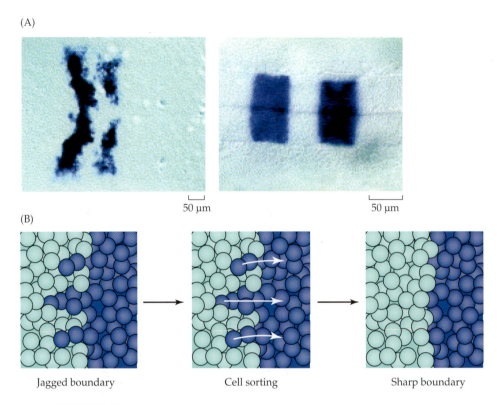

50 µm   50 µm

(B)

Jagged boundary   Cell sorting   Sharp boundary

FIGURE 2.24 **Sharpening of rhombomere boundaries** (A) Expression of the transcription factor *Krox20* in zebrafish rhombomeres 3 and 5 is diffuse at first (10 hours after fertilization), but 8 hours later it is much sharper. (B) The sharpening of expression does not come from the death of cells that are not properly sorted across the border, but from both the migration of cells to the proper side of the boundary and the change in gene expression of individual cells. This sharpening of rhombomere boundaries requires ephrin signaling. (From Cooke & Moens, 2002.)

that are not, will either stop expressing the homeobox gene or migrate to join other cells that are (**FIGURE 2.24**). These steps result in a sharp boundary in gene expression between rhombomeres, which in turn will result in distinct fates for the cells that make up the rhombomeres.

Given the dramatic nature of Hox mutations in flies, you may be wondering why such mutations in humans don't result in a wholesale swapping of body parts—a hand where a foot should be, ears where the eyes should be, and so on. Of course no one knows the answer for sure, but the absence of such dramatic changes probably reflects, in part, the much more complex body plans of mammals versus insects. Put simply, when there are so many more cells, leading to more body parts—more sub-sub-sub regions of an arm, an eye, or a brain—then ever more specific plans are needed to make each variant. This is surely one reason why mammals have four sets of Hox genes, rather than the single set our common ancestor with flies had. Other homeobox genes in flies also typically have more than one homologue in vertebrates. Vertebrates need more variant genes to make more subassemblies of cells that make up various structures. This variety of activity of several homeobox genes also means there is some redundancy of effort, hence more than one of the genes must be knocked out to produce an obvious phenotype (see Figure 2.18).

This more complex body plan also means that when things do go wrong, if loss of a homeobox gene results in the failure of a cell assembly to make some structure, this will have more far-reaching consequences for the assembly of neighboring body parts. In fact, there may be such cases of development in humans, but they would be lethal. If such embryos were lost very early in development, the woman might not have even been aware that she was pregnant. Embryos lost a little later in development, so-called miscarriages (technically known as **spontaneous abortions**), are still microscopic and therefore typically go unexamined. Probably the most misshapen embryos are lost at such microscopic stages. Those Hox mutations in humans that are not embryonically lethal, that modify the protein's function only slightly, tend to show distortions in the shape or number of late-developing structures that are not crucial for survival, such as digits and external ears (Quinonez & Innis, 2014), rather than absolute transformations of one body part into another. For example, mutations of HoxA or HoxD genes can result in formation of additional digits on the feet and/or the hand (**FIGURE 2.25**).

How's It Going?

What are ephrins and how do they play a role in segmentation of the brainstem?

■ 2.8 ■
SEVERAL SIGNALS DESIGNATE THE CAUDAL END OF THE BODY AND NERVOUS SYSTEM

FIGURE 2.25 **Hox gene mutations can result in additional fingers.** © Evaristo SA/ AFP/Getty Images.)

The developing vertebrate nervous system benefits from several signals informing cells about their position in the anterior-posterior axis (i.e., from **rostral** [head end] to **caudal** [tail end]). The gradients of these signals are initially established by the same BMP antagonists (organizer signals like noggin and chordin) that induce neural development, which are most concentrated in the rostral end. Eventually these signals determine the order of homeobox gene expression that we've discussed so far.

Meanwhile, the posterior end of the body has a high concentration of four factors that also direct neural fate: the BMP proteins themselves (which are antagonized by organizer signals concentrated at the anterior end of the neural plate), and three other factors—FGF, Wnt, and retinoic acid (**FIGURE 2.26**). All three serve as "posteriorizing" agents, directing cells to take on a fate appropriate for the caudal end of the animal. We'll discuss each in turn.

- **FGF** The portions of the neural plate that will give rise to the spinal cord are exposed to more BMP signaling between cells than are those at the head end for two reasons—the BMPs are more concentrated at the posterior end, and the organizers antagonizing BMP

spontaneous abortion The accidental loss of an embryo, sometimes called a miscarriage.

rostral Referring to the head end.

caudal Referring to the tail end.

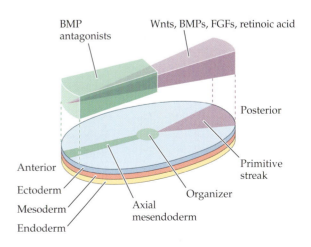

FIGURE 2.26 **Anterior-posterior cues in the neurula** (After Robb & Tam, 2004.)

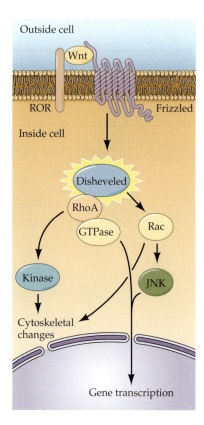

Outside cell

Wnt

ROR Frizzled

Inside cell

Disheveled

RhoA

GTPase Rac

Kinase JNK

Cytoskeletal changes

Gene transcription

FIGURE 2.27 **One of several Wnt signaling pathways** (After MacDonald et al., 2009.)

signaling are more concentrated at the anterior end. Given that, you might expect the posterior neural plate cells to be in danger of becoming epithelial cells rather than neural cells, but that fate is averted by exposure to fibroblast growth factor (FGF), a family of secreted heparin-binding proteins that bind to a family of FGF receptors on the cell membrane surface. FGF signaling often induces cells to divide (i.e., they often act as **mitogens**, factors that encourage mitosis), but it can also have a wide variety of other effects. FGF is first secreted by the endoderm to induce cells to become mesoderm, but it also acts on overlying ectoderm to make those cells less sensitive to BMP signaling, so the cells of the presumptive spinal cord continue developing a neural fate despite lesser exposure to the organizer signals from the head. The gradient of FGF signaling in the rostral end also establishes the opposite gradients of homeobox genes *Emx2* and *Pax6*, which direct the specifications of different cortical regions, which we described earlier (Cholfin & Rubenstein, 2007) (see Figure 2.22).

• **Wnt** Another protein secreted from the posterior end of the neural plate is Wnt, which acts on a family of cell surface receptors called **Frizzled** to have a diverse range of effects by regulating gene expression (**FIGURE 2.27**). There are over a dozen Wnt genes in vertebrates (Nusse & Varmus, 2012), but we'll lump them together as simply **Wnt**. The posterior concentration of cells secreting Wnt is itself a result of maternal polarity genes.

• **Retinoic acid** Finally, the posterior neural plate also has a high concentration of cells producing **retinoic acid** (**RA**), a steroid hormone (**FIGURE 2.28**). RA is not a protein, so there is no gene for the molecule; rather, RA production in cells is determined by the expression of enzymes that derive the steroid from vitamin A (**FIGURE 2.29A**). Like all steroids, retinoic acid is lipophilic, so it readily crosses the cell membrane to enter cells. In

mitogen A substance that promotes mitosis.

Frizzled A cell surface receptor protein that responds to Wnt.

Wnt A gene that encodes the secreted protein Wnt, which is concentrated in the posterior end of vertebrate embryos.

(A) (B)

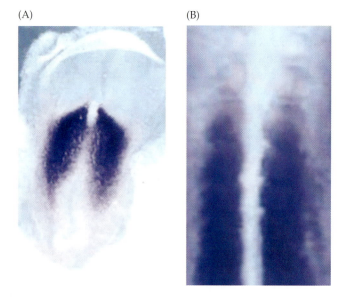

FIGURE 2.28 **Expression of the enzyme required to produce retinoic acid** (A) Looking down on a chick neural plate where the node is anterior, you can see that the enzyme required to produce retinoic acid (purple) is expressed posterior to the node, where the brain will develop. (B) Later in development, the enzyme is not expressed in the neural tube (white region in center) but is in the mesoderm on either side, with highest levels expressed posteriorly (bottom of photo). (From Maden, 2006.)

(A) Retinoic acid

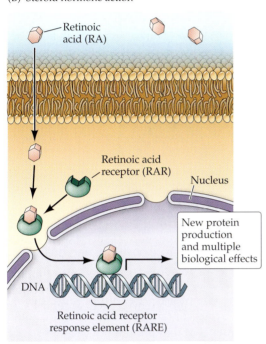

(B) Steroid hormone action

FIGURE 2.29 **Steroid receptor signaling** (A) The steroid retinoic acid (RA) acts through (B) the retinoic acid receptor (RAR) to inform cells in the developing nervous system where they are along the anterior-posterior axis. Note this diagram is not to scale, as the RA molecule is much smaller than the RAR protein.

target cells, retinoic acid binds to an intracellular protein, the **retinoic acid receptor** (**RAR**), then the steroid-receptor complex binds to DNA to regulate gene expression. Which genes are regulated? Those genes in which the promoter region contains a **retinoic acid response element** (**RARE**), a specific sequence of nucleotides that recognizes the steroid-receptor complex (**FIGURE 2.29B**). (The abbreviations can be confusing, so note that the steroid itself is RA, its receptor is RAR, and the promoter regions they bind are RAREs). RAREs are found in the promoter regions of several homeobox genes. We'll discuss other steroid hormones and their receptors in Chapter 7 when we discuss sexual differentiation of the nervous system in vertebrates.

In addition to being highly concentrated at the posterior end of the embryo, FGF, Wnt, and retinoic acid have something else in common—they all regulate expression of Hox genes, which then direct cellular differentiation in a direction that is appropriate for each cell's position in the anterior-posterior axis. This regulation of Hox genes is most important for specifying cell fate in the spinal cord, but it also plays a role in the brainstem. For example, treating embryos with additional RA can "posteriorize" rhombomeres to express more posterior Hox genes, and therefore to adopt the fate of more posterior rhombomeres (Kessel, 1993).

Because RA is a steroid, it readily passes through cells, including skin cells, to circulate throughout the body, and so it is a powerful **teratogen** ("monster producer"), a chemical that interferes with fetal development to produce birth defects. Embryos exposed to exogenous RA illustrate its role as a marker for

retinoic acid (RA) A steroid-like molecule, concentrated in the posterior end of vertebrate embryos, that promotes development of posterior structures. It is a powerful teratogen.

retinoic acid receptor (RAR) A member of the steroid receptor superfamily that serves as a receptor for retinoic acid.

retinoic acid response element (RARE) A specific sequence of DNA nucleotides that is bound by the retinoic acid–retinoic acid receptor complex, thereby regulating expression of the associated gene.

teratogen A substance that causes malformations in development.

FIGURE 2.30 Retinoic acid (RA) is a powerful teratogen. (A) Mouse mothers fed a miniscule amount of RA (0.00025 mg/g) produce misshapen embryos (right). (B) Embryos exposed to RA form fewer anterior structures, including brain structures. (A from Anchan et al., 1997; B from Linney & LaMantia, 1994.)

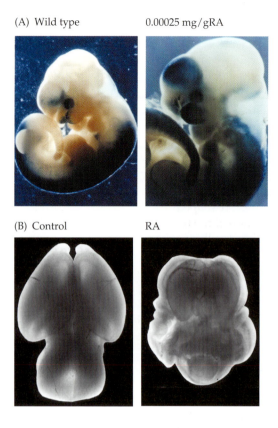

(A) Wild type   0.00025 mg/gRA

(B) Control   RA

posterior fate—the larger the dose of RA, the more posterior nervous system is produced, so little or no brain develops (**FIGURE 2.30**).

A form of RA called isotretinoin (trade name Accutane) is a very effective treatment for acne. But because of RA's teratogenic capacity, drugs containing isotretinoin are boldly labeled warning women that they should not use these drugs if they are pregnant or *might become* pregnant. Unfortunately, since women of an age to seek acne treatment also tend to be young enough to be fertile, and since many pregnancies in the United States are unplanned, many babies have been born with birth defects caused by the drug, including absence of ears or jaws and malformed brains (Lammer et al., 1985). This is, of course, yet another reason to educate teenagers about effective birth control methods and to make those methods widely available.

> Since women of an age to seek acne treatment also tend to be young enough to be fertile, many babies have been born with birth defects.

--- **How's It Going?** ---

What are the four signals that direct the nervous system to take on a caudal fate?

■ **2.9** ■

CONTINUED GRADIENTS IN BMP SIGNALING ESTABLISH THE DORSAL-VENTRAL AXIS IN THE NERVOUS SYSTEM

So far we've talked about factors that affect development of the nervous system in the anterior-posterior axis. Other factors are needed to direct neural cells to differentiate in the manner that's appropriate for their position in the dorsal-ventral axis. In Chapter 1 we saw that ectodermal cells secrete bone morphogenetic protein (BMP) to induce each other to follow an epidermal fate.

When the underlying mesodermal tissues (which had formed the dorsal lip of the blastopore at the start of gastrulation) serve as an organizer to disrupt that BMP signaling, the ectodermal cells are induced to follow a neural fate, forming the neural plate. After the neural plate has formed a neural tube, a sheet of ectodermal cells closes overhead, continuing to release BMP onto the dorsal portion of the tube. Thus the concentration of BMP informs cells in the neural tube how far dorsal or ventral they happen to be.

Meanwhile, some underlying mesodermal cells condense at the midline to form a rod-shaped structure, the **notochord**, under the ventral portion of the neural tube (see Figure 1.17). Eventually the notochord will contribute to forming the bony vertebral column, and it is the embryonic presence of the notochord that distinguishes all the *chordates*, including the vertebrates (see Figure 1.1). But during the early development that this chapter is concerned with, the notochord plays a crucial inductive role in guiding differentiation of the neural tube, specifically guiding development in the ventral portion of the neural tube.

> The notochord plays a crucial inductive role in guiding development in the ventral portion of the neural tube.

As development proceeds, the antagonism of the ectodermal cells that closed dorsal to the neural tube (which will eventually become epidermal cells, that is, skin) and the notochord lying ventral to the neural tube will direct cellular differentiation to establish the dorsal-ventral axis. The establishment of this dorsal-ventral axis is most obvious in the spinal cord, where eventually sensory cells will predominate in the dorsal spinal cord while motor-related neurons, including **motor neurons**, the neurons that innervate muscles, will predominate in the ventral spinal cord.

The discovery of the factor from the notochord that induces the development of the ventral spinal cord, including the differentiation of cells into motor neurons, illustrates the complementary contributions of work in flies and vertebrates. A gene discovered in flies was named *hedgehog* because some mutations of the gene resulted in spiky hairs sticking up so they looked rather like a hedgehog (Nusslein-Volhard & Wieschaus, 1980). The sequencing of the gene revealed that it included the code for a **signal peptide**, a sequence of N-terminal amino acids that directs the full protein to the secretory pathway, which meant the product was a secreted protein. Eventually it was found that hedgehog is part of an evolutionarily conserved signaling system for regulating gene expression in many different systems (**FIGURE 2.31**).

Vertebrates were soon found to have several homologous genes (Riddle, Johnson, Laufer, & Tabin, 1993). Note that, as in the case with the Hox genes, the difference between vertebrates and flies is not in which genes are present, or even their function, broadly speaking. Rather, vertebrates appear to have become more complex by duplicating those genes, which then have diverged in structure and function to produce more complex bodies. One of the vertebrate homologues was playfully named **Sonic hedgehog (Shh)**, after a character in a SEGA video game that was popular at the time. Shh was found to be produced and secreted by the embryonic notochord (Roelink et al., 1994). The discovery that notochord secretes Shh provoked the question of whether this protein could play a role in the induction of the **floor plate**, a morphological and functionally distinct ventral portion of the spinal cord. Earlier, studies in chick embryos had shown that transplanting a bit of notochord alongside the neural tube, or even dorsal to the tube, could induce the formation of an additional floor plate, as long as the notochord was within 25 μm or so of the tube (van Straaten, Hekking, Thors, Wiertz-Hoessels, & Drukker, 1985; van Straaten, Hekking, Wiertz-Hoessels, Thors, & Drukker, 1988). Was Shh the factor that notochord used to induce the floor plate? We'll look at that question next.

notochord An embryonic rod-shaped structure that is derived from mesodermal tissue in all vertebrates and that induces formation of the ventral neural tube above.

motor neurons Neurons that send axons out to the periphery to innervate and control muscles.

signal peptide A particular sequence of N-terminal amino acids that directs the full protein to the cell's secretory pathway.

Sonic hedgehog (Shh) A gene that encodes the signaling protein that is secreted by the notochord and induces formation of the floor plate and the differentiation of motor neurons in the vertebrate neural tube.

floor plate The ventral portion of the vertebrate neural tube, the developing spinal cord.

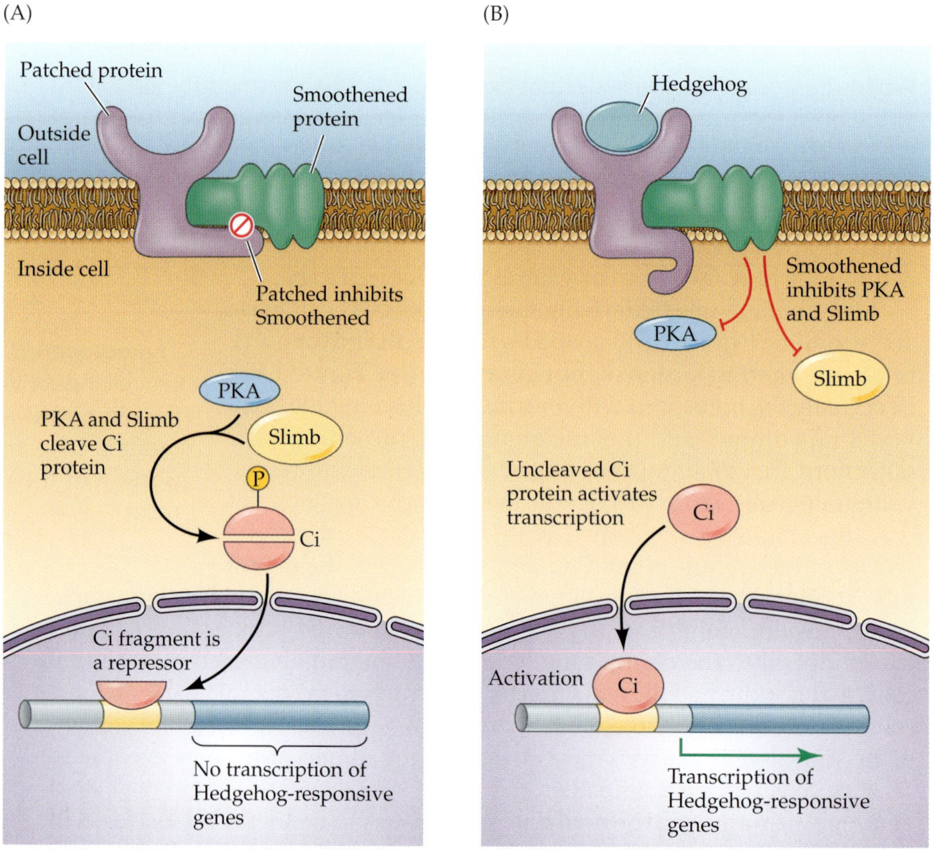

(A)

(B)

FIGURE 2.31 Hedgehog signaling pathway (A) Normally a protein fragment suppresses expression of a set of genes. (B) Binding of Hedgehog to the Patched/Smoothened receptor halts that suppression so that a set of genes is expressed. (After Johnson & Scott, 1998.)

RESEARCHERS AT WORK

What Notochord Factor Induces the Floor Plate and Motor Neurons?

■ **QUESTION**: What factor(s) secreted by the notochord induce the ventral neural tube to form a floor plate with motor neurons on either side?

■ **HYPOTHESIS**: If Shh is the factor secreted by notochord to induce formation of the spinal cord floor plate, then supplying Shh to another part of the neural tube should induce the formation of an additional, ectopic (misplaced) floor plate there.

■ **TEST**: Transfect cells of a tumor line that grows readily in vitro, in this case COS cells, so that they secrete Shh, and place such cells next to the neural tube to see if a floor plate forms.

■ **RESULT**: COS cells that had not been transfected to express Shh had no discernible influence on the neural tube, but COS cells secreting Shh induced the neural tube to form a floor plate, much like transplants of notochord (Roelink et al., 1994). Eventually such manipulations were shown to be capable of inducing differentiation of motor neurons on both sides of the ectopic floor plate (Placzek, Jessell, & Dodd, 1993; Tanabe & Jessell, 1996) (**FIGURE 2.32**).

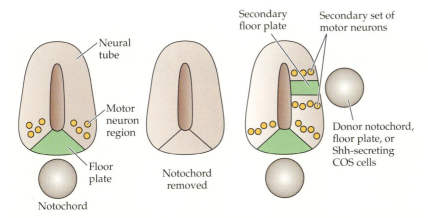

FIGURE 2.32 **Notochord induces motor neuron differentiation** (After Placzek et al., 1990.)

■ **CONCLUSION**: Shh is at least one of the important factors secreted by the notochord to direct the ventral half of the neural tube to differentiate into floor plate. This is an initial step in the production of motor neurons just dorsal to the floor plate on either side of the ventral spinal cord.

Meanwhile, the dorsal half of the neural tube, which receives relatively little Shh stimulation and considerably more BMP stimulation (from the overlying ectoderm), differentiates in another fashion, involving different patterns of gene expression, into the **roof plate**. Those portions of the neural tube sandwiched between the roof plate and floor plate differentiate along yet another pathway, involving a different pattern of gene expression (**FIGURE 2.33**).

roof plate The dorsal portion of the vertebrate neural tube.

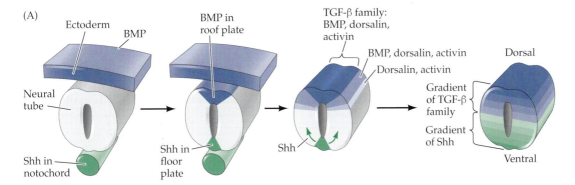

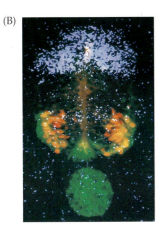

FIGURE 2.33 **Antagonism between BMP and Shh help establish the dorsal-ventral axis.** (A) The gradients of BMP from the ectoderm dorsally and of Shh from the notochord ventrally are maintained as the roof plate and floor plate begin expressing those signals. (B) In this photomicrograph, Shh is stained green, dorsalin (like BMP, a member of the TGFβ family) is light blue, and the motor neurons are stained orange. (B courtesy of Dr. T. M. Jessell.)

FIGURE 2.34 Failure to split the telencephalon (A) This severely distorted brain, a single, crescent-shaped cortical lobe, in a case of holoprosencephaly, came from a baby who stopped breathing 18 minutes after birth. (B) Cyclopia in a lamb. Note the severely distorted snout, which is one reason such individuals typically die at birth or shortly after. This case was caused by a teratogen, but defects in Shh signaling can also cause cyclopia. (A from Tantbirojn et al., 2008; B courtesy of Dr. L. James and USDA Poisonous Plant Laboratory.)

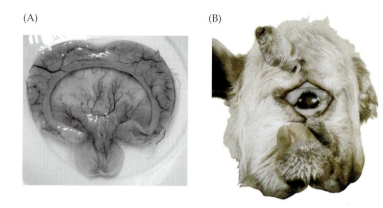

(A) (B)

Shh similarly guides differentiation of the ventral neural tube at the anterior end, directing formation of structures appropriate for ventral brainstem, including motor neurons in various cranial nerve nuclei. Another function of Shh is inhibiting outward growth at the far rostral end of the telencephalon. This has two obvious effects on brain development. One is to inhibit the rostral growth of the cortex at the midline, forming a notch in the developing telencephalon (see Figure 2.19B). This is the beginning of cortical development into two hemispheres, one on the left and another on the right. Also, as cells grow out to form the eye field, Shh from the underlying notochord at the midline inhibits this growth, so the eye field splits to form the left and right optic cups (the subjects of Spemann's groundbreaking work on induction we discussed in Chapter 1). If something interferes with Shh secretion or signaling at this stage, then it fails to split the developing telencephalon and eye field, so a single telencephalon and a single eye cup form. Thus in severe cases, the individual will be born with only a single forebrain, a condition known as *holoprosencephaly* ("single prosencephalon") (**FIGURE 2.34A**), and a single eye, *cyclopia* (**FIGURE 2.34B**) (Chiang et al., 1996; Tantbirojn, Taweevisit, Sritippayawan, & Uerpairojkit, 2008). The improper development of the eyes typically distorts other facial features, so these individuals may develop severe facial distortions, which in addition to very misshapen brains is probably why the condition is usually lethal before or shortly after birth. The stillbirth of such babies may have inspired prehistoric myths about Cyclops that inspired Homer's *Odyssey*.

In Chapter 1 we learned that the same mechanism of regulating BMP signaling establishes the dorsal-ventral axis in *Drosophila*, too. Recall that a fly analogue of chordin interferes with the signaling of a BMP homologue, so neural tissue develops (see Figure 1.32; because dorsal and ventral are reversed in vertebrates versus invertebrates, the neural tissues develop in what we call the ventral portion of flies).

How's It Going?
What are some of the gradients that inform neural tube cells of their dorsal-ventral position?

■ 2.10 ■

FIND OUT WHERE YOU ARE TO COORDINATE YOUR FATE WITH THAT OF YOUR NEIGHBORS

Thanks to pioneering work in fruit flies that guided research in vertebrates, we can understand, at least in principle, how development of the brain and spinal cord is organized. Thus various cortical regions develop at the anterior end, the midbrain develops behind that, and the various fates unfold for the eight rhombomeres, including those giving rise to the cerebellum. Farther posterior, other factors direct the neural tube to form the spinal cord. While the homeobox genes control differentiation along the anterior-posterior axis, the continuing action of BMP from overlying ectoderm and Shh from the mesodermal notochord below control differentiation along the dorsal-ventral axis, including the formation of the earliest neurons to develop in vertebrates, the motor neurons.

Remember that this whole complex program of development requires cell-cell interaction, each cell both responding to signals from its neighbors and providing signals in turn to guide differentiation. Recall Brenner's analogy contrasting this system with the mosaic program of development. He likened the cell-cell interaction program of development to an "American plan," where any cell can take on any role, based not on what its ancestors were (as in mitotic lineage in *C. elegans*), but on the environment around it.

I like to offer a metaphor about putting together a symphony orchestra, where traditionally string instruments are placed in the front, woodwinds in the middle, horns and drums in the back, etc. One way to construct such an orchestra is to interview individual musicians and, depending on what instruments they play and who plays each instrument best, seat them as first violin, second violin, first flute, second flute, etc. But another way would be to just seat people at random in the chairs and then say to the person in the first violin chair, "You're going to be the best violinist, so I'm going to teach you that instrument and spend more time with you than the second violinist." Just so, the body makes a bunch of cells, and each one of them will take on whatever highly specialized function is needed if it ends up in the right place at the right time. While we may not be able to randomly assign people to acquire varying degrees of skill (they must have the right aptitude and attitude to excel on a particular instrument), the body can indeed randomly assign cells to a particular fate because every cell has the requisite genes, so it's just a matter of directing the cell to express the right genes in the right order.

In the next several chapters we will see many more examples of how cell-cell interactions guide further development, determining how many neurons are made, where the neurons migrate to, what fates they take on there, where they send their growing axons, how the axons make synapses, what neurotransmitters they release, and so on. Each of these many choices, which are absolutely critical in order for the individual cell to take on the appropriate fate, will be decided as a result of communication with other cells.

The point in evolution when our ancestors began using such cell-cell interactions to determine cell differentiation was a crucial step in the development of the human brain. Understanding the consequences of this first step will help you understand the organization of this book. This watershed event, relying on cell-cell interactions to coordinate development, specifically cell fate, has taken up our time in this and the preceding chapter, as it will in several more chapters to come. But reliance on cell-cell communication to determine cell fate eventually led to situations where it was the *electrical activity of neurons* that affected the fates of other cells, often other neurons. Chapter 8 will discuss those cases where neural activity regulates neural development. Once neural activity began directing development, it was only a matter of time before that neural activity would sometimes originate in sensory cells, at which point information from the outside world, *sensory experience*, could start guiding neural development, which is the subject of Chapter 9. Then once sensory processes began directing neural development, it was only a matter of time before *social experience*, the exposure to another individual, would start guiding neural development. Thus Chapter 10, the final chapter of the book, will be about social guidance of neural development, which is how, for example, human culture came to be a major force shaping the human brain and therefore the human mind.

> Using cell-cell interactions to determine cell differentiation was a crucial step in the development of the human brain.

SUMMARY

■ The evolution of divergent species is driven primarily by changes in later development, so phylogenetically closely related species appear very similar to one another as embryos. Thus the embryos of all vertebrates appear rather similar to one another and share many stages of development and genetic mechanisms to guide development. **See Figure 2.1**

■ In *Drosophila*, *maternal factors* result in the establishment of a basic anterior-posterior axis, with *bicoid* concentrated at the head end of the *syncytium*, and *nanos* concentrated at the tail end. **See Figures 2.3 and 2.4**

■ Placing *bicoid* mRNA in various parts of the embryo will result in the formation of head structures at those positions. **See Figure 2.5**

■ *Maternal polarity genes* are transcription factors that organize anterior-posterior gradients in several other transcription factors within the syncytium. **See Figure 2.6**

■ Maternal polarity genes regulate the expression of another set of transcription factors, the *gap genes*, such as *hunchback*. Because maternal polarity genes often antagonize each other's effects, the resulting spatial distribution of gap gene products tends to be more sharply defined than the distribution of maternal polarity genes. **See Figures 2.7 and 2.8**

■ Gap genes in turn regulate the expression of transcription factors called *pair-rule genes*. Each of the pair-rule genes is expressed in alternating stripes along the anterior-posterior axis. **See Figure 2.9**

■ Pair-rule genes regulate the expression the *segment polarity genes*, which encode cell-cell signaling systems to regulate gene expression as *cellularization* of the embryo takes place. Because each of the segment polarity genes is expressed in a stripe about one segment in width, each of the newly separated cells has been exposed to a sequence of transcription factors that is unique for the particular segment where it resides. **See Figure 2.10**

■ Gap genes, pair-rule genes, and segment polarity genes all regulate the expression of *homeotic selector genes*, called *Hox genes*, which all contain a *homeobox*. **See Figures 2.11 and 2.12**

■ Hox gene mutant flies may have whole body parts transformed, such as a leg constructed where an antenna should be. **See Figures 2.13 and 2.14**

■ Hox genes in *Drosophila* show *colinearity*: their sequence on the chromosome aligns with their expression in the anterior-posterior axis. **See Figure 2.16**

■ Hox genes in mammals also display colinearity, despite forming complexes on four different chromosomes, with the homologous genes on the various chromosomes diverging in function. **See Figures 2.17 and 2.18**

■ As in insects, Hox genes and other homeobox genes play a role in segmentation in mammals, including in the nervous system. For example, the homeobox gene *Otx2* guides the fate of the developing midbrain and forebrain, while other homeobox genes are expressed only in the spinal cord. *Engrailed* is expressed only in the posterior midbrain, which seems to make the midbrain/hindbrain junction a "local organizer" capable of using *FGF* to induce a wide range of neural tissue to form a midbrain and cerebellum. **See Figures 2.19–2.21**

■ Homeobox genes *Emx2* and *Pax6* are expressed in opposite gradients in the developing cortex, and together they regulate the size of various cortical regions. **See Figure 2.22**

■ The hindbrain forms eight *rhombomeres*, each of which takes on a separate fate, under the direction of homeobox genes and *ephrins* that sharpen the demarcation between rhombomeres. **See Figures 2.23 and 2.24**

■ There are four factors concentrated at the posterior end of the embryo to direct differentiation along the anterior-posterior axis: BMP proteins, FGF, *Wnt*, and *retinoic acid (RA)*. Each regulates Hox genes and other homeobox genes to guide differentiation appropriate to a particular segment of the nervous system. **See Figures 2.26 and 2.27**

■ RA is a steroid hormone and *teratogen*, so exogenous RA causes embryos to develop more caudal structures, for example, "posteriorizing" the development of rhombomeres, or preventing the formation of the brain. **See Figures 2.28–2.30**

■ The dorsal-ventral axis is formed by secretion of BMP from ectodermal cells above the neural tube and *Sonic hedgehog* (*Shh*) from the notochord below. **See Figures 2.31 and 2.32**

■ BMP stimulation induces the dorsal neural tube to form a *roof plate*, while Shh induces the ventral tube to form a *floor plate* and, just dorsal to the floor plate on each side, *motor neurons*. **See Figure 2.33**

Go to the Companion Website
sites.sinauer.com/fond
for animations, flashcards, and other review tools.

CHAPTER

3

Upward Mobility
NEUROGENESIS AND MIGRATION

A VERY LATE BLOOMER After a normal, uneventful boyhood, at age 14 Neil began wondering why his body wasn't going through the changes he saw in his peers reaching puberty. Certainly they noticed. Neil lost confidence and started dropping out of teenage activities altogether. When he consulted doctors, they just assured him he was a "late bloomer." By age 17 he still had no pubic hair, no beard development, and no interest in sex, so a doctor prescribed a low dose of testosterone, again telling him puberty would start soon. It didn't (Smith & Quinton, 2012). Neglected by his physicians, Neil stopped taking the testosterone and just assumed he was never going to change. At college he studied biomedical sciences, in part hoping to understand what was happening. At 22 years of age, working as a scientist, Neil decided to approach an endocrinologist and tell him his story. One of the first questions the endocrinologist asked Neil was, "Do you have a sense of smell?" which no doctor had ever asked before. As a matter of fact, Neil did not have a sense of smell, and he was also partially deaf.

The endocrinologist immediately had a guess about Neil's condition, which tests soon confirmed. His circulating levels of two hormones, the gonadotropins called luteinizing hormone (LH) and follicle-stimulating hormone (FSH), were very low. MRI scans of the brain revealed an absence of olfactory bulbs, and a bone scan showed low bone density. These seemingly unrelated symptoms—absence of puberty, lack of olfactory sensitivity, partial deafness, and disinterest in sexual relations—were all the result of a failure of cells to migrate into Neil's brain before birth. We'll learn more about cell migration in brain development, including the cause of Neil's condition, in this chapter.

CHAPTER PREVIEW

In the phases of brain development we've covered so far, cell-cell interactions regulating gene expression have induced some parts of the ectoderm to become the nervous system (Chapter 1) and have established the basic dorsal-ventral and rostral-caudal polarities within that neural ectoderm (Chapter 2). Now the developing brain faces a population problem—not enough cells. To make the nearly 100 billion neurons in the human brain (Herculano-Houzel, 2012), the fetus reaches a peak of adding 250,000 new neurons per minute! There was a time when scientists thought new neurons were generated only early in life, but the past few decades have shown that, to a limited extent, we make new neurons throughout life. We'll review evidence that these additions in adulthood really do matter and that we can make choices to encourage real brain growth.

mitosis The process of cell division in which both resulting cells receive the full complement of genetic material.

pleiotropy The phenomenon in which a single gene plays a role in several, seemingly unrelated, traits.

In the embryo and fetus, this **mitosis**, the splitting of cells so that each of the resulting cells has its own nucleus and separate cytoplasm, producing this monstrous number of cells, takes place in a relatively restricted number of places, following a well-established pattern, with some cells dividing continually while others leave the mitotic cycle and begin taking on their future fates. We'll see that some cells differentiate into glia but continue to divide, giving rise in some cases to more glia. Other postmitotic cells will eventually become neurons. With all those divisions taking place in one zone, the postmitotic cells must crawl away to establish brain structures. During these extensive migrations, the cells behave in many ways like autonomous organisms, sometimes shinnying along glial fibers or avidly crawling along the adhesive surfaces of some cells while actively avoiding the surfaces of others. We'll see that in both the cerebral cortex and the cerebellum, there is an orderly addition of newly generated cells that offers the first hint of what fate each cell must take on. We'll conclude with a fascinating instance of cell migration where seemingly disparate brain structures and functions depend on the same migratory pathway and therefore depend on the same set of genes. In fact, it is commonplace in development of the nervous system (or any other aspect of biology) that a single gene may be involved in many different processes, a fundamental finding we consider to kick off this chapter.

■ 3.1 ■
THE SAME GENE MAY PLAY A ROLE IN MANY DIFFERENT DEVELOPMENTAL EVENTS

Neil's condition is one of many human conditions that are caused by a mutation in a single gene. In fact, so far 11 different genes have been identified that when disabled lead to Neil's syndrome (Smith & Quinton, 2012). Depending on how much you already know about endocrinology, you may see a connection between some of Neil's symptoms—the lack of gonadotropin hormones would forestall puberty, which in turn would prevent secondary sex characters like voice change, pubic hair, and libido. But how would those connect to a lack of sense of smell? And what's the connection between any of those symptoms and deafness? This case illustrates an important concept in genetics. **Pleiotropy** occurs when one gene influences many different traits, especially when those traits seem unrelated to one another (Paaby & Rockman, 2013). We saw one example of pleiotropy in Chapter 2, when we noted that the *Sonic hedgehog* (*Shh*) gene plays a role in the development of the floor plate, the differentiation of motor neurons, and the splitting of the optic plate to produce two optic cups. At first glance, the development of motor neurons and of two eyes seem unrelated to one another, so we might not expect one gene to be crucial for both. Of course, while those may *appear* to be unrelated developmental events, in reality they are all dependent on Sonic hedgehog signaling (**FIGURE 3.1A**). Similarly, by the end of this chapter, we'll see how Neil's seemingly disparate symptoms are in fact related to a failure of a particular group of cells to migrate.

Some other famous examples of pleiotropy are that about half of all cats with white fur and blue eyes are also deaf (an example Darwin notes in *On the Origin of Species*) and, in Mendel's pea plants, that the gene that affected seed coat texture also affected blossom colors (Fairbanks & Rytting, 2001). Another example is phenylketonuria (Paul, 2000), a single gene disorder in humans that can, if left unchecked, lead to mental disability, eczema, sparse hair, and light-colored skin.

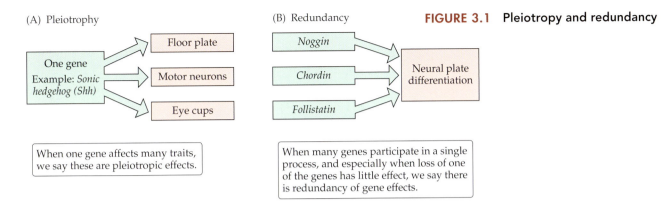

(A) Pleiotrophy

One gene
Example: *Sonic hedgehog (Shh)*

Floor plate

Motor neurons

Eye cups

When one gene affects many traits, we say these are pleiotropic effects.

(B) Redundancy

Noggin

Chordin

Follistatin

Neural plate differentiation

When many genes participate in a single process, and especially when loss of one of the genes has little effect, we say there is redundancy of gene effects.

FIGURE 3.1 **Pleiotropy and redundancy**

In this book, the names of several genes we've encountered so far provide glimpses of pleiotropy. In Chapter 1, we saw that the gene for bone morphogenetic protein (BMP) plays a crucial role in directing ectodermal cells to become epithelia rather than neural tissue, but the name tells you that this same gene must be important in bone development, too. By convention, whichever name is first applied to a gene becomes its official name, and the research group studying bone development that first named BMP (Urist & Strates, 1971) likely had no idea the gene would also have a crucial role in the development of the nervous system.

It's probably fair to say that with respect to gene influences, pleiotropy is the norm, not the exception (Johnson & Barton, 2005; Stearns, 2010; Wagner & Zhang, 2011). A human engineer who tried to build an organism would probably have each protein act in a dedicated fashion, doing a single task at a single point in development. But natural selection, using trial and error, seems to take a different strategy, sometimes co-opting existing proteins into new processes. So if you're surprised that humans may have only 10,000 genes to build such a complicated organism, comfort yourself with the knowledge that almost all of their gene products play multiple roles (Hodgkin, 1998), sometimes at very different times in development.

Keep in mind that pleiotropy, the involvement of a single gene in many traits, is a separate concept from **redundancy**, when several genes contribute to producing a single trait (**FIGURE 3.1B**). In Chapter 1 we saw how several proteins from the dorsal lip of the blastopore "organize" the induction of the neural plate. In Chapter 2 we saw how there tends to be more redundancy in vertebrates than in insects, as the duplication of *Hox* genes in mammals indicated. In both cases, loss of one of the genes involved may have only a modest effect (see Figure 2.15). The fact that many genes contribute to a process seems reasonable when you consider how complicated developmental mechanisms can be. To return to an earlier example, anything that interferes with Sonic hedgehog signaling, like mutations in the *Sonic hedgehog* gene itself or its receptor or any of the proteins participating in the second messenger systems activated by the receptor, will affect all the disparate processes where Shh signaling is involved.

redundancy   Here, the phenomenon in which several genes play a role in a process such that loss of one may have a relatively minor effect.

■ **3.2** ■

THE DEVELOPING BRAIN GENERATES NEURONS AT A TREMENDOUS RATE

By the nineteenth century scientists understood that cells are the basic building blocks of all living things, so they understood that growth requires the multi-

FIGURE 3.2 **Early observations of mitotic figures** Note that in both drawings, the mitotic figures are seen only next to the ventricles, not the pial surface. (Left, image from His, 1887; right, image from Ramón y Cajal, 1894.)

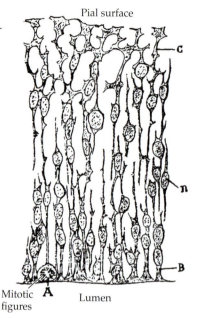
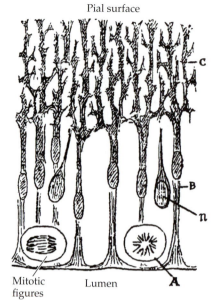

Pial surface

Pial surface

Mitotic figures

Lumen

Mitotic figures

Lumen

mitotic figures The tangled threads of duplicated chromosomes being pulled apart that are seen in cells undergoing mitosis.

meninges The three layers of tissue protecting the vertebrate central nervous system.

pia mater Also called simply *pia*. The innermost layer of the vertebrate meninges, found along the outer surface of the brain.

ventricular zone The regions adjacent to the ventricles of the brain and central canal of the spinal cord, where cell division continues throughout life.

neuroblast A cell that will divide to produce neural cells.

plication of cells through mitosis. Today we also know that mitosis provides both resulting cells with a complete copy of the individual's genes, as opposed to *meiosis*, where cell divisions eventually produce a gamete (egg or sperm) with half the individual's genes. Although there wasn't yet the technology to watch cells divide in real time, nineteenth-century microscopic examination of dead tissue slices revealed distinctive profiles called **mitotic figures**—dark, tangled threads of chromosomes in the process of separating so both resulting cells would have a complete set of genes. One clue that mitotic figures were a snapshot of cells about to divide was that they were usually seen only in tissues that were known to be growing, which were only a few tissues in adult specimens but just about everything in embryos.

When nineteenth-century anatomists like Wilhelm His (1831–1904) and Santiago Ramón y Cajal (1852–1934) examined the developing neural tube, they noticed that the mitotic figures were always seen next to the interior surface of the tube (next to the hollow interior of the tube, called the lumen), not toward the outer surface (**FIGURE 3.2**). As the neural tube develops further, that inner surface will delimit the brain's ventricles, while the outer surface will be protected by the **meninges**, innermost of which is the **pia mater** (or pia). Thus we talk of the inner, ventricular surface and the outer, pial surface. Even in adult brains, mitotic figures are found almost exclusively lining the ventricles, in the so-called **ventricular zone**.

Based on such static snapshots of cell division in action, early biologists pieced together an understanding of neural mitosis. At first, the neural tube is only one cell thick, each cell bordering both the interior and exterior surfaces of the tube. As the neural tube grows in size, getting thicker, the cells retain their connection to both surfaces, getting stretched out like a victim on a medieval torture rack. As these early cells, called **neuroblasts**, are dividing, their nuclei shuttle back and forth within the cytoplasm, from ventricular surface to pial surface, displaying mitotic figures only when close to the ventricular surface (see Figure 3.2).

As the tube grows thicker, some cells continue to contact both upper and lower boundaries of the tube, continue to shuttle their nuclei up and down, and continue to divide to produce two cells stretched across the width of the tube. In other words, a neuroblast may divide to produce two

> As the neural tube grows in size, getting thicker, some cells retain their connection to both surfaces, getting stretched out like a victim on a medieval torture rack.

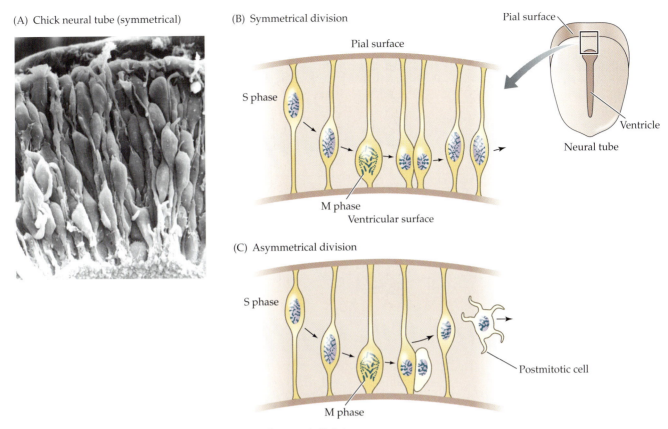

(A) Chick neural tube (symmetrical)

(B) Symmetrical division

Pial surface

S phase

M phase

Ventricular surface

(C) Asymmetrical division

S phase

M phase

Postmitotic cell

Pial surface

Ventricle

Neural tube

FIGURE 3.3 Symmetrical and asymmetrical neural divisions
(Part A courtesy of Dr. K. Tosney.)

neuroblasts. We say these cells are *dividing symmetrically* (**FIGURE 3.3A,B**). But soon some neuroblasts *divide asymmetrically*, such that one daughter cell remains attached to both the ventricular and pial surfaces and prepares to divide again, while the other daughter cell has no attachment to either surface and migrates away from the ventricular zone. This postmitotic daughter cell will populate the width of the neural tube, differentiating into a neuron or glia, as we'll detail soon (**FIGURE 3.3C**). Eventually there will be enough cells detached from the ventricular zone that they will accumulate near the pial surface to form the **marginal zone**.

Some cells retain their connections to both pial and ventricular surfaces as the brain grows tremendously. These incredibly stretched cells are radial glia, which play a crucial role in guiding neuronal migration, as we'll describe later in this chapter. Other cells release their hold on the pial surface and stay in the ventricular zone, capable of mitosis, throughout life. Right now, along the ventricular zone of your brain, cells are continually dividing to provide glial cells and, we now know, a few neurons, as we'll also discuss later. The ventricular zone is sometimes referred to as a germinal zone because one meaning of *germinal* is "providing material for future development" (as in a germinal idea, like Darwin's theory of natural selection).

The process of dividing to produce cells that differentiate into neurons is called **neurogenesis**. It's important to keep in mind that neurogenesis is *not* neurons dividing to form new neurons—neurons do not normally divide. Rather, neurogenesis is the process of nonneuronal cells in a germinal zone dividing to produce daughter cells that leave the germinal zone and change into neurons. The production of new glia is called **gliogenesis**, and we've known for some time that gliogenesis happens throughout life. Unlike

marginal zone The outermost layer of the developing vertebrate brain. By adulthood it will form the molecular layer of the cerebral cortex.

neurogenesis The mitosis of cells that will give rise to neurons.

gliogenesis The mitosis of cells that will give rise to glia.

Neurogenesis is *not* neurons dividing to form new neurons— neurons do not normally divide.

(A)

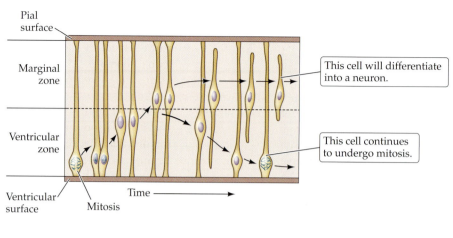

FIGURE 3.4 **Cell nuclei shuttle from ventricle to pia in mitosis.** (A) In asymmetrical divisions, one resulting cell becomes postmitotic while the other continues to divide. (B) The first postmitotic cortical cells settle in the outermost marginal zone that will become layer I. (C) Later arriving cells will form just beneath the marginal zone to form the other five layers of cortex, and the intermediate zone will become the region filled with axons coming and going from cortex.

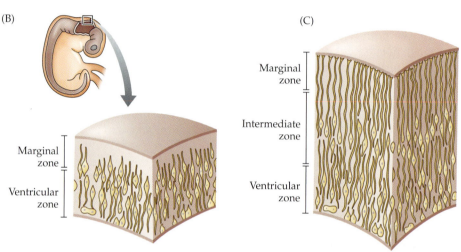

intermediate zone The layer between the ventricular zone and marginal zone of the developing vertebrate brain.

neurons, differentiated glial cells can themselves divide to produce more glia (Ge, Zhou, Luo, Jan, & Jan, 2009). This retained capacity for mitosis must carry some risk, since sometimes things go wrong and glial cell division gets out of hand to become cancerous. Most brain tumors, especially malignant ones, are *gliomas*, masses of astrocytes or oligodendrocytes that are rapidly dividing.

As more and more cells break out of the mitotic cycle and differentiate into neurons and glia, they accumulate beneath the outermost marginal zone and above the ventricular zone in a region called, reasonably enough, the **intermediate zone** (also sometimes called the mantle zone; **FIGURE 3.4**).

How's It Going?

1. Compare and contrast pleiotropy and redundancy.
2. Where does most mitosis take place in the developing nervous system, and what are the three initial layers that form in the brain?

■ **3.3** ■

SHORTLY AFTER DIVISION, NEURAL CELLS DIVERGE TO BECOME NEURONS OR GLIA

Once cells become postmitotic, they can begin to differentiate into specific cell fates, to serve different functions in the nervous system. One basic distinction is whether the cell will become a neuron or a glial cell. In at least some cases, this basic decision is made even before the cell has left the germinal layer: a proportion of cells in the cortical ventricular zone express the gene for **glial fibrillary acidic protein** (**GFAP**), a structural protein normally expressed in radial glia and astrocytes but not neurons. In monkeys, that proportion of ventricular cells expressing GFAP rises as fetal development proceeds (Levitt, Cooper, & Rakic, 1983), presumably because most of the cells destined to become neurons have already been produced by the end of gestation, and most of the remaining neuroblasts are going to produce glia.

What determines whether a cell will become a neuron or a glia? Recall from Chapter 1 that in the worm *Caenorhabditis elegans*, there is a strict relationship between mitotic lineage and the fate of each cell, whether it will become a neuron, a glia, or some other type of cell. We learned in that chapter that, unlike worms, the embryos of other model organisms display self-regulation, which seems incompatible with the idea of mitotic lineage affecting cell fate. But note that all the manipulations we discussed that affected cell fate were, by definition, abnormal situations: a limb bud was removed, an mRNA was injected, a gene was knocked out or overexpressed. Scientists had to resort to such manipulations because, in an organism with millions of embryonic cells, there's simply no way to do the bookkeeping needed to identify every cell and watch its fate from last division to final differentiation. So, it might be possible that normally mitotic lineage also specifies cell fate in mammals, but we can't keep track of all the cells involved to observe it. Maybe some dividing cells are already specified to produce only neurons as progeny, while others are already progressing down the path of becoming glia and will produce only glial progeny. In that case, the difference between worms and mice might not be the importance of mitotic lineage, but the compensatory ability of self-regulation in mice.

> In an organism with millions of embryonic cells, there's simply no way to do the bookkeeping needed to identify every cell and watch its fate from last division to final differentiation.

In trying to evaluate this idea that some dividing cells produce only neurons while others produce only astrocytes and yet others produce only oligodendrocytes, several labs used methods to label cells approaching their last mitosis and keep track of the fate of their progeny, as we'll see next.

RESEARCHERS AT WORK

Labeling of Dividing Cells Disputes the Idea That Lineage Determines Fate

■ **QUESTION**: Do dividing cells start narrowing in on cell fates, each producing progeny that all develop into only one particular type of cell—neuron, astrocyte, or oligodendrocyte?

■ **HYPOTHESIS**: All the progeny produced by the final few rounds of mitosis will be alike, either all neurons, or all astrocytes, or all oligodendrocytes.

■ **TEST**: In developing chick spinal cord, infect cells that are nearing their final division with a retrovirus that inserts the gene *lacZ*, which produces an enzyme normally made only in bacteria called **beta-galactosidase** (**β-Gal**). Any progeny of that first infected cell will inherit the *lacZ* and so make β-Gal, which we can make visible with histochemistry so that all the progeny cells are stained blue. We say the *lacZ* is a *reporter gene*, because its product will report to us which cells descended from the originally infected cell (**FIGURE 3.5A**).

continued

glial fibrillary acidic protein (GFAP) A structural protein normally expressed in radial glia and astrocytes but not neurons.

lacZ A bacterial gene that encodes the enzyme beta-galactosidase.

beta-galactosidase (β-Gal) A bacterial enzyme, encoded by the *lacZ* gene, that often serves as a reporter gene or marker in studies of neural development.

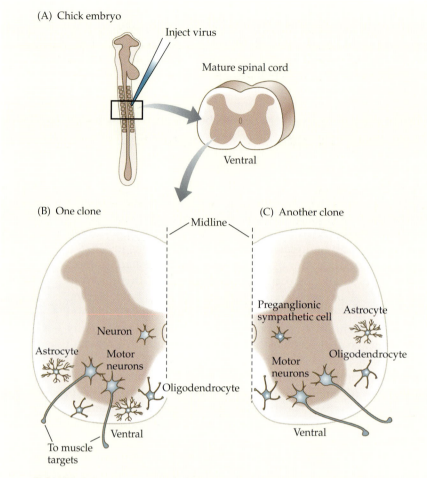

(A) Chick embryo
Inject virus
Mature spinal cord
Ventral

(B) One clone
Midline
(C) Another clone
Neuron
Astrocyte
Motor neurons
Oligodendrocyte
Ventral
To muscle targets
Preganglionic sympathetic cell
Astrocyte
Oligodendrocyte
Motor neurons
Ventral

FIGURE 3.5 Two typical clones (After Leber et al, 1990.)

■ **RESULT**: In fact, there was no apparent relationship among the cells produced by the infected cell. In many cases, both neurons and glia were produced as progeny. Despite being "siblings" or "cousins" in terms of mitotic lineage, their fates could be quite different. For example, the descendants of one cell produced several large motor neurons, several astrocytes, and a number of oligodendrocytes (**FIGURE 3.5B**). Another cell divided to produce motor neurons innervating two different muscles, another neuron that was a sympathetic preganglionic cell, and a smattering of glia (**FIGURE 3.5C**) (Leber, Breedlove, & Sanes, 1990).

■ **CONCLUSION**: There is no discernible pattern of offspring from a single infected cell. Thus there is no evidence that neural cells in their final stages of mitosis are becoming specified to produce only one cell type or another. Thus even in their final divisions, neural cells in vertebrates remain multipotent, capable of producing progeny that pursue a variety of fates.

A similar analysis of cell fate in the retina reached the same conclusion—an infected cell might produce many different cell types found in the retina, or even *all* of them (Livesey & Cepko, 2001; Turner, Snyder, & Cepko, 1990). So just as mitotic lineage doesn't seem to play a role in specifying cell fate in early embryos of insects and vertebrates, there's no evidence that it plays a role in deciding whether cells become neurons or glia in the developing nervous system in those species. What, then, controls whether postmitotic cells become neurons or glia?

In Chapter 4 we'll learn that the differentiation of neural cells into particular neuronal or glial fates is primarily influenced by cell-cell interactions of the sort we discussed in Chapter 2. For now, let's see how cell division populates the growing intermediate zone in the cerebellum and cerebral cortex.

■ **3.4** ■

THE CEREBELLUM AND CEREBRAL CORTEX FORM IN LAYERS

So far we've spoken of the neural tube as if it's all alike, as indeed it is in early development. But as the anterior end develops into the brain, additional layers form between the inner ventricular zone and the outer pial surface. In the cerebellum, the intermediate zone retains relatively few cell bodies and becomes filled with axons coming to and fro. Above that, glia and neurons accumulate, including huge **Purkinje cells** that form a single layer, with tiny granule ("grains," as of grains of sand) cells forming layers on either side: the **internal granule layer** next to the intermediate zone and the **external granule layer** outside the marginal zone (**FIGURE 3.6A**).

Purkinje cells   The large, multipolar neurons that form a single layer in the vertebrate cerebellar cortex.

internal granule layer   A layer of small neurons ventral to the Purkinje cell layer in the vertebrate cerebellar cortex.

external granule cell layer   A layer of granule neurons that migrate to the top of the developing cerebellum before migrating ventrally to form the internal granule cell layer in adulthood.

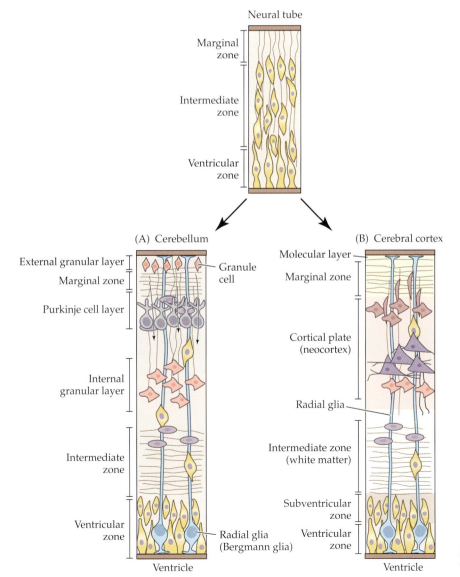

FIGURE 3.6   Developing layers in the cerebellum and cerebral cortex

subventricular zone (SVZ) The region just next to the ventricular zone, where many cells divide to provide neurons and glia to the developing vertebrate brain and, in at least some brain regions, new neurons in adulthood.

FIGURE 3.7 The developing cerebral cortex (A) Some dividing cells retain connections from ventricle to pia, eventually becoming radial glia. Other cells migrate from the ventricular zone to the subventricular zone, to divide symmetrically, while others continue asymmetric divisions. (B) Glial cells arise after most neurons have been produced, and some glia themselves divide. (After Noctor et al., 2004, 2008.)

The internal granule layer is formed when cells in the external granule layer migrate below the Purkinje cells, as we'll detail later in this chapter. Eventually the outermost marginal zone will contain very few cell bodies, consisting almost exclusively of axons and dendrites.

In the cerebral cortex, several additional layers join the three basic layers of ventricular zone, intermediate zone, and marginal zone. Immediately above the ventricular zone, a thin layer of cells detach from the ventricular lining but continue to divide to provide precursor cells to populate the region. In other words, they are neuroblasts that are close to, but not lining, the ventricular surface (**FIGURE 3.6B**). These cells form a region called the **subventricular zone** (**SVZ**), which is confusing since it is in fact above the ventricular zone (sorry). Sometimes the ventricular zone and SVZ are together called the *proliferative zone* because almost all divisions giving rise to brain cells happen there. Of course, they are also both examples of germinal zones.

As in the ventricular zone, cells in the SVZ continue to divide throughout life, but while divisions in the ventricular zone continue to be asymmetric (one daughter cell becoming postmitotic while the other divides again), mitosis in the SVZ can be symmetric, with both cells becoming postmitotic or both dividing (Noctor, Martinez-Cerdeno, Ivic, & Kriegstein, 2004; Noctor, Martinez-Cerdeno, & Kriegstein, 2008) (**FIGURE 3.7A**). The SVZ is a potential source of

(A) Asymmetric/symmetric divisions differ within proliferative zones

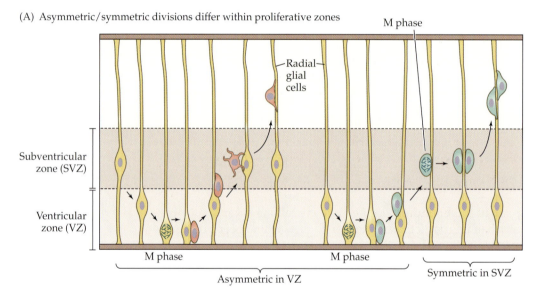

(B) Distribution and mode of neural stem and progenitor cell divisions

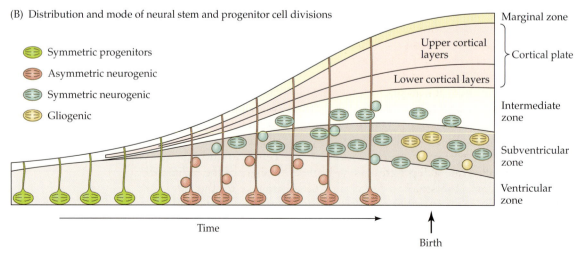

adult neurogenesis (Alvarez-Buylla & Garcia-Verdugo, 2002) that we'll discuss later in this chapter. By adulthood, the ventricular zone will shrink to form the ependymal layer that lines the cerebral ventricles.

The cortical intermediate zone never accumulates many cell bodies, filling up with axons that, upon myelination, form the inner **white matter** of the brain. Above that layer, neurons and glia accumulate in what is called the **cortical plate**, which will eventually form **gray matter**. The small marginal zone remains outside the cortical plate, and eventually dendritic and axonal processes reach past the marginal zone to form the **molecular layer** (so called because it is devoid of cell bodies), immediately next to the pia (**FIGURE 3.7B**). The molecular layer and marginal zone together will form the outermost layer, layer I, of the six-layered **neocortex** (Rakic, 2009), with the cortical plate providing the other five layers, rich with neuronal cell bodies. As the cortical plate grows, the marginal zone/molecular layer becomes a relatively thin outermost layer of the cortex, filled with axons and dendrites (see Figure 3.6B).

Before we describe in further detail the development of the six neocortical layers in the cerebrum, we need to talk about the birthdating of cells, next.

How's It Going?

1. What do experiments labeling a late-dividing cell and its progeny tell us about the use of mitotic lineage to determine neural fate?

2. Compare symmetric and asymmetric divisions in the subventricular zone.

3. What are the beginnings of the two main types of tissue (hint—they are different in color) in the cortex?

■ 3.5 ■

WE CAN LABEL NEWLY SYNTHESIZED DNA TO DETERMINE THE BIRTHDATES OF CELLS

When we speak of the **birthdate** of a cell, we are talking about the point in development when this particular cell stopped dividing and began differentiating into a particular fate. The earliest methods to determine when cells went through their final mitoses took advantage of the fact that one nucleotide, **thymidine**, is found in DNA but not RNA. Therefore, when we introduce thymidine into an animal (sometimes by injecting it into a pregnant animal to reach embryos), most of it will be taken up by cells in the synthetic phase (S phase) of mitosis: duplicating chromosomes in preparation for division. The thymidine will not be taken up for RNA synthesis, because RNA molecules substitute the nucleotide uridine in place of the thymidine found in DNA. We can radioactively tag molecules of thymidine, typically by replacing either hydrogen atoms or carbon atoms with radioactive isotopes (tritium [^{3}H in the case of one, carbon-14 [^{14}C] in the other). When the radiolabeled thymidine is injected into the animal, all cells in S phase take it up, so it is depleted from circulation quickly. That means only cells in S phase within a few hours of the injection will take up the labeled thymidine. Those cells that undergo many cell divisions after this will progressively dilute their share of the radiolabel, which will become too small to detect. But cells that take up the label and either stop dividing altogether or divide only a few more times will have sufficient radiolabel in their nuclei that we will be able to detect it weeks or months later when the animal is sacrificed (**FIGURE 3.8A**). The typical way

white matter   The inner portion of the vertebrate brain, consisting primarily of myelinated axons coursing to or from the cerebral cortex, hence light in color in postmortem preparations.

cortical plate   In developing cortex, the expanding layer of postmitotic cells that settle beneath the marginal zone and above the intermediate zone. It will form layers II–VI.

gray matter   The outer portion of the vertebrate brain, predominated by neuronal and glial cell bodies rather than myelin, hence dark in color in postmortem preparations. It is organized in six layers in mammals.

molecular layer   The outermost layer of the vertebrate cerebral cortex, consisting primarily of dendrites and axons with relatively few cell bodies.

neocortex   The six-layered outer region of the mammalian cerebral cortex.

birthdate   Here, the time during development when a given cell underwent its final mitosis before differentiating into a neuron or glial cell.

thymidine   A nucleotide used in the synthesis of DNA. Because thymidine is not used in RNA, it can serve as a DNA-specific marker.

FIGURE 3.8 Birthdating cells
(A) Labeled thymidine or BrDU taken up by a cell in the final or nearly final round of mitosis will still be apparent in adulthood. (B) In this autoradiogram of five motor neurons in the spinal cord of an adult rat, one has only a single silver grain over the nucleus (arrow). The other motor neurons have many silver grains, indicating they underwent their final mitosis about the 12th day of gestation, when the radioactive thymidine was injected into the rat's pregnant mother. (B courtesy of Dr. Cynthia L. Jordan.)

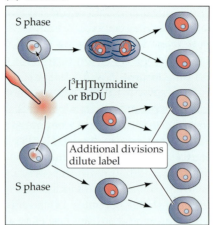

(A) Birthdate

S phase

[³H]Thymidine or BrDU

Additional divisions dilute label

S phase

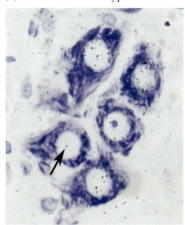

(B) Differentiated cell type

---

autoradiography A process in which a tissue "takes its own picture" when photographic film is exposed to radioactively labeled markers.

5-bromo-2′-deoxyuridine (BrdU) A synthetic nucleotide that can serve as a substitute for thymidine in the synthesis of DNA but can be readily distinguished from thymidine by the use of antibodies.

to detect labeled cells is to slice the brain into thin sections, lay those next to photographic film, and store the sections in the dark. Over time, radiolabeled thymidine molecules will emit particles that affect the photographic film. Eventually (typically after weeks to months), we will be able to chemically develop the film to find those cell nuclei that had contained the labeled thymidine: they will have small black silver grains on the adjacent film (**FIGURE 3.8B**). This is an instance of the method called **autoradiography**: the tissue "takes a picture of itself." We know that those labeled cells underwent their final mitosis, or near-final mitosis, back when we injected the animal weeks or months before sacrifice.

While thymidine autoradiography was enormously useful for determining the various birthdates of neurons and glia in different brain regions, it has several disadvantages. The biggest is that you have to get the tissue sections onto photographic film with little or no light (very discomfiting because you also have to slice the tissue with a very sharp blade, in the dark). But there is also the problem of having to wait weeks or months for the isotopes to emit their radioactive particles to expose the film—tritium has a half-life of 12 years! Which brings to mind another disadvantage—working with radioactive materials. Finally, you also have to chemically develop the film in a darkroom.

Thus when another method of birthdating cells arose, it quickly spread (Balthazart & Ball, 2014). The molecule **5-bromo-2′-deoxyuridine (BrdU)** is enough like thymidine that cells in S phase will take it up and incorporate it in DNA in place of thymidine. That same similarity between thymidine and BrdU means the latter will not be taken up into RNA. But BrdU is sufficiently different from thymidine that scientists can make antibodies that bind only to BrdU, not thymidine or uridine or other nucleotides. This means we can use the antibodies to indicate where BrdU is found in tissue sections. This method is much faster than thymidine autoradiography: you inject BrdU at one stage of development, where it is taken up by cells in the S phase preparatory to division. After the animal grows up, you sacrifice the animal, section the brain, and instead of putting those sections on photographic film in a darkroom for weeks or months, in just a few days you can chemically react the tissues to produce a colored chemical product wherever BrdU is found, using a process called immunohistochemistry, which I'll describe next.

> BrdU is sufficiently different from thymidine that scientists can make antibodies that bind only to BrdU, not thymidine or uridine or other nucleotides.

We can detect cells containing BrdU in their nuclei by **immunohisto-chemistry** (**IHC**), the use of antibodies to leave a discernible chemical reaction product in tissues. IHC is widely used in the life sciences, including experiments described in several future chapters, so we'll discuss it in some detail here. While we'll talk about using it to detect BrdU, the general method can be used to detect any molecule that you can make an antibody recognize, which means almost any molecule, including any protein the body might make.

For immunohistochemistry, you must have **antibodies** (also called immunoglobulins), large Y-shaped proteins produced by the immune system that specifically bind the molecule of interest, in this case BrdU. Typically the antibodies are made by injecting the molecule of interest into an animal, which then produces the antibodies, and you can harvest them by, say, taking some serum from the animal and purifying it to get the antibodies you want.

So you can take thinly sliced brain tissue and expose it to a solution of antibodies directed at BrdU. After allowing time for the antibodies to attach to any BrdU in the tissue (a few hours or days), you rinse off any unbound antibodies and then run the tissue through a series of solutions to form a visible, colored reaction product (called a chromogen) wherever the antibodies were, a process that takes a few hours (**FIGURE 3.9A**). Now you have a visible signal wherever the BrdU molecules bound by the antibodies were: in the nuclei of cells that were undergoing their final or near-final mitosis back when the BrdU was injected into the animal, hours or months before sacrifice.

Thus you can inject a pregnant rat **dam** (a mother of a domesticated animal species) with BrdU to expose the embryos at a particular stage, then let the pups be born and grow up. Then you can sacrifice the pups as adults and use BrdU IHC to examine the brain. Any cells that underwent a final division around that particular stage of embryonic development should still have BrdU in their nuclei (see Figure 3.8). For major types of neurons, there are particular times in development when they are generated (**FIGURE 3.9B**). For example, spinal motor neurons have undergone their final divisions by the 14th day of development in rats (Breedlove, Jordan, & Arnold, 1983); injections of thymidine or BrdU after that stage of development will label many spinal neurons, but never label spinal

immunohistochemistry (IHC)
A method for detecting a particular protein in tissues in which an antibody recognizes and binds to the protein and then chemical methods are used to leave a visual reaction product around each antibody.

antibodies Large, Y-shaped proteins produced by the immune system that recognize and bind to particular shapes in molecules.

dam Here, a mother of a domesticated animal.

(B) BrdU

cFos

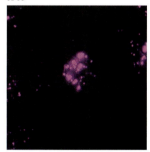

Merge

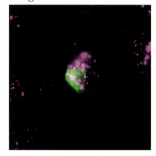

10 μm

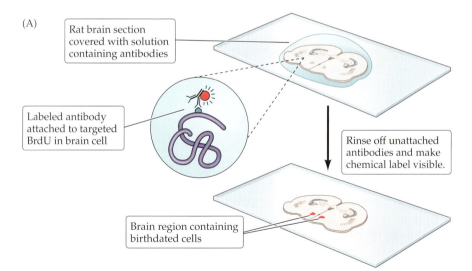

(A)

Rat brain section covered with solution containing antibodies

Labeled antibody attached to targeted BrdU in brain cell

Rinse off unattached antibodies and make chemical label visible.

Brain region containing birthdated cells

FIGURE 3.9 Immunohistochemistry (A) Using antibodies to detect BrdU in brain tissue. (B) The green label for BrdU tells us this hypothalamic neuron in an adult rat was born when the BrdU was administered at puberty, and the cFos label in purple tells us it was activated in response to hormone treatment to elicit ovulation. (B courtesy of Dr. Margaret A. Mohr.)

For major types of
neurons, there are
particular times in
development when
they are generated.

motor neurons. By the way, the fact that exposure to mitotic markers in adulthood does not label neurons reassures us that our birthdating procedure is valid. If the marker were being incorporated into DNA outside of mitosis, as part of ongoing DNA repair for example, then it might get taken up by adult postmitotic cells.

Whenever you're using immunohistochemistry, there are several issues of concern. First and foremost is whether your antibody really is attaching to the molecule you targeted. An antibody does not recognize the entire molecule. Rather, it binds to an **epitope**, a particular shape found on some part of the targeted molecule. Perhaps there is some other molecule in the tissue you're examining that offers that same epitope that your antibody is binding. In the case of IHC targeting BrdU, if you examine tissue from animals that were not injected with BrdU, then you should not see any staining. When using IHC to find endogenous proteins, the best way to be sure your procedure is revealing only the targeted molecule and not some other molecule sharing that epitope is to examine tissue from an animal in which the gene for that molecule has been "knocked out" (see Chapter 2). If your procedure stains tissues from wild-type animals but not knockout animals, it is likely it really does reveal the molecule of interest.

Now that we know how to birthdate cells, we can return to the story of the development of the cerebral cortex, which involves an orderly migration of cells to form layers.

epitope The particular shape of a molecule that a given antibody recognizes and binds.

radial glial cells Also called simply *radial glia*. Long, slender glial cells that stretch from the ventricular surface to the pial surface in the vertebrate cerebral cortex.

■ 3.6 ■
NEWBORN CELLS SHINNY UP GLIAL POLES

As the neural tube expands, dividing to produce more cells, some of the proliferating neuroblasts in the inner part of the tube, the ventricular surface, become detached from both the inner and outer surfaces. No longer simply one cell thick, the neural tube now has postmitotic cells from the ventricular zone to the pial surface. However, there are some cells that maintain processes attached to both the ventricular surface and the pial surface. These cells are still dividing, but they have also begun differentiating, producing proteins typically seen in glia rather than neurons, including glial fibrillary acidic protein (GFAP) we mentioned earlier. As the neural tube continues to grow many layers thick, these glial cells get stretched out until most of each cell's membrane forms a thin fiber. If you look at a cross section of the neural tube at this point, these glia look like spokes on a wheel and so are called **radial glial cells** (**FIGURE 3.10A**), which serve as guiding lines for migrating cells.

As cells in the
ventricular zone and
subventricular zone
become postmitotic,
they latch on to
radial glial cells and
shinny outward like a
gymnast climbing up
a rope.

As cells in the ventricular zone and subventricular zone become postmitotic, they latch on to radial glial cells and shinny outward like a gymnast climbing up a rope. The migrating cell extends a small part of itself, usually called a process, along the glial rope; the process attaches to the fiber, then contracts, pulling the cell body along the line (**FIGURE 3.10B**).

These migrating precursor cells, some of which will differentiate into other sorts of glia while others become neurons, seem like independent amoeba-like animals, and indeed we'll see that they exhibit considerable autonomy, making individual choices about where to go. As the neural tube grows even larger, forming the embryonic brain, the radial glia stretch out even farther, offering postmitotic cells a route to the outer layers. Although there are times when migrating cells reach from one radial glia to another, to move across the radiating lines, like Tarzan moving from one vine to another, they mostly seem to stick to one glial cell and move out. Because this outward migration

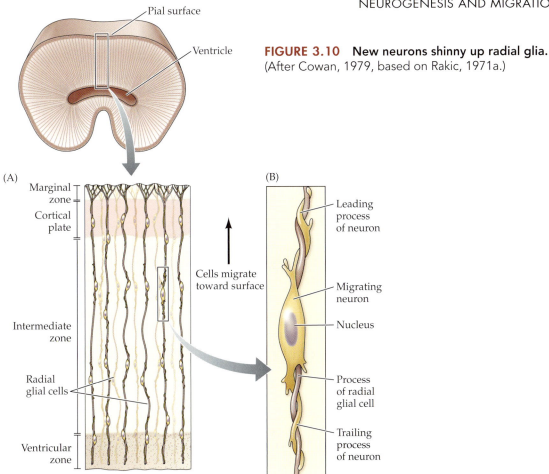

FIGURE 3.10 **New neurons shinny up radial glia.** (After Cowan, 1979, based on Rakic, 1971a.)

on a single radial glia is more common, and more readily studied in vitro, we know the most about that.

In the developing cerebral cortex, the first postmitotic cells departing from the ventricular zone stay just below the pial surface to form the **cortical preplate** (so called because it is present before the cortical plate), which will make up the marginal zone, with relatively few cell bodies. As we noted earlier, a molecular layer of dendrites and axons will form beyond this marginal zone, and together marginal zone and molecular layer will become thinner relative to the whole cortex, forming the outermost layer of neocortex, layer I. The next wave of postmitotic cells will nestle below the preplate and form the cortical plate. These cells will form the innermost of the six layers, layer VI. As more cells migrate up from the proliferative zone, the cortical plate will expand to form the other five layers of neocortex. Beneath the cortical plate, the intermediate zone will eventually be filled with axons, including myelinated axons, making the white matter of the cortical interior. The SVZ lies below the white matter, with the ventricular zone beneath the SVZ, forming the boundary with the fluid-filled ventricles (see Figure 3.7B).

As cortex thickens, radial glia in the ventricular zone continue asymmetric divisions to provide postmitotic cells to join the layers overhead. In the subventricular zone (which, remember, is paradoxically above the ventricular zone), some cells continue to divide but do so in a symmetrical manner, where both daughter cells may stop dividing and differentiate into neurons or glia (see Figure 3.7A). In general, the first postmitotic cells differentiate into neurons, while the later waves of postmitotic cells differentiate into glia to populate the neocortex.

cortical preplate The region between the ventricular zone and marginal zone in developing vertebral cerebral cortex, which develops into the gray matter of the neocortex.

Each wave of neurons first migrates past their predecessors to settle farther out the glial highway, closer to the pial surface.

You could imagine that the waves of neurons migrating into neocortex might do so in more or less random order, each cell looking for an available slot in any layer. But in fact, each wave of neurons first migrates past their predecessors to settle farther out the glial highway, closer to the pial surface. Let's see how birthdating revealed this "inside-out" pattern of development.

RESEARCHERS AT WORK

The Cortex Develops in an Inside-Out Manner

■ **QUESTION**: Is there any order in which newborn neurons settle into the six layers of neocortex?

■ **HYPOTHESIS**: Postmitotic neuronal precursors will establish themselves in an orderly way as the cortical plate thickens, either settling beneath the neurons that preceded them into the cortical plate or settling above the preceding neurons.

■ **TEST**: Inject a DNA marker such as tritiated thymidine into rhesus monkeys at different prenatal stages of development, let them grow up, and then determine where neurons with different birthdates are found in the cortex.

FIGURE 3.11 **Later-born neurons settle in upper cortex.** (After Rakic, 1974.)

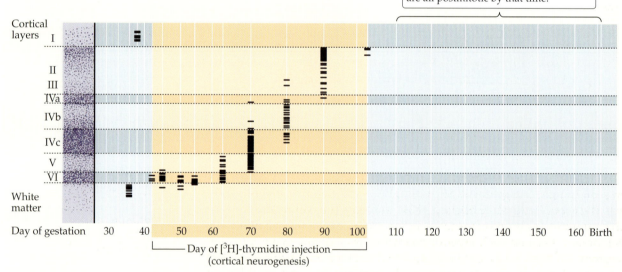

Injections of thymidine after embryonic day 110 never label any neurons, so they are all postmitotic by that time.

■ **RESULTS**: The earliest born cells, those labeled by thymidine injections earliest in gestation (day 30–40) settled in two layers: either in the intermediate zone just above the ventricular zone or in the marginal zone (layer I) closest to the external surface. Cells born after that were found *in between* these two layers, in cortical layers II–VI. Injections a bit later in gestation, on day 50 or so, labeled only cells in the deepest layer, layer VI. Injections at successively later stages of development labeled cells at progressively higher cortical layers. For example, thymidine injected on gestational day 105 labeled exclusively neurons in the uppermost portion of layer II (**FIGURE 3.11**). Injections on gestational day 110 or later labeled glia and endothelial cells in the cortex, but no neurons (Rakic, 1974).

■ **CONCLUSION**: After those first few neurons settle in layer I, the cortical plate is settled in an inside-out fashion, with neurons first established in layer VI and subsequently born neurons migrating past their predecessors such that the latest-born neurons end up at the top of layer II. Thus there is indeed an orderly progression of the age of neurons in the cortex.

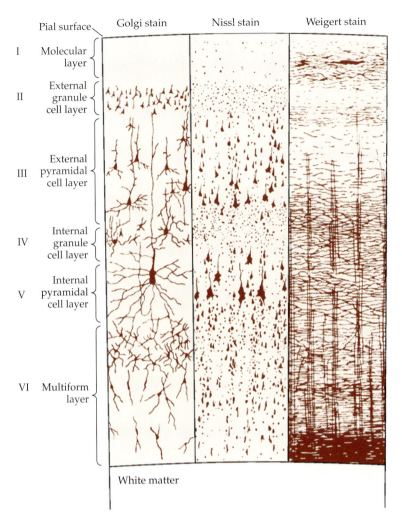

Pial surface

I Molecular layer

II External granule cell layer

III External pyramidal cell layer

IV Internal granule cell layer

V Internal pyramidal cell layer

VI Multiform layer

White matter

Golgi stain · Nissl stain · Weigert stain

FIGURE 3.12 **The six layers of neocortex** (From Heimer, 1994.)

A similar inside-out addition of newborn neurons to cerebral cortex has been seen in mice (Angevine & Sidman, 1961), rats (Berry & Rogers, 1965), ferrets (Jackson, Peduzzi, & Hickey, 1989), and cats (Luskin & Shatz, 1985), so this migration pattern seems to be a characteristic of mammalian cortical development. Because each layer of neocortex has a distinctive population of neurons (**FIGURE 3.12**), decisions about which layer to inhabit also affect what particular phenotype the neuron will adopt. In Chapter 4, we'll look at the question of whether cells are specified to take on this particular phenotype before they leave the proliferative zone or are induced to take on a particular phenotype after they settle into a layer.

Recall we said that some neurons don't migrate straight out along radial glia, but take tangential routes across neighboring radial glia. Many of these cells arise from one of several proliferative zones lining the lateral ventricles. These other germinal zones are the **ganglionic eminences** (**GEs**), transient structures in the ventral telencephalon. There are three GEs: a medial and a lateral GE seen in rostral telencephalon (Corbin, Nery, & Fishell, 2001) and a caudal GE in posterior telencephalon (C. Wonders & Anderson, 2005). Postmitotic cells from these zones contribute interneurons, especially GABA-ergic neurons, to several cortical regions and give rise to the neurons in the basal ganglia and amygdala (Parnavelas, 2000). Unlike the cortical neurons we've discussed so far, which tend to go straight out radially along a radial glia, the neurons from the ganglionic eminences migrate tangentially, moving from one radial glia to another to reach their destinations, which include the basal ganglia and other limbic structures, as well as neocortex (**FIGURE 3.13**).

ganglionic eminences (GEs) Transient bumps along the lateral ventricles of the developing telencephalon, from which neurons migrate tangentially across radial glia.

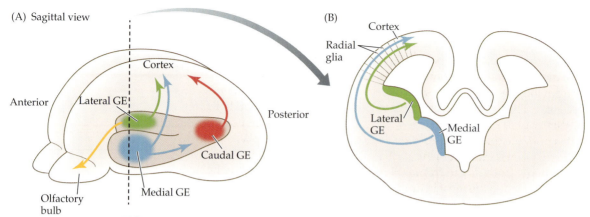

(A) Sagittal view

Cortex

Anterior

Lateral GE

Posterior

Caudal GE

Olfactory bulb

Medial GE

(B)

Cortex

Radial glia

Lateral GE

Medial GE

FIGURE 3.13 **Neuronal migration from ganglionic eminences** (A) The three ganglionic eminences provide a wave of newly born cells to populate the cortex. (B) Cells from the ganglionic eminences first migrate tangentially across radial glia. These will become neurons of the basal ganglia and amygdala and interneurons in the cortex. (After Parnavelas, 2000, and Wonders & Anderson, 2006.)

The neurons from the ganglionic eminences migrate tangentially, moving from one radial glia to another to reach their destinations.

Because of their varied destinations and tangential migration path, we know less about how these cells are informed about where to go or what controls the fate they adopt when they arrive (C. P. Wonders & Anderson, 2006).

There is recent evidence of a large wave of neurons, seen only in humans, that are still migrating into the frontal cortex after birth (Paredes et al., 2016). Observations of postmortem brain slices from infants revealed these migrating neurons forming an arc along anterior cingulate cortex, just above the corpus callosum (**FIGURE 3.14**). It's not clear whether these arc neurons are newly born or had been born prenatally

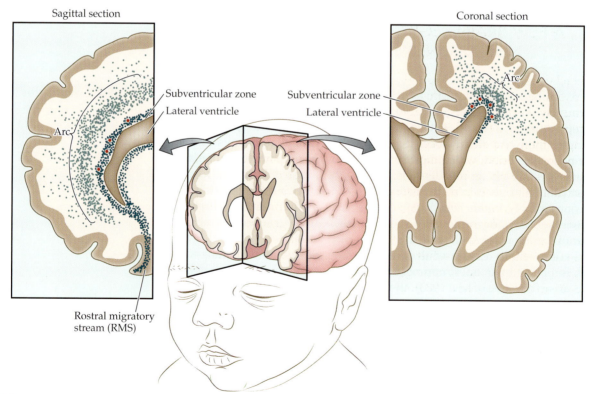

Sagittal section

Subventricular zone

Lateral ventricle

Arc

Rostral migratory stream (RMS)

Coronal section

Arc

Subventricular zone

Lateral ventricle

FIGURE 3.14 **Late migration into human frontal cortex** The number of migrating arc neurons diminishes over the first months of life and is complete by 2 years of age. The rostral migratory stream (RMS), which provides new neurons for the olfactory bulb, will continue throughout life. While the late migrating arc neurons have been seen only in humans, the RMS is evident in other mammals, too. (After Paredes et al., 2016.)

and were simply waiting to participate in the late migration, which is most prominent in the first few months of life. Like the neurons that had migrated earlier from the GEs, most of these neurons will become inhibitory interneurons using GABA. The arc neurons first move tangentially across radial glia, then follow them out to settle into various cortical layers. Because they arrive after birth, it is possible that arc neurons provide additional neural plasticity as the infant learns from social interactions. If indeed the late arc migration happens only in humans, and only in frontal cortex, that might explain why the brain region that expanded the most in human evolution was of the prefrontal cortex, a topic we'll return to in Chapter 4.

How's It Going?

1. What do we mean by the "birthdate" of a cell, and how can we determine that?
2. What are the advantages of BrdU over tritiated thymidine for birthdating?
3. What are radial glia, and what role do they play in migration?
4. What is the relationship between birthdate of neurons and their position in neocortical layers?
5. What are the ganglionic eminences, and how do migrating cells originating there move relative to radial glia?
6. What is the potential significance of the arc of late-migrating neurons in the human frontal cortex?

■ **3.7** ■

A FEW BRAIN REGIONS DISPLAY CONTINUING NEUROGENESIS THROUGHOUT LIFE

When I was in graduate school in the late 1970s, everyone knew that we make all of our neurons early in life and that by adulthood, life is a downhill slide because while neurons might die, you have no way to replace them. This firmly accepted (if grim) fact of life was taught in every neuroanatomy class. Funny thing is, there had already been published reports of neurogenesis well past the neonatal period, which most of the field (including most neuroanatomy instructors) simply ignored. Joseph Altman used tritiated thymidine to reveal postnatal neurogenesis in the hippocampus, olfactory bulb, and neocortex of rats and cats (Altman, 1963, 1969; Altman & Das, 1965), findings that were mostly ignored for over a decade. Even when other groups replicated the findings (Graziadei & Graziadei, 1979; Kaplan & Hinds, 1977), still "everyone knew" there was no adult neurogenesis. Then when Fernando Nottebohm reported that adult neurogenesis was widespread in the canary brain (Goldman & Nottebohm, 1983), there was abundant public skepticism that the new cells were really neurons (and not glia, which were already known to be produced throughout life). Finally, in a brute-force approach, recording from random neurons until some proved, upon postmortem examination, to be newborn, it was proved that at least some adult-born neurons participate in auditory processing (Paton & Nottebohm, 1984). You might think the field would have reconsidered Altman's earlier findings in mammals, but in fact adult neurogenesis was then dismissed as something restricted to songbirds.

Today it's widely accepted that neurons are continually generated in the olfactory system and the dentate gyrus of the hippocampal formation (Ming & Song, 2011) and that a very few new neurons are added to the adult neocortex

THE CONTROVERSY OF NEUROGENESIS IN ADULTHOOD

After Fernando Nottebohm's report of neurogenesis in the adult songbird forebrain (Paton & Nottebohm, 1984), Pasko Rakic, the great scientist who made seminal contributions to almost every chapter in this book, was vocally skeptical that adult neurogenesis happens in mammals, or at least primates. In the high-impact journal *Science*, he reported that neurogenesis in monkeys was completed in the brain, including the hippocampus and olfactory bulb, by puberty (Rakic, 1985).

But various labs "rediscovered" Altman's findings of neurogenesis in the hippocampus in adult rats and mice, showing that the rate of neurogenesis and/or survival of newborn neurons could be altered by factors like hormones (Gould et al., 1992) and environmental enrichment (Kempermann et al., 1997). Adult neurogenesis in the hippocampus was replicated and expanded to other species (Gould, McEwen, Tanapat, Galea, & Fuchs, 1997), including monkeys (Gould, Tanapat, McEwen, Flugge, & Fuchs, 1998) and humans. The findings in humans were made possible when several cancer patients who had received intravenous BrdU in adulthood as part of their treatment donated their brains for postmortem examination. Sure enough, there were neurons in the dentate gyrus of the hippocampal formation that contained BrdU in their nuclei (Eriksson et al., 1998). Eventually evidence of adult neurogenesis was also found in the human olfactory bulb (Bedard & Parent, 2004). So more than 30 years after Altman's report, there was finally a consensus that adult neurogenesis

occurs in the hippocampus (Ming & Song, 2011) and olfactory bulb of mammals.

But if adult neurogenesis in the hippocampus and olfactory bulb occurs in humans, why hadn't Rakic seen evidence of it in monkeys? Soon Rakic reexamined the question, using BrdU rather than thymidine, and reported that indeed there is neurogenesis in the hippocampus and olfactory bulb of adult monkeys (Kornack & Rakic, 1999) that apparently his work with thymidine had missed. In other words, he was retracting two of the claims he made in his 1985 *Science* report.

The next frontier came with reports of adult neurogenesis in the neocortex of monkeys (Gould, Reeves, Graziano, & Gross, 1999). Again, earlier reports of neurogenesis in the neocortex of rats (Altman, 1969; Kaplan, 1981) had simply been ignored. And again, Rakic published an article in *Science* proclaiming that there was absolutely no sign of neurogenesis in the monkey neocortex (Kornack & Rakic, 2001). This history illustrates the problem with negative findings, when you fail to detect something or fail to find a difference between two groups. Just because you don't *see* something, that doesn't prove it doesn't *exist* ("absence of proof is not proof of absence"). In the meantime, there have been ample reports of adult neurogenesis in the neocortex and other regions, including the amygdala and striatum (Gould, 2007), in addition to ongoing adult neurogenesis in the hippocampus and olfactory bulb.

> Keep in mind that the number of neurons added in adulthood is a very, very small proportion of the brain.

(Gould, 2007), despite decades of resistance in the scientific community (**BOX 3.1**). But keep in mind that the number of neurons added in adulthood is a very, very small proportion of the brain. On the other hand, just because adult-born neurons are the minority, that doesn't mean they are unimportant, as we'll see.

Adult-born neurons originate in three germinal zones that we will discuss in turn: (1) the olfactory neuroepithelium, (2) the subventricular zone, and (3) the subgranular zone.

The *olfactory neuroepithelium* contains cells that continue to divide to provide new olfactory sensory neurons. The work of primary olfactory sensory neurons is hazardous; because their job is to detect chemicals, including some that are harmful, they sometimes encounter very toxic chemicals and die. In industry, someone exposed to fumes from organic solvents can sometimes be rendered **anosmic**, unable to detect smells. This happens because the chemicals have killed the primary olfactory sensory neurons in the epithelium lining the nasal cavity. We've known for some time that most of these people (if kept away from the solvents) recover their sense of smell in a few months (Hastings, 1990), which was the first clue that primary

anosmia The condition of being unable to detect odors.

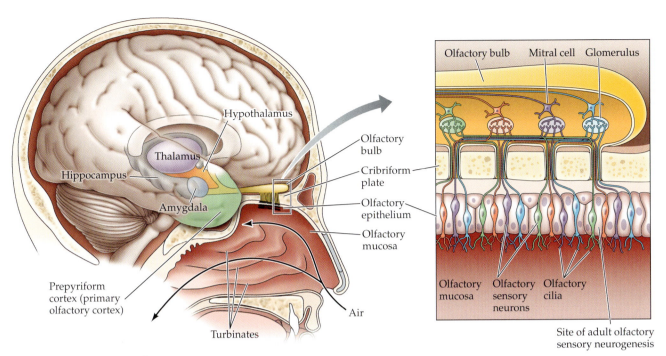

FIGURE 3.15 **The olfactory system**

olfactory neurons can be replaced in adulthood. Similarly, a blow to the head can sometimes shear olfactory sensory axons as they pass through the cribriform plate (**FIGURE 3.15**), killing the receptors, and in these cases people sometimes recover their sense of smell months or years later. Birthdating experiments in animals (Graziadei & Graziadei, 1979) and humans (Bedard & Parent, 2004) confirmed that olfactory sensory neurons can be replaced throughout life and that they may undergo continuous replacement even without trauma (Magrassi & Graziadei, 1995). The cells that divide to provide the new olfactory receptor cells reside within the olfactory epithelium alongside the receptor cell bodies (Brann & Firestein, 2014).

> Birthdating experiments in animals and humans confirmed that olfactory sensory neurons can be replaced throughout life and that they may undergo continuous replacement even without trauma.

The *subventricular zone* (SVZ) of the lateral ventricle produces cells that migrate rostrally to continuously replace interneurons in the adult olfactory bulb. You might expect that these olfactory interneurons would also arise from the nearby olfactory epithelium. Instead, cells from the SVZ in the anterior-most horn of the lateral ventricles migrate along a pathway called the **rostral migratory stream** (**RMS**) (Altman, 1969; Curtis et al., 2007) (**FIGURE 3.16A**). Apparently there are no radial glia to guide this migration, but astrocytes form a tunnel surrounding the migrating cells (**FIGURE 3.16B**) (Kaneko et al., 2010), presumably restricting their wandering and thereby forming a "pipeline" to the olfactory bulb. Extensive contact among migrating cells suggests they follow one another, making a continuous chain of cells. Fascinatingly, some of the cells continue to divide as they migrate along the RMS (Menezes, Smith, Nelson, & Luskin, 1995). Upon arrival in the olfactory bulb, the cells differentiate into granular and periglomerular interneurons (**FIGURE 3.16C**) (Pencea, Bingaman, Freedman, & Luskin, 2001). As we noted earlier, there is growing evidence that a very few adult-born neurons are found in the neocortex (see Box 3.1), and they are assumed to arise from the SVZs of the lateral ventricles, but they are so sparse in number that no one has been able to trace the path(s) they take.

The *subgranular zone* of the dentate gyrus, part of the hippocampal formation, was the first known germinal zone for adult-born neurons. In the earliest

rostral migratory stream (RMS) A collection of cells that migrate from the anterior horn of the lateral ventricles to the olfactory bulbs in adult mammals.

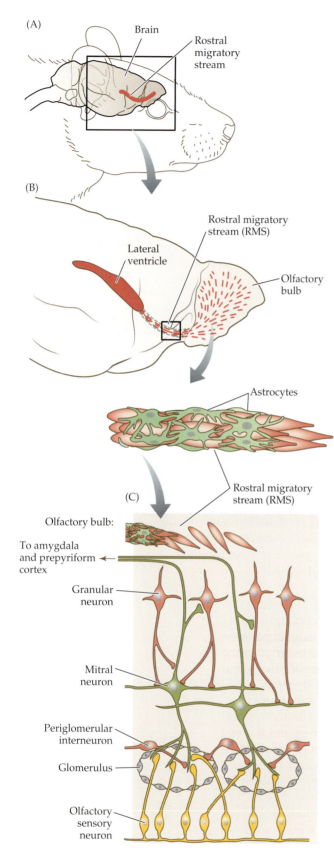

(A) Brain — Rostral migratory stream

(B) Rostral migratory stream (RMS) — Lateral ventricle — Olfactory bulb

Astrocytes — Rostral migratory stream (RMS)

(C) Olfactory bulb:

To amygdala and prepyriform cortex

Granular neuron

Mitral neuron

Periglomerular interneuron

Glomerulus

Olfactory sensory neuron

FIGURE 3.16 **The rostral migratory stream (RMS)** Astrocytes (green) are thought to ensheath the RMS, thereby guiding the migrating cells. (After Lennington et al., 2003.)

reports of postnatal neurogenesis, the sites showing the greatest numbers of new neurons were the dentate gyrus granule cell layers (Altman & Das, 1965). These cells were arising from a layer just below the granule cell layer called the **subgranular zone** (**FIGURE 3.17**). If the only change were that granule neurons were added to the dentate gyrus throughout life, it would get progressively larger with age, but it doesn't. So there appears to be some turnover, new neurons replacing old neurons. The higher frequency of neurogenesis in the hippocampus is probably why this region was the first to be accepted as a site for adult neurogenesis. Also, the idea of continuing neurogenesis in the hippocampus may have seemed more plausible because we know this brain region is important for memory and learning. Since we continue to learn throughout life, maybe we need new hippocampal neurons to support that function.

In fact, testing whether hippocampal neurogenesis is causally related to learning is not easy (Leuner, Gould, & Shors, 2006). But there are several correlations that suggest the neurogenesis is functionally significant. First, animals put into situations where they are expected to learn more, so-called enriched environments with many toys and other animals to interact with, display more new hippocampal neurons (Kempermann, Kuhn, & Gage, 1997; York, Breedlove, Diamond, & Greer, 1989), indicating either that they produce more neurons than animals in standard conditions, or that the new neurons, once produced, survive longer in the enriched conditions. Second, administering radiation or toxins to inhibit neurogenesis interferes with memory functions that are known to rely upon the hippocampus (Saxe et al., 2006; Winocur, Wojtowicz, Sekeres, Snyder, & Wang, 2006). Finally, neurogenesis appears to slow down in older animals at about the time, or even before, their learning capacity diminishes (Leuner, Kozorovitskiy, Gross, & Gould, 2007; Seib & Martin-Villalba, 2014). So while the case isn't airtight, there's good reason to think adult neurogenesis in the hippocampus is functionally significant for learning (Aimone et al., 2014; Opendak & Gould, 2015).

Adult neurogenesis may also be clinically significant. For example, the proportion of newly generated neurons found in the hippocampal granule cell layer is increased in mice that are treated with antidepressants such as Prozac (fluoxetine) (Santarelli et al., 2003), so perhaps depression can result from too little hippocampal neurogenesis (Miller & Hen, 2015; Serafini et al., 2014). Even if that's not the original cause of depression, perhaps antidepressants work by boosting hippocampal neurogenesis. There is also evidence that hippocampal neurogenesis is important in the stress response (Cameron & Glover, 2015). Indeed, one of the earliest manipulations shown to affect the rate of neurogenesis was treatment with

(A) Newborn cells in the dentate gyrus

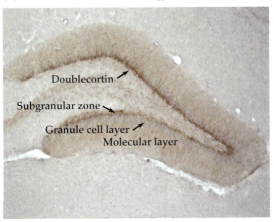

(B) Close-up view of newborn granule cells

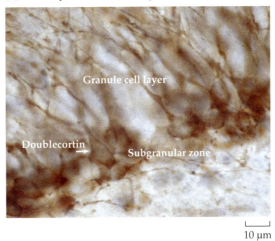

FIGURE 3.17 Continuous neurogenesis in the adult brain (A) The subgranular zone of the dentate gyrus generates new neurons in adulthood. (B) Doublecortin serves as a marker for newborn neurons. (From Oomen et al., 2009.)

adrenal steroids (Gould, Cameron, Daniels, Woolley, & McEwen, 1992), which are normally released as part of the stress response.

How's It Going?

1. Where does neurogenesis take place in the adult brain (hint—there are three widely accepted sites)?
2. What is the evidence that adult neurogenesis is behaviorally significant?
---

subgranular zone The portion of the dentate gyrus where cells divide in adulthood to contribute new neurons to the overlying granular layer.

peripheral nervous system (PNS) The entire nervous system other than the central nervous system; it includes the enteric and autonomic nervous system. It is derived from neural crest cells.

■ 3.8 ■
NEURAL CREST CELLS MIGRATE TO POSITIONS THROUGHOUT THE BODY

While most of neuroscience research is focused, reasonably enough, on the brain, sometimes we can learn a lot by studying (relatively) simpler portions of the nervous system. A wonderful example is the neural crest, comprised of those ectodermal cells that pinch off from the sheet of cells forming the neural tube to sit for a time alongside its dorsal side (see Figure 1.17). It is precisely because these cells lie outside the hubbub of activity in the neural tube itself that we can more readily keep track of them. Perhaps the most dramatic activity of the neural crest cells is their extensive migration to take up residence in disparate parts of the body to take on a remarkably diverse range of functions, including those of the **peripheral nervous system** (**PNS**), which is the entire nervous system outside the CNS (**TABLE 3.1**). We'll see that neural crest cells migrate along well-marked pathways to reach their destinations, and that the fates they take on are due to gene regulatory influences they encounter en route and at journey's end (Simoes-Costa & Bronner, 2015).

Recall that neural crest cells are found alongside the entire rostral-caudal extent of the neural tube. From one individual to another, you can predict what migration route(s) and fate(s) crest cells will take, based on where they fall on the rostral-caudal axis.

From one individual to another, you can predict what migration route(s) and fate(s) crest cells will take, based on where they fall on the rostral-caudal axis.

TABLE 3.1
Some Derivatives of the Neural Crest

Derivative	Cell type or structure derived
Peripheral nervous system (PNS)	Neurons, including sensory ganglia, sympathetic and parasympathetic ganglia, and plexuses Schwann cells and other glial cells
Endocrine and paraendocrine derivatives	Adrenal medulla Calcitonin-secreting cells Carotid body type I cells
Pigment cells	Epidermal pigment cells
Facial cartilage and bones	Facial and anterior ventral skull cartilage and bones
Connective tissue	Corneal endothelium and stroma Tooth papillae Dermis, smooth muscle, and adipose tissue of skin, head, and neck Connective tissue of salivary, lachrymal, thymus, thyroid, and pituitary glands Connective tissue and smooth muscle in aoritic arteries

For example, crest cells at the head end will form the embryonic pharyngeal arches, which will migrate out to form the face and the cranial nerve ganglia (**FIGURE 3.18**). All vertebrate embryos have six pharyngeal arches at some point in development, but in humans the fifth arch degenerates, leaving five arches that form a variety of neural and nonneural structures, as detailed in **TABLE 3.2**.

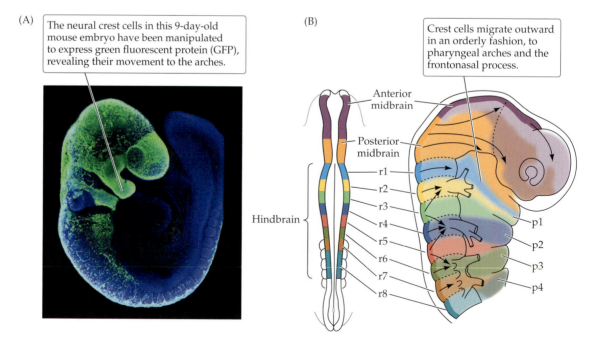

(A) The neural crest cells in this 9-day-old mouse embryo have been manipulated to express green fluorescent protein (GFP), revealing their movement to the arches.

(B) Crest cells migrate outward in an orderly fashion, to pharyngeal arches and the frontonasal process.

Anterior midbrain

Posterior midbrain

Hindbrain

r1
r2
r3
r4
r5
r6
r7
r8

p1
p2
p3
p4

FIGURE 3.18   Neural crest cells colonize pharyngeal arches. (Part A courtesy of Drs. P. Trainor and A. Barlow; B after Le Douarin, 2004.)

TABLE 3.2
Some Derivatives of the Pharyngeal Arches

Pharyngeal arch	Skeletal elements (neural crest plus mesoderm)	Cranial nerves (neural tube)
1	Incus and malleus; mandible, maxilla, and temporal bone regions (from neural crest)	Maxillary and mandibular divisions of trigeminal nerve (V)
2	Stapes bone of the middle ear; styloid process of temporal bone; part of hyoid bone of neck (all from neural crest)	Facial nerve (VII)
3	Lower rim and greater horns of hyoid bone (from neural crest)	Glossopharyngeal nerve (IX)
4	Laryngeal cartilages (from lateral plate mesoderm, not neural crest)	Superior laryngeal branch of vagus nerve (X)
6[a]	Laryngeal cartilages (from lateral plate mesoderm, not neural crest)	Recurrent laryngeal branch of vagus nerve (X)

Source: After Larsen 1993.
[a] The fifth arch degenerates in humans.

Along those portions of the neural tube that will become spinal cord, neural crest cells take a variety of migratory pathways to reach their diverse destinations. Some crest cells migrate a very short distance to stop just short of the **somites** lined up along either side of the neural tube (somites are mesodermally derived blocks of cells that will form bone, muscle, cartilage, and skin). The crest cells making this short trip will develop into neurons of the **dorsal root ganglia** (**DRG**), primarily bipolar neurons receiving sensory information from the periphery and transmitting it to neurons in the dorsal spinal cord (**FIGURE 3.19**). Other crest cells migrate ventrally between the somites and the neural tube, stopping near the notochord to differentiate into **autonomic ganglia**, collections of neurons that receive input from the CNS and in turn innervate various organs of the body. A subset of crest cells

somites   Paired blocks of mesoderm found on either side of the neural tube.

dorsal root ganglion (DRG)
A collection of neuron cell bodies embedded in vertebrate dorsal roots of the spinal cord that provides neurites to gather sensory information from the periphery and transmit it to the dorsal horn of the spinal cord.

autonomic ganglia   Collections of neurons outside the central nervous system that provide autonomic innervation of body organs.

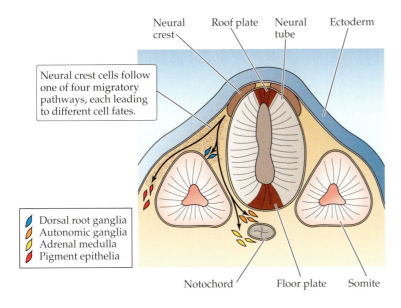

Neural crest cells follow one of four migratory pathways, each leading to different cell fates.

Neural crest   Roof plate   Neural tube   Ectoderm

Dorsal root ganglia
Autonomic ganglia
Adrenal medulla
Pigment epithelia

Notochord   Floor plate   Somite

FIGURE 3.19   **Four pathways for neural crest cells** (After Sanes, 1989.)

FIGURE 3.20 **Rostral-caudal differences in neural crest fate** (After Le Douarin, 1982.)

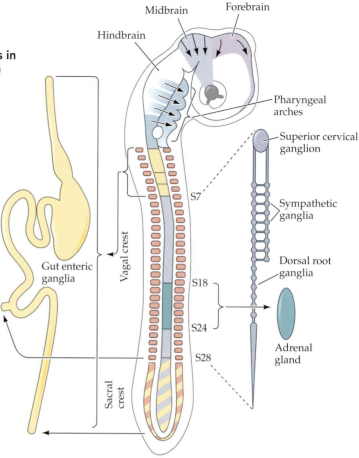

melanocytes Neural crest-derived pigment cells that provide color to the skin.

cell adhesion molecules (CAMs) A class of molecules, found in extracellular regions, that adhere to some cells and not others.

neural cadherin (N-cadherin) Also called *NCad*. A family of calcium-dependent adhesion molecules.

taking this path settle near the aorta (the large artery coming from the heart) and form the inner portion of the adrenal gland, the adrenal medulla that secretes epinephrine (also called adrenaline) when activated by sympathetic fibers. (The outer layer of the adrenal gland, the cortex that secretes steroid hormones, is derived from mesodermal tissue.) In the fourth major pathway, neural crest cells flow dorsally over the somites to eventually differentiate into **melanocytes** ("pigment cells"), providing color to the skin. The neural crest contribution to skin pigmentation, while not directly relevant to neural development, will prove useful in experiments we'll describe in Chapter 4.

Knowing which spinal segment the neural crest cells came from allows us to predict which of several possible fates they may take on. For example, some crest cells from every segment will form the dorsal root ganglia, but the sympathetic ganglia will arise exclusively from crest cells in the thoracic and lumbar regions (**FIGURE 3.20**). Crest cells from the cervical spinal region will contribute to cranial parasympathetic ganglia, while cells from the most caudal spinal cord, the sacral region, will form parasympathetic ganglia elsewhere in the body.

■ **3.9** ■

CELL ADHESION MOLECULES ATTRACT AND REPEL MIGRATING CELLS

To begin their migration from the neural tube to the rest of the body, neural crest cells, which have formed a cohesive column on either side of the neural tube until now, must first loosen their hold on one another. They down-regulate the expression of a **cell adhesion molecule** (**CAM**) called **neural cadherin** (**N-cadherin**) (the cadherins are a family of calcium-dependent adhesion

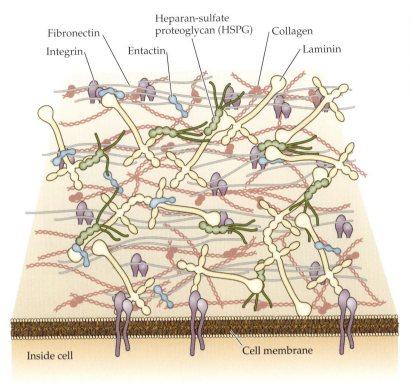

Fibronectin   Heparan-sulfate proteoglycan (HSPG)   Collagen

Integrin   Entactin   Laminin

Inside cell   Cell membrane

FIGURE 3.21   **Basal lamina** First, neural crest cells must dissolve such basal lamina with metalloproteases to commence their migration. Then they migrate along the surfaces to reach their destinations. (After Alberts et al., 2007.)

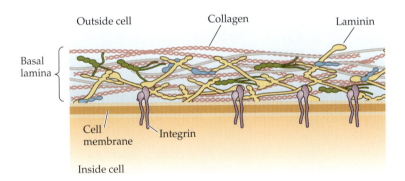

Outside cell   Collagen   Laminin

Basal lamina

Cell membrane   Integrin

Inside cell

collagen A long-chained filamentous protein that contributes to the formation of extracellular matrix and is especially prominent in connective tissue.

laminin A long-chained glycoprotein that readily binds other molecules and is a major component of the extracellular matrix and connective tissue.

fibronectin A long-chained glycoprotein that contributes to the formation of the extracellular matrix.

basal lamina A particularly thick layer of extracellular matrix that surrounds mature muscle fibers and many organs.

metalloproteases Enzymes that break down proteins and require atoms of a metal such as zinc to function.

molecules [Basu, Taylor, & Williams, 2015]), which caused them to stick to one another before migration (McKeown, Wallace, & Anderson, 2013; Taneyhill & Schiffmacher, 2013). They also down-regulate the expression of genes for proteins forming tight junctions with one another, again loosening their bonds before striking out. Finally, they must dissolve, and break free of, a dense network of extracellular matrix (ECM) proteins—molecules like the proteins **collagen** and **laminin** and glycoproteins like **fibronectin**. These long, sticky molecules are also CAMs, and they help form the **basal lamina** ("base layer"), which anchors and attaches cells together, maintaining the integrity of tissues (Brownell & Slavkin, 1980) and therefore the body as a whole (**FIGURE 3.21**). To break free of the basal lamina between the neural tube and overlying epithelium, the neural crest cells secrete a class of proteases (or *proteinases*, enzymes that dissolve proteins) called **metalloproteases** (the *metallo* indicates they use a metal such as zinc to help break down proteins). Having used the metalloproteases to dissolve an opening in the basal lamina (Christian, Bahudhanapati, & Wei, 2013; Duong & Erickson, 2004),

To begin their migration from the neural tube to the rest of the body, neural crest cells must first loosen their hold on one another.

FIGURE 3.22 **Neural crest cells migrate through anterior somites to avoid ephrin.** (A,B from Krull et al., 1997; C after O'Leary & Wilkinson, 1999.)

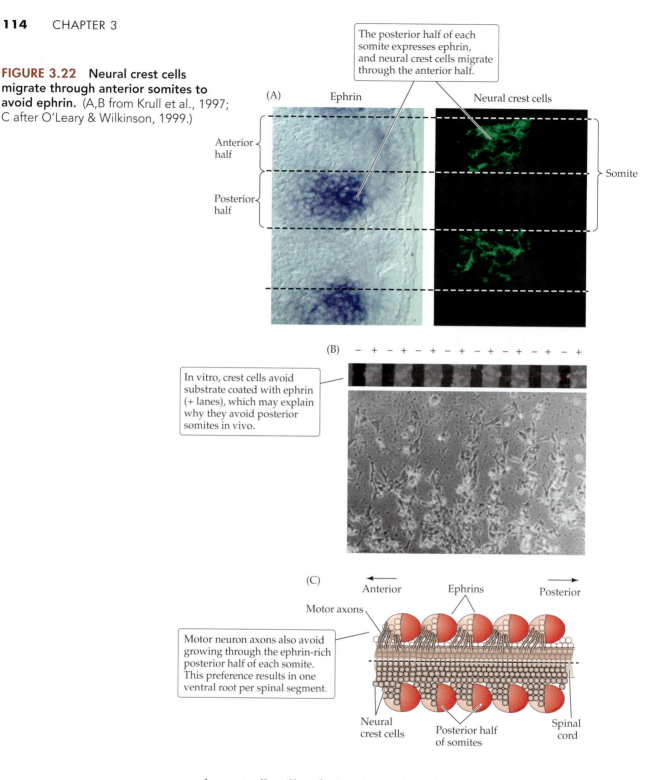

(A) Ephrin   Neural crest cells

The posterior half of each somite expresses ephrin, and neural crest cells migrate through the anterior half.

Anterior half

Posterior half

Somite

(B) − + − + − + − + − + − + − + − +

In vitro, crest cells avoid substrate coated with ephrin (+ lanes), which may explain why they avoid posterior somites in vivo.

(C)

← Anterior   Ephrins   Posterior →

Motor axons

Motor neuron axons also avoid growing through the ephrin-rich posterior half of each somite. This preference results in one ventral root per spinal segment.

Neural crest cells   Posterior half of somites   Spinal cord

the crest cells will prefer to migrate along those same extracellular matrix molecules—collagen, laminin, and fibronectin—as we'll see shortly.

Once the neural crest cells break free from the neural tube, they do not form a continuous wave of cells from the rostral end to the caudal end. Rather, they go out in discrete streams, each from a different spinal segment. One factor that breaks the cells up into these streams is the fact that they avoid passing over or through the caudal half of each of the somites lying alongside the neural tube (**FIGURE 3.22A**). This choice of the anterior half of each somite might reflect a preference for some factor in the anterior somite or an avoidance of some factor in the posterior end. In fact, this seems to be a case of guidance by avoidance, herding the cells into discrete streams. The posterior half of each somite richly expresses **ephrins**, the membrane-bound proteins that interact with ephrin receptors (part of the large receptor tyrosine kinase [RTK] family; see Figure 4.3). We encountered ephrin signaling in Chapter 2 as being important for sharpening the boundaries of gene expression in the rhombomeres. Here the ephrins seem to repulse migrating neural crest cells, because when crest cells are placed in culture, they avoid strips of the dish coated with ephrins (**FIGURE 3.22B**).

We'll discuss axonal outgrowth in more detail in Chapter 5, but for now note that the first axons to grow out of the spinal cord are from motor neurons, which will seek out their muscle targets. These axons also avoid the posterior half of the somite (**FIGURE 3.22C**). This is why motor axons leave the spinal cord in discrete ventral roots, one for each spinal cord segment, rather than in a continuous sheet. That is important because when the bony vertebral column forms to surround and protect the spinal cord, axons must pass through discrete channels lying between adjacent vertebrae.

Returning to the neural crest cells, once they start moving, they prefer to migrate along particular pathways marked by basal lamina molecules, primarily fibronectin (Dufour et al., 1988). We noted that the crest cells stop expressing one CAM, NCad, in order to break free to commence migration. Next they start expressing another CAM, **neural cell adhesion molecule** (**NCAM**), which helps them grip surfaces to pull themselves along. Once they near their eventual target, the crest cells cease expressing NCAM and again express NCad (Akitaya & Bronner-Fraser, 1992), coalescing with one another at their new home.

Upon reaching its final destination, a neural crest cell begins differentiating into one of the many possible cell fates, as is appropriate for its final resting place. The story of how crest cells are directed to their final cell fates will be taken up in Chapter 4. As a preview, I'll say now that the eventual differentiation of neural crest cells results from a combination of the factors they encounter along the way, such as Shh from the notochord (see Figure 3.19), and those they encounter at their final destinations. Indeed, in vitro studies treating neural crest cells with various factors that they normally encounter show that, to a remarkable degree, we can direct their fate by controlling the sequence of factors they are exposed to (**FIGURE 3.23**). In Chapter 4 we'll

ephrins A family of membrane-bound signaling molecules that bind to ephrin receptors, which are part of the receptor tyrosine kinase (RTK) superfamily.

neural cell adhesion molecule (NCAM) Also called *neural-cam*. An adhesive molecule expressed by many neurons.

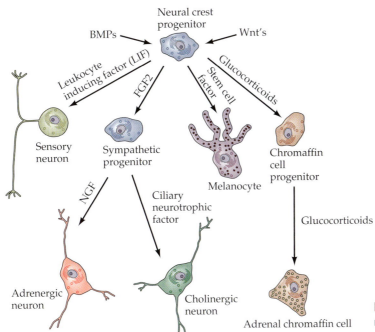

FIGURE 3.23 Factors that direct neural crest cell fate

The eventual differentiation of neural crest cells results from a combination of the factors they encounter along the way, and those they encounter at their final destinations.

describe experiments showing that when neural crest cells are transplanted into a new position in the rostral-caudal axis, they take on a migratory route and cell fate that is appropriate for their new position, further supporting the idea that their fate is guided by other cells. Then we'll see that for those neural crest cells that become postganglionic neurons in the sympathetic and parasympathetic systems, another factor controlling their differentiation is their interaction with synaptic targets. For now, let's consider another remarkable feat of migration, this time in the cerebellum.

Bergmann glia   Long, slender glial cells in cerebellar cortex that guide neurons migrating from the external granule cell layer to the internal granule cell layer.

parallel fibers   The long axons from granule neurons of the cerebellum that innervate Purkinje neuron dendrites.

How's It Going?

1. What are the four main migratory pathways of neural crest cells, and what fates do they take when migration is over?

2. What steps do neural crest cells take to break free and commence migration?

3. How do adhesive factors play a role in guiding migrating neural crest cells?

■ 3.10 ■

CEREBELLAR GRANULE CELLS PARACHUTE DOWN FROM ABOVE

Early in cerebellar development, the Purkinje cells migrate dorsally from the fourth ventricle along radial glia, settle in a single layer, and begin growing massive dendrites toward the cerebellar pial surface (Yuasa, Kawamura, Ono, Yamakuni, & Takahashi, 1991). By adulthood, all the dendrites of each Purkinje cell are restricted to a single plane, and the dendrites of neighboring Purkinje cells are all aligned in the same plane (**FIGURE 3.24**).

Meanwhile, a population of small cells departs from the edge of the pons, a region called the rhombic lip, and they move along the outer surface of the cerebellar cortex until they aggregate along the dorsal surface, forming the external granule layer. There these cells continue to divide, in part because the Purkinje cells beneath them are secreting our old friend Sonic hedgehog, which promotes mitosis in granule cells (Wallace, 1999). This proliferation continues into the early postnatal period. Starting late in fetal development, postmitotic granule cells begin migrating ventrally into the cortex (**FIGURE 3.25**), climbing down extensions of specialized astrocytes called **Bergmann glia**, in a migration that will extend into the early postnatal period (Rakic, 1971b). Shortly after they begin their descent into the cortex, the granule cells grow processes, which later differentiate into axons, extending out in a straight line perpendicular to the Purkinje cell dendrites. Because these axonal processes all grow perpendicular to the Purkinje dendritic array, they are all

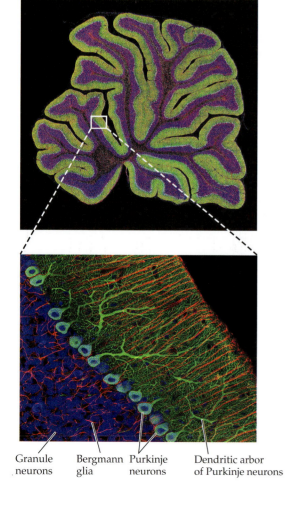

Granule   Bergmann   Purkinje   Dendritic arbor
neurons   glia   neurons   of Purkinje neurons

FIGURE 3.24   The remarkable organization of cerebellar cortex The large Purkinje cells form a single layer beneath the entire cerebellar cortex. Each has a massive dendritic tree extending toward the pial surface. (Courtesy of Drs. T. Deerinck and M. Ellisman, University of California, San Diego.)

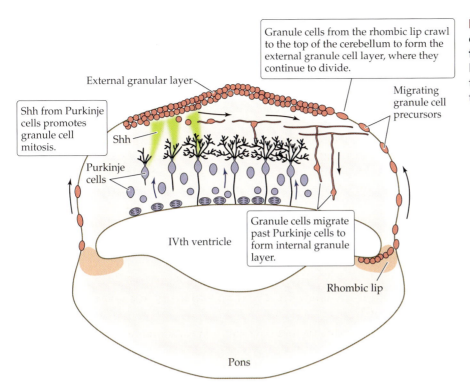

Granule cells from the rhombic lip crawl to the top of the cerebellum to form the external granule cell layer, where they continue to divide.

External granular layer

Shh from Purkinje cells promotes granule cell mitosis.

Shh

Purkinje cells

Migrating granule cell precursors

Granule cells migrate past Purkinje cells to form internal granule layer.

IVth ventricle

Rhombic lip

Pons

FIGURE 3.25 **Granule cells crawl over the cerebellar surface.** Granule cells shinny down Bergmann glia until they reach the inner granule cell layer ventral to the Purkinje cells.

parallel to one another and hence are referred to as the **parallel fibers** of the cerebellum. While the parallel fibers continue to grow, the cell bodies of the granule neurons dive deeper into the cerebellum, trailing their connections to the still-growing parallel fibers like a spider descending on its web (Kumada & Komuro, 2004) (**FIGURE 3.26**). Eventually the granule cell bodies will pass below the Purkinje cell bodies to settle into their adult placement in the internal granule layer. The granule cells will receive excitatory input from the so-called mossy fibers from pontine neurons and will make excitatory synapses in turn upon Purkinje cell dendrites via those parallel fibers they spooled out behind them during their migration.

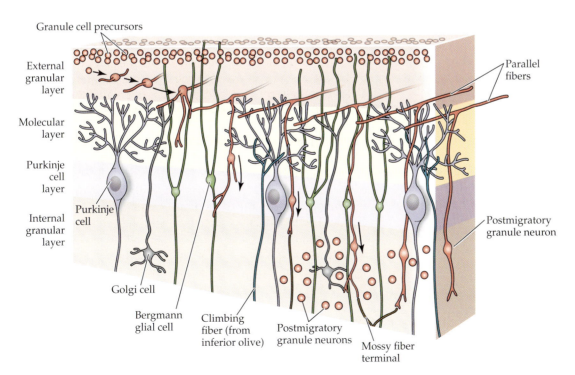

Granule cell precursors

External granular layer

Molecular layer

Purkinje cell layer

Internal granular layer

Purkinje cell

Golgi cell

Bergmann glial cell

Climbing fiber (from inferior olive)

Postmigratory granule neurons

Mossy fiber terminal

Parallel fibers

Postmigratory granule neuron

FIGURE 3.26 **Granule cell paratroopers** (After Kumada & Komuro, 2004.)

> Late in the nineteenth century, people in China and Japan who bred mice for a hobby isolated several strains that were said to "dance," weaving, reeling, or whirling about as they walked.

Thus as a result of these choreographed migrations, the mature cerebellar cortex normally presents a highly ordered structure, a single layer of massive Purkinje cell bodies, their dendrites strictly aligned, and a layer of tiny granule cells below, each sending axons up to stimulate the dendrites of many Purkinje cells in a row. But things don't always go smoothly in development. Late in the nineteenth century, people in China and Japan who bred mice for a hobby isolated several strains that were said to "dance," weaving, reeling, or whirling about as they walked. These so-called Japanese dancing mice were studied and described by Robert Yerkes (founder of the Yerkes National Primate Research Center), who concluded they originated from China several centuries before (Yerkes, 1907).

Even in Yerkes' day, it was established that the dancing behavior of these various strains was the result of recessive mutations (Falconer, 1951; Gruneberg, 1947). By the late 1950s and early 1960s, several different genes were found to cause various forms of "dancing" and were named after their behavioral effects: *weaver*, *reeler*, *staggerer*, and so on (Sidman, 1965). When the brains of these mice were examined, they almost all showed obvious anatomical disorder in a shrunken, misshapen cerebellum (Sidman, 1968; Sidman, Lane, & Dickie, 1962). Rather than being in orderly arrays of Purkinje cells and granule layers as seen in normal mice, cells in these mutant cerebellums were in disarray (**FIGURE 3.27**).

The disarray in these cerebellar mutants suggested a breakdown in the orderly process of cell migration. To better understand the mechanisms that might underlie this failure, scientists turned to cultures, trying to get neurons and glia to survive in vitro (Gasser & Hatten, 1990; Hatten & Mason, 1990). In cultures from wild-type mice, the glia formed long processes and the granule cells would clasp on and migrate along them. The granule cell migration appeared rather aimless—once they reached the end of the "glial monorail," they would reverse and migrate back in the other direction (Hat-

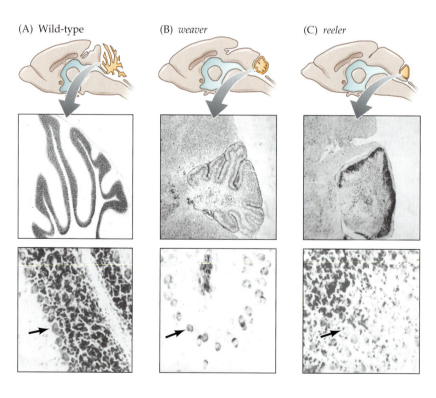

FIGURE 3.27 Abnormal cerebellums in dancing mice (A) In wild-type mice, Purkinje cells (arrow) align in a single layer with their dendrites growing toward the pial surface. (B) In *weaver* mutant mice, the Purkinje cells are properly aligned in a layer, but the granule cells are missing. (C) In *reeler* mice, the Purkinje cells are not aligned in a single layer and the dark granule cells are scattered all over the place. In the middle panel, note that the nearby cerebral cortex of reeler looks abnormal, too. (From Dr. A. L. Leiman, unpublished observations.)

(A) Wild-type   (B) *weaver*   (C) *reeler*

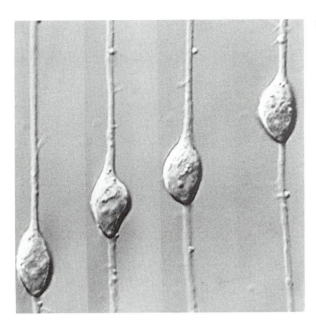

FIGURE 3.28 **Riding the glial monorail** Cerebellar granule cells shinny along the rodlike processes of astrocytes in vitro. You can watch a video of the neurons in motion on the website for this text. (From Hatten, 1990.)

ten, 1990) (**FIGURE 3.28**). Then the investigators tried isolating either glia or neurons from different mice and reintroducing them in vitro. Interestingly, glia would not form the long processes unless cocultured with neurons. This result indicated that the neurons provide some inductive effect, triggering the glia to take on the correct morphology. Even more interestingly, granule cells from weaver mice were not as effectively inductive as granule cells from wild-type mice. In contrast, even weaver glia would respond to wild-type neurons, so the weaver neurons seemed deficient in their ability to affect their glial colleagues. The success of this in vitro approach made it possible to ask whether the defect in weaver mice was a problem in the neurons or the glia, or both, as we'll see next.

RESEARCHERS AT WORK

Weaver Neurons Fail to Grasp Glial Fibers

- **QUESTION**: What is the nature of the defect in migration in *weaver* mutant mice?

- **HYPOTHESIS**: The genetic mutation either affects granule cell neurons, making them unable to migrate, or affects the glia, so they cannot support granule cell migration.

- **TEST**: Place granule cells and glia together in vitro to observe migration. In some cultures, place wild-type granule cells with glia from weaver mice, while in other cultures mix weaver granule cells with wild-type glia. Is migration disrupted in one or both types of mixed cultures?

- **RESULT**: Granule cells from wild-type mice migrated equally well along glia from either wild-type or weaver glia. However, granule cells from *weaver* mutants failed to migrate along glia from either wild-type or mutant mice (Hatten, Liem, & Mason, 1986). The weaver granule cells never seemed to attach to glia (**FIGURE 3.29**).

continued

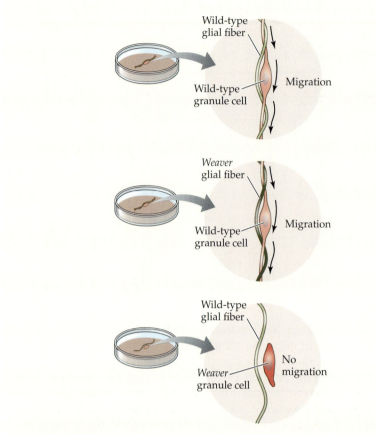

FIGURE 3.29 **Which cell is responsible for the failure of migration in weaver mice?** Mixed cultures of either wild-type granule cells and weaver glia, or weaver granule cells combined with wild-type glia, demonstrated that mutant glia could support migration, but mutant granule cells could not effectively attach to wild-type glia.

■ **CONCLUSION**: The defect caused by the gene mutation in weaver mice does not affect the ability of their glia to support migrating neurons. Rather, the failure of cerebellar migration in weaver mice is due to the inability of their granule cells to grip glial fibers in order to climb along them.

astrotactin   A gene that is expressed by cerebellar granule cells to help them to grip glial fibers for migration.

Later the gene responsible for the weaver phenotype was isolated and found to encode for a membrane-bound potassium channel (Hess, 1996). Interestingly, just aggregating cultures of wild-type granule cells with weaver granule cells allows even the mutant cells to migrate (Gao, Liu, & Hatten, 1992), so it appears that this potassium channel is needed in at least some granule cells in a group so that they all can grasp and migrate along glial fibers.

Another gene, **astrotactin**, encodes for a protein expressed by granule cells to help them grip glia; antibodies to the protein disrupt migration in vitro (Fishell & Hatten, 1991). Knocking out the *astrotactin* gene slows and disrupts cerebellar migratory patterns (Adams, Tomoda, Cooper, Dietz, & Hatten, 2002).

The migratory failure in another dancing mouse, reeler, is much more profound than that seen in weaver mice. This is obvious even in the gross structure of the cerebellum in the two mutants: the reeler cerebellum is even smaller and more disorganized than that of weaver (see Figure 3.27). Once *reeler* was isolated and sequenced, it was found to code for a protein, named

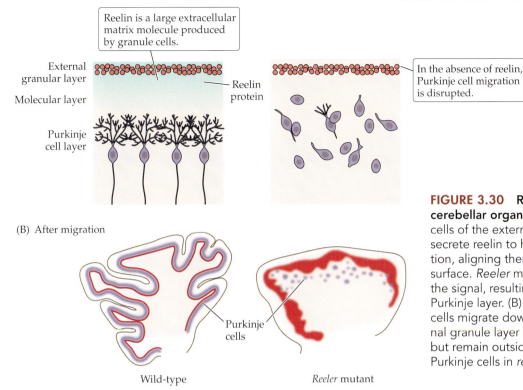

(A) Before migration

Reelin is a large extracellular matrix molecule produced by granule cells.

External granular layer

Molecular layer

Purkinje cell layer

Reelin protein

In the absence of reelin, Purkinje cell migration is disrupted.

(B) After migration

Purkinje cells

Wild-type

Reeler mutant

FIGURE 3.30 **Role of reelin in cerebellar organization** (A) Normally cells of the external granule cell layer secrete reelin to halt Purkinje migration, aligning them beneath the pial surface. *Reeler* mutants don't produce the signal, resulting in a disorganized Purkinje layer. (B) The external granule cells migrate down to form the internal granule layer in wild-type mice, but remain outside the disorganized Purkinje cells in *reeler* mutants.

reelin, that is secreted into the extracellular matrix around cells expressing the gene (D'Arcangelo et al., 1995). In the developing cerebellum, granule cells normally secrete reelin while they are massed in the external granule cell layer on the pial surface of the cerebellum. This provides a uniform gradient of reelin along the outer surface, and the Purkinje cells migrating from the ventricular zone appear to stop once they reach a particular concentration of the protein. Normally this results in all the Purkinje cells coming to rest in a single layer just beneath the surface of the cerebellar cortex. However, in reeler mice, the absence of functional reelin means the Purkinje cells migrate much longer than usual, coming to rest at random, typically in clumps (**FIGURE 3.30A**). The absence of an orderly array of Purkinje cells results in neurons in the external granule cell layers that seem to lose the call to migrate and so stay bunched on the surface of the cerebellum (**FIGURE 3.30B**). Reelin also plays a role in modifying migration in the cerebral cortex, which causes a more modest, but still obvious, disarray there in *reeler* mutants (Pearlman & Sheppard, 1996; Rice & Curran, 2001).

How's It Going?

1. What is the normal pattern of cell migration in the developing cerebellum?

2. What underlies the disrupted migration in weaver and *reeler* mutant mice?

▪ 3.11 ▪

CELLS CRUCIAL FOR SMELL AND REPRODUCTION MIGRATE INTO THE EMBRYONIC BRAIN

Another instance of cell migration can help us understand what happened to Neil, whom we met at the start of the chapter. Well after the telencephalon has formed from the rostral end of the neural tube and become thick with cells, there is an invasion of cells from outside the neural tube. These cells come from the **olfactory placodes**, plate-shaped (that's what *placode* means)

reelin A gene that encodes the protein reelin and when dysfunctional results in very a disorganized cerebellum and cerebrum in mice.

olfactory placode A plate-shaped collection of cells outside the developing brain, from which cells migrate through the olfactory bulb and to the rest of the brain.

FIGURE 3.31 Invasion of the GnRH neurons

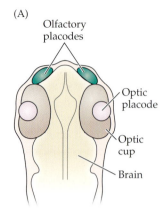

(A)

Olfactory placodes

Optic placode

Optic cup

Brain

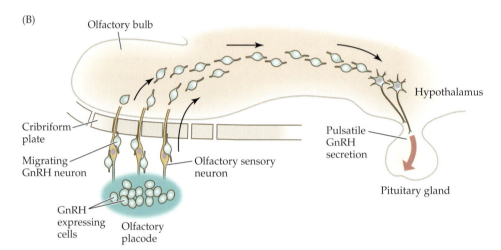

(B)

Olfactory bulb

Hypothalamus

Cribriform plate

Migrating GnRH neuron

Olfactory sensory neuron

Pulsatile GnRH secretion

GnRH expressing cells

Olfactory placode

Pituitary gland

olfactory sensory neurons (OSNs) The primary sensory neurons of the olfactory epithelium, which contact and recognize particular odorants.

olfactory bulb An anterior projection of the brain that terminates in the upper nasal passages and, through small openings in the skull, receives axons from olfactory sensory neurons.

olfactory ensheathing glia (OEGs) Glia found in the olfactory epithelium that guide newly generated olfactory sensory neurons into place.

gonadotropin-releasing hormone (GnRH) A protein released by neurons in the hypothalamus into the median eminence, which directs anterior pituitary cells to secrete the gonadotropins: follicle-stimulating hormone (FSH) and luteinizing hormone (LH).

collections of ectodermal cells that amass just outside the budding olfactory bulbs of the developing telencephalon (**FIGURE 3.31**). (An optic placode forms across from the opening of each developing optic cup that will form a lens, and otic placodes next to the developing brainstem will form the ears and vestibular system.)

The incoming cells from the olfactory placode migrate toward the young olfactory bulb and take on several different fates (Schwanzel-Fukuda & Pfaff, 1989; Wray, Grant, & Gainer, 1989). Some stop just outside the olfactory bulb and become primary **olfactory sensory neurons** (**OSN**) forming a sheet called the olfactory epithelium. These neurons extend axons through holes in the bony cribriform plate to synapse in the **olfactory bulb** and eventually express olfactory receptor proteins to detect odorants. Other incoming cells become specialized glial cells called **olfactory ensheathing glia** (**OEG**), which also populate the olfactory epithelium and the outermost layers of the olfactory bulb.

Yet another subset of invading cells shinny along the axons of the olfactory sensory neurons, entering the olfactory bulb and then plunging into the brain. Curiously, even before they leave the olfactory placode, these cells are already producing the neuropeptide, **gonadotropin-releasing hormone** (**GnRH**), that they will release from their axons someday. Thus we can track the migrating cells using immunohistochemistry directed at GnRH. These GnRH-expressing cells eventually settle in the hypothalamus, where they will differentiate into neurons that send axons to the hypothalamic-pituitary blood portal system. When the neurons release their GnRH into the portal system, it will reach the pituitary and stimulate cells there to release two hormones, follicle-stimulating hormone (FSH) and luteinizing hormone (LH). FSH and LH are called gonadotropins because they stimulate development of the gonads. Although their names are derived from their action on the ovary, where they were first studied, FSH and LH are also crucial for development of the testes. In males, the release of FSH and LH at puberty induces the testes to start sperm production and to release steroid hormones, including testosterone. The testosterone in turn promotes development of sexual interests and secondary sexual characters, which in humans include pubic hair and beard growth, changes in voice, enlargement of the penis, and greater muscle mass.

> Although their names are derived from their action on the ovary, where they were first studied, FSH and LH are also crucial for development of the testes.

Several things are curious about this migration that begins with cells passing through the olfactory bulb. First, there is evidence that the invading cells include not only ectodermal cells from the olfactory placode, but also neural crest cells that first invade the olfactory placode and then mix with the ectodermal cells there to contribute to the migration (Forni, Taylor-Burds, Melvin, Williams, & Wray, 2011; Whitfield, 2013). In fact, some neural crest cells take on each of the three fates we've mentioned—OSNs, OEGs, and GnRH cells. Another curious thing is that the GnRH cells don't settle down together in any particular part of the hypothalamus. Rather than gathering in a nucleus, the cell bodies remain scattered about. All they have in common is that they make GnRH, and they send their axons to the base of the hypothalamus to release the GnRH into the portal system to reach the pituitary. I don't know of another instance in the neurosciences where a group of neurons that all play a very specific physiological function are not aggregated together in a single place to form a nucleus. Apparently, they don't need to be in close contact with each other in order to coordinate the release of GnRH.

By the way, there is evidence that OEGs may be useful in promoting regeneration within the CNS (Li, Field, & Raisman, 1997) and therefore may be useful for many therapies, perhaps even including the repair of spinal cord injuries (Granger, Franklin, & Jeffery, 2014).

You may have guessed that this migration is related to Neil's symptoms described at the start of the chapter. **Kallmann syndrome** is a failure of the pituitary to release gonadotropins to stimulate the gonads at puberty (technically called hypogonadotropic hypogonadism) *plus* lack of a sense of smell (Smith & Quinton, 2012). Mutations in several different genes can cause Kallmann, but what they all have in common is a failure to produce the OEGs and the primary olfactory sensory neurons with axons leading from the olfactory epithelium into the olfactory bulb. This is why the person will be anosmic: he or she has no olfactory sensory neurons. Furthermore, without these olfactory sensory neurons, there are no axons for the GnRH cells to climb into the olfactory bulb and thence to the hypothalamus. Consequently the person has no GnRH neurons to stimulate the pituitary to release gonadotropins, so the gonads don't develop and puberty never comes. Some of the genes involved in this process are also involved in development of the ear (Maier, Saxena, Alsina, Bronner, & Whitfield, 2014), hence people with Kallmann may be partially or wholly deaf as well.

Nothing can be done at present to allow people with Kallmann to detect odors, but hormone treatment can produce puberty and the changes associated with it. Typically, male patients are treated with testosterone, which, if presented early enough (ideally before age 16), will produce the male secondary sex characters and stimulate libido. Another approach that is being tried is treatment with gonadotropins, which in addition to stimulating the testes to produce testosterone for the above effects, can also stimulate sperm production if the patient wants to father children. Canadian poet Brian Brett has written a poignant and revealing memoir, *Uproar's Your Only Music*, about growing

Kallmann syndrome A condition in which individuals fail to reach puberty and are unable to detect odors.

up with Kallmann syndrome (Brett, 2004). "Little" Jimmy Scott, a jazz singer with a unique voice, which is neither male nor female, that probably resulted from his never going through puberty, talks about his condition on YouTube. Unfortunately for Neil, his treatment did not begin in earnest until he was 22 (Smith & Quinton, 2012), at which point testosterone is less effective at building up secondary sex characters and strengthening bone. Neil found being labeled a "late bloomer" in his 20s very humiliating and feels it has had a long-lasting effect on his life. To avoid having this happen to other people, Neil urges for a greater awareness of Kallmann syndrome so that physicians know to ask the next 16-year-old late bloomer, "Can you smell odors?" If the answer is no, then further testing for Kallmann syndrome is in order.

How's It Going?

1. Describe the migration of GnRH neurons.
2. What causes the constellation of seemingly unrelated symptoms in Kallmann syndrome?

SUMMARY

■ *Pleiotropy* is a condition in which a single gene contributes to many different traits, as when a failure of cell migration causes the many disparate symptoms of *Kallmann syndrome*. **See Figure 3.1**

■ In the neural tube, *mitosis* occurs when the cell nuclei are near the ventricular surface. As the tube thickens, cells may span the width of the tube, but the nuclei will still shift to the ventricular surface during mitosis. **See Figure 3.2**

■ Symmetrical divisions provide more *neuroblasts*, including some cells that will later serve as *radial glial cells*. Asymmetric divisions produce one postmitotic daughter cell that joins the outermost *marginal zone* or, later, the *cortical plate* to differentiate into a neuron (*neurogenesis*) or glia (*gliogenesis*). **See Figures 3.3 and 3.4**

■ Studies using cell lineage markers in chicks and mice show that the progeny of a single neuroblast may take on many different fates, indicating little or no role of mitotic lineage in cell differentiation. **See Figure 3.5**

■ In the cerebral cortex, some neuroblasts detach from the *ventricular zone* to form the *subventricular zone* (*SVZ*), continuing to divide to produce neurons and glia. **See Figures 3.6 and 3.7**

■ A few of the earliest-generated neurons settle in the sparsely populated layer I, but thereafter each cohort of neurons migrates past their predecessors, settling in an "inside-out" order, with innermost layer VI containing the oldest neurons and layer II the youngest. **See Figures 3.10–3.12**

■ Some neurons arise from one of three *ganglionic eminences* and migrate tangentially across radial glia

rather than simply out a single glial fiber. In humans, an arc of new neurons is still migrating into the frontal cortex in the first few months of life. **See Figures 3.13 and 3.14**

■ Neurogenesis extends into adulthood in three regions: the olfactory epithelium producing replacement *olfactory sensory neurons*, the SVZ of the lateral ventricles that supply the *rostral migratory stream* of *olfactory bulb* interneurons, and the *subgranular zone* of the dentate gyrus in the hippocampal formation. **See Figures 3.15 and 3.16**

■ Although it is not universally accepted, there is considerable evidence that a few new neurons are also added to the adult *neocortex* and that experience affects the production or survival of these new neurons. **See Figure 3.17**

■ Neural crest cells contribute to the pharyngeal arches that make up cranial nerve ganglia, and in the spinal regions they will migrate out four major pathways to form *dorsal root ganglia*, *autonomic ganglia*, the adrenal medulla, and melanocytes. **See Figures 3.18–3.20**

■ Crest cells regulate expression of *cell adhesion molecules* and secrete proteases to start their migration, then follow paths of cell adhesion molecules to reach their destination. They display a strong preference to migrate along pathways rich in the glycoprotein *fibronectin*. **See Figure 3.21**

■ Migrating crest cells are also herded into one stream per spinal segment by avoidance of the *ephrins* expressed in the posterior half of each *somite*. **See Figure 3.22**

■ Cerebellar *Purkinje cells* arise from the ventricular layer to form a single layer. Granule cells arise from the rhombic lip to migrate over the dorsal surface of the cerebellar cortex, continue dividing in that *external granule layer*, and then migrate down along *Bergmann glia* fibers past the Purkinje cells to form the *internal granule layer*. **See Figures 3.24–3.26**

■ Migrating granule cells are also guided by adhesive signals. The *weaver* mutation makes the cells incapable of grasping glia in order to migrate properly. In *reeler* mutants, the lack of *reelin* secretion from the external granule layer means Purkinje cells don't align in a single layer, severely disrupting the subsequent migration of cells to the internal granule cell layer. **See Figures 3.28–2.30**

■ Cells from the *olfactory placode* provide primary olfactory sensory neurons that send axons through the cribriform plate to synapse in the *olfactory bulb*, as well as cells that become *olfactory ensheathing glia* along those axons, plus cells that migrate along those axons to enter the hypothalamus and become *GnRH* neurons. Kallmann syndrome results when the primary olfactory sensory neurons don't form, depriving the person of a sense of smell and denying presumptive GnRH neurons a pathway into the brain. **See Figure 3.31**

Go to the Companion Website
sites.sinauer.com/fond
for animations, flashcards, and other review tools.

4

Seeking Identity
NEURAL DIFFERENTIATION

A LIVING GRAVEYARD   For thousands of years, a disfiguring disease inspired fear and repulsion. Long known as "leprosy," it is characterized by pale lumps on the skin caused by peripheral nerve damage that eventually leads to numbness, especially of the face, hands, and feet. The Bible's instructions to avoid the spread of leprosy were clear: "And the leper in whom the plague is, his clothes shall be rent, and his head bare, and he shall put a covering upon his upper lip, and shall cry, Unclean, unclean … he shall dwell alone; without the camp shall his habitation be" (Leviticus 13:45–46). Being called a "leper" became the same as being identified as someone to be avoided and reviled. Because of the stigma attached to that term, today the disease is called **Hansen's disease**, after Norwegian physician G. Armauer Hansen, who in 1873 identified the bacterium *Mycobacterium leprae*, the first bacterium to be identified as the cause of a human disease (Irgens, 1984).

Until well into the twentieth century, people suffering from the disease were typically banished to "leper colonies," with little or no resources to support them. In Hawaii, the disease was known as "the sickness that is a crime," as victims were declared civilly dead, their spouses free to remarry (Tayman, 2010), and forcibly outcast to a harsh, remote peninsula on the island of Molokai without shelter or medical support, and almost no food. When Belgian missionary Damien De Veuster (1840–1889) arrived at the colony in 1873, some 800 men, women, and children with Hansen's disease were living in shacks, slowly starving in what was called a "living graveyard." The energetic Father Damien soon erected housing, organized farming, established schools, dressed wounds, and far too often, made caskets and dug graves. Damien refused the safeguards that were hoped to protect him, visiting each colonist at least once per week, and welcoming all into his home. Originally, Damien was to be the first of four volunteer priests taking turns ministering to the colony, to minimize the risk of contracting the disease. But shortly after arriving, Damien petitioned his superiors to let him stay. When they agreed, Damien wrote farewells to his brother, parents, and friends, knowing he would never see them again. Damien would spend the final 16 years of his life tending the lepers of Molokai.

CHAPTER PREVIEW

So far in this book we have discussed the induction of the nervous system from the ectoderm (Chapter 1), the laying out of a comprehensive body plan (Chapter 2), and the migration of neural cells to populate the brain and periphery (Chapter 3). By the end of that migration, most cells have already taken steps toward either a glial or neuronal fate, but it is only after they have reached their final destinations that most differentiation takes place, and that is the topic of this chapter.

In Chapter 1 we saw that in *Caenorhabditis elegans* differentiation is guided by mitotic lineage, as maternal factors are divided and subdivided until each of the thousand or so cells expresses a unique set of genes and thereby takes up a unique fate. Recall this is called mosaic control of differentiation. In

Father Damien De Veuster (1840–1889) This photograph, taken the year of his death, shows the scarring on his face and hands caused by Hansen's disease. (Photo by William Brigham, 1889.)

Hansen's disease Formerly called leprosy. A chronic bacterial disease that causes nerve damage and disfiguring sores on the limbs and face.

ommatidium A cluster of photoreceptors and supporting cells that form a single visual unit in the compound eye of an insect. Many ommatidia make up the eye.

contrast, the other model organisms, with bodies made up of many more cells, rely upon *cell-cell interactions* to direct each cell to a particular fate. It is this method of differentiation that we'll focus on now.

Trying to chart the differentiation of all the different neuronal phenotypes would be very daunting indeed, as one hallmark of the brain is the bewildering array of different shapes of cell bodies, dendritic trees, and axonal projections. As far as we can tell, of the nearly 100 billion neurons in the human brain, no two are exactly alike in appearance, so they are probably not exactly alike in function, either. Instead, research has focused on understanding neuronal differentiation in a few model systems where we can readily keep track of the cells during development. The theme that emerges from them all is that cells influence their neighbors, nudging them to express particular sets of genes to differentiate into particular fates. Sometimes this intercellular communication occurs at the point of contact between two cells; in other cases a cell may release a diffusible chemical to alter its neighbors. In other cases, the cell may expose itself to a series of inductive influences, both cell-cell contact and diffusible factors, as it migrates through the brain and body. By the end of its journey, the cell's fate has been specified.

The first such system we'll consider is the *Drosophila* eye, where there is a very orderly sequence of neuronal differentiation of photoreceptors and a powerful array of genetic approaches to understand that sequence. We'll see that the orderly differentiation of photoreceptors comes about because they communicate with one another by direct cell-cell contact, coordinating each other's fates in a particular sequence. Next we'll learn how cells in the vertebrate spinal cord respond to diffusible factors to communicate with each other to determine which will become motor neurons versus interneurons or glia.

Then we'll consider the vertebrate neural crest where, as we learned in Chapter 3, a population of cells migrate long distances to assume a wide variety of fates. Here we'll see that the cells they encounter during and after their migration direct crest cells to take on one fate versus another. We'll also learn that neural crest cells can sometimes "dedifferentiate," returning to an earlier, more plastic stage such that they can follow another fate, which turns out to be important for the progression of Hansen's disease. Finally we'll turn to the mammalian neocortex, where again migrating cells are directed to one neural fate or another by communication and competition both before and after they wander. The chapter will close with the observation that because natural selection tends to vary events later in development rather than early in development, the additional "brain power" that has been gained across vertebrates was primarily due to augmentation of those last-born, outermost layers of cortex.

■ 4.1 ■

THE FRUIT FLY RETINA DEVELOPS THROUGH AN ORDERLY PROGRESSION OF GENE EXPRESSION AND SIGNALING

One of the earliest demonstrations of the control of neuronal differentiation again emphasized the importance of cell-cell communication and lateral inhibition. The compound eye of the fruit fly (**FIGURE 4.1A**) develops from a relatively uniform-looking disc of cells into a highly organized array of tightly packed hexagons, each of which will give rise to a single **ommatidium** (plural *ommatidia*), a cluster of eight photoreceptors and supporting

(A)

(B)

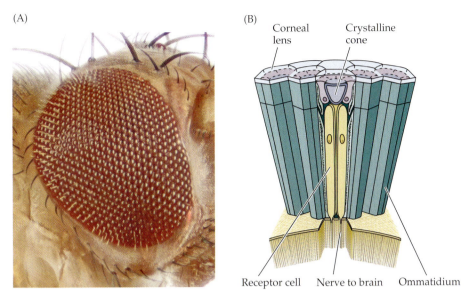

FIGURE 4.1   The *Drosophila* compound eye (A courtesy of David McIntyre; B after Rubin, 1989.)

cells (**FIGURE 4.1B**). Early twentieth-century researchers had speculated that each ommatidium might arise from successive divisions of a single precursor, but analysis of the many mutations known to affect eye structure, as well as direct microscopic observations of developing larvae and pupae, refuted this view (Landis, 1996; Ready, Hanson, & Benzer, 1976). In the late larval stages the optic disc that will form the adult's eye is marked by a prominent furrow, a physical indentation as cells contract and expand. This furrow starts at the posterior end of the eye and then moves anteriorly. In front of this advancing furrow the cells of the eye field are homogenous and dividing, but behind the furrow, cells have stopped dividing and begin differentiating until they form a packed hexagonal field. Because it seems to organize the eye, leaving an orderly array of differentiated cells in its wake, the furrow is known as a **morphogenetic furrow** (**FIGURE 4.2A**). The differentiation left behind by the morphogenetic furrow includes differentiation of photoreceptors in the ommatidia (Spratford & Kumar, 2014) (**FIGURE 4.2B**).

The forces behind this morphogenetic furrow came to light from studies of the gene **hedgehog** (**hh**), a *Drosophila* homologue of our old friend Sonic hedgehog (Shh). Mutations of *hh* were known to cause the compound eye to have a rough texture. When the gene was cloned and sequenced, it was revealed to produce a secreted protein (because it included a *signal peptide* directing the product to the cell's secretory pathway, Chapter 2), which was found to be required for the furrow to form and move anteriorly. Early in the process, *hh* is expressed and secreted by cells adjacent to the posterior eye disc and induces the cells there to stop dividing and start differentiating into ommatidia. Then the cells within each budding ommatidium start expressing and secreting Hh, effectively "pushing" the morphogenetic wave forward to complete the process, like a line of dominoes, each knocking down its neighbor. One effect of the secreted Hh is to induce expression of **decapentaplegic** (**dpp**, the gene for a bone morphogenetic protein homologue), which halts mitosis in the cells in front of the furrow and induces them to start expressing proneural genes. If the photoreceptors are genetically prevented from expressing either *hh* or *dpp*, the morphogenetic furrow stalls (Rogers et al., 2005). Such genetic manipulations were once only possible in invertebrates,

morphogenetic furrow   The prominent indentation of the insect optic disc that moves anteriorly, marking the differentiation of photoreceptors.

hedgehog (hh)   A gene in *Drosophila* that is a homologue of *Sonic hedgehog* (*Shh*). It encodes the secreted protein hedgehog, which influences cell differentiation.

decapentaplegic (dpp)   The gene that encodes dpp, a bone morphogenetic protein homologue in insects that halts mitosis in the optic disk and promotes differentiation of photoreceptors.

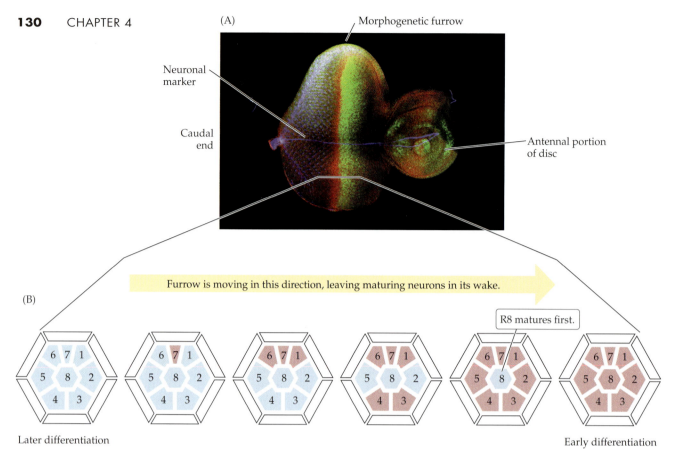

(A)

Morphogenetic furrow

Neuronal marker

Caudal end

Antennal portion of disc

Furrow is moving in this direction, leaving maturing neurons in its wake.

(B)

R8 matures first.

Later differentiation

Early differentiation

FIGURE 4.2 **The morphogenetic furrow leaves differentiated ommatidia in its wake.** (A courtesy of Drs. N. Brown, S. Paddock, and S. Carroll; B after Tomlinson, 1988.)

transgenic Referring to an organism in which foreign DNA has been deliberately inserted.

construct Here, a stretch of DNA that has been manipulated before being inserted into an organism's genome to create a transgenic animal.

knockout Here, an animal in which a particular gene has been deliberately removed or disabled.

knockin Referring to an animal in which an endogenous gene has been deliberately replaced with another allele, often an allele associated with a human disorder.

sevenless A gene that encodes a receptor tyrosine kinase (RTK) and that when stimulated triggers differentiation of photoreceptor 7 in *Drosophila*.

receptor tyrosine kinase (RTK) A family of membrane-bound signaling proteins that, when activated by a ligand, phosphorylate tyrosine sites on particular proteins.

but today it is possible to knockout genes in many other species, including mammals, as described in **BOX 4.1**.

In addition to some pigment cells to absorb stray photons, and cells to secrete a lens overhead, each of the 800 or so *Drosophila* ommatidia contains eight different photoreceptor neurons, each with a unique morphology and function in the adult. These are designated simply as cells R (for receptor) 1 through 8. The earliest cell to begin expressing neuron-specific genes is R8, which is in the center of the array. Then R2 and R5 differentiate on either side of R8. Next, R1, R3, R4, and R6 differentiate, leaving R7, which will detect UV light (Harris, Stark, & Walker, 1976), to mature last. This sequence is seen in each developing ommatidium, so the farther you are from the morphogenetic furrow, the more time the ommatidium has had to go through this sequence, hence the further along receptor maturation has gone (see Figure 4.2B).

The sequence of receptor differentiation is invariant because the first receptor, R8, directs the fate of the other receptors in a cascade of transcription factors. In Chapter 1 we learned that in insects there is a competition between neighboring ectodermal cells using the Delta-Notch system such that one cell, typically the one in the middle of those expressing proneural genes, delaminates and becomes a neuron while laterally inhibiting its neighbors from doing so. In each young ommatidium the same Delta-Notch competition results in one cell, typically the one in the middle becoming the first neuron, R8 (Baker & Zitron, 1995). But rather than inhibiting its neighbors from becoming neurons, R8 sends signals that regulate gene expression in two of its neighbors, causing them to become R2 and R5. These cells then team up with R8 to induce differentiation of the other receptors (Treisman, 2013). Last to develop is R7, which is controlled exclusively by R8, as we'll see next.

BOX 4.1

TRANSGENICS, KNOCKOUTS, AND KNOCKINS

Much of the explosion of understanding about neural development has come from the ability to alter genes in organisms to see the roles they play. For most of the twentieth century such things were possible only with invertebrate species like fruit flies and worms, where random mutations could be induced in a large number of individuals that could then be readily screened to look for interesting phenotypes. That approach simply wasn't practical (or affordable!) with vertebrate models such as rats, mice, and monkeys. In those days, the only interesting mutations one could study were the very few spontaneous mutations that had been isolated, like those in the "dancing mice" with abnormal cerebellums we discussed in Chapter 3. But over the past few decades scientists have worked out ways to manipulate the genome in a variety of ways, primarily in laboratory mice, and these advances have revolutionized the field.

The first wave of genetically modified mice were the **transgenics**, organisms in which an extraneous stretch of DNA has been inserted into the genome. To make a transgenic mouse, you first use recombinant techniques to assemble the desired length of DNA, usually called a **construct**, that you want to insert. The construct typically consists of the gene of interest along with an upstream length of DNA to serve as the promoter. To maximize the chances that the gene would be expressed in the final organism, in the past researchers typically used a "universal" promoter like that found before the beta-actin gene, which is normally expressed in nearly all cells. With growing confidence in the technique, these days researchers may use a tissue-specific promoter so that the gene will be expressed only in the nervous system, or only in astrocytes.

Typically, the gene of interest has been deliberately modified, for instance, mutated to mimic a genetic disorder. Once the construct is assembled, it can be injected into zygotes, often 100 at a time. In a percentage of zygotes, the construct is incorporated into a chromosome at random. Alternatively, you can put the construct into a stem cell and inject that into an early embryo, as we described in Figure 1.20. In either case, the embryos are surgically imported into the uterus of pseudopregnant mouse dams and allowed to develop and be born. After birth, the pups are screened, typically with polymerase chain reaction (PCR) reagents, to see which mice actually incorporated the construct into their DNA and therefore may express the transgene, and potentially pass along the gene to their offspring.

In other cases, you may want to *disable* the gene of interest to see what happens to development, as a way to infer the gene's role in that process. In this case you make a **knockout** organism, one in which the gene of interest has been disabled—either it is missing or it includes a premature stop codon—so no functional protein is made. It used to be much more difficult to make a knockout than a transgenic, because you can't just insert a construct at random; rather, you have to replace a very specific sequence of DNA that occurs only once in the genome. To do this, you made a construct that includes nucleotide sequences that are complementary to much of the targeted gene, so that it will bind to the target gene and possibly replace it, but that has been otherwise disabled so it cannot produce a functional product. More recently, disabling specific genes has been made easier by use of the CRISPR system of using synthetic RNAs and an enzyme called Cas to target and disable a particular gene (Shalem, Sanjane, & Zhang, 2015), as described in detail in the Appendix. An important advantage of CRISPR is that it can be used in many different species, not just fruit flies and mice. It can also knock out several genes at the same time.

Knockin animals are those in which the endogenous gene has not been disabled, but has been replaced with a modified allele, typically an allele that has been associated with a disease or disorder. The idea is to see whether this allele, which causes a disorder when it occurs in humans, will also cause a disorder in mice. If so, then it might be possible to study the knockin mice to better understand the human disorder and to test possible therapies for the disorder.

A mutation in flies called *sevenless*, because R7 was missing, provided the first insights into this signaling between the photoreceptors. Sequencing the *sevenless* gene revealed that it encodes a **receptor tyrosine kinase (RTK)**, meaning it has several hydrophilic peptide sequences that embed the protein in a membrane (i.e., *transmembrane domains*), plus an intracellular domain that includes a *kinase,* an enzyme that adds phosphate groups to proteins (i.e., it *phosphorylates* proteins),

FIGURE 4.3 **Receptor tyrosine kinase (RTK) family of receptors** The *Drosophila* gene *sevenless* is a part of this family. We will encounter several of these receptors in future chapters. (After Lemmon & Schlessinger, 2010.)

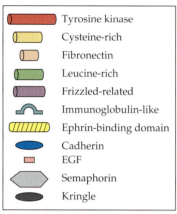

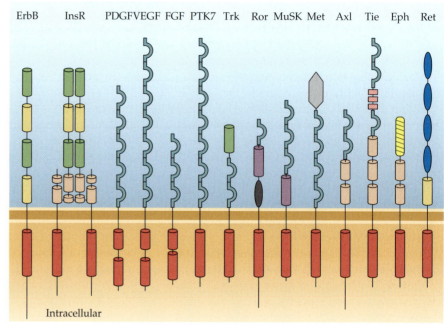

bride of sevenless (boss)
A gene in *Drosophila* that encodes the membrane-bound protein boss, which is expressed in photoreceptor 8 and binds sevenless in a nearby cell, inducing it to become photoreceptor 7.

in this case adding phosphates exclusively to the amino acid tyrosine (Hafen, Basler, Edstroem, & Rubin, 1987). This would turn out to be one in a large family of RTK signaling systems (Lemmon & Schlessinger, 2010), which includes the ephrin receptors we discussed in Chapters 2 and 3, as well as receptors for EGF and fibroblast growth factors, plus signals we'll discuss in future chapters (**FIGURE 4.3**). Normally R7 expresses the *sevenless* gene, but if the protein isn't functional, the cell fails to receive the signal to differentiate into a functional UV receptor. Later another gene was found to also be crucial for R7 differentiation, and it was called ***bride of sevenless (boss)*** because its protein is normally expressed by the R8 cell to activate the sevenless receptor in R7 (Reinke & Zipursky, 1988). Thus the sequential interaction of the eight photoreceptors, capped by R8's expression of boss to activate the sevenless receptor in R7, provides each ommatidium with the full complement of photoreceptors (**FIGURE 4.4**).

The wave of differentiating photoreceptors in the eye also has consequences for a wave of maturation in the optic lobe as the precisely timed arrival of photoreceptor axons organize layers of neuronal maturation (Sato, Suzuki, & Nakai, 2013). Now let's return to the developing neural tube in vertebrates to consider some early steps in neural differentiation.

How's It Going?

1. How do cell-cell interactions start and propagate the morphogenetic furrow of the *Drosophila* eye?
2. What is the sequence of events in the specification of the eight photoreceptors in fruit flies?
3. How does R8 direct the fate of R7?

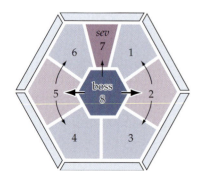

FIGURE 4.4 **Photoreceptor R8 directs the development of the other photoreceptors.**

■ 4.2 ■

SEVERAL FACTORS INFLUENCE WHETHER A CELL WILL BECOME A NEURON OR A GLIA

We noted in Chapter 3 that some dividing cells in the ventricular zone are already moving toward differentiation as glia, as indicated by expression of glial fibrillary acidic protein (GFAP) (Levitt, Cooper, & Rakic, 1983), which is not expressed by neurons. We also saw that the cohort of offspring cells created by the last few cell divisions can take on a wide variety of cell fates—neurons, astrocytes, oligodendrocytes. In other words, there was no evidence of mitotic lineage affecting cell fate in vertebrates.

Whether a cell becomes a neuron or glia depends on the balance of many different forces.

What controls whether postmitotic cells become neurons or glia? In vitro studies have made it possible to expose cells from the developing nervous system to various factors and observe whether they start expressing genes typical of neurons, or astrocytes, or oligodendrocytes. One influence on whether cells become neurons or glia is bone morphogenetic protein (BMP). Recall from Chapter 1 that BMPs are secreted proteins that induce ectodermal cells to take on an epithelial fate. The organizer proteins from the dorsal lip of the blastopore interfere with that signaling so that some ectodermal cells take on a neural fate instead. After the neural tube has formed, BMP signaling induces cells to become astrocytes rather than neurons (D'Alessandro, Yetz-Aldape, & Wang, 1994; Mabie et al., 1997). It's as though BMP is still acting to deflect cells from a full neuronal phenotype. (By the way, if you haven't noticed yet, the adjective *neuronal* designates that we're talking about neurons only, not glia, whereas the adjective *neural* could be referring to neurons, or glia, or both.)

In addition to encountering many secreted factors, postmitotic cells are also communicating with each other directly by cell-cell contact to control neural fate. Recall from Chapter 1 the Delta-Notch system of signaling. These membrane proteins interact such that when Delta from one cell binds to Notch in its neighbor, Notch inhibits proneural gene expression, pushing the cell toward differentiating into epidermis. Hence, among a group of cells in contact with each other, whichever one initially expresses the most Delta suppresses a neural fate in its neighbors, a case of lateral inhibition. In postmitotic neural cells, vertebrate homologues of the Delta-Notch signaling mediate a similar competition to determine which of these neural cells will become neurons rather than glia (**FIGURE 4.5**) (Casarosa, Fode, & Guillemot, 1999; Fode et al., 2000; Horton,

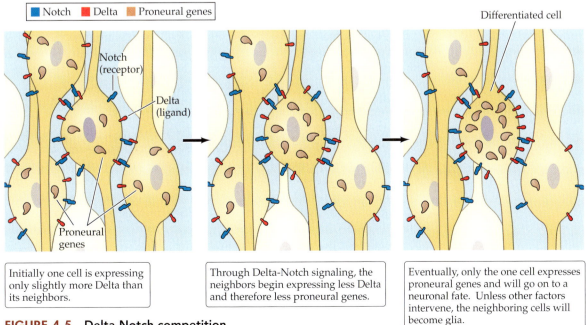

Initially one cell is expressing only slightly more Delta than its neighbors.

Through Delta-Notch signaling, the neighbors begin expressing less Delta and therefore less proneural genes.

Eventually, only the one cell expresses proneural genes and will go on to a neuronal fate. Unless other factors intervene, the neighboring cells will become glia.

FIGURE 4.5   Delta-Notch competition

(A)

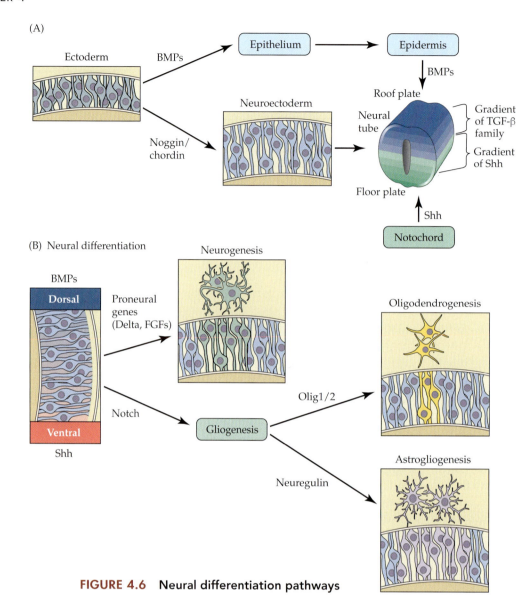

(B) Neural differentiation

FIGURE 4.6 Neural differentiation pathways

fibroblast growth factors (FGFs)
A family of proteins that act upon a family of receptor tyrosine kinase (RTK) receptors. They are concentrated in the posterior portions of the developing vertebrate nervous system and demarcate the midbrain/hindbrain junction.

Meredith, Richardson, & Johnson, 1999). The cell expressing more Delta will suppress expression of Delta and proneural genes in its neighbors and go on to differentiate into a neuron. By suppressing Delta expression in its neighbors, this neuron-to-be consigns the other cells to become glia. (**FIGURE 4.6A**).

In contrast, another family of secreted proteins that we encountered in Chapter 2, the **fibroblast growth factors** (**FGFs**), push cells to differentiate into neurons (Guillemot & Zimmer, 2011; Hegarty, O'Keeffe, & Sullivan, 2013; Lee & Jessell, 1999). The receptors to FGF represent another family of RTK receptors (see Figure 4.3). BMPs and FGFs are just two of many identified factors (there are probably others that haven't been identified yet) that can induce neural cells to become either neurons or glia. The picture we see is that cells are exposed to a blend of many factors secreted by various neighboring cells, and whether the cell becomes a neuron or glia depends on the balance of these many different forces.

The cells that are thus repressed from expressing Delta, and therefore the proneural genes, will differentiate into glia. While we don't know as much about the signals that direct these cells into the various types of glia, we do know that those destined to become oligodendrocytes will start expressing two related oligodendrocyte transcription factors, Olig1 and Olig2 (**FIGURE 4.6B**). Those presumptive glia that begin expressing a gene called *neuregulin* are more likely to differentiate into astrocytes (see Figure 4.6B). Any factors that boost proneural gene expression in these cells will push them away from the pathway to glia.

So you can see how this is a stochastic process, where a particular pattern of outcomes is reliably seen but you cannot predict precisely what will happen to any particular individual cell. Thus, some cells express slightly more proneural genes at the start and so become neurons, while most of their neighbors become glia. This sorting out of cells is going on throughout the neural tube, in many different local arenas of competition, where cells are using lateral inhibition with their immediate neighbors to determine which will take on a neuronal fate.

Next let's consider a case where cells are directed to a particular neuronal fate well before they have a chance to innervate their targets, specifically motor neurons.

Isl-1, Isl-2 Homeotic genes that encode the transcription factors Isl-1 and Isl-2, which are expressed in developing motor neurons.

Lim domains Domains found in various transcription factors that form zinc fingers to bind DNA in the process of gene regulation.

zinc fingers Relatively small stretches of amino acids found in various transcription factors that bind to DNA in the process of gene regulation.

◼ 4.3 ◼
THE MOLECULAR DIFFERENTIATION OF MOTOR NEURONS IS ORDERLY

Recall from Chapter 2 that both FGF and retinoic acid induce a more posterior fate to the vertebrate neural tube. Next, Shh from the notochord beneath the neural tube induces the ventral region to form the floor plate, where cells soon begin making their own Shh. Soon after that the notochord is displaced ventrally, so next it will be Shh from the floor plate itself that induces cells in the adjacent neural tube on both sides to start differentiating into motor neurons and spinal interneurons. One action of Shh that promotes this neuronal differentiation is the repression of homeotic genes *Pax* (specifically *Pax3* and *Pax7*) and *Msx* (specifically *Msx1* and *Msx2*) (Goulding, Lumsden, & Gruss, 1993; Liem, Tremml, Roelink, & Jessell, 1995). Some of these cells will go on to become motor neurons while others will become spinal interneurons. It's not clear what determines which of these two fates a given cell will take on, but small changes in Shh concentration can make a big difference in cell differentiation in vitro (Tanabe & Jessell, 1996), so it may be tiny regional differences in Shh or it may be some competitive process between cells that determines which will be motor neurons and which interneurons.

Small changes in Shh concentration can make a big difference in cell differentiation in vitro.

At this point, those cells that will become motor neurons begin expressing a homeotic gene called **Isl-1** (for insulin gene enhancer, because it was first identified in relation to insulin function, another example of pleiotropy in gene function). The Isl-1 protein contains two **Lim domains** that form so-called **zinc fingers** to bind DNA. This clues you in that this protein is a transcription factor, and in the case of motor neurons, Isl-1 turns on expression of related gene **Isl-2** (Tsuchida et al., 1994). Thus the developing spinal cord has cells expressing a variety of different transcription factors, effectively sorting themselves out into different future fates (**FIGURE 4.7**). Once the developing motor neurons start sending axons out the ventral roots, they can take on one of four different fates. Some will become part of the medial motor column (MMC) of the ventral spinal cord and will innervate axial muscles (i.e., those muscles attached to the main body trunk), while others will become part of the lateral motor column (LMC) innervating limb muscles. Within the LMC there are two more divisions: the medial half of the LMC tends to innervate ventral limb muscles, while the lateral half tends to innervate dorsal limb muscles. Finally, some motor neurons will become preganglionic sympathetic cells that will innervate neurons in the sympathetic ganglia

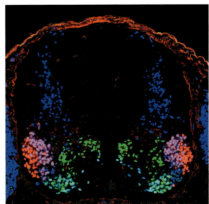

FIGURE 4.7 Early differentiation of motor neurons The various colors in the bottom half of the image mark the different Hox genes being expressed by the motor neurons there. (Courtesy of Dr. J. S. Dasen.)

FIGURE 4.8 **Sequential expression of genes in different motor neuron pools** Note that ventral muscles tend to express one signaling molecule, ephrin, while dorsal muscles express another, semaphorin (Sema). (After Tanabe & Jessell, 1996.)

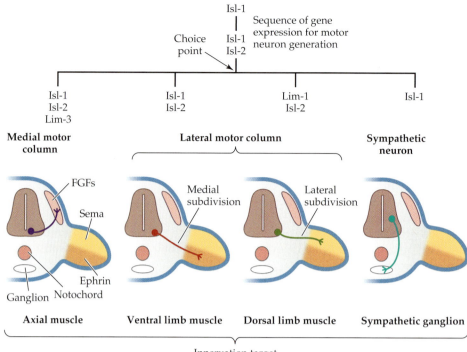

(recently formed by the arrival of neural crest cells) that lie along either side of the spinal cord. While all four of these motor neuron populations began by expressing Isl-1 followed by Isl-2, at this point each population expresses a unique combination of transcription factors (**FIGURE 4.8**) (Miri, Azim, & Jessell, 2013).

The particular expression pattern of these genes interacts with factors signaling the motor neurons' positions in the anterior-posterior axis (FGF and retinoic acid among them), so the four types of motor neurons at different levels of the spinal cord begin expressing particular combinations of Hox genes (Philippidou & Dasen, 2013) (**FIGURE 4.9**). These Hox genes in turn direct each motor neuron to express the combination of genes required for it to take on its unique role.

Having mentioned the neural crest cells that form the sympathetic ganglia, let's turn now to how neural crest cell differentiation is controlled.

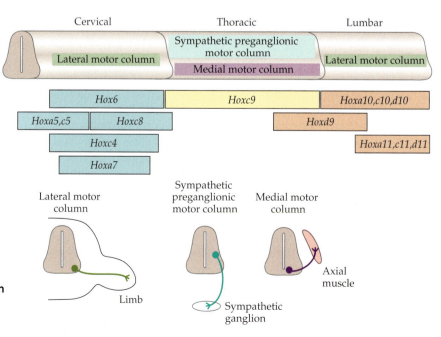

FIGURE 4.9 **Hox gene expression patterns in the spinal cord** (After Philippidou & Dasen, 2013.)

┌───┐
│ **How's It Going?** │
│ │
│ 1. How does Delta-Notch–like signaling result in some cells │
│   becoming neurons while others become glia? │
│ │
│ 2. What factor directs some cells to become motor neurons, and │
│   what genes will those cells express to follow that path? │
│ │
│ 3. What are the four main types of motor neurons that arise, and │
│   what is the evidence that their patterns of gene expression differ │
│   from one another? │
└───┘

■ **4.4** ■

NEURAL CREST CELLS ARE AFFECTED BY THEIR MIGRATION AND DESTINATION

Recall from Chapter 3 that neural crest cells migrate long distances to take on a wide variety of cell fates, and that crest cells originating from different regions of the rostral-caudal axis take up different subsets of those fates. Are the fates of the crest cells settled before they start migrating, or does the process of migration and settling down direct their differentiation? That question can be most readily addressed by considering neural crest contributions to the two divisions of the autonomic nervous system.

> Are the fates of the crest cells settled before they start migrating, or does the process of migration and settling down direct their differentiation?

The neural crest cells that contribute to the autonomic nervous system take on two rather different fates that we can readily detect. Some will become the postganglionic neurons of the sympathetic nervous system, forming those sympathetic ganglia that will receive innervation from a subset of spinal motor neurons, and be noradrenergic (meaning they will release **norepinephrine [NE]** as a neurotransmitter). Other neural crest cells will become postganglionic neurons of the parasympathetic nervous system and therefore be cholinergic (meaning they will release **acetylcholine [ACh]**). In fact, all of them will initially be cholinergic, but the sympathetic cells will switch from using ACh to NE (then, as we'll see later in this chapter, a select few will be induced by their synaptic targets to switch back to a cholinergic fate). The crest cells with these two different fates are anatomically segregated (Le Douarin, Renaud, Teillet, & Le Douarin, 1975). Those that will become parasympathetic cholinergic neurons arise from crest at the rostral end near the brain (segments 1–7 in chicks) and from the far caudal end along sacral spinal cord. Those that will become noradrenergic sympathetic neurons arise from neural crest along the middle spinal segments, numbered 7–28 (see Figure 3.20). Now we can ask, How do the crest cells know which neurotransmitter to use? More formally, what directs differentiation of the cells to either a cholinergic or noradrenergic fate?

We can readily think of two possibilities. Maybe some process at work before the cells migrate away from the neural tube designates them as thoracic or sacral, and thereby instructs them where to go and, once there, which neurotransmitter to make. Alternatively, maybe when the crest cells leave the neural tube, they are still undifferentiated, capable of becoming either cholinergic or adrenergic, but are induced to take on the appropriate fate once they reach the periphery.

A famous experiment indicates that this latter hypothesis is correct. Nicole Le Douarin and colleagues transplanted tissue between the embryos of chicks and quail, relying on the distinctive microscopic appearance of the cells (Le Douarin, 1973) to note which cells were from the graft and which from the host (the same method used to find the midbrain/hindbrain organizer, as described in Chapter 2). The grafted donor tissue was accepted, and through the process of embryonic self-regulation (see Chapter 1), it melded with host cells. Because

norepinephrine (NE)   A catecholaminergic neurotransmitter released by most sympathetic fibers to activate the "fight or flight" response in various organs.

acetylcholine (ACh)   The cholinergic neurotransmitter released by parasympathetic fibers as well as vertebrate motor neurons.

chimera Here, an individual made up of cells displaying more than one genotype, formed from the combination of cells from two separate zygotes. Typically, such an individual has received genes from more than two parents.

the chicks used were from the white leghorn strain, while melanocytes (pigment cells) from quail crest produce brown feathers, the resulting individuals provided striking evidence that the cross-transplants worked. They displayed a patchwork of brown feathers and white, depending on which rostral-caudal length of crest came from which species. Individuals such as these that are derived from cells originating from two separate zygotes are called **chimeras**, after the monstrous combined animals of Greek myth. Le Douarin and colleagues used these chimeras to test hypotheses about neural crest fate, as we show next.

RESEARCHERS AT WORK

Neural Crest Cells Adopt New Fates after Transplantation

■ **QUESTION**: What factors direct neural crest cells to become cholinergic or noradrenergic?

■ **HYPOTHESIS**: The crest cells are instructed before migration about where they should go and what neurotransmitter they should use upon arrival.

■ **TEST**: Transplant embryonic neural crest (along with the neural tube) from the head, which normally becomes cholinergic cells, to the thoracic region, where crest cells normally become noradrenergic (**FIGURE 4.10A**). In other animals, transplant crest from the thoracic region (which normally becomes noradrenergic) to the head region (where cholinergic crest cells normally originate). Conduct cross-transplants of neural tube and associated neural crest between chickens and quail (**FIGURE 4.10B**). If the crest cells know where to go and what phenotype to adopt before they migrate, then cells from the head, if transplanted to the thoracic region, should take a new migratory pathway to join the parasympathetic nervous system and use ACh.

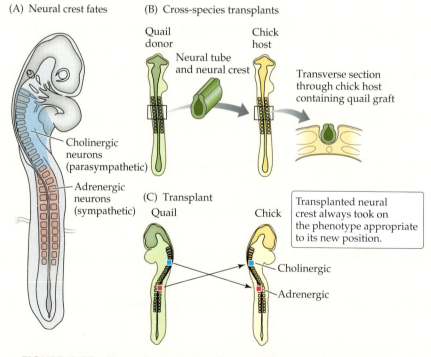

FIGURE 4.10 Transplanted neural crest cells take on the fates appropriate to their new positions. (After Le Deouarin, 1980.)

- **RESULT**: In fact, neural crest cells that were moved from one region to another migrated to the destination appropriate to their *new* position, and they took on the phenotype appropriate for their new destination. Crest from the head, which would normally become cholinergic parasympathetic neurons, became noradrenergic when transplanted more caudally in the thoracic region. Conversely, thoracic crest cells, when transplanted to the head region, became cholinergic cells. In other words, the cells changed their fates to coincide with their new places in the rostral-caudal axis (Le Dourain, 1980) (**FIGURE 4.10C**).

- **CONCLUSION**: The neural crest cells are not specified before they begin their migration to reach a particular goal and take on a particular fate. Rather, if they are moved from one part of the rostral-caudal axis to another, they will take on the migratory route and the neural fate that is appropriate to their new position. By elimination, they must be guided along a migratory route by local cues along the pathway (see Chapter 3), and they must be induced by factors encountered along the way or at their final destination to use either norepinephrine or acetylcholine as a neurotransmitter.

An alternative explanation for these results arose from a third hypothesis that relied on the fact that a normal part of development is the widespread death of cells (the topic of Chapter 7). Maybe all neural crest, from head to tail, has populations of *both* types of cells: those fated for a noradrenergic fate *and* those fated for a cholinergic fate. If so, then maybe in the normal case one population dies off in some regions of the rostral-caudal axis, and the other population dies in other regions. In that case, then the transplants simply switched which population would die. This was a difficult possibility to rule out from transplants alone, but happily a series of in vitro studies made it clear that neural crest cells do indeed follow local cues to follow particular pathways and that their fate is determined by the combination of factors they encounter en route and at their final destination (Takahashi, Sipp, & Enomoto, 2013).

Some of these neural crest cells differentiate into **Schwann cells**, which provide myelination for axons in the peripheral nervous system. This decision to take on a Schwann cell fate is induced by contact with axons (Takahashi et al., 2013). The axons express a signaling molecule, *neuregulin*, which binds ErbB receptors (part of the RTK family; see Figure 4.3) on Schwann cells to induce them to myelinate, as we'll discuss in more detail in Chapter 6.

Sometimes when injury causes a withdrawal of distal nerves, the abandoned Schwann cells may dedifferentiate (Adameyko et al., 2009), and may then resume migrating. Such dedifferentiation of Schwann cells is a critical link in Hansen's disease, which we learned about in the opening of this chapter. *Mycobacterium leprae* initially infects Schwann cells, where it replicates for years until the immune system recognizes the invader and initiates acute inflammation. That swelling of the nerve within its sheath damages the axon (Britton, 1998), which retracts, so Schwann cells lose contact with the axon. Freed from the inductive signal that kept them Schwann cells, the cells dedifferentiate to their former migrating stage. The infected cells then migrate to muscles and to skin, where they are induced to become melanocytes (Masaki et al., 2013). The infected melanocytes trigger reactions in the skin that produce *granulomas*, the inflammatory swellings that mark the disease (**FIGURE 4.11**).

In fact, Hansen's disease is not very contagious, because 95% of people are immune (Britton & Lockwood, 2004), and it is spread, not by skin contact, but by airborne droplets of saliva from coughing or sneezing (which may be

Schwann cells Glial cells that provide myelin sheaths for axons in the peripheral nervous system and cap neuromuscular junctions.

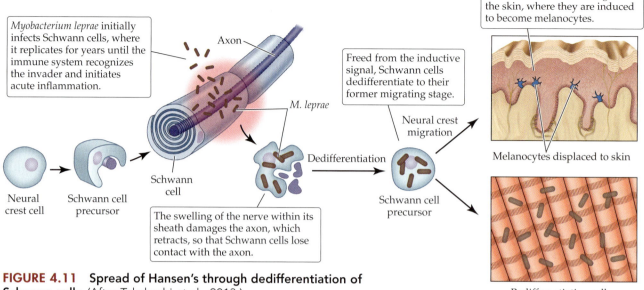

FIGURE 4.11 Spread of Hansen's through dedifferentiation of Schwann cells (After Takahashi et al., 2013.)

In fact, Hansen's disease is not very contagious, because 95% of people are immune.

why the Bible says sufferers should put a covering upon their upper lip). Father Damien was among the unlucky 5% who are susceptible, because he contracted Hansen's disease and died of it on Molokai 16 years after his arrival. A few months before his death, an American nun, Mother Marianne Cope, came to the colony to continue his work. After tending to Damien at his deathbed, she spent 35 years caring for other Hansen's patients without ever getting the disease, dying of natural causes at 80. Eventually the Catholic Church declared both Father Damien and Mother Marianne saints.

Today Hansen's disease is curable with antibiotics, but its symptoms are so variable that the main problem is diagnosing and treating it before much damage has been done. Clinical trials are under way for a vaccine (Boer, Joosten, & Ottenhoff, 2015; Richardus & Oskam, 2015).

Even after neural crest cells have taken on a neuronal fate, they are still subject to inductive reactions to other cells they encounter, including, as we'll see next, the targets they innervate.

■ **4.5** ■

THE NEUROTRANSMITTER PHENOTYPES OF AUTONOMIC NEURONS ARE GUIDED BY THEIR TARGETS

The two parts of the autonomic nervous system, the sympathetic and parasympathetic branches, typically exert opposing effects on a given target organ. For example, the sympathetic fibers release NE to accelerate heart rate while parasympathetic fibers release ACh to slow it down. The neural crest transplant studies we reviewed above tell us that whether the crest cells become sympathetic or parasympathetic is not predetermined by where they arise in the rostral-caudal extent of the spinal cord, but determined by exposure to various factors along their migratory pathways and final resting places.

But, in one of those twists of natural selection that seem designed to vex students of neuroscience trying to memorize simple rules, there are a few exceptions to that rule that sympathetic fibers are noradrenergic. For example, **sweat glands**

sweat glands Specialized structures that release sweat onto the skin surface to reduce body temperature. They are activated by the release of acetylcholine from sympathetic fibers.

are stimulated by axons from the sympathetic nervous system (as you'd expect since the sympathetic nervous system is active in "fight or flight" situations), but they act by releasing ACh, not NE. Interestingly, early in development all postganglionic neurons in the autonomic nervous system manufacture and release ACh, then those from the sympathetic trunks switch transmitters from ACh to NE.

There are a few exceptions to the rule that sympathetic fibers are noradrenergic.

Oddly enough, the sympathetic fibers growing out to innervate sweat glands go through the transition from making ACh to NE, as the other sympathetic fibers do, but after innervating the sweat glands, these particular sympathetic fibers revert to making ACh rather than NE (Guidry & Landis, 1998). It seems unlikely that this change in phenotype is due to their migration and final resting place, because most of their neighbors in the sympathetic ganglia, which innervate other targets, remain noradrenergic. So, are these particular sympathetic neurons preprogrammed to go back to a cholinergic fate at a particular stage of development, just in time to provide ACh to activate sweat glands, or are they being cued by some other cell to switch transmitters? It turns out that the sweat glands themselves direct sympathetic fibers to revert to producing ACh, as we'll see next.

RESEARCHERS AT WORK

Targets Can Regulate the Neurotransmitter Phenotype of Afferents

- **QUESTION**: What directs sympathetic neurons innervating sweat glands to switch neurotransmitters from NE to ACh?

- **HYPOTHESIS**: Perhaps the sweat glands themselves instruct the fibers to switch phenotypes, and only this minority of sympathetic neurons become cholinergic.

- **TEST**: Transplant target tissues, providing sweat glands to sympathetic fibers that would normally remain noradrenergic, to determine whether they change from noradrenergic to cholinergic. Conversely, determine whether sympathetic fibers that would normally provide cholinergic innervation to sweat glands go back to releasing NE if they are directed to innervate a different target. In rats, which have sweat glands only in the hairless pads of their feet, this means transplanting a foot pad with sweat glands onto the back, and transplanting skin from the back onto a foot pad. Transplant the skin in newborn rats from the Lewis strain of inbred rats (that are nearly identical genetically, so you don't worry about tissue rejection), allow them to grow up, and then determine whether the sympathetic fibers innervating the transplanted "islands" of skin are cholinergic or noradrenergic.

- **RESULT**: Those sympathetic fibers that innervated foot pad tissue transplanted onto the back reverted from the noradrenergic fate they would have followed innervating the back to providing, instead, cholinergic neurotransmitters appropriate for sweat glands. In the converse situation, normally cholinergic sympathetic fibers, upon innervating skin from the back transplanted to the foot pad, changed to a noradrenergic fate (Schotzinger & Landis, 1988) (**FIGURE 4.12**).

continued

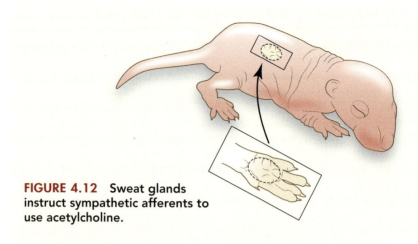

FIGURE 4.12 Sweat glands instruct sympathetic afferents to use acetylcholine.

■ **CONCLUSION**: The small group of sympathetic fibers that end up innervating sweat glands are instructed by their target to change neurotransmitters from NE back to ACh. Follow-up experiments working in culture later implicated several factors coming from the sweat glands, including some proteins that are similar to leukemia inhibitory factor (LIF) and ciliary neurotrophic factor (CNTF), that direct sympathetic fibers to change transmitters (Landis, 1996).

Interestingly, this developmental mechanism of regulating neurotransmitters means that if natural selection alters the distribution of sweat glands in the skin, sympathetic fibers will adjust, altering neurotransmitter phenotype to provide the ACh stimulation those glands need to function. A similar process of target-regulated switching from noradrenergic to cholinergic transmission occurs in sympathetic neurons that will innervate connective tissue surrounding bone (Asmus, Parsons, & Landis, 2000). So target regulation of neurotransmitter phenotype may be a general property of sympathetic development.

Now let's consider the enormous question of how cells in the cerebral cortex are directed to differentiate to a particular fate.

■ **4.6** ■

THE FATE OF A CORTICAL NEURON IS INFLUENCED BOTH BEFORE AND AFTER MIGRATION

In Chapter 3 we learned about the "inside-out" development of the mammalian neocortex, with the innermost layer VI cells born before those of layer V, which precede those in layer IV, and so on (see Figure 3.12). Because of this orderly pattern of neuronal addition to the cortex, knowing the birthdate of a neuron will allow you to predict with reasonable accuracy what layer it will reside in. Furthermore, neurons in different cortical layers take on different characteristics, so you can also use birthdates to reasonably predict the particular fates of cortical neurons. For example, large pyramidal neurons tend to be found in layers III and V, while layers II and IV tend to have smaller neurons (see Figure 3.13). So knowing a cortical neuron's birthdate offers some prediction about its differentiation, which is simply a reflection of what genes the cell is expressing and in what order.

But that still doesn't tell us what mechanisms caused the cell to express those particular genes in that particular order to take on that particular structure and function. Is it birthdate alone that controls gene expression, such that mitosis on this particular day stimulates expression of particular genes and inhibits the expression of others? If so, how would the cell know how far development had progressed when it was born? Does the cell itself keep track? Or do its neighbors in the ventricular and subventricular zones inform the postmitotic cell: today's the 63rd day of gestation, so go to layer V and become a big pyramidal cell; or, it's day 90, so go to layer II and become a tiny granule neuron?

Conversely, maybe when the postmitotic neuronal precursor leaves the subventricular zone, it has received no instruction at all but simply shinnies up a radial glia, passes the layers that have already been settled, then gets off at the end of the line (or, just before the marginal layer). Depending on which layer the cell ends up in, cell-cell interactions with its new neighbors would control gene expression, directing the cell to take up a fate appropriate to this layer. An experiment testing these hypotheses found that both are correct, as we see next.

> Knowing the birthdate of a neuron will allow you to predict with reasonable accuracy what layer it will reside in.

RESEARCHERS AT WORK

Cortical Neuron Fate Is Specified after the S Phase

■ **QUESTION**: In ferrets, as in other mammals that have been examined, cortical cells settle in an inside-out manner, with the cells generated last settling in the outer layer II. What directs an individual cell to settle in a particular layer?

■ **HYPOTHESES**: Cells migrating along radial glia may simply settle in the next "available" layer (i.e., the outermost layer). Alternatively, they may target a particular layer that's appropriate to their birthdate.

■ **TEST**: Cross-transplant ventricular zone cells from a young ferret cortex, which would normally settle in an inner layer (VI) (**FIGURE 4.13A**), into an older brain, where dividing cells are settling in an outer layer (II) (**FIGURE 4.13B**). Prelabel the younger cells with thymidine to birthdate them and also so they can be distinguished from the host cells. Let the labeled cells migrate, then see whether they settle in layer VI, as they would have done if left in the donor (and therefore would be appropriate for their birthdate in the donor), or bypass layer VI to settle in layer II, as their neighboring host cells do.

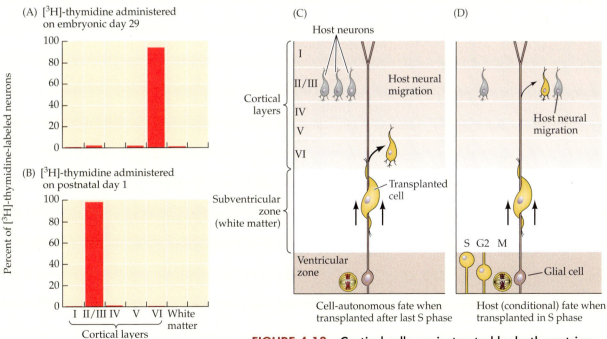

FIGURE 4.13 **Cortical cells are instructed by both ventricular neighbors and their migratory route.** (After McConnell & Kaznowski, 1991.)

continued

■ **RESULT**: In fact, the transplanted cells tended to settle in *either* layer VI or layer II, not layers in between. In other words, some went to the layer appropriate to their birthdate back in the donor, while others went to the next available layer at the end of the radial glial guide. It turned out that cells that were transplanted when they were in the S (synthesis) phase of mitosis (duplicating the chromosomes preparatory to division) followed the fate of their older host neighbors, settling in outer layer II (**FIGURE 4.13C**). But cells that were transplanted *after* the S phase seemed to have received some instruction from their ventricular neighbors back in the donor about where to go, because they got off the radial glial cell early to settle in the inner layer VI, just as they would have done if they had been left in their donor's brain (McConnell & Kaznowski, 1991) (**FIGURE 4.13D**).

■ **CONCLUSION**: While ventricular zone cells are dividing, they are pluripotent, able to settle wherever the end of the radial glial line takes them. On the other hand, once they have become postmitotic, they seem to pick up some signal from the ventricular zone that instructs them about which cortical layer they should settle into. For all these cells, once they settle into a particular cortical layer, they take on a fate that is appropriate for that layer, presumably under the influence of their neighbors.

In some ways, this outcome is in keeping with a persistent theme we've seen in development. The older the embryos are, the less they are able to self-regulate, overcoming the loss of cells, for instance. And, in general, the younger an embryo is, the more pluripotent the cells are, such that at the early stages of development, the cells are totipotent. With the above experiments, the younger the ventricular cells are, the more pluripotent they are. Thus cells that are still in S phase, that have not yet divided, can take on whatever fate will work in their environment. But once they have become postmitotic, they appear less flexible and will settle into a deeper layer, even if that layer is already well populated. Note that another way to describe this same concept is to say that younger cells respond to many environmental cues from other cells, but older cells become more specified. Just as younger people tend to be more physically and mentally flexible than older people, younger cells tend to be more plastic than older cells. This may be why young brains are more plastic, better able to learn new material and to compensate for brain damage, than older brains.

> *Younger cells respond to many environmental cues from other cells, but older cells become more specified.*

By now you probably have a sense for how differentiation works, how each cell is influenced—by other cells, mostly nearby neighbors that make physical contact as in the Delta-Notch system; by diffusion, as in the spinal motor neurons and neural crest migrators; or by synaptic contact, as in sympathetic innervation of sweat glands. The final common pathway by which these various influences direct differentiation (and therefore fate) is expression of transcription factors, which in turn affect expression of many other genes, including other transcription factors. Unfortunately for us as students of development, it can be difficult to comprehend how these various transcription factors, each affecting the expression of the others, actually result in a particular fate. **FIGURE 4.14**, for example, illustrates one particular network of gene regulation that has been worked out for mouse embryonic stem cells (Zhou, Chipperfield, Melton, & Wong, 2007). If nothing else, this network demonstrates that natural selection does not necessarily land upon the simplest solution to a problem, but rather favors whatever solution works to help the individual reproduce. I show you this figure, not to discourage you from studying gene regulation, but to give you a sense of how complicated these interactions can be, and how important it is to try to isolate one particular part of the

> *Natural selection does not necessarily land upon the simplest solution to a problem, but rather favors whatever solution works to help the individual reproduce.*

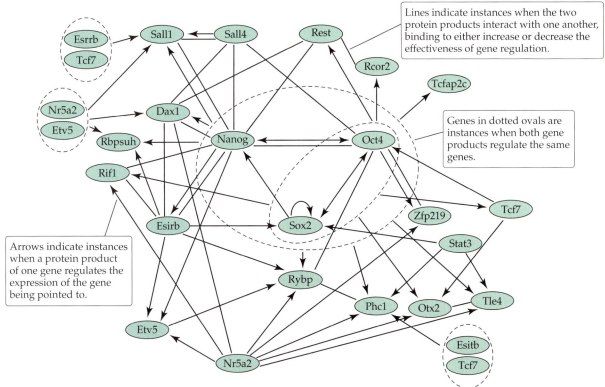

FIGURE 4.14 Gene regulation is not simple. This is just an example of interactions that have been identified in a set of genes coding for transcription factors. In this case, the genes are those known to be regulated in mouse embryonic stem cells. It is not thought to be a complete snapshot of gene regulation in this cell, nor is it thought to be a particularly complex network. (After Zhou et al., 2007).

pathway to try to understand differentiation. Essentially we have done that in this chapter by carefully choosing well-understood models.

There is one important sense in which the evolution of the human brain has exploited the increased plasticity of the younger brain. As we'll see next, we all seem to be having a long, extended childhood, which has made all the difference.

How's It Going?

1. What is the evidence that neural crest cells follow local cues to migrate and that they are induced to a specific fate by taking that pathway?

2. What is the role of neural crest cells in the progression of Hansen's disease?

3. How do synaptic targets sometimes shift the fate of innervating sympathetic neurons?

4. What factors specify the fates of cells in various layers of the cerebral cortex?

■ 4.7 ■

LATER EVENTS IN DEVELOPMENT ARE MORE EVOLUTIONARILY LABILE

In Chapter 2 we noted that phylogenetically related species may greatly resemble each other in embryonic development, such that all vertebrate species look much alike as very early embryos. Echoing a theme in Darwin's *On the Origin of Species*, we noted that if a feature is adaptive only for the adult,

(A)

FIGURE 4.15   A tight squeeze

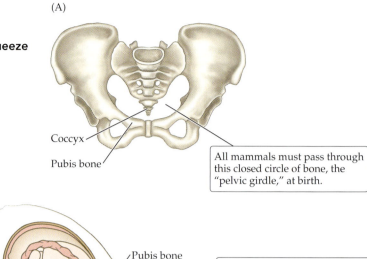

Coccyx

Pubis bone

All mammals must pass through
this closed circle of bone, the
"pelvic girdle," at birth.

(B)

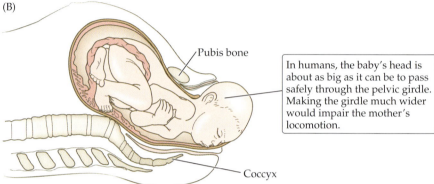

Pubis bone

In humans, the baby's head is
about as big as it can be to pass
safely through the pelvic girdle.
Making the girdle much wider
would impair the mother's
locomotion.

Coccyx

then there is no particular reason why natural selection would have to resort
to changes early in development to gain that advantage. We also speculated
that, in fact, natural selection may tend to result primarily in changes in later
development, not only because a later development may be sufficient to provide an adaptive advantage, but also because changes earlier in development
may have further ramifications that prove lethal.

One prominent example of changes late in ontogeny regards the tremendous
growth of the human brain after birth. Because newborn mammals must pass
through the mother's pelvic girdle at birth (**FIGURE 4.15A**), and because the
head is the single largest structure that makes this passage, there is a limit on
how large the brain can be at birth (**FIGURE 4.15B**). Women have wider hips,
and therefore larger pelvic girdles, than men, probably to accommodate larger
newborn heads, but making the hips any wider might impair locomotion. The
bones of the newborn's skull are not yet rigidly affixed to one another, allowing
some deformation of the head to make the passage, but any further squeezing
would risk brain damage. There are times when the newborn's head cannot
readily pass through the mother's pelvic girdle, which today can be remedied
by cesarean section but in past days resulted in injury or death of the child,
the mother, or both. Thus the newborn human head is probably as large as it
can possibly be without making reproduction an even riskier business than it
already is. We simply cannot be safely born with bigger heads.

Many writers have noted that evolution seems to have circumvented this
limit on brain size by having the incredibly rapid pace of fetal
development simply prolonged so that it continues after birth
(Montagu, 1989). Indeed, human brain weight, relative to body
weight, continues growing at a feverish pace after birth, unlike
brain weight in other great apes (**FIGURE 4.16A**) (Bogin, 1998). This
continued postnatal growth of the brain seems to account for the
larger brain in humans versus other primates (**FIGURE 4.16B**). It is
as if we continue being fetuses after we are born. Thus we have a
period of intrauterine fetal development followed by a period of

Human brain weight,
relative to body
weight, continues
growing at a feverish
pace after birth, unlike
brain weight in other
great apes.

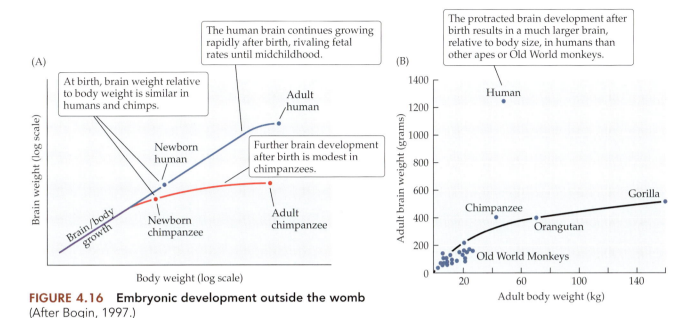

(A)

At birth, brain weight relative to body weight is similar in humans and chimps.

The human brain continues growing rapidly after birth, rivaling fetal rates until midchildhood.

Adult human

Newborn human

Further brain development after birth is modest in chimpanzees.

Brain weight (log scale)

Brain/body growth

Newborn chimpanzee

Adult chimpanzee

Body weight (log scale)

(B)

The protracted brain development after birth results in a much larger brain, relative to body size, in humans than other apes or Old World monkeys.

Adult brain weight (grams)

Human

Chimpanzee

Gorilla

Orangutan

Old World Monkeys

Adult body weight (kg)

FIGURE 4.16 **Embryonic development outside the womb** (After Bogin, 1997.)

*extra*uterine fetal development. One consequence of this circumstance is that while all other mammals undergo fetal development in the dark, relatively quiet, and isolated world of the uterus, we undergo much of our "fetal" development out in the world, exposed to sights, sounds, smells, and so on. Interestingly, the majority of our additional synapses are acquired after birth (note the much greater increase in cortical dendritic complexity, and therefore additional synapses, acquired after birth in Figure 8.23). In the final two chapters of this book, we'll talk about evidence that sensory experiences and social experiences play a crucial role in this extrauterine phase of fetal brain development, which may be responsible for the unique complexity of the human brain and therefore the unique ability of humans to alter our environment (for better or worse).

There is also a popular idea that not only are we born with fetal brains, but in fact the rest of the human body undergoes "arrested development" resulting in mostly hairless bodies, large foreheads, and short jaws that rather resemble those of juvenile chimpanzees (**FIGURE 4.17**). Stephen Jay Gould

FIGURE 4.17 **Did humans arise through neoteny?** Note that a newborn chimp has a larger forehead and larger eyes than adults but has a much shorter jaw and less hair. In this way, the young chimp resembles adult humans much more than adult chimps do. Are adult humans simply juvenile chimps that are sexually mature? (From Naef, 1926.)

FIGURE 4.18 **Later becomes larger** Natural selection seems to be enlarging brain size primarily by prolonging the later stages of development. The dashed line represents the border of the underlying white matter. (From Hill & Walsh, 2005.)

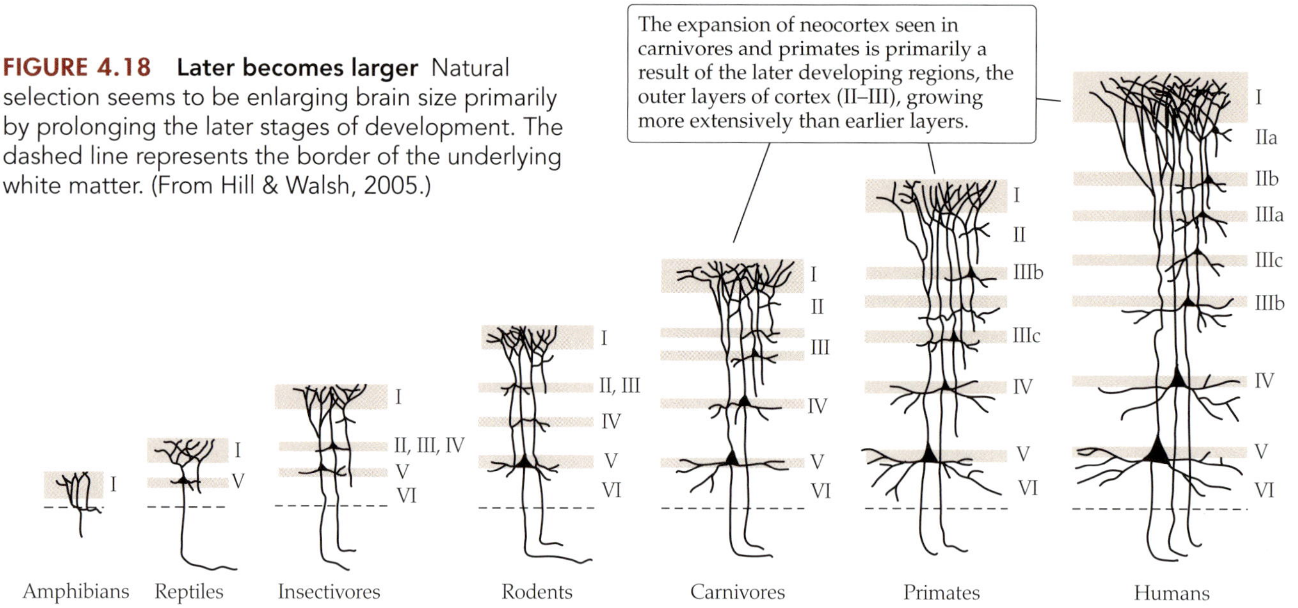

The expansion of neocortex seen in carnivores and primates is primarily a result of the later developing regions, the outer layers of cortex (II–III), growing more extensively than earlier layers.

Amphibians   Reptiles   Insectivores   Rodents   Carnivores   Primates   Humans

neoteny An instance when a descendant species halts development at what was a juvenile stage of its ancestral species and becomes sexually mature at that stage.

proposed the term **neoteny** to denote instances when the juvenile stages of an ancestor become the adult stage of a descendant (Gould, 1977b). It is as though our ancestors halted the developmental process for the entire body, including keeping the brain juvenile and plastic into adulthood, while allowing the sexual organs to fully mature. In this light, Gould also likened humans to "permanent children" who are, obviously, capable of reproducing. Perhaps we are all Peter Pan, and a childlike ability to wonder may be a core attribute of the human condition.

This propensity to grow larger by expanding later stages of development may explain a characteristic that has been noted repeatedly in the comparison of the cerebral cortex across vertebrate species: specifically, as you compare species that may be considered progressively more intelligent—say from rodents to carnivores to nonhuman primates to humans—the greatest proportional changes seem to be in the outermost cortical layers. We have seen that these outer layers are the last to be added to the cortex, so natural selection seems to have expanded primate brains primarily by extending the last stages of cortical development, resulting in thicker layers II, III, and IV, rather than V and VI (**FIGURE 4.18**). This trend has sometimes been summarized as "later becomes larger," meaning larger brains evolved by having later-added structures enlarge and thereby expanding the later-arising structures relatively more than early-arising structures (such as cerebral cortex versus brainstem). We can roughly gauge how late a brain structure arises by simply asking when neurogenesis is complete for each structure (Finlay, Darlington, & Nicastro, 2001), or considering when axons start growing out, or when synapses or myelin sheaths form (Workman, Charvet, Clancy, Darlington, & Finlay, 2013). In this way, later-developing brain regions expand more as natural selection increases total brain power.

As you compare species that may be considered progressively more intelligent the greatest proportional changes seem to be in the outermost cortical layers.

In fact, if you examine the various cortices in Figure 4.18, you'll see that the layer designated III in rodents must be subdivided into anatomically distinct layers IIIa, IIIb, and IIIc in humans, while layer II is also subdivided, in this case into two distinct layers. In turn, the big difference between carnivores and the living species thought to come closest to resembling the common ancestor of the mammals, the insectivores, is that a single cortical layer in insectivores is subdivided into layers II, III, and IV in carnivores.

Squirrel monkey
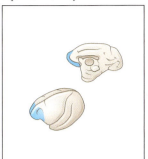

Cat
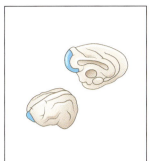

Rhesus monkey
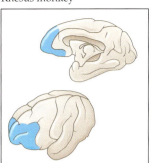

Chimpanzee
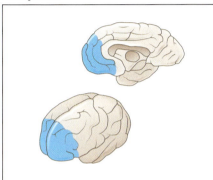

Human
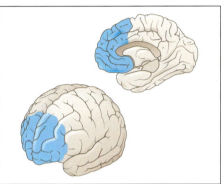

FIGURE 4.19 **The expansion of prefrontal cortex in humans** The last region of cortex to grow and mature is also the region that expanded most in humans compared with primates and other mammals.

The "later becomes larger" principle also seems to apply to whole brain regions, since the greatest difference in brain volume between us and the other great apes, and therefore presumably between us and our common ancestor with the great apes, is the expansion of the prefrontal cortex (**FIGURE 4.19**), which, as we'll see in Chapter 8 (Figure 8.25), is the last to fully mature in terms of the thinning of the gray matter.

> The idea that greater intelligence is always a good thing is itself a value judgment.

By the way, if you are used to thinking that increased intelligence is always an advantage, you may be wondering why rats, for example, haven't evolved to have larger brains. The idea that greater intelligence is always a good thing is itself a value judgment, and natural selection does not favor development of "best" species, merely species that successfully reproduce. Presumably the reason rats have not evolved bigger brains is because, given the ecological niche they occupy, the brains they have are sufficient to maximize their chances of reproducing new rats to take their place in their niche. Surely some slightly brainier rats have arisen now and then over the millennia, but perhaps any improvement in occupying their ecological niche wasn't enough to compensate for the additional energy needed to make that bigger brain. In other words, perhaps brainier rats do not, in fact, produce more offspring than rats with brains of a species-typical size. Likewise, while many have speculated that one day humans will have enormously expanded brains, maybe we're already as smart as we need to be to occupy our niche. In fact, maybe we're already too smart for our own good.

How's It Going?

1. Why would protracted fetal development have to occur postnatally in humans?
2. Describe the idea of "later becomes larger" in evolution of the brain, and give some examples that seem to fit that notion.

SUMMARY

■ Cells release *hedgehog* (*hh*) to trigger and propagate the *morphogenetic furrow* across the developing fruit fly eye. Within each *ommatidium*, one cell outcompetes others to become photoreceptor 8 (R8), which then directs differentiation of the other photoreceptors, culminating in the expression of *bride of sevenless* (*boss*) to activate *sevenless*, a *receptor tyrosine kinase* (*RTK*) in photoreceptor R7. **See Figures 4.2–4.4**

■ In the developing vertebrate neural tube, cell-cell interactions include the release of BMP that pushes cells to become glia, the release of *FGFs* that push cells to become neurons, and Delta-Notch signaling between cells in contact with one another. These result in a mix of neurons and glia. **See Figures 4.5 and 4.6**

■ Spinal motor neurons are first induced by exposure to moderate levels of Sonic hedgehog (Shh) from the notochord and floor plate. All the cells that will become motor neurons express homeotic genes *Isl-1* and *Isl-2* during early differentiation, then express different combinations of transcription factors to differentiate into different subpopulations innervating various targets. They also express different Hox genes at anterior-posterior levels. **See Figures 4.7–4.9**

■ Neural crest cells migrate to many different parts of the body to take on different roles. Their differentiation is guided by exposure to factors from other cells along their routes, at their destinations, and from their innervation targets. Crest cells transplanted from one segment of the anterior-posterior axis to another take on the fate appropriate to their new position. **See Figure 4.10**

■ If crest cells encounter axons in the periphery, that contact induces them to differentiate into *Schwann cells* providing myelin. In *Hansen's disease*, infected Schwann cells dedifferentiate to revert to migratory behavior, thereby spreading infection to the skin, which forms granulomas in response. **See Figure 4.11**

■ The subset of sympathetic postganglionic neurons that innervate *sweat glands* are directed by their target to switch from using *norepinephrine* (*NE*) to using *acetylcholine* (*ACh*) as a neurotransmitter. **See Figure 4.12**

■ In mammalian neocortex, cells migrate from the ventricular zone to none of six layers, differentiating into the appropriate neuronal phenotypes. Transplant studies show that shortly after becoming postmitotic, a cell is instructed by the ventricular zone to migrate to a particular layer; prior to that time, it can migrate to the end of the radial glia and then take on the fate appropriate to whatever layer is forming there. **See Figure 4.13**

■ The many transcription factors that direct neural differentiation often interact, affecting each other's transcription and/or protein activity, which is probably responsible for the extremely wide diversity of structure and function in neurons and glia. **See Figure 4.14**

■ Larger brains tend to evolve as a result of prolonged later embryonic stages that result in more cells and/or more elaborate neural processes found in later-arising structures, such as the outer layers of neocortex in mammals and the prefrontal cortex in primates. **See Figures 4.15 and 4.16**

■ The evolution of humans from an ancestor in common with the other apes may have been driven by *neoteny*, extending fetal-typical brain development after birth and stopping development of the body at a stage resembling the juvenile stage in other apes. **See Figures 4.17–4.19**

Go to the Companion Website
sites.sinauer.com/fond
for animations, flashcards, and other review tools.

5

Feeling One's Way
AXONAL PATHFINDING

MAKING CONNECTIONS All his life, Joseph G. felt that he had a harder time "connecting the dots like other people." Kath M. always stood out for the unusual pauses in her speech. As a child, Sarah M. had a hard time learning to walk, although she's an assistant dance teacher today. What these three have in common is they have no **corpus callosum** (**CC**), the thick bundle of myelinated axons connecting the two cerebral hemispheres in eutherian mammals (but is absent in other vertebrates, including marsupials and monotremes). Their condition, **agenesis of the corpus callosum** (**AgCC**), is present at birth but, because there may be no obvious symptoms, can go undetected. That means it is sometimes discovered accidentally, in MRI scans or routine autopsies (Hunter, 2005). First described in 1812 (Reil, 1812), AgCC is rare but not unheard of: about 1–3 people out of a 1,000 lack a CC (Larsen & Osborn, 1982). Joseph didn't learn he had AgCC until he was 45; Kath was 59 (Wolf, 2013).

Perhaps the most famous person with AgCC was Kim Peek, the model for the title character in the movie *Rain Man*, who was also a **savant**, a person with an unusually well-developed skill or ability. Unable to walk until he was 4, Kim was reading at age 3 and apparently never forgot what he read (Treffert & Christensen, 2005). Before he died of a heart attack at age 59, Kim memorized about 9,000 books, mostly nonfiction, each taking about an hour. Given the novel *The Hunt for Red October*, Kim read it in 75 minutes. When asked four months later about a minor character, Kim knew the page number where the character appeared, and quoted text from that page verbatim!

Obviously something went wrong during development in these people such that the usual connections between the left and right cortex failed to form. By now you have a sense of how complicated brain development is, so in some sense the wonder is that the CC develops fine about 99.9% of the time. What processes typically guide formation of the CC, why do some people not develop the structure, and how can they function as well as, or better than, most people who have it? Is the only reason we have a CC to keep the two hemispheres from flopping around in our skull?

CHAPTER PREVIEW

So far we've described the basic formation of a body plan (Chapters 1 and 2), the production and migration of neurons and glia (Chapter 3), and the factors that direct them to differentiate in their final destination (Chapter 4). Now these newly differentiated neurons distributed throughout the nervous system must produce axons and send them to the appropriate targets, which will be different for each type of neuron. That means each different type of axons must be guided to grow in the right direction, which is the topic of this chapter.

We'll start by describing the remarkable mobile tip of growing axons, the growth cone, which relies on complex chemical signaling systems to explore its environment, being attracted to some directions and repulsed from

Kim Peek

corpus callosum (CC) The large bundle of axons that communicate between the two cerebral hemispheres in placental mammals.

agenesis of the corpus callosum (AgCC) The condition of being born with the corpus callosum either absent or severely reduced.

savant A person with extraordinary talent in a specific endeavor, such as calculations, music, or memory.

neuron doctrine The early twentieth-century proposal that neurons are structurally distinct from one another and communicate across gaps called synapses.

growth cones The extensions of dendrites and axons that grow away from the cell body to make synaptic contacts.

lamellipodia The sheetlike extensions of membrane produced by growth cones.

filopodia The slender, rodlike extensions of membrane produced by growth cones.

tubulin The specialized protein that assembles to form microtubules providing structure to the cytoplasm.

microtubules The long filamentous materials formed from tubulin that provide internal structure for cells.

others. As the apparently shapeless growth cone advances, it leaves a highly organized axon in its wake. We'll see that while basic adhesion is necessary to permit growth cone movement, there are also several families of signaling systems to attract or repel growth cones, guiding different axons to different targets, using both cell-cell contact and factors that are diffused through the body like a bread crumb trail. A whole class of neurons needs to send their axons across the body midline to reach their target, and we'll learn quite a bit about the system that first lures them across the midline but then ensures they don't cross back over again. Finally, we'll see how these systems direct development of the corpus callosum, that enormous band of axons connecting the two cortical hemispheres, which surprisingly enough may not be that crucial for behavioral function.

■ 5.1 ■
RAMÓN Y CAJAL DESCRIBED GROWTH CONES AND DISCERNED THEIR SIGNIFICANCE

At the end of the nineteenth century, a lively debate raged in neuroscience about whether the nervous system consisted of an incredibly intricate, continuous network of fibers, or whether it was, as we know now, made up of billions of separate cells, each offering fine processes that touched, but didn't quite merge with each other. In other words, were neurons "continuous or contiguous"? The latter position, which came to be known as the **neuron doctrine**, was championed by the great Spanish anatomist Santiago Ramón y Cajal (1852–1834). Using Golgi's method to impregnate the entirety of individual neurons with a dense black silver precipitate, Cajal observed for the first time the tips of elongating fibers, which he called **growth cones** (Ramón y Cajal, 1890) (**FIGURE 5.1A,B**). For Cajal, growth cones provided strong evidence that neurons were separate from one another, each sending out processes to connect with other cells. In his typically vivid prose, Cajal later described the growth cone as a "battering ram, endowed with exquisite chemical sensitivity, with rapid ameboid movements" (Ramón y Cajal, 1904/1983). Cajal's inferences were based on static views of dead growth cones, so it would be some time before anyone would confirm his description of this hotbed of activity in living growth cones. Although American biologist Ross Harrison did not use the term *growth cone*, he would describe living tadpole growth cones in culture, saying, "We therefore cannot escape the conclusion that the extension of the fiber is due to the activity of the enlargement at its end" (Harrison, 1910, p. 819) (**FIGURE 5.1C**).

Cajal's "ameboid" description is apt, as the growth cone, like an amoeba, constantly changes shape, advancing by extending out a portion of itself, a "foot," latching onto something, and then contracting to pull the rest of the structure forward. In fact, growth cones typically have two different kinds of "feet": the broad, sheetlike **lamellipodia** ("sheet feet," singular *lamellipodium*), and the long, slender, rodlike **filopodia** ("thread feet," singular *filopodium*). While the growth cone itself looks otherwise amorphous, the axon it trails behind it is highly organized. Molecules of the protein **tubulin** assemble to form the **microtubules** that provide a sleek structural core to the axon (**FIGURE 5.2**) as well as a scaffold for extensive **axonal transport**, in which additional building components are brought from their manufacturing site in the cell body to the growth cone (**anterograde transport**), and chemical messages are sent from the growth cone back to the cell body (**retrograde**

> Cajal later described the growth cone as a "battering ram, endowed with exquisite chemical sensitivity, with rapid ameboid movements."

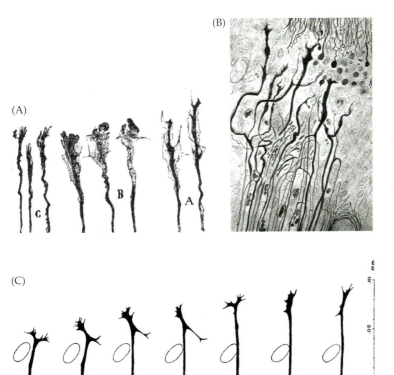

(B)

(A)

FIGURE 5.1 **Early portraits of growth cones** (A) Cajal's drawings show growth cones in the embryonic chick spinal cord. (B) Regenerating axons from severed dorsal root ganglion neurons form growth cones attempting to reenter the spinal cord. (C) Harrison traced the progress of living growth cones in culture over the course of 47 minutes. (A from Ramón y Cajal, 1890; B from Ramón y Cajal, 1928; C from Harrison, 1910.)

(C)

transport). The retrograde transport can, among other things, allow growth cones to call for alterations in gene expression back at the cell body. This means that the growth cone can call for a change in the supplies being manufactured for shipment downstream. As we'll see later, such signaling can change the behavior of the growth cone so that it can turn from one pathway to another, or even change from being attracted to a pathway to avoiding it.

In this chapter, we will concentrate on the role of growth cones in the guidance of growing axons, but you should know that when dendrites grow, they do so by forming growth cones, too. We know much more about the growth of axons than dendrites, in part because axons typically appear first in development and so are easier to track and manipulate in a relatively unpopulated nervous system. By the time dendrites start growing, there are already millions of cells and axons in place, and it can be harder to tell one neuronal extension from the other. Plus, because axons typically travel farther than dendrites, it's easier to study cues that guide turning of the growth cone from developing axons than dendrites. Nevertheless, you should keep in mind that these same forces that attract and repel axons to their targets are also at work to connect dendrites to their synaptic partners.

An important component of growth cones, the contractile protein **actin**, provides the force to push the tips of filopodia out and pull them back again (Lowery & Van Vactor, 2009) (**FIGURE 5.3A**). If a filopodium latches onto some adhesive portion of the environment, then the subsequent contraction of the filopodium will pull the entire growth cone in that direction. If the filopodium tip does not adhere to the substrate, then the contraction simply brings the filopodium back so it can make another try later. In this way, the growth cone "sniffs" the environment, following the

axonal transport The process by which materials are moved within axons, in both directions, along microtubules.

anterograde transport The transport of materials within axons in the direction of the axonal terminals.

retrograde transport The transport of materials within axons in the direction of the cell body.

actin A protein that mediates contraction of cell parts, such as filopodia of growth cones.

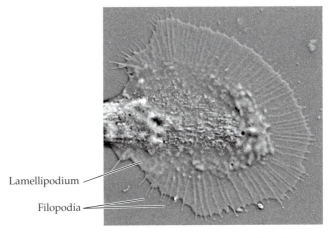

Lamellipodium

Filopodia

FIGURE 5.2 **Growth cone in culture** (From Munnamalai & Suter, 2009, courtesy of Dr. Daniel Suter.)

FIGURE 5.3 Growth cones in action (A) The different components of the moving growth cone are found in different compartments. (B) This schematic view shows an active growth cone assembling microtubules in response to an attractive signal and avoiding a repulsive one. (C) Calcium imaging reveals a sudden spike in Ca^{2+} in a single filopodium (white arrowheads) in response to application of a chemical signal (netrin). (A from Dent & Gertler, 2003, courtesy of Dr. E. Dent; C from Gomez & Zheng, 2006.)

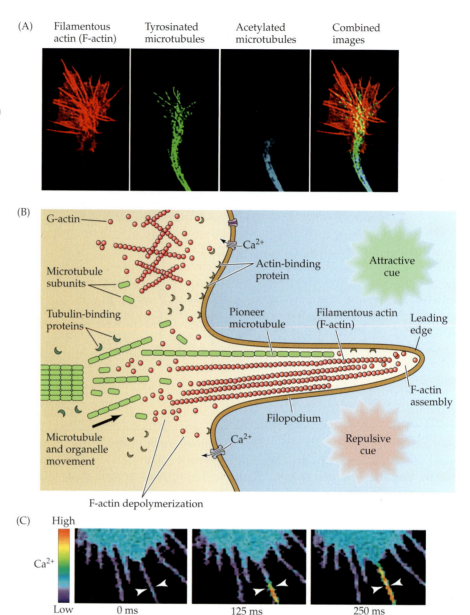

guidance cues we'll discuss shortly and leaving a "tail" of an assembled axon (**FIGURE 5.3B**). Individual actin molecules polymerize to form filaments that attach to actin-binding proteins anchored in the membrane of the filopodia and lamellipodia at one end, and attach to myosin molecules anchored to immobile microtubules at the other. The myosin proteins pull on actin filaments to reel them in, either retracting the filopodium back or pulling the growth cone forward. These processes are modulated by release of internal stores of Ca^{2+} and influx of Ca^{2+} (**FIGURE 5.3C**) (Gomez & Zheng, 2006).

How's It Going?

1. Name the parts of a growth cone.
2. How do growth cones advance?
3. How are microtubules assembled and what function(s) do they serve?

■ **5.2** ■

IN VITRO APPROACHES REVEAL PRINCIPLES OF AXONAL GROWTH AND ADHESION

One of the early questions experimenters had about growing axons might not have occurred to you: does the axon grow by adding membrane close to the cell body, pushing the entire axon forward (like pushing a plumber's "snake" down a drain), or by adding membrane down at the growth cone end of the axon, assembling additional axon behind it? The microtubule core of axons was exploited to answer this question. Fluorescently labeled tubulin molecules were introduced into a neuron in vitro, and these were gradually added to the growing axon's core. Then a portion of the fluorescently labeled axon was deliberately exposed to enough light to "bleach out" the fluorescent tags on tubulin molecules in that one spot. If the entire axon was being shoved forward, then that bleached spot should advance as the growth cone moved forward. But instead, the bleached spot stayed put as the growth cone advanced, and the distance between the spot and the growth cone lengthened (Okabe & Hirokawa, 1990). So in fact, the axon is extended by continual addition of more microtubules to the end of the structure, sometimes called "tip growth" (**FIGURE 5.4**).

Of course, eventually axons *must* grow by other means, by simple expansion of both the length and diameter of the process. Think of a motor neuron in your lumbar spinal cord, for example, with an axon reaching out to muscles in the calf

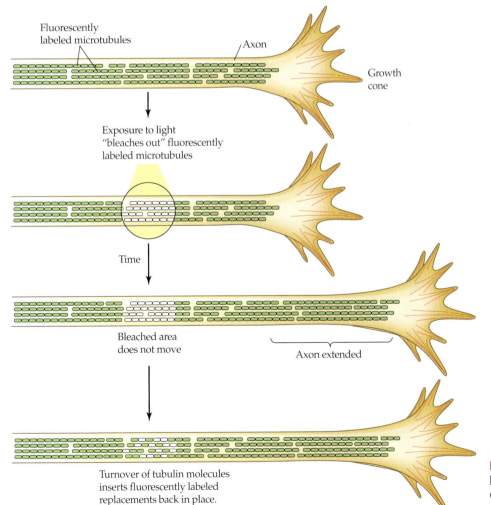

Fluorescently labeled microtubules

Axon

Growth cone

Exposure to light "bleaches out" fluorescently labeled microtubules

Time

Bleached area does not move

Axon extended

Turnover of tubulin molecules inserts fluorescently labeled replacements back in place.

FIGURE 5.4 **Axons grow by adding to the tip.** (After Okabe & Hirokawa 1990.)

of your leg. That axon connected those two structures well before you were born, and it has grown tremendously in length from that time until today, without the need for any growth cone. In the experiment we just described, for example, eventually that spot of faded fluorescence grew brighter again, as more fluorescently tagged tubulin molecules were inserted (see Figure 5.4). That result showed that there is turnover of the tubulin molecules within the organized microtubules, as would be required for the entire axon to expand in length. Presumably this much slower growth process in development is due to gradually extending the microtubule core and adding lengths of membrane at every point along the axon.

> Picture the cell body anterogradely transporting unassembled building materials down the axon, to have them sort of gush out.

So the picture we have is of the cell body anterogradely transporting unassembled building materials down the axon, to have them sort of gush out, all the pieces disassembled, to add membrane and be organized locally to extend the tip of the axon's organized core. Among those building materials, tubulin molecules arrive individually, without organization, and then organize, with the help of local enzymes that encourage either assembly or disassembly, onto the tip of the previously assembled microtubules.

The *direction* of that axonal extension is determined by the growth cone as it probes its environment with filopodia, to see where they can get the best grip to pull the growth cone along. Thus for the growth cone to advance, filopodia have to be able to adhere to something. Indeed, when neurons explanted into a dish send out axonal growth cones, they make very little progress along the surface of glass or plastic alone. So scientists typically provide a coating of some sort along the glass so that the growth cone can gain purchase to move. In the early days, there was something of an art of making the dish surface sticky enough for filopodia to adhere, but not so sticky that the rest of the growth cone would stay put. These findings naturally led to the idea that maybe growth cones simply follow whichever pathway is stickiest. So you can imagine them simply moving along a gradient of least adhesive to most adhesive surfaces, an adhesive gradient. But when experimenters deliberately tested this hypothesis, they found that things aren't that simple, as we'll see next.

RESEARCHERS AT WORK

Getting a Grip: The Role of Adhesion in Axonal Growth

■ **QUESTION**: Do axonal growth cones simply follow the most adhesive surface, moving up an adhesive gradient to find their target?

■ **HYPOTHESIS**: Growth cones in a dish will preferentially grow along whichever pathway they find most adhesive.

■ **TEST**: Culture retinal ganglion cell explants on surfaces coated with five different substrates: the **cell adhesion molecules (CAMs)**, specifically two called **L1** and **N-cadherin**, as well as **polylysine** (molecules of the amino acid lysine that have polymerized with one another), and the extracellular matrix protein **laminin** on a polylysine surface, and laminin on a nitrocellulose surface. For each preparation, determine how firmly growth cones adhere to the surface by directing a gentle flow of culture solution at the structure and measuring how long it takes to dislodge the growth cone. In separate cultures, note how rapidly the growth cones grow on each of the surfaces. If more-adhesive substrates encourage faster growth, then there should be a positive correlation between adhesivity and growth rate (**FIGURE 5.5A**).

---

cell adhesion molecules (CAMs) A class of molecules, found in extracellular regions, that adhere to some cells and not others.

L1 cell adhesion molecule (L1) Also called *L1CAM*. An important cell adhesion molecule.

neural cadherin (N-cadherin) Also called *NCad*. A cadherin that is important in neural development.

polylysines Long-chain molecules consisting of many lysines strung together.

laminin A long-chained protein that is a major component of the extracellular matrix.

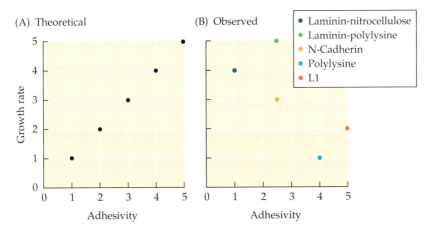

FIGURE 5.5 **Adhesion and growth do not correlate.** (After Lemmon et al., 1992.)

- **RESULTS**: In fact, there was no correlation between adhesivity of the growth cone for a substrate and its growth rate (**FIGURE 5.5B**). The most adhesive substrate was the CAM L1, yet axons grew slowly on that surface. Only polylysine produced slower growth than L1. In a follow-up experiment, when the explants were grown on alternating stripes of different substrates, they did not preferentially grow along the stickiest substrate, L1 (Lemmon, Bturden, Payne, Elmslie, & Hlavin, 1992).

- **CONCLUSION**: Adhesive surfaces are certainly required for growth cones to advance, but growth requires the *right degree* of adhesivity. Too much adhesion can impede growth, presumably because it becomes difficult to dislodge the lamellipodia to move the entire growth cone forward. The authors concluded that differences in adhesion might play a *permissive* role, allowing particular growth cones to grow along particular pathways, but not, at least in this system, an *instructive* role, directing growth cones in a particular direction or to choose a particular pathway among several available alternatives.

As we'll see next, study of more specific systems would eventually reveal instructive cues, too. Thus growth cone guidance is not simply a matter of following the stickiest pathway or the pathway that has just the right amount of adhesivity. Instead, there must be other mechanisms that direct the axonal growth cone to advance in a particular direction and for a growth cone that has been growing in one direction to turn in another direction. Rather, the growth cone is making active choices, following guidance cues, which we discuss next.

> Growth cone guidance is not simply a matter of following the stickiest pathway.

How's It Going?

1. What is the role of adhesivity in the movement of growth cones?
2. What is the disadvantage for the function of a growth cone if a surface is too adhesive?

■ **5.3** ■

GUIDANCE CUES MAY BE ATTRACTIVE TO ONE TYPE OF GROWTH CONE AND REPULSIVE TO OTHERS

contact guidance   The process by which growth cones are guided upon direct contact with the membranes of other cells.

concentration gradient   The condition in which a particular substance, such as a CAM, is more concentrated at one end of a structure than at another.

chemotropism   The tendency to follow along a particular chemical trail.

chemotaxis   The process of displaying chemotropism.

neurotropic   Referring to materials that attract neuronal growth cones.

attractive   Here, referring to materials to which growth cones readily attach.

repulsive   Here, referring to materials to which growth cones will not attach. Repulsive signals often cause the collapse of filopodia that contact them.

Of course, for the exterior of the growth cone to be selective about which adhesive trails to follow, it needs yet another important component that we haven't mentioned yet: proteins embedded within the filopodium membrane that serve as surface receptors to recognize various external molecules, to which the filopodium will either adhere or not. In fact, we'll see shortly that there are several families of receptors and their ligands. The direction of growth cone advance will then depend on which of these proteins are studding the outside of its membrane. This permits growth cones to detect, and make a selection among, a wide range of guidance cues.

There are two classes of these cues that growth cones can use for guidance:

1. Short-range cues depend on **contact guidance**, as the ligand embedded in the membrane of one cell engages a receptor embedded in the membrane of another cell. Either the growth cone has the membrane-bound ligand that interacts with a receptor, or it has a receptor to interact with a ligand. Note that in such cases, which membrane-bound molecule we consider to be the receptor and which the ligand is somewhat arbitrary.

2. Long-range cues are at work when *diffusible molecules* released by a cell or group of cells spread out so that a **concentration gradient** of the molecule may be detected some distance from its origin, a process known as **chemotropism** ("following a chemical [trail]"). Here the diffusible molecule is smaller and so is easily understood as the ligand that binds the membrane-bound receptor in the growth cone. In situations like this, where axonal growth cones are moving toward a specific chemical, we say they are displaying **chemotaxis**. The specific substances they are attracted to, which we'll discuss shortly, are said to have a **neurotropic** (the *tropic* is pronounced with a long *o*; there's nothing tropical about it) effect, meaning they draw neuronal growth cones closer.

To warn you about some confusing terminology, in Chapter 7 we'll discuss proteins that are important for keeping neurons alive, said to be *neurotrophic* ("neuron-feeding"). I'll try to always use the correct term, but you should know that researchers sometimes seem to use the terms *neurotropic* and *neurotrophic* interchangeably. To make matters worse, in Chapter 7 we'll see that several proteins have both a neuro*tropic* effect (attracting axons of particular cells) and a neuro*trophic* effect (keeping the neuron alive after the axon arrives).

> Filopodia that can't stick to something won't be able to pull the growth cone in that direction.

As we'll see in the remainder of this chapter, both types of cues, contact guidance and chemotropism, can be either **attractive** or **repulsive**, either leading the growth cone closer or diverting it away (Tessier-Lavigne & Goodman, 1996). A cue is attractive when the filopodia of a growth cone adhere to the substance and pull the growth cone in that direction. A cue is repulsive either because the filopodia can't stick to it or because the repellent cue causes the filopodia to collapse altogether, falling back into the body of the growth cone (**FIGURE 5.6A**). Filopodia that can't stick to something won't be able to pull the growth cone in that direction. Overall then, the growth cone extends in the direction being called for by the majority of the filopodia, a sort of navigation by consensus.

Whether an external cue is attractive or repulsive depends on which molecules the growth cone has embedded in its membrane, and in this way each growth cone is in charge of its own destiny (Huber, Kolodkin, Ginty, & Cloutier, 2003). Because the growth cone determines which cues it finds attractive and repulsive,

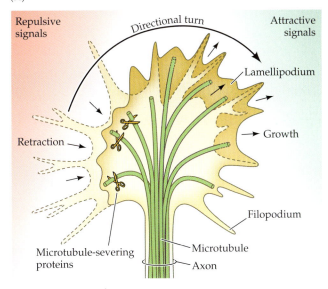

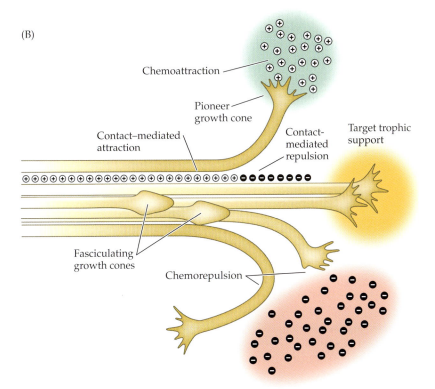

FIGURE 5.6   Growth cones actively navigate. (A) Close-up view shows a growth cone changing directions. (B) Such responses result in some growth cones turning one way while others turn the other way. (B after Huber et al., 2003.)

we'll see that the same cue may be attractive to one growth cone and repellent to another (**FIGURE 5.6B**). There are also cases when the same growth cone is initially attracted to a cue and then, after passing some milestone, becomes repelled by that same cue. Often that switch in preference is mediated by a decrease in cyclic AMP (cAMP) levels within the growth cone (Ming et al., 1997), which alters the intracellular signaling systems of the membrane-bound receptors when they encounter ligands (Murray, Tucker, & Shewan, 2009).

Next let's consider some of the options growth cones have for the receptors they may embed in their membrane and the factors those receptors may encounter on their journey.

─────────── How's It Going? ───────────

1. What are the two main types of cues that guide growth cones?
2. How does a filopodium respond to contact with an attractive surface versus a repulsive surface?

■ **5.4** ■

FAMILIES OF RECEPTORS OFFER A MULTITUDE OF GUIDANCE CUES

neuronal cell adhesion molecule (NCAM) Also called *neural-CAM*. An adhesive molecule expressed by many neurons.

homophilic binding The property of a material that readily binds to itself.

heterophilic binding The property of two different materials that readily bind to one another.

calcium-dependent cell adhesion molecules (cadherins) A class of transmembrane proteins, the adhesive properties of which are sensitive to local levels of calcium.

ephrins A family of membrane-bound signaling molecules that bind to ephrin receptors, which are part of the receptor tyrosine kinase (RTK) superfamily.

ephrin receptors A family of receptor tyrosine kinase (RTK) molecules that bind to ephrins.

integrins A family of adhesive molecules that are both membrane bound and secreted into the extracellular matrix.

extracellular matrix (ECM) The collection of various long-chain molecules that loosely bind one another to form a layer outside many cell membranes.

collagen A long-chained filamentous protein that contributes to the formation of extracellular matrix and is especially prominent in connective tissue.

fibronectin A long-chained glycoprotein that contributes to the formation of the extracellular matrix.

basal lamina A particularly thick layer of extracellular matrix that surrounds mature muscle fibers and many organs.

First we'll consider the receptors a growth cone might insert to recognize short-distance cues from contact guidance. Then we'll consider growth cone receptors that can be either attracted or repulsed by long-distance cues, diffusible factors.

We will discuss in some detail five families of receptors underlying *contact guidance* of growth cones:

1. Cell adhesion molecules (CAMs) include **neural cell adhesion molecule (NCAM)** (Walsh & Doherty, 1997) and L1 cell adhesion molecule (L1 or L1CAM) (Kamiguchi & Lemmon, 1997), which we discussed above. CAMs can either bind to one another (**homophilic binding**) or bind to other CAMs (**heterophilic binding**) (**FIGURE 5.7A**). Both NCAM and L1 are in a superfamily of immunoglobulin genes crucial for the immune system.

2. **Calcium-dependent cell adhesion molecules (cadherins)** are, as the name indicates, dependent on the presence of Ca^{2+} in order to bind to one another (Stepniak, Radice, & Vasioukhin, 2009). In other words, they display homophilic binding (**FIGURE 5.7B**). As with the Wnts we discussed in previous chapters (see Figure 2.27), the intracellular portion of cadherins interacts with the catenins.

3. Our old friends from Chapter 2, the **ephrins**, are cell surface proteins that bind a family of **ephrin receptors**, which are members of the receptor tyrosine kinase (RTK) family that we discussed in Chapter 4 (see Figure 4.3) (Flanagan & Vanderhaeghen, 1998). In other words, ephrins are the membrane-bound ligands, and they bind the ephrin family of RTK receptors (**FIGURE 5.7C**). In Chapter 2 ephrins were responsible for guiding migrating cells to cluster together, sharpening up the boundaries of the rhombomeres (see Figure 2.24). When bound, both ephrin receptor and ligand can activate kinases and/or Rho GTP-ase activating proteins (Rho-GAPS) (Dickson, 2001).

4. The **integrins** are a family of receptors that recognize various molecules in the **extracellular matrix** (**ECM**) (Hynes, 1992), including **collagen**, laminin, and **fibronectin**, which polymerize to form a local extracellular "coating" around a cell (see Figure 3.21). We discussed ECM molecules in Chapter 3 in their role in guiding cell migration. Outside the CNS, some tissues and organs are surrounded by organized layers of ECM called **basal lamina** (Bunge, Bunge, & Eldridge, 1986), which plays a role in muscle innervations, as we'll discuss in Chapter 6. Every integrin receptor is composed of two parts, an alpha subunit and a beta subunit, which heterodimerize to form the functional receptor that, when bound by an ECM ligand, can activate kinases and/or Ca^{2+} channels inside the cell (Albelda & Buck, 1990) (**FIGURE 5.7D**). In humans, there are 18 different alpha subunits and eight different beta subunits (Hynes, 2002), so there is the potential for a staggering number of different integrin receptors that can be assembled, depending on which alpha subunit is paired with which beta subunit.

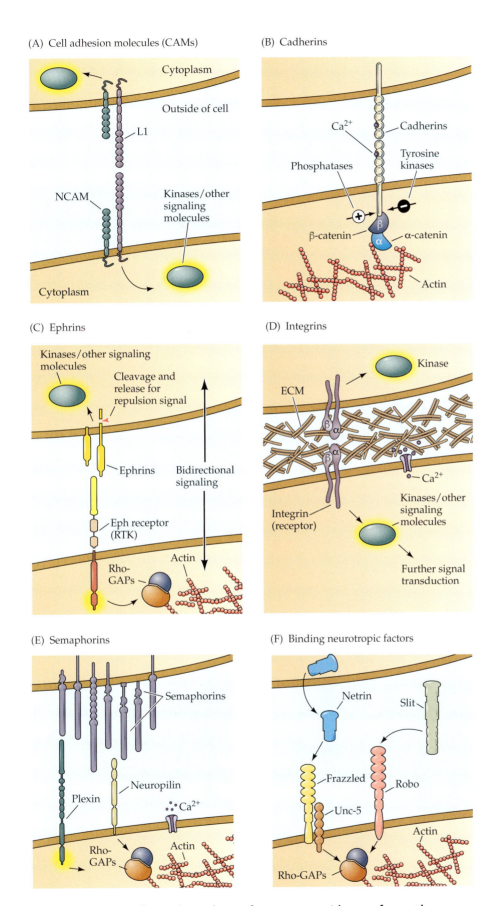

(A) Cell adhesion molecules (CAMs)

(B) Cadherins

(C) Ephrins

(D) Integrins

(E) Semaphorins

(F) Binding neurotropic factors

FIGURE 5.7 Membrane-bound cues for contact guidance of growth cones

5. There are eight different classes of **semaphorins**, representing 20 different genes in humans (Yazdani & Terman, 2006). Semaphorins most often serve to inhibit growth cones and so deflect them from wandering in the wrong direction (**FIGURE 5.7E**). They interact with a family of receptors called the **plexins** (Yoshida, 2012) and another family of receptors called the **neuropilins** (Zachary, 2014). Both plexins and neuropilins act through Rho-GAPs, typically causing growth cone collapse. When serving as contact cues, the semaphorins remain embedded in the membrane of the cell that produced them, but as we'll see, semaphorins can be secreted to serve as a chemotropic cue, too.

Remember, these are five *families* of receptors and their ligands, so there are several versions of each, and some, like the integrins, are composed of several subunits that can be combined in very many different ways to provide a wide variety of choices for growth cones. We will also discuss three families of receptors for neurotropic factors, diffusible substances that provide chemotactic cues for growth cones:

1. The **unc-5** family of receptors (Serafini et al., 1994) were first discovered in *Caenorhabitis elegans*, while the **Frazzled** family of receptors (Bradford, Cole, & Cooper, 2009; Keino-Masu et al., 1996) were first studied in humans (and called DCC). Both are receptors for a family of secreted proteins called **netrins**, named for the Sanskrit word for "guide" (Kennedy, Serafini, de la Torre, & Tessier-Lavigne, 1994) (**FIGURE 5.7F**). Among other things, the netrins play an important role in attracting axons to the midline and then keeping them from crossing again, as we'll discuss below.

2. The **roundabout** (**Robo**) family of receptors (Dickson & Gilestro, 2006) bind to diffusible **Slit** proteins, both of which were first discovered in *Drosophila* (see Figure 5.7F). Robo and Slit also play a role in axonal crossing of the midline in both flies and humans, as discussed below.

3. The same plexin and neuropilin receptors that bind semaphorins in contact guidance also bind them when they are secreted to serve as chemotactic cues (see Figure 5.7E).

Whew, that is a lot of different guidance cues! First, of course, this gives you an idea of how important axonal guidance is, that natural selection has come up with so many systems to participate. The various receptors and their ligands are summarized in **TABLE 5.1** for review, but you will get to know

> The integrins are composed of several subunits that can be combined in very many different ways to provide a wide variety of choices for growth cones.

semaphorins A family of secreted and membrane-bound molecules that often serve to repulse growth cones, including those that express neuropilin or plexins, and so direct them away from a boundary.

plexins A family of receptors that bind and respond to semaphorins.

neuropilins A family of membrane receptor proteins that govern growth cone guidance. They bind and respond to semaphorins.

unc-5 A family of membrane-bound receptors that respond to netrins.

Frazzled A family of membrane-bound receptors that respond to netrins.

TABLE 5.1
Families of Proteins for Axonal Guidance

Family	Receptor	Transduction
CAMs (NCAM, L1)	CAMs (homophilic/heterophilic)	Activate kinases
Cadherins	Cadherins (homophilic)	β-catenin
Ephrins	Ephrin receptors (heterophilic)	Tyrosine kinase/Rho-GAPs
Extracellular matrix (ECM) components (collagen, laminin, fibronectin)	Integrins	Activate kinases/Ca^{2+} channels
Semaphorins (membrane bound or secreted)	Plexins, neuropilins	Rho-GAPs
Netrins (secreted)	Frazzled, Unc-5	Rho-GAPs
Slits (secreted)	Robo	Rho-GAPs

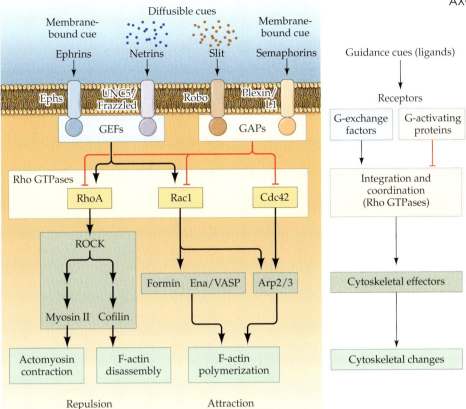

FIGURE 5.8 **Growth cones control their destiny.** The growth cone determines which cues it will find attractive or repulsive by controlling which molecules it embeds in its membranes, and therefore which cues the filopodia will find adhesive. Those cues are then integrated, funneling down two common final molecular pathways to either destabilize or stabilize the growth cone. (After Lowery & Van Vactor, 2009.)

each of them a bit better for their roles in various models we'll encounter in this chapter. The variety of guidance cues also reminds you that there is a bewildering variety of different axonal projections even in the *Drosophila* nervous system, never mind the human brain.

Eventually, of course, this wide variety of guidance cues must either attract or repulse growth cones by either encouraging stabilization of polymerized actin in the filopodium and growth cone in the case of attraction, or destabilizing actin polymerization in the case of repulsion. The four best-studied guidance cues tap two molecular pathways inside the growth cone to either stabilize or destabilize the structure (**FIGURE 5.8**) and therefore attract or repulse the growing axon.

Now let's see an example of how an axon can use contact guidance and chemotaxis to find its way.

netrins A family of secreted, diffusible proteins that attract some growth cones while repelling others.

roundabout (Robo) A family of membrane-bound receptors that respond to diffusible Slit proteins.

Slit A family of diffusible proteins that mark the midline in the developing nervous system.

How's It Going?

1. Name five classes of molecules involved in contact guidance of growth cones.

2. Name three classes of diffusible factors that can guide growth cones.

3. Make a table of the contact guidance cues and diffusible factors, listing the receptor(s) for each.

4. Why might natural selection favor the evolution of more than one receptor for a given membrane-bound ligand?

■ 5.5 ■
PIONEER NEURONS AND GUIDEPOST CELLS ESTABLISH PATHWAYS FOR LATER AXONS

pioneer neurons   Neurons that are the first in a region to send out axonal growth cones to establish a path that many later-arising axons will follow.

guidepost cells   Certain cells that seem to serve as targets for axonal growth cones establishing a particular pathway.

An example of how growth cones rely on contact guidance to navigate is seen in grasshopper embryos, where sensory neurons differentiate out in the periphery and then sprout axonal growth cones that must find their way to the nervous system, to form the connections necessary to communicate sensory information. These first-arising neural cells that must blaze a trail to the nervous system are called **pioneer neurons**. Their growth cones manage this journey by reaching first one **guidepost cell**, and then another. For example, at the tip of the embryonic limb bud, two pioneer neurons in the embryonic tibia, destined to be sensory neurons, differentiate and send out growth cones. At first, the filopodia appear to branch out in every direction. But once a filopodium encounters a particular guidepost cell in the femur, it attaches to that cell and pulls the rest of the growth cone there (Bentley & Keshishian, 1982). Again the growth cones send filopodia in every direction until one encounters another guidepost and pulls the growth cone there. Now the filopodia reach out in every direction until encountering and pulling the rest of the growth cone to the next guidepost cell, thereby turning the developing axon that much closer to the segmental ganglion, which is its final destination (**FIGURE 5.9A**). In this way, the developing axons navigate several turns that are required to reach their target, as a blind man using a cane does to detect first one street corner then another.

Of course, although the growth cones *seem* to latch onto first one guidepost cell and then another, it's always possible there is some other cue at work. But using a laser to kill a guidepost before the filopodia can encounter it results in the growth cone stalling at the previous guidepost, filopodia continuing to stretch out in every direction, with any progress seemingly at random (**FIGURE 5.9B**).

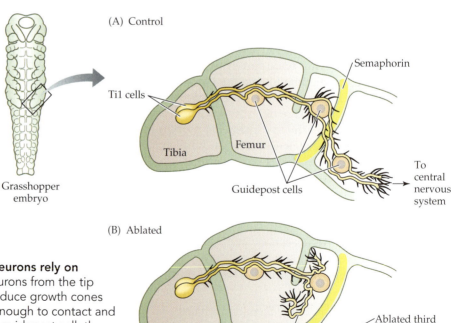

(A) Control

Semaphorin

Ti1 cells

Tibia   Femur

Guidepost cells

To central nervous system

Grasshopper embryo

(B) Ablated

Ablated third guidepost cell

Growth cone searches randomly for guidepost cell

FIGURE 5.9   Pioneer neurons rely on guideposts. Sensory neurons from the tip of the embryonic leg produce growth cones with filopodia just long enough to contact and pull the axon first to one guidepost cell, then another. If the third guidepost cell is ablated with a laser, the growth cone seems to wander about. (After Bentley & Caudy, 1983.)

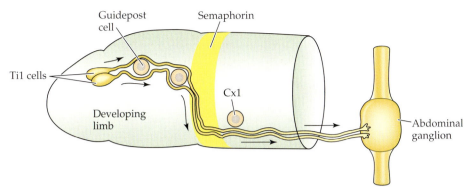

FIGURE 5.10 **Guideposts and road barriers** At one point, the filopodia are repelled by a membrane-bound protein called semaphorin, so they turn. Eventually the growth cone reaches a point at which some filopodia contact another guidepost cell, dragging the trailing axon across the semaphorin-expressing cells and toward the abdominal ganglion. (After Kolodkin et al., 1993.)

The axons of these pioneer neurons are directed by more than just guidepost cells. They are attracted to some substrates and repelled by others. In the case of the grasshopper sensory pioneers, the growth cone is repelled by semaphorins that are expressed by a band of cells at the transition from outer leg to inner leg (Kolodkin, Matthes, & Goodman, 1993) (**FIGURE 5.10**). When the pioneer growth cone filopodia touch a cell expressing semaphorin on its surface, the filopodia collapse (which is why the people who discovered semaphorin in chicks first called it "collapsin") (Luo, Raible, & Raper, 1993). Eventually a filopodium happens to bypass semaphorin-expressing cells and touches the third guidepost cell. Then it sticks to the guidepost and drags the growth cone across the "no-man's land" of semaphorin. Semaphorin proved important for the guidance of sensory axon growth cones in vertebrates, too (Koropouli & Kolodkin, 2014). For example, antibodies to the protein in chick spinal cord neutralized its ability to collapse growth cones and led to aberrant projections (Shepherd, Luo, Lefcort, Reichardt, & Raper, 1997).

The axons of later-arising neurons, either motor axons from the ganglion growing out toward muscle targets, or sensory neurons from the periphery headed to the ganglion, will simply adhere to these pioneer fibers, in a process known as **fasciculation** (making a *fasces*, which in Latin means "a bundle of rods bound together"; fascism celebrates the idea of everyone agreeing and behaving the same). Typically, growth cones that are advancing by fasciculation are almost arrow shaped, with few filopodia (see Figure 5.6B), as there's no need to "sniff around." This means they grow more rapidly than axons that must find their way. Interestingly, once many later-arising axons have fasciculated with the pioneer fibers and found their various targets, some of those pioneering cells will die before the embryo hatches (Klose & Bentley, 1989), as if they have already served their purpose. In Chapter 7 we'll see that such instances of cell death are a normal part of development throughout the body, including the nervous system.

When there are no guidepost cells available, axons may rely on chemotropic cues, and several systems make it clear that something in the extracellular space provides cues for the three spatial axes: the rostral-caudal, dorsal-ventral, and medial-lateral axes. For example, many fish and amphibians have two giant neurons, called **Mauthner cells**, in the brainstem, each of which sends an axon across the midline and caudally toward the spinal cord. When a segment of the embryonic brainstem containing the Mauthner cells

> Growth cones that are advancing by fasciculation are almost arrow shaped, with few filopodia.

fasciculation Here, the tendency of axonal growth cones to adhere to and follow along a preexisting bundle of axons.

Mauthner cells Large identifiable neurons found in the brainstem in many fishes and amphibians.

FIGURE 5.11 Which way is caudal? (A,B after Hibbard, 1965; C,D from Constantine-Paton & Caprianica, 1975.)

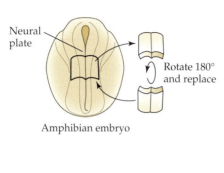

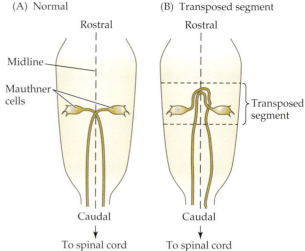

(A) Normal

(B) Transposed segment

(C)

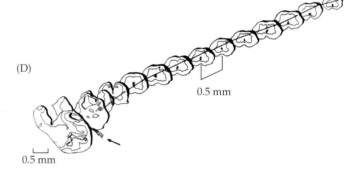

(D)

0.5 mm

0.5 mm

is cut out and reversed before it is put back, the Mauthner cells first send their axons toward what would have been the posterior portion of the segment but is now actually toward the head (Hibbard, 1965). However, once the growth cones reach the unmanipulated part of the brain, something must signal that this is the wrong way, because they reverse direction and eventually reach the spinal cord after all (**FIGURE 5.11A,B**). Similarly, when an embryonic eye from one tadpole was grafted more posteriorly into another tadpole (**FIGURE 5.11C**), the axons entered the brain and grew caudally, as they normally would. However, because they were already "downstream" from their normal brainstem target, they simply grew all the way down the spinal cord (Constantine-Paton & Caprianica, 1975). Interestingly, these retinal axons that normally grow in the dorsal-lateral portion of the brainstem grew only in the dorsal-lateral quadrant of the spinal cord (**FIGURE 5.11D**). Experiments such as these indicate that there must be some kinds of indicators of general directions in the developing nervous system such that, anywhere along the length of the nervous system, a growth cone can tell which way is rostral, which way is dorsal, and which way is lateral. Later we'll see that when amphibian eyes are transplanted to a less radical position in the body, they can find their way toward the correct target with uncanny accuracy. Those experiments indicate that, in addition to local cues about the direction of rostral, dorsal, and lateral, there are sometimes cues about where particular targets are, as though they diffuse some substance that serves as a "beacon" to say, "Here I am." If axons are close enough to pick up those cues, they can travel novel routes, in novel directions, to reach them.

> There must be some kinds of indicators of general directions in the developing nervous system such that, anywhere along the length of the nervous system, a growth cone can tell which way is rostral, which way is dorsal, and which way is lateral.

─────────────── **How's It Going?** ───────────────

1. How do growth cones from the early-arising sensory neurons in the grasshopper leg reach the central nervous system? What experimental procedure demonstrated the importance of guidepost cells?

2. What is the role of semaphorin in this process?

3. What is fasciculation? Do growth cones move faster or slower when fasciculating?

4. What experimental procedure with Mauthner neurons indicated the presence of local cues signaling directions in the vertebrate nervous system?

▪ 5.6 ▪

MANY AXONAL GROWTH CONES HAVE TO DEAL WITH CROSSING THE MIDLINE

An important problem for the developing nervous system is that the two sides of the body must act in coordination. This means that information must cross the body's midline, and so while many neurons remain entirely on one side of the body or the other, some neurons must send their axons across the midline to deliver information needed to coordinate movement, sensory processing, and so on. A bundle of axons crossing the midline is called a commissure, so we call any neuron that sends its axons across the midline a **commissural neuron**.

A series of interesting mutants in flies revealed a key system for keeping noncommissural neurons from crossing the midline, and at the same time guiding commissural neurons across the midline, then making sure they don't cross back over. The ganglia running along the ventral midline in insects are connected to one another through a pair of longitudinal connectives, which consist of axons passing information between ganglia. Within each ganglion, there are two commissures, consisting of axons transmitting information across the midline. Some projection neurons send their axons along the longitudinal connectives between ganglia, while other neurons are commissural, sending their axons across the midline, and then often continuing along one of the longitudinal pathways (**FIGURE 5.12A**).

A genetic screen of many mutants revealed one that came to be named *roundabout* (*robo*) because in this mutant some commissural neurons sent their axons across the midline again and again, basically going in circles like an errant car stuck in a traffic roundabout. Also, longitudinally projecting neurons, which were not supposed to cross the midline, sometimes crossed and recrossed the midline in these mutants (**FIGURE 5.12B**). This mutant illustrates a problem about commissural neurons that might not have occurred to you, namely that whatever mechanism is used to entice the axonal growth cone across the midline must be turned off soon afterward or the axon will just cross over again.

Later analysis revealed that Robo was a receptor for a diffusible protein called Slit, which is normally found only along the midline (**FIGURE 5.13A,B**). Axons in mutants with a disabled Slit protein reached the midline, but seemed unable to ever leave it again (**FIGURE 5.12C; FIGURE 5.13C,D**). Conversely, a mutation called *commissureless* caused neurons, including those that would

commissural neuron   A neuron that extends its axon across the body midline.

> Whatever mechanism is used to entice the axonal growth cone across the midline must be turned off soon afterward or the axon will just cross over again.

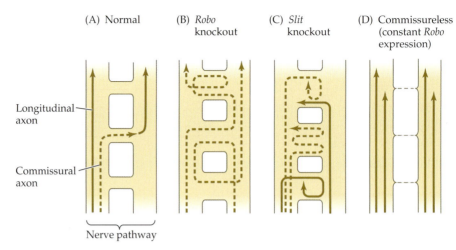

(A) Normal   (B) *Robo* knockout   (C) *Slit* knockout   (D) Commissureless (constant *Robo* expression)

Longitudinal axon

Commissural axon

Nerve pathway

FIGURE 5.12   Once you cross the border, you're not supposed to come back. (A) Normally, longitudinal axons that will simply grow to the next ganglion express Robo from the start, and so are repelled by the midline. Growth cones from commissural neurons do not express Robo at first, and so are attracted by the midline, but then express Robo, so they won't cross again. (B) In *roundabout* knockouts, neurons are not repelled by the midline, and so they may cross and recross many times (like a car stuck going round and round a traffic roundabout). (C) If Slit is deficient, both longitudinal and commissural axons enter the midline, but neither ever leaves again as they are never repulsed by the midline. Therefore, the commissures clog up with these arrivals that never leave. (D) In *commissureless* mutants, all neuronal growth cones express Robo, so none cross the Slit boundary at the midline, resulting in no commissures. (After Dickson & Gilestro, 2006.)

floor plate   The ventral portion of the vertebrate neural tube, the developing spinal cord.

have been commissural neurons, to overexpress Robo. In that case, no neurons crossed the midline, and so no commissures developed (hence the name) (**FIGURE 5.12D**).

In vitro studies established that growth cones expressing Robo are repelled by Slit, which explained the phenotype in those mutants. Neurons destined to project longitudinally express Robo from the start, so their axonal growth cones never cross the midline. In contrast, commissural neurons do not insert Robo into their membranes at first, and so nothing prevents them from crossing the midline. However, once the growth cone crosses the midline, *then* they insert Robo into their membranes so now they find Slit as repellant as do the longitudinal axons, and so rather than cross the midline again, they move on to their target.

By now you'll not be surprised to learn that this system of getting axons to cross the midline once and only once in flies is also at work in other species. As researchers explored this system in vertebrates, another player was discovered, another diffusible factor that serves to attract axonal growth cones to cross the midline, as we'll see next. This additional factor explains why normally commissural axons head toward the midline in the first place.

In vitro studies revealed that the axons of commissural neurons in the vertebrate dorsal spinal cord would grow toward explants of neural **floor plate** (see Chapter 2). In fact, the commissural neurons would move around an intervening explant of roof plate to reach a floor plate behind it. Because the growth cones moved toward the floor plate explants before their filopodia could possibly reach them, there had to be some diffusible substance that the floor plate was releasing to attract the axons.

There had to be some diffusible substance that the floor plate was releasing to attract the axons.

(A) Slit protein

(B) Robo protein

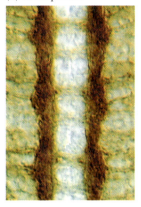

FIGURE 5.13   The center cannot hold. (A) Slit protein is normally restricted to the midline. (B) Robo is expressed only when axons are avoiding the midline, so Robo protein is found only in the ganglia and the longitudinal nerves connecting them. (C) Staining of all axons reveals the ladderlike structure of paired ganglia running from top to bottom, with a pair of commissural pathways connecting each. (D) In *slit* knockout flies, all axons are free to enter the midline and, in the absence of Slit to repel them away, the axons just stay there. (From Kidd et al., 1999, courtesy of Dr. C. S. Goodman.)

(C) Wild-type

(D) *Slit*$^{-/-}$

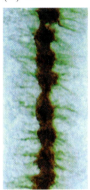

Scientists extracted two related brain proteins, which they named netrins, that could also attract commissural axons. When they cloned and sequenced the genes for netrin-1 and netrin-2 (Serafini et al., 1994), they learned they were homologous to a gene in *C. elegans, unc-6*, which had been implicated in axon guidance. They set about asking if netrins are the diffusible beacon for commissural neurons, as we'll see next.

RESEARCHERS AT WORK

What Makes the Floor Plate so Attractive?

■ **QUESTION**: Are netrins preferentially expressed in the floor plate, and do they attract commissural neuron growth cones?

■ **HYPOTHESIS**: Floor plate expression of netrin attracts commissural neuron growth cones.

■ **TEST**: First use in situ hybridization to see where netrin-1 and netrin-2 are expressed. Then transfect the netrin genes into COS cells (a cell culture line that normally does not express netrins) and see if commissural growth cones are now attracted to these cells.

■ **RESULTS**: Netrin-1 was indeed expressed exclusively in the floor plate region, while netrin-2 was expressed in the ventral two-thirds of the spinal cord (**FIGURE 5.14A,B**). Commissural growth cones were not attracted to normal COS cells, but when the COS cells were transfected with the netrin-1 gene, commissural axons grew toward them (**FIGURE 5.14C**) (Kennedy et al., 1994).

continued

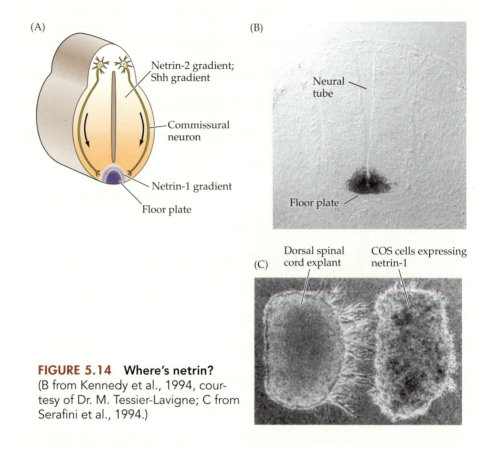

FIGURE 5.14   Where's netrin?
(B from Kennedy et al., 1994, courtesy of Dr. M. Tessier-Lavigne; C from Serafini et al., 1994.)

■ **CONCLUSION**: The ventral spinal cord makes and diffuses netrin-2 while the floor plate region makes and diffuses netrin-1, and the two together attract commissural neuron growth cones to the ventral midline.

Later studies would show that netrins are detected by a Frazzled receptor, (see Figure 5.7F) which is expressed by commissural neuron growth cones. Once these growth cones have crossed the midline, they will begin expressing Robo, which means they will now grow *away* from the midline for two reasons. First, because they will be expressing Robo, they will be repelled by the Slit protein concentrated in the midline in vertebrates, just as it is in flies. What's more, Slit stimulation of the Robo receptor also *suppresses* Frazzled activity, so the growth cones will no longer be attracted by netrins at the ventral midline (i.e., the floor plate) (**FIGURE 5.15**). The suppression of Frazzled by Robo activation is another reason why longitudinal axons, which express Robo all along, avoid the midline. They are repelled by Slit, plus they are not attracted to netrins.

The ability of growth cones to change their adhesive preferences while growing out brings up the question of how they get the new proteins they need to insert in their membrane to cause that change in preference. In some cases, it's a matter of the growth cone removing proteins from the membrane that had been available, or now inserting proteins that had been made earlier back in the cell body and held in reserve within the growth cone. In other words, there may not be a change in protein synthesis (Roche, Marsick, & Letourneau, 2009), just in how previously synthesized proteins are

The ability of growth cones to change their adhesive preference while growing out brings up the the question of how they get the new proteins they need to insert in their membrane.

(A) Commissural axon before crossing midline

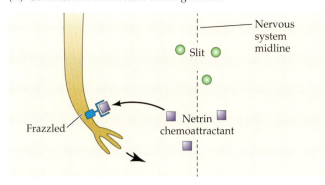

FIGURE 5.15   **Come closer, closer, closer... now go away.** (A) Commissural axons express Frazzled but not Robo, and so are attracted to the midline. (B) Once they cross the midline, the growth cone starts expressing Robo, which inhibits Frazzled. Thus the axon tip, now repelled by Slit at the midline and no longer attracted to netrin, is free to follow other cues to find its target cell for innervation. (C) Longitudinal axons express Robo continuously and so are repelled by Slit at the midline and do not express Frazzled, meaning they are not attracted by netrins, either.

(B) Commissural axon after crossing midline

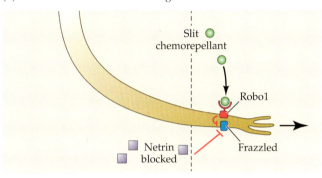

(C) Longitudinal axon

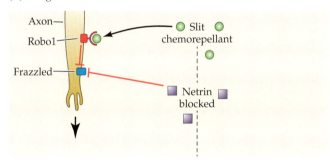

deployed. But there is also growing evidence that some mRNAs are present in growth cone and that a limited amount of translation to make proteins de novo occurs within the growth cone itself. There's even evidence that the growth cone may translate transcription factors that can be retrogradely transported back to the cell body to alter gene expression, such as making new proteins and/or mRNAs to send down to the axon (E. Kim & Jung, 2015; Lin & Holt, 2008). So despite the relative autonomy of the growth cone, there is also communication between the growth cones and the cell body.

How's It Going?

1. What are the phenotypes of fly mutants when robo, slit, and commissureles are disabled?

2. How does expression of robo differ in commissural neurons versus non-commissural neurons in flies?

3. How does netrin expression in the vertebrate spinal cord influence commissural neuron growth cones?

4. What is the receptor for netrin, and how does it interact with robo?

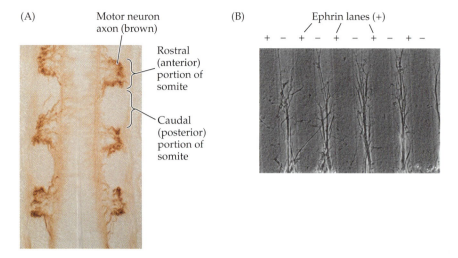

(A) Motor neuron axon (brown)

Rostral (anterior) portion of somite

Caudal (posterior) portion of somite

(B) Ephrin lanes (+)

+ − + − + − + − + −

FIGURE 5.16   I don't wanna go there. (A) Motor neuron axons grow only through the rostral compartment of each somite. (B) They probably prefer the anterior compartment because the posterior compartment expresses ephrins; motor neuron axons in vitro avoid substrate coated with ephrins ("+" lanes). (From Wang & Anderson, 1997, courtesy of the authors.)

■ 5.7 ■

MOTOR NEURONAL AXONS MUST FIND THE CORRECT TARGET MUSCLES

In Chapter 3 we noted that neural crest cells migrate only through the rostral half of each somite, and that later the motor neurons send their axons only through that same rostral compartment (**FIGURE 5.16A**). Recall that the neural crest cells chose the rostral compartments because they were repelled by ephrins expressed in the caudal compartments (see Figure 3.25). Similarly, motor neuron axons are also repelled by ephrins, which we can show by culturing motor neurons in dishes where the bottom is coated with lanes of ephrins interspersed between lanes without them. The motor neurons grow out almost exclusively on the ephrin-free lanes in between (**FIGURE 5.16B**) (Wang & Anderson, 1997).

We learned in Chapter 4 that shortly after cells differentiate into spinal motor neurons, they begin expressing different patterns of transcription factors, depending on both their position in the ventral horn and the eventual target muscles they will innervate (see Figure 4.8). Do those differences in gene expression program the growing axons to seek out particular muscles? A classic experiment indicates they do, as we'll see next.

RESEARCHERS AT WORK

Can You Navigate Your Way Home?

■ **QUESTION**: Are motor neurons already specified to innervate a particular target muscle when their axons exit the ventral roots?

■ **HYPOTHESIS**: Motor neuron axonal growth cones follow local cues to innervate whatever muscle in need of innervation is closest.

■ **TEST**: Cut out a section of embryonic chick spinal cord about four segments in length and reverse it so that motor neurons originally in the last thoracic segment are transposed and now exit the spinal cord several segments more caudally, in the third lumbar segment (**FIGURE 5.17A,B**). Will they follow the new local cues to innervate muscles appropriate for the third lumbar segment, or will they chart a novel route to reach their original target?

(A) Motor axons exiting the spinal cord

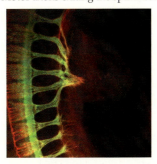

FIGURE 5.17   Charting a new course (A from Huettl et al., 2011, courtesy of Dr. A. Huber-Brosamie; B–D after Lance-Jones & Landmesser, 1980.)

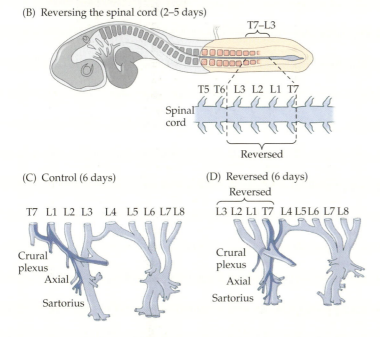

(B) Reversing the spinal cord (2–5 days)

(C) Control (6 days)

(D) Reversed (6 days)

■ **RESULT**: When the motor neurons that would normally innervate axial muscles and the sartorius muscle were transposed caudally, their axonal growth cones grew out and *took a new pathway* to reach those muscles (**FIGURE 5.17C,D**). In other words, the transposition did not change their final target muscle. They were not simply growing out in a predetermined trajectory, but navigating, taking a new path to reach an intended target (Lance-Jones & Landmesser, 1980).

■ **CONCLUSION**: Even before their growth cones exit the spinal cord, motor neurons are at least somewhat specified to innervate a particular muscle or type of muscle. Their axons can navigate new routes to reach those targets, suggesting that there are diffusible cues to help them orient and grow toward those targets. The muscles appear to release a signal, serving as a beacon for axons to zero in on their target.

In Chapter 4 we noted that as motor neurons differentiate, they express particular patterns of transcription factors that vary depending on which muscle groups they will innervate (see Figure 4.9), and those cues may play a role in specifying which target muscles they will seek out. That in turn could play a role in motor neurons' ability in this experiment to send their axons along a new pathway to reach those targets.

Another well-studied model where axonal growth cones seem to be able to navigate novel pathways to find their appropriate targets is the retinotectal system of fish and amphibians, which we take up next.

How's It Going?

1. What cues direct motor neuron axons to grow through the rostral half of a somite?

2. Describe the classic experiment that indicated motor neuron axons are seeking a specific muscle target, not just the nearest uninnervated muscle.

■ 5.8 ■

THE AXONS OF RETINAL GANGLION CELLS MUST REACH THE MIDBRAIN

Just in case you're proud of yourself for being a mammal and think you're therefore superior to mere fish and amphibians, you should know that those "lower" vertebrates enjoy a remarkable advantage over mammals in at least one way. If the optic nerves connecting your eyes to your brain were ever severed, you would immediately be blind and remain blind for life, because the retinal ganglion cells would never be able to reinnervate your brain. However, when an optic nerve is surgically severed in a fish, frog, or salamander, it is, of course, blind in the affected eye immediately, but its retinal ganglion cells, unlike yours or mine, can regenerate. Thus a month or so after the surgery, the animal will again be able to see with the healed eye (Sperry, 1949). We can tell that, because if we offer it a tasty worm or fly, the animal will direct itself toward the treat and eat it. This remarkable regenerative ability of the optic system in fish and amphibians led to an equally remarkable line of research that is a venerable branch of modern neuroscience. You really can't consider yourself a neuroscientist unless you know about this work.

> The remarkable regenerative ability of the optic system in fish and amphibians led to an equally remarkable line of research that is a venerable branch of modern neuroscience.

You may have heard of Roger Sperry's famous work with "split-brain" patients, but before doing that, he conducted a highly respected line of research in the regeneration of connections between the retina and the tectum, the roof of the midbrain where most visual information enters the brain in fish and amphibians. There were thought to be three possible explanations for why the animals could see after the regeneration took place:

1. Perhaps the regenerating retinal ganglion cells simply grew toward the brain together and, because of mechanical barriers and their spacing with one another, ended up innervating the first tectal cells they came to and, because of their spacing with one another, ended up restoring the original connections between the eye and brain.

2. Alternatively, maybe the regenerating retinal ganglion cells did not simply innervate the first tectal cells they encountered, but actually selected the proper tectal target and reestablished the original connections that way.

(A)

FIGURE 5.18 **Which way is up?** When the eye is reversed 180°, frogs' behavior makes it clear that their vision is reversed. When food is offered above the animal, it directs its tongue downward in an attempt to retrieve it. (After Sperry, 1963.)

(B)

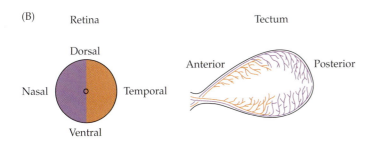

3. Or maybe the retinal ganglion cells innervated the tectum willy-nilly, but higher centers in the brain, receiving the information from the tectum, were able to adapt and make sense of the newly patterned information reaching them.

In a springboard from the known capacity of amphibian and fish visual systems to regenerate, Sperry added an important logical and literal twist: after severing the optic nerve of a frog, he also rotated the eye 180° before sewing it back in place (Sperry, 1963). This offered an opportunity to test different explanations for why the recovered animals were able to see. Eventually vision returned to the rotated eye, but the frog's behavioral responses to stimuli in the eye were reversed: if a treat was dangled on the upper left, the frog shot its tongue out to the lower right, and vice versa (**FIGURE 5.18A**). Thus the frog's vision was reversed by the rotation, which meant the retinal ganglion cells must have reconnected to their original tectal targets, not the targets appropriate to their new position in the visual field. Follow-up studies confirmed that the original pattern of innervation was reestablished: ganglion cells in nasal retina (the part closest to the nose) innervated posterior tectum, while ganglion cells in temporal retina (the part closest to the temple) innervated anterior tectum (**FIGURE 5.18B**).

These frogs with reversed vision also called into dispute the idea that retinal ganglion cells might innervate the tectum at random and then higher brain centers would sort out the input to make sense of the visual world. If that had happened, the animal would eventually be able to properly orient to stimuli in that eye. Some animals were kept for years, but they were never able to make sense of the reversed vision.

These experiments were seen as a powerful example of "nature" being more important than "nurture," because the retinal cells seemed to know where they should go and could take new pathways to get there, and the animal seemed unable to learn to overcome that wiring if the eye was reversed. Sperry proposed the **chemoaffinity hypothesis**, that each retinal neuron had a particular chemical "key" that was looking for a particular "lock" in a tectal cell, such that it would only innervate a particular part of the tectum. Sperry's hypothesis was supported by many experimental findings in the coming years, including one we'll describe next.

> Sperry added an important logical and literal twist: after severing the optic nerve of a frog, he also rotated the eye 180° before sewing it back in place.

chemoaffinity hypothesis Roger Sperry's proposal that axonal growth cones seek out a particular target cell based on chemical signals that mark both. It explains how regenerating retinal cells manage to reestablish their original synaptic contacts with tectal cells.

RESEARCHERS AT WORK

I'd Rather Walk over Here

■ **QUESTION**: Do the axons of retinal ganglion cells entering the tectum prefer innervating some target neurons over others?

■ **HYPOTHESIS**: Neurons from different parts of the tectum differ in their membranes, being attractive to the appropriate retinal ganglion cell axons and/or repulsive to others.

■ **TEST**: Gather cells separately from anterior tectum and posterior tectum of chick embryos. Homogenize both samples and extract just the fragments of cell membranes from each. Coat a filter with alternating stripes of a "carpet" of membranes; that is, stripes of membrane from anterior tectum should be separated by stripes of membrane from posterior tectum. Now culture onto the filter explants of embryonic chick retinas, from either the nasal region (which normally innervates posterior tectum) or the temporal region (which normally innervates anterior tectum). Observe whether axonal growth cones from the retina prefer growing over membranes from one region of tectum or the other.

■ **RESULT**: The axons from nasal retina displayed no preference in growing along membranes. However, axons from temporal retina grew preferentially on membranes from anterior tectum, which they would normally innervate (Walter, Henke-Fahle, & Bonhoeffer, 1987). When a strip of retina, from temporal to nasal, was allowed to choose either anterior or posterior tectum membranes, only temporal retinal cells showed any preference (**FIGURE 5.19**).

Temporal retina   Nasal retina

FIGURE 5.19   Repulsion at work (From Knöll et al., 2007.)

■ **CONCLUSION**: Temporal retina axons can distinguish between the membranes of posterior and anterior tectal cells and prefer the latter. This means that they are either attracted to anterior membranes or repulsed by posterior membranes. To distinguish between these two explanations, the experimenters used high temperatures to "denature" the membranes, presumably by destroying whatever membrane molecules the axons can detect. They found that heating up the posterior membranes alone was enough to cause the temporal retinas to no longer show a preference. Thus the axons are not *attracted* by anterior membranes, but rather are *repelled* by posterior membranes.

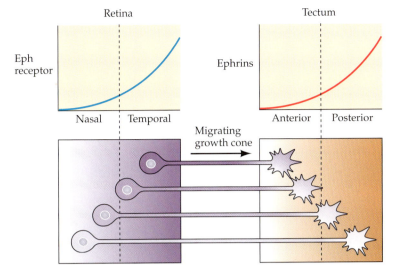

FIGURE 5.20 Growth cones seeking a balance There are opposing gradients of Eph expression in the retina and tectum, such that each retinal ganglion cell grows until repulsion reaches a threshold level. Temporal retinal neurons express the most Eph receptor and so stop in anterior tectum. This results in a gradient of retinal space projected onto the tectum. (After Wilkinson, 2001.)

The membrane-bound proteins responsible for retinal axons' preference for one type of tectal membrane or another are the ephrins, which we've encountered before (see Figure 3.22). Part of a large family of proteins, ephrins can be attractive or repulsive, depending on which pairing of ephrin receptor (which you may recall are receptor tyrosine kinases, RTKs; see Figure 4.3) and ephrin ligands are being expressed by the two interacting cells. Ephrin (Eph) receptors are expressed by retinal cells in a gradient, higher in temporal retina than nasal retina (**FIGURE 5.20**). The ephrin ligands for these receptors are expressed in a gradient in the tectum, higher in posterior tectum than anterior tectum. Because the ephrin receptors in growth cones are inhibited by the ephrin ligands, the temporal retina axons that enter anterior tectum will be inhibited from growing farther. Conversely the relative absence of Eph receptors in nasal retina growth cones means they will grow farther into tectum, eventually reaching the posterior end. Growth cones arising from the midpoint of the retina grow to the midpoint of the tectum, and so on. Thus growth cones from each region of the retina grow until they reach an equilibrium point, where their concentration of Eph receptor just inhibits further progress (**FIGURE 5.21**). In this way, there is

> Growth cones from each region of the retina grow until they reach an equilibrium point, where their concentration of Eph receptor just inhibits further progress.

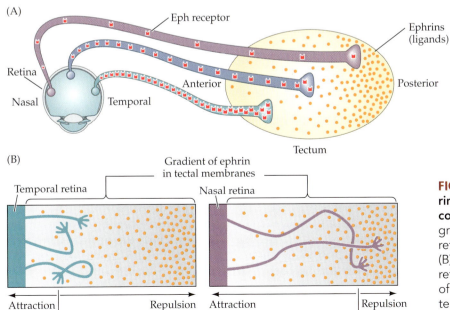

FIGURE 5.21 Gradients of ephrin result in gradients of growth cone movement. (A) The normal gradient of ephrin receptors in retina and of ephrin in the tectum. (B) In vitro, explants from nasal retina grow further up a gradient of ephrin than do explants from temporal retina. (After Baraniga, 1995, and Hansen et al., 2004.)

a gradient of retina-to-tectum innervation. Note that because the retina also represents a two-dimensional map of visual space, this means that there will be a "map" of information about visual space projected onto the tectum.

We will revisit experiments in regenerating retinotectal projections in Chapter 9, where we'll learn that after regenerating axons reach the roughly correct portion of the tectum, visual experience will fine-tune the topographic projection of the retina to the brain. So while the initial results of these experiments were seen as a vindication of the importance of "nature," it turns out that "nurture," specifically visual experience, is also important for the developing visual system, even in frogs.

Retinal innervation of tectum involves membrane-bound guidance cues for growing axons, where the growth cone must come into contact with a cell to be either attracted or repelled. Sometimes repulsion can help axons project in a new direction, as in formation of the corpus callosum, which we describe next.

How's It Going?

1. Describe the result of Sperry's experiment of severing the optic nerve and reversing the position of the eye in frogs. How did those results indicate that retinal axons had re-established their original connections in the tectum?

2. How was it demonstrated that some retinal axons preferred membranes from a particular region of the tectum?

3. How do the concentrations of ephrins and their receptors result in a gradient of innervation from retina to tectum?

■ 5.9 ■

THE CORPUS CALLOSUM IS DIRECTED ACROSS THE MIDLINE BY A GLIAL BRIDGE

Normally the fibers crossing through the corpus callosum consist of axons arising from the cortex of one hemisphere that project to homologous cortex in the other hemisphere (**FIGURE 5.22A**). There are over 100 medically recognized syndromes associated with a totally or partially missing corpus callosum (CC) (Hunter, 2005). Despite that, about 75% of cases of agenesis of the CC (AgCC) have no identified cause. It is most commonly seen along with major malformations of the brain, but when AgCC happens by itself, without other brain abnormalities (**FIGURE 5.22B**), it is essentially asymptomatic (Jeeves & Temple, 1987) and so may go undetected. Six people accidentally found to have AgCC were extensively tested without uncovering any behavioral impairment (Meyer, Roricht, & Niehaus, 1998). In some reports, people with AgCC have a slightly slower reaction time if you present a signal to one cerebral hemisphere and require them to push a button in response using the hand controlled by the opposite hemisphere. But the difference in reaction time is pretty subtle: 400 ms versus 360 (M. C. Lassonde, Sauerwein, & Lepore, 2003). These individuals without a corpus callosum do not show the sorts of deficits that are seen in adults who have the CC surgically severed, the famous "disconnection syndrome" seen in the so-called split-brain patients that Sperry and others studied (Gazzaniga, 2000). Presumably it is because the people with AgCC *grew up without a corpus callosum* that their developing brains compensated, forming alternative, subcortical pathways to connect cortex on both sides (Barr & Corballis, 2002). Indeed, when the corpus callosum is surgically severed in *childhood*, the patients as adults do not show the disconnection syndrome seen in those who had the surgery in adolescence or adulthood (M. Lassonde, Sauerwein, Chicoine, & Geoffroy, 1991). Some, but not all, people with AgCC

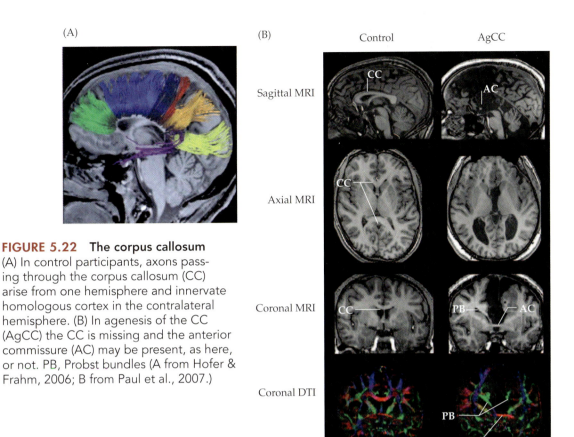

FIGURE 5.22 The corpus callosum (A) In control participants, axons passing through the corpus callosum (CC) arise from one hemisphere and innervate homologous cortex in the contralateral hemisphere. (B) In agenesis of the CC (AgCC) the CC is missing and the anterior commissure (AC) may be present, as here, or not. PB, Probst bundles (A from Hofer & Frahm, 2006; B from Paul et al., 2007.)

may have a larger than typical **anterior commissure** (Hetts, Sherr, Chao, Gobuty, & Barkovich, 2006), which might provide an alternative path across the brain's midline. But the majority of people with AgCC have normal-sized anterior commissures, so we don't know what alternate routes their hemispheres developed for communication.

The individuals with AgCC we met at the start of the chapter each report different "symptoms," so in fact it is possible that the difficulties they describe have *nothing* to do with the absence of a CC. After all, most of us have some difficulties in one arena of life or another, and the fact that these folks report different symptoms makes it hard to see how absence of a CC could be behind them all. Similarly, the astonishing savant ability that Kim Peek had is very rare, but other savants have no notable brain abnormalities. AgCC is commonly accompanied by **autism spectrum disorder (ASD)** (Paul et al., 2007), which is characterized by impaired social interactions, problems communicating, and severely limited behavior and interests. Kim definitely had autism, as do most savants, but most people with autism are not savants. Autism is a "spectrum" disorder because severity can range from severe to mild. At the mild end, Asperger's syndrome is not uncommon in engineers (or professors).

AgCC is usually associated with another unusual structure, called **Probst bundles** (Hetts et al., 2006; Probst, 1901), axons from each cerebral cortex that seem to approach the midline without being able to cross it and so grow in the rostral-caudal axis instead. There are strains of mice in which AgCC is common, and they also display Probst bundles (Richards, Plachez, & Ren, 2004). Experiments with those mice offer clues to what might be going on in human AgCC.

The first cortical axons to cross the midline in mice are from the cingulate cortex that sits on the midline. At about embryonic day 15.5 in mice, axons from neurons found in cingulate layers II, III, and V grow ventrally beneath the gray matter, then grow medially across the midline (Rash & Richards, 2001). Like pioneer neurons found in insects, these pioneer axons provide a platform upon which later-arriving axonal growth cones fasciculate and cross the midline. The mouse axonal growth

> The majority of people with AgCC have normal-sized anterior commissures, so we don't know what alternate routes their hemispheres developed for communication.

anterior commissure A relatively small collection of axons, found in the ventral portion of the brain, that communicate between the two cerebral hemispheres.

autism spectrum disorder (ASD) A disorder of social interaction that may be accompanied by problems in communication and severely limited behavior and/or interests.

Probst bundles Malformations caused by growing axons that fail to produce a corpus callosum.

cones cross the midline using the same Robo-Slit signaling system (Fothergill et al., 2014; Shu, Sundaresan, McCarthy, & Richards, 2003) we discussed earlier, first attracted to Slit and then, after crossing the midline, expressing Robo so they are repelled by it, and therefore they don't cross the midline again.

But for the cortical callosal neurons, glia also play a critical role for axons to cross the midline: if the fibroblast growth factor (FGF) receptor gene is disabled in the glia at the midline, the CC does not form, while knocking out the FGF receptor in neurons has no effect (Smith et al., 2006). What are glia doing to help formation of the CC? A dramatic demonstration that glia can help CC axons cross the midline was the reversal of AgCC in mice, as we'll see next.

RESEARCHERS AT WORK

Glia Can Help Axons Cross a Border

- **QUESTION**: What factors guide early callosal axon growth cones across the midline? Are glia simply providing a mechanical barrier that forces the growth cones across the midline, or do the axons prefer growing across such glia?

- **HYPOTHESIS**: As tightly packed glia form in the midline before the first callosal fibers arrive, they play a role in guiding growth cones.

- **TEST**: First use a surgical procedure to prevent CC formation, cutting the midline "bridge" of glia in mice just before or just after birth. Then in some mice surgically implant, at about the position of the severed bridge, a slab of nitrocellulose that glia, probably astrocytes, will readily grow upon and coat. Examine the brains a few days later to see whether this implant will rescue formation of the CC.

- **RESULT**: In fact, cortical axons grew across the glial-coated bridge at the midline forming a CC rather than Probst bundles, but only in those rostral-caudal regions where the glial bridge was present (**FIGURE 5.23**). Surgical introduction of materials that glia will not coat, including various kinds of plastic, were ineffective at guiding axons across the midline (Silver & Ogawa, 1983).

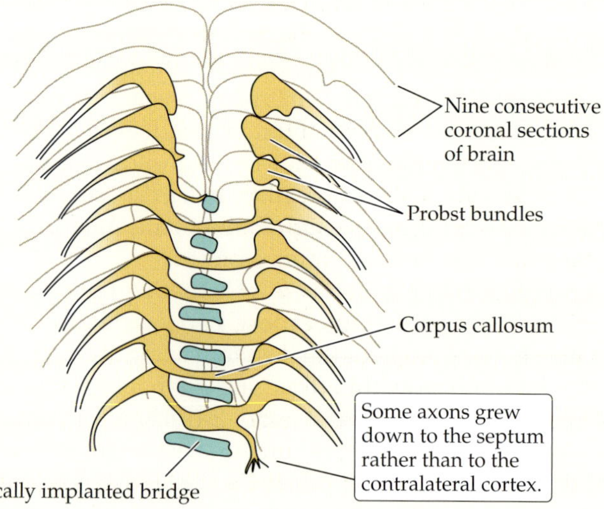

FIGURE 5.23
Restoring the corpus callosum in mice (After Silver & Ogawa, 1983.)

- Nine consecutive coronal sections of brain
- Probst bundles
- Corpus callosum
- Some axons grew down to the septum rather than to the contralateral cortex.
- Surgically implanted bridge

- **CONCLUSION**: Developing callosal axons are not simply forced across the midline by a mechanical barrier, but need the guidance of glia to reach the other side.

(A) Wild-type

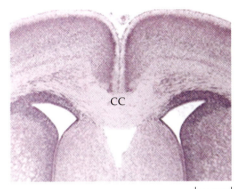

Ephrin-B1 knockout

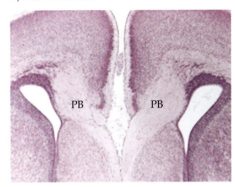

200 µm

(B) Wild-type

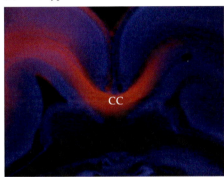

Ephrin mutant

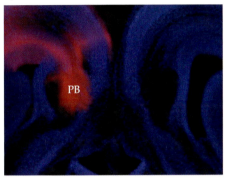

FIGURE 5.24 *Ephrin* **mutants display AgCC.** (A) Compared with wild-type mice (left), mice with the *ephrin-B1* gene disabled do not form a CC, forming Probst bundles (PB) instead. (B) Fluorescent tracers confirm that cortical axons fail to cross the midline and make up the Probst bundles. (From Bush & Soriano, 2009.)

One genetic cause of AgCC is a mutation in the gene for L1 cell-adhesion molecule, which suggests that this protein may play a role in formation of the CC. You might expect that dysfunction of a cell adhesion molecule would have widespread effects in the developing brain, and indeed the L1 mutation, in addition to inducing AgCC, is accompanied by mental impairment and hydrocephalus (Fransen, Van Camp, Vits, & Willems, 1997). Ephrins also play a role in the development of the CC, as some mutations of ephrins can result in AgCC in mice (**FIGURE 5.24**).

Looking ahead to Chapter 8, you should also know that many of the axons that cross the midline through the corpus callosum early in development will disappear by adulthood (Innocenti, Manger, Masiello, Colin, & Tettoni, 2002; Innocenti & Price, 2005; J. H. Kim & Juraska, 1997; Olavarria & Van Sluyters, 1995), a "pruning" of connections that we'll learn is commonplace in axonal connections throughout the developing brain.

Before we leave this topic, you should know that there is one well-established cause of AgCC: prenatal exposure to alcohol. AgCC is seen in as much as 6% of children with fetal alcohol syndrome (FAS; **FIGURE 5.25**) (Roebuck, Mattson, & Riley, 1998), who usually have mental impairment that can range from mild to quite severe (Bookstein, Sampson, Connor, & Streissguth, 2002).

There is one well-established cause of AgCC: prenatal exposure to alcohol.

(A) Control infant   Corpus callosum   (B) Infant with FAS

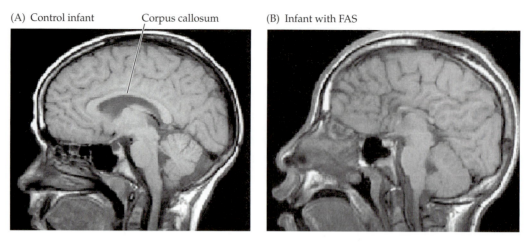

FIGURE 5.25 **Fetal alcohol syndrome can cause AgCC.** (Courtesy of Dr. E. Riley.)

How's It Going?

1. What are the symptoms of people who have AgCC and no other brain malformations?
2. What might explain why people with AgCC do not behave like the famous "split-brain" patients that were later studied by Roger Sperry?
3. What molecular system normally direct axons of the corpus callosum across the midline?
4. Describe the experimental surgery that can restore the corpus callosum in mice.

SUMMARY

■ *Growth cones* use *lamellipodia* and *filopodia* to move in an amoeba-like fashion, trailing the growing axon (or dendrite) behind them. The cell body sends building materials *anterogradely* to the growth cone, which sends *retrograde* signals back to the cell body. **See Figures 5.1 and 5.2**

■ Molecules of *tubulin* polymerize to add to the tip of the *microtubule* core of the axon, and they are polymerized or depolymerized within the growth cone as it responds to external cues. **See Figures 5.3 and 5.4**

■ An adhesive substrate is necessary for growth cone movement, but several families of signaling molecules direct growth cones through either short-range *contact guidance* or longer-range *chemotropism*. Both types of cues can be either *attractive* or *repulsive*. **See Figure 5.5**

■ *Actin* provides force to push filopodia forward and retract them again. If the filopodium adheres firmly enough to an external factor, contraction will pull the rest

of the growth cone in that direction. External signals that cause the filopodium to collapse will turn the growth cone away. **See Figure 5.6**

■ Families of guidance cues and their receptors mediate contact guidance, including *cell adhesion molecules* like *NCAM* and *L1* that bind to each other (by either *heterophilic* or *homophilic binding*), *cadherins* (homophilic binding), *ephrins* (heterophilic binding), and the *integrins*, which bind molecules in the *extracellular matrix*. **See Figure 5.7**

■ Families of *chemotactic* guidance cues include the *netrins* (binding *unc-5* and *Frazzled* receptors), *Slit* (binding *Robo*), and *semaphorins* (binding *plexin* and *neuropilin* receptors). **See Figure 5.8 and Table 5.1**

■ The earliest axons from *pioneer neurons* may rely on *guidepost cells* and the other cues to establish a pathway that later axons may join by *fasciculation*. **See Figures 5.9 and 5.10**

■ Multiple studies indicate that there are general orientation cues for all three body axes available for growth cone navigation, such that they can orient, for instance, caudally in the dorsal lateral quadrant of the nervous system. **See Figure 5.11**

■ Axons of *commissural neurons* expressing Frazzled are attracted to netrin molecules in the midline, but after crossing the midline, they begin expressing Robo and are then repulsed by midline Slit and stop expressing Frazzled. Longitudinal axons express Robo from the start and so are repulsed by midline Slit, never crossing the midline. **See Figures 5.12–5.15**

■ Different groups of motor neurons express different transcription factors (as detailed in Chapter 4), and their axons navigate to reach the proper muscle targets. If transposed, motor neuronal axons can take novel pathways to reach their original targets, indicating there are long-range cues to guide them. **See Figures 5.16 and 5.17**

■ Amphibian and fish retinal ganglion cells can also take novel pathways to reach their appropriate targets in the tectum. Their growth cones are attracted to some contact guidance cues and repulsed by others. The *concentration gradient* of nasal-to-temporal retinal axons innervate the tectum in an anterior-to-posterior gradient, establishing a two-dimensional visual field on the tectum. **See Figures 5.18–5.21**

■ Cortical axons establishing the *corpus callosum* (*CC*) are attracted to the midline, using the Robo-Slit signaling system, and are then deflected across the midline by a glial "bridge" to then innervate homologous portions of the contralateral cortex. All eutherian mammal species have a CC, but some people born without a CC (*AgCC*) show little or no behavioral deficits. On the other hand, fetal alcohol exposure, in addition to raising the risk of AgCC, often results in some degree of mental impairment. **See Figures 5.22–5.24**

Go to the Companion Website
sites.sinauer.com/fond
for animations, flashcards, and other review tools.

6

Making Connections
SYNAPSE FORMATION AND MATURATION

"STICKS AND STONES" In the first months after Kathleen gave birth to her second child, Oscar, she noticed that he seemed to keep his eyes closed a lot, but her pediatrician wasn't worried. When Oscar was 8 months old, the staff at his day care suggested Kathleen take the baby to a specialist because he seemed developmentally delayed. Teams of evaluators came to Kathleen's home several times to observe Oscar. By the final visit, they had reached a consensus and the speech pathologist broke the news. She told Kathleen that Oscar was either autistic or "globally delayed." A social worker herself, Kathleen immediately dismissed the idea of autism, because Oscar was so quick to smile at people and engage with them, but she asked the woman, "Global delay? Does that mean my son is mentally retarded?" The woman whispered her reply, "We're not allowed to say that anymore, but yes."

After months of additional testing Kathleen finally got a more definitive diagnosis, but it brought no comfort. Oscar had fragile X syndrome, a condition caused by a gene on the X chromosome, which, among other things, meant he inherited it from Kathleen. Googling to learn more about the condition only intensified Kathleen's fear, anguish, and guilt. She kept thinking that her beautiful baby boy was mentally retarded and it was her fault. Kathleen came to despise the terms "mentally retarded," "retarded," and "retard." Struggling to get Oscar the help he needed, looking for an outlet for her frustrations, Kathleen started a blog called *My Son's a Retard!* to draw attention to how hurtful the term can be. The blog starts, "Is the name of my blog offensive to you? I hope so, because it sure as hell offends me... Do you GET that it is my son you are talking about? My BABY? STOP IT!" It turns out, words can hurt you just as surely as sticks and stones.

What is fragile X syndrome, how is it inherited, and how does it affect brain development? To understand the answers to those questions, we need to learn about the origin and maturation of synapses.

CHAPTER PREVIEW

Chapter 5 covered the guidance of axonal growth cones toward their appropriate targets. Now we come to the process by which an initial contact, if maintained, develops into a fully mature synapse. Continuing the theme of the previous chapter, we'll see that adhesive molecules play a role in determining whether an initial contact is retracted or maintained and allowed to grow. If the contact is maintained, another familiar theme is revealed as each of the two neurons provides signals to induce the other to start assembling synaptic machinery. Presynaptic terminals use both secreted signals and cell-cell contact to induce the target cell to produce the postsynaptic assembly of neurotransmitter receptors and associated proteins needed to receive neuronal messages. Conversely, the target cell uses signals to induce the presynaptic axon to assemble synaptic vesicles full of neurotransmitter and the complex of proteins needed to release them when an action potential arrives. Both cells also secrete molecules to form the extracellular matrix filling the synaptic cleft between them.

We'll learn about several of the proteins and their receptors that have been identified (so far) to mediate this reciprocal development on either side of the synaptic cleft. A normal part of this process is a competition between neighboring synapses, which we'll expand upon in Chapter 8, for the resources needed to survive until adulthood. Contrary to what you might have expected, retaining too many of the synapses formed in the juvenile period may not be a good thing. We'll go into some detail about the development of the synapses between motor neurons and their target muscles, because that is where the story is most complete.

We'll also discover that while the electrical activity of embryonic synapses is rather slow and sluggish, alterations in the expression of ion channels during ontogeny sharpen synaptic responses so that they become rapid and brief, maximizing the amount of information a given synapse can convey over a short period of time. Changes in ion channel expression can also convert neurotransmitter receptors such that they have an excitatory effect rather than an inhibitory one.

So far I've spoken of only the neuron and its target, but glia are also crucial in synaptic development. The chapter closes with the signaling between neurons and glial cells that cap axon terminals to insulate the synapses from outside influences, as well as glia that form myelin, which also greatly sharpens neuronal signaling. Unfortunately, the same glia that provide this essential hastening of neural communication during normal development of the brain may impede regeneration in adulthood. What's more, the highly specialized proteins that make up myelin also make us susceptible to specific attacks, as happens in multiple sclerosis.

To begin with, let's consider how synaptic transmission evolved in the first place, by considering the precursor molecules that animals without nervous systems use, which natural selection modified to create the elaborate machinery needed on both sides of a synapse.

■ 6.1 ■

CALCIUM REGULATORS AND ENVIRONMENTAL SENSORS EVOLVED TO MEDIATE SYNAPTIC SIGNALING

Being single-celled organisms, of course prokaryotes have no need of synapses. But they do possess many of the evolutionary precursors to synapses. For example, yeast express calcium (Ca^{2+}) transporters and the Ca^{2+}-binding protein **calmodulin**, to regulate internal Ca^{2+} levels, as well as Ca^{2+}/calmodulin-dependent protein kinase II (CaMKII) and a variety of other **kinases** and **G proteins** that are regulated by extracellular signals (Cyert, 2001; Ryan & Grant, 2009). This machinery to respond to extracellular signals allows the single-celled organism to sense outside conditions and respond to them. The homologues of these kinases and G proteins in multicellular organisms (metazoans) have been adapted to provide the postsynaptic response to neurotransmitters (**FIGURE 6.1A**). Single-celled flagellates, thought to be the closest living relatives to the ancestors of metazoans, have numerous **tyrosine kinases** that respond to environmental conditions (Ryan & Grant, 2009). These consist of proteins that are embedded within the membrane, with separate branches extending into the

calmodulin   A protein that binds Ca^{2+}, regulating intracellular concentrations of the ion.

kinases   Enzymes that promote the addition of phosphates to particular sites on proteins, a process called phosphorylation.

G proteins   A family of proteins, named for their binding to GDP and GTP, that transmit intracellular signals.

tyrosine kinases   Enzymes that phosphorylate particular tyrosine molecules found in particular proteins.

transmembrane proteins   Proteins that, due to having several hydrophobic domains, are embedded in a cell membrane. They include ion channels and neurotransmitter receptors.

scaffolding proteins   A family of proteins that anchor intracellular portions of transmembrane proteins, such as neurotransmitter receptors in postsynaptic sites.

(A)

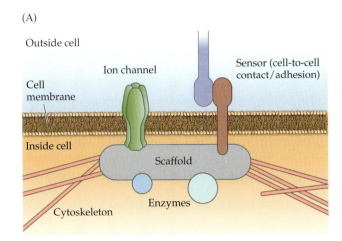

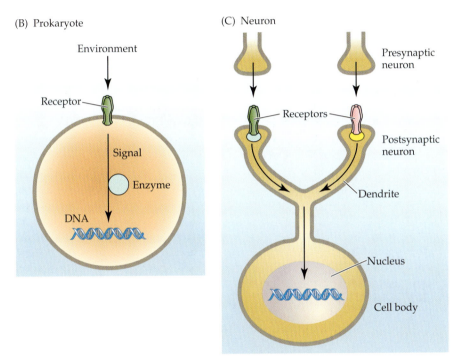

FIGURE 6.1 **Precursors to synapses** (A) Even prokaryotes make transmembrane proteins with extracellular receptors to respond to environmental cues as well as intracellular mechanisms to alter internal chemical reactions, including changes in gene expression. (B) They also make scaffolding proteins to anchor the intracellular portion of proteins to the internal cytoskeleton. (C) The metazoan homologues of these proteins participate in synaptic signaling. (After Emes & Grant, 2012.)

extracellular and intracellular space, so called **transmembrane proteins**, and include ion channels and sensors to detect extracellular conditions. Associated with these transmembrane proteins are **scaffolding proteins** to anchor the assembly within the cytoskeleton and interact with intracellular enzymes to alter the internal chemistry of the cell (**FIGURE 6.1B**). The homologues of these proteins in metazoans have been co-opted to play roles in the assembly of the postsynaptic machinery (**FIGURE 6.1C**), as we discuss in more detail below.

The most primitive metazoans, such as sponges, have no recognizable synapses, but they stick together by utilizing adhesion molecules, including **calcium-dependent adhesive molecules (cadherins)**, the calcium-dependent adhesive molecules that rely on homophilic binding, which we discussed in the previous chapter. Sponges also use cell contact signaling systems, as well as precursors to other synaptic components, including potassium (K^+) channels, and glutamate and GABA receptors (Emes &

calcium-dependent cell adhesion molecules (cadherins) A class of transmembrane proteins, the adhesive properties of which are sensitive to local levels of calcium.

The duplication of many synaptic genes about 500 million years ago, followed by divergence in their functioning, gave rise to the bewildering array of neurotransmitter receptors that we see in mammals today.

(A)

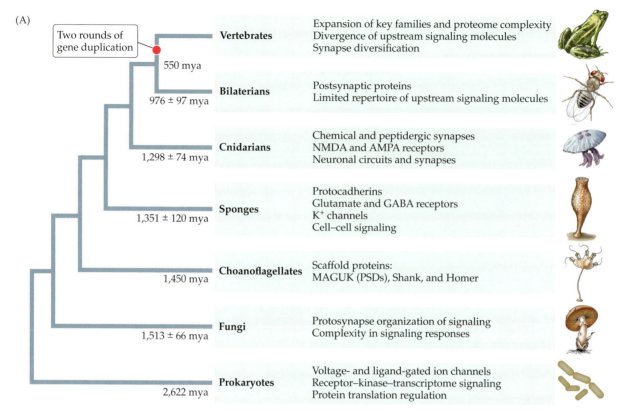

Two rounds of gene duplication

Vertebrates	Expansion of key families and proteome complexity Divergence of upstream signaling molecules Synapse diversification	
550 mya		
Bilaterians	Postsynaptic proteins Limited repertoire of upstream signaling molecules	
976 ± 97 mya		
Cnidarians	Chemical and peptidergic synapses NMDA and AMPA receptors Neuronal circuits and synapses	
1,298 ± 74 mya		
Sponges	Protocadherins Glutamate and GABA receptors K⁺ channels Cell–cell signaling	
1,351 ± 120 mya		
Choanoflagellates	Scaffold proteins: MAGUK (PSDs), Shank, and Homer	
1,450 mya		
Fungi	Protosynapse organization of signaling Complexity in signaling responses	
1,513 ± 66 mya		
Prokaryotes	Voltage- and ligand-gated ion channels Receptor–kinase–transcriptome signaling Protein translation regulation	
2,622 mya		

(B)

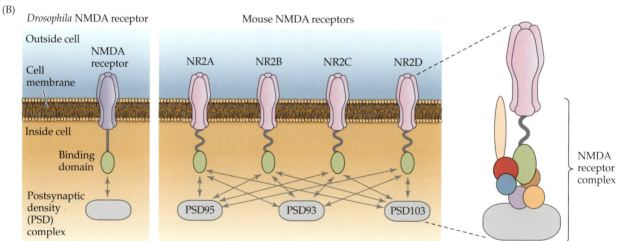

FIGURE 6.2 Evolution of synapses (A) A major event in vertebrate synaptic evolution was the duplication and divergence of many synapse-related genes about 500 million years ago. (B) The proliferation of receptor subunits in vertebrates allowed far more varied and complex synaptic signaling. (A after Emes & Grant, 2012; B after Ryan & Grant, 2009.)

neurotransmitter receptors
Proteins embedded in the postsynaptic membrane that bind to neurotransmitter in the cleft and trigger a response in the target cell.

duplication Here, replication of a gene in the course of evolution.

divergence Here, the gradual change in the structure and function of the different versions of a gene that had been duplicated.

Grant, 2012) (**FIGURE 6.2A**). Cnidarians such as jellyfish and hydras exhibit true synapses, including several peptidergic **neurotransmitter receptors** and the separate precursors to NMDA versus AMPA receptors for glutamate. Synapses in jellyfish and *Drosophila* basically work the same as in vertebrates, but an important event in vertebrate evolution was the **duplication** of many synaptic genes about 500 million years ago (Hedges, Dudley, & Kumar, 2006), followed by **divergence** in their functioning, giving rise to the bewildering array of neurotransmitter receptors that we see in mammals today (**FIGURE 6.2B**). We saw a similar duplication of *Drosophila* genes followed by divergence in vertebrates in Chapter 2, where the Hox gene array that occupies a single chromosome in flies has been duplicated across four chromosomes in mammals and has diverged in structure and function (see Figure 2.17).

■ 6.2 ■
WE CAN DIVIDE SYNAPSE STRUCTURE AND DEVELOPMENT INTO THREE PARTS

So what needs to be assembled for a fully functional synapse? We can divide the synapse into three components. The **presynaptic terminal** contains neurotransmitter-filled **synaptic vesicles** and the machinery to release them, a specialized subcellular region called the **active zone**. To respond to the neurotransmitter, the **postsynaptic region** includes one or more **postsynaptic densities** (**PSDs**), so called because they appear dark in electron micrographs and are typically opposite the presynaptic terminal's active zone. Between these two components lies the **synaptic cleft**, which is filled with an **extracellular matrix** (**ECM**), composed of a variety of long-chain molecules that polymerize with one another, including **collagen, laminin**, and **fibronectin** (Singhal & Martin, 2011) (see Figure 3.21).

We can divide the maturation of the synapse into three stages (**FIGURE 6.3**):

1. The *initial contact* between the presynaptic terminal and the target cell is mediated by the action of adhesive molecules, such as the homophilic binding of cadherins we mentioned before and discussed in Chapter 5. Both pre- and postsynaptic partners have only the rudiments of the structural specializations needed for a synapse, but each begins guiding development of those specializations in the other.

presynaptic terminal The region of the axon terminal at a synapse.

synaptic vesicles Roughly spherical containers of neurotransmitter found in presynaptic terminals.

active zone The portion of the presynaptic terminal that actively releases neurotransmitter.

postsynaptic region The portion of a neuronal target specialized to respond to a presynaptic terminal.

postsynaptic densities (PSDs) Dense regions of a postsynaptic site that are specialized to detect and respond to neurotransmitter from the presynaptic terminal.

synaptic cleft The gap between the presynaptic terminal and postsynaptic region of a synapse.

extracellular matrix (ECM) The collection of various long-chain molecules that loosely bind one another to form a layer outside many cell membranes.

collagen A long-chained filamentous protein that contributes to the formation of extracellular matrix and is especially prominent in connective tissue.

laminin A long-chained glycoprotein that readily binds other molecules and is a major component of the extracellular matrix and connective tissue.

fibronectin A long-chained glycoprotein that contributes to the formation of the extracellular matrix.

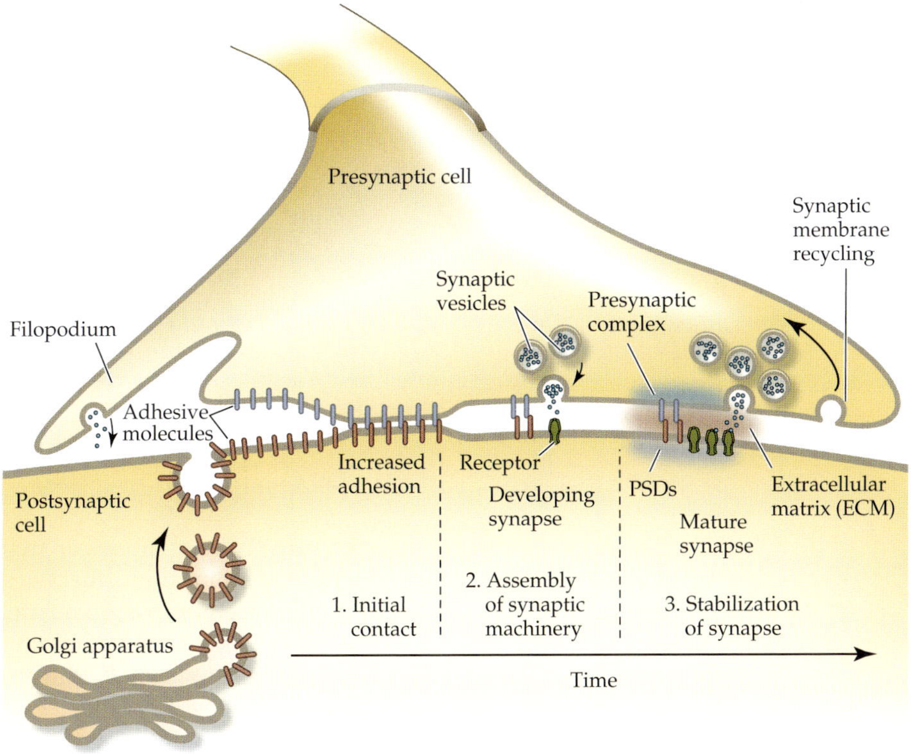

FIGURE 6.3 Overview of synaptic maturation

2. The *assembly of synaptic machinery* to release vesicles from the presynaptic terminal (Ziv & Garner, 2004) and to respond to that signal in the postsynaptic region (Bresler et al., 2004) is the second stage. The machinery is fairly minimal at first, and so transmission may be weak or unreliable. Indeed, the two partners sometimes detach to abort the beginning synapse. Otherwise, the synapse is strengthened and moves on to stabilization.

3. In the *stabilization of the synapse* the pre- and postsynaptic apparatus become more elaborate, resulting in more rapid and crisp communication. Young synapses are slow to respond and slow to recover, while mature synapses respond quickly and recover quickly. This rapid, discrete response provides more effective synaptic communication.

> Synapses that can provide brief, discrete signals can transmit more information per second than signals that, once activated, last a long time.

It's probably obvious why you want the postsynaptic cell to respond very soon after neurotransmitter is released. If a message is important (look out for that lion!), you want it to arrive soon. But it may be less obvious why, once the postsynaptic response begins, it is better for it to be short-lived rather than long lasting. It's a question of how much information you can transmit across a synapse in a short period of time. Synapses that can provide brief, discrete signals can transmit more information per second than signals that, once activated, last a long time. Think of how worthless a doorbell would be if, once you pushed the button, it rang forever. That doorbell could send only one signal, once. Likewise, a doorbell that rang for a minute would be inefficient (and annoying). If a doorbell rings very briefly, then we can send a wider range of information by, say, pushing it once a minute versus 60 times per minute (also pretty annoying, but it sends a message). Likewise, neural signals that are short and discrete, delivering crisp bursts of signal, can provide a wide range of information depending on how rapidly the signals arrive.

As synaptic development continues, chemical signals from each partner, either secreted or through cell-cell contact, induce further development in the other. We'll discuss the molecular mechanisms underlying these signals in each of the three stages in turn.

How's It Going?

1. What precursors to synaptic transmission are seen in unicellular organisms?
2. What happened about 500 million years ago to increase the diversity of synaptic signaling?
3. What are the three main divisions of any synapse?
4. Name the three main stages in synapse development.

■ 6.3 ■

A SYNAPSE BEGINS WITH ADHESION

neural cell adhesion molecule (NCAM) Also called *neural-CAM*. An adhesive molecule expressed by many neurons.

ephrins A family of membrane-bound signaling molecules that bind to ephrin receptors, which are part of the receptor tyrosine kinase (RTK) superfamily.

The same adhesive molecules that play a role in axon guidance, bringing the two cells into contact in the first place, also serve to keep them together long enough for other processes to begin that may cement the synapse into place. Some of the adhesive molecules at work in budding synapses are **neural cell adhesion molecule (NCAM)** and the **ephrins** (Dalva, McClelland, & Kayser, 2007; Halbleib & Nelson, 2006) that we mentioned in the previous chapter on axon guidance. Foremost of those adhesive factors holding the new synapse together are the cadherins we mentioned earlier, part of a superfamily of proteins called *protocadherins*, altogether consisting of over 100 different genes (Junghans,

Haas, & Kemler, 2005; Morishita & Yagi, 2007). Because cadherins and protocadherins tend to stick together in a homophilic fashion, like to like, the wide variety of members of these families provides a way for axons and their targets to display some discrimination, preferring to bind to partners expressing the same specific variety of cadherin.

When the axonal growth cone and its target first encounter each other, neither has much in the way of specializations for forming a synapse. The growth cone releases a little neurotransmitter more or less constantly, and the target cell has only diffusely distributed receptors. But once they remain stuck together by cadherins long enough, each will induce the other to develop larger and more differentiated machinery to communicate synaptically.

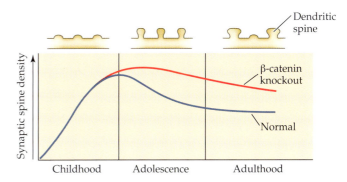

FIGURE 6.4 **The density of synaptic spines in cortical neurons changes with age.** (After Bian et al., 2015.)

One of the first steps after initial homophilic binding of cadherin receptors in a presynaptic axon and a potential target is the formation of a complex between the intracellular domain of cadherin with our old friend from Chapter 1, β-**catenin**, which links the intracellular domain of the cadherin receptor to actin filaments inside the cell (Calabrese, Wilson, & Halpain, 2006). This linkage serves to anchor the membrane-bound cadherin receptor to the cell's cytoskeletal interior and allows the assembly of scaffolding proteins to complete the postsynaptic structures. The importance of cadherins in the maintenance of synapses was demonstrated in research on dendritic spines in the somatosensory cortex of mice (Bian, Miao, He, Qiu, & Yu, 2015). The number of spines increases dramatically after birth, but many of them disappear during the transition from adolescence to adulthood (**FIGURE 6.4**). This "pruning" of spines, and therefore the synapses upon those spines, is part of an overall process of synapse elimination that we'll discuss in detail in Chapter 8. When the researchers knocked out the gene for β-catenin, that prevented the pruning of synaptic spines (see Figure 6.4, red line). This result suggested that the cadherin–β-catenin complex was important for the loss of synaptic spines and therefore played a role in synaptic stabilization. Bian and colleagues tested this idea in several ways, as we'll see next.

β-**catenin** A transcription factor that also regulates cell adhesion and plays a role in several stages of neural development.

RESEARCHERS AT WORK

Dendritic Spines Compete for Survival

■ **QUESTION**: What determines which synaptic spines will be retained and which will be lost?

■ **HYPOTHESIS**: The cadherin–β-catenin complex stabilizes the synapse at retained spines, while insufficient cadherin–β-catenin complexes make spines vulnerable to pruning.

■ **TEST 1**: The authors coated plastic beads with cadherin and introduced them to cultured neurons that express β-catenin. Would those spines that contact cadherin-coated beads be more likely to remain?

■ **RESULT 1**: Indeed, those dendritic spines that contacted a cadherin-coated bead grew more than spines contacting control, uncoated beads (Bian et al., 2015) (**FIGURE 6.5A**). This result suggested that contacting a cadherin-expressing surface encourages spine growth. However, another important finding was that when a spine contacted a cadherin-coated bead, its *neighboring* spines were more likely to disappear than the neighbors of spines contacting control beads (Bian et al., 2015) (**FIGURE 6.5B**). This result suggested that when one spine

continued

is stabilized, its neighbors are destabilized, and therefore likely to disappear. Perhaps neighboring spines compete for a limited supply of β-catenin to stabilize postsynaptic structures and therefore the spines. Thus when a presynaptic cell expressing cadherin contacts a spine, the spine gathers β-catenin to stabilize postsynaptic structures, depriving neighboring spines of β-catenin. This hypothesis suggested another test.

FIGURE 6.5 Dendritic spines compete for β-catenin to survive. (After Bian et al., 2015.)

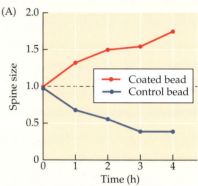

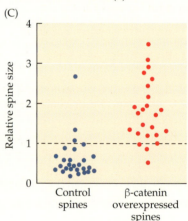

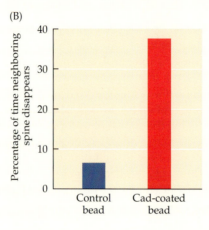

■ **TEST 2**: Induce a few neurons in the somatosensory cortex of mice to overexpress β-catenin, and then follow the fate of their spines versus the spines of neurons that do not overexpress β-catenin.

■ **RESULT 2**: Spines overexpressing β-catenin grew larger than control spines without extra β-catenin (Bian et al., 2015) (**FIGURE 6.5C**).

■ **CONCLUSION**: Dendritic spines of a given neuron compete with one another for survival. When a cadherin-expressing axon terminal contacts a cadherin-expressing spine (via homophilic binding), the spine gathers β-catenin to anchor the intracellular portion of its cadherin in place, which allows further maturation of the postsynaptic structures to proceed. Apparently there is a limited supply of β-catenin in a given neuron, such that when one of its spines accumulates β-catenin to stabilize the synapse, less β-catenin is available for other nearby spines. Spines with sufficient β-catenin remain, while spines with insufficient β-catenin disappear. This means spines on a given dendrite normally compete for survival, with those receiving contact from a cadherin-positive axon terminal more likely to persist, while their near neighbors are more likely to disappear. If there is no β-catenin, as in the knockout mice (see Figure 6.4, red line), then all spines are on an equal basis in the competition (none of them have any β-catenin) and so they all remain.

In the competition between synaptic spines for survival, presumably the spines that are contacted first, and therefore start gathering β-catenin first, have an advantage over those contacted later. Also, because homophilic binding of cadherins is required for a spine to accumulate β-catenin, spines contacted by an axon expressing the matching cadherin will be favored over those receiving contact from an axon expressing a different cadherin. Given all the different combinations of adhesive factors that neurons and their targets can use (see Chapter 5), the two participants in any potential synapse can be quite specific in choosing their partners. For that specificity to be implemented, there has to be some system like this one, where several different initial synapses compete for some limited resource (like β-catenin), for the appropriate synapse to be retained while inappropriate, or perhaps *less* appropriate, synapses disappear. In Chapter 8 we'll return to the issue of synapses that disappear in development, and we'll find that neural activity is also crucial for the survival of synapses—those that are active are better able to gather those limited resources to survive, while inactive synapses next door fade away.

The loss of some initially formed synapses may sound like a bad thing, or may sound inefficient, but losing some of those synapses is almost certainly a good thing, as we'll see next.

How's It Going?

1. What are three classes of adhesive molecules that regulate initial synaptic contact?
2. Describe the experiment that indicated that synaptic spines compete for intracellular β-catenin to determine which synapses remain.
3. How is variation in the adhesiveness achieved to promote specificity of connections?

---

fragile X syndrome (FXS) A disorder caused by extended repeats of CGG trinucleotides in the gene named *fragile X mental retardation 1* (*FMR1*).

intellectual disability A lifelong impairment in intellectual function and adaptive behavior, formerly known as mental retardation.

■ 6.4 ■
FRAGILE X SYNDROME SUGGESTS THERE CAN BE TOO MUCH OF A GOOD THING

Fragile X syndrome (**FXS**) is caused by extended repeats of the trinucleotide CGG in a gene on the X chromosome, which is called *fragile X mental retardation 1* (*FMR1*). The repeats effectively silence the gene (Kazdoba, Leach, Silverman, & Crawley, 2014). "Fragile X" stems from the fact that when the CGG repeats are very long, they form a constriction in the chromosome that can be seen in karyotypes. The gene name also illustrates a hazard of nomenclature in biology, as the term *mental retardation* has come to have a pejorative connotation (Ansberry, 2010), as the mother Kathleen related in the introduction to this chapter. There is a movement afoot to replace *mental retardation* with the term **intellectual disability**, a lifelong impairment of intellectual function and adaptive behavior. As with most X-linked disorders, FXS is less common in females than in males, who have only one copy of the gene available to produce the FMR1 protein (Lozano, Rosero, & Hagerman, 2014). Thus Kathleen, whom we met at the start of this chapter, may have had more than a typical number of repeats in *FMR1* on one of her X chromosomes, but the allele on her other X probably had few repeats, so she could still make

Coping with fragile X syndrome This young man with fragile X syndrome hugs his mother on his way to see his tutor. (© Chris Walker/Chicago Tribune/Getty Images.)

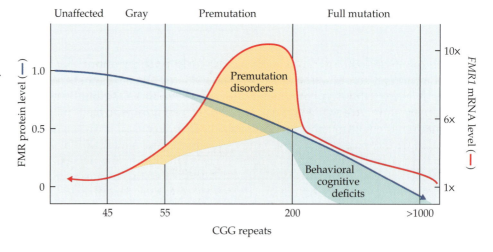

FIGURE 6.6 Trinucleotide repeats in the gene for fragile X syndrome (FXS) (After Lozano et al., 2014.)

ataxia   Lack of motor coordination.

functional FMR1 protein. Many children with FXS show symptoms of autism, such as reduced speech and poor social skills (although Kathleen's son had difficulties with language, he was fairly social and easily made eye contact with others). FXS is often accompanied by an elongated face, protruding ears, hyperflexible ("double-jointed") fingers, and low muscle tone, another illustration of genetic pleiotropy.

The most commonly inherited cause of intellectual disability (Lyons, Kerr, & Mueller, 2015), FXS results when there are more than 200 CGG repeats. In parents carrying more than 50 or so CGG repeats in *FMR1*, there is sometimes an expansion of the repeats in gamete production, so the child exhibits the syndrome. In addition to potentially passing on an extended repeat, people with 55 to 200 CGG repeats may show more subtle symptoms of **ataxia** (lack of motor coordination) and tremor. Interestingly, as the number of repeats increases, there is initially an increase in *FMR1* mRNA levels, as if the brain is trying to compensate for a less effective protein (**FIGURE 6.6**, red line), yet protein levels still drop (Figure 6.6, blue line) (Kenneson, Zhang, Hagedorn, & Warren, 2001), suggesting that the transcript with extended repeats is not translated efficiently to produce enough of the protein, or that extended repeats result in alternate transcripts that are dysfunctional (Peschansky et al., 2015). The greater number of transcripts may overcome reduced translation until there are so many repeats that protein levels dip below a certain level and symptoms result.

The FMR1 protein is an RNA-binding protein that typically acts by repressing translation of a variety of products, including scaffolding proteins (such as PSD95), as well as kinases, potassium channels, and neurotransmitter receptors (Richter, Bassell, & Klann, 2015). So the problems in FXS are caused by an *overabundance* of several neural proteins that would normally be in short supply, and therefore subject to competition between synapses, when FMR1 is fully functional. Mouse models of FXS have been generated by knocking out the *Fmr1* gene (Kazdoba et al., 2014), and one of the most prominent neural phenotypes is an *increased* number of dendritic spines in cortical neurons. It may be that there are too many scaffolding proteins and therefore not enough competition of the sort we saw in the elimination of dendritic spines above (see Figure 6.4). Thus some pharmacological treatments that are being explored include trying to interfere with the overabundance of particular types of synapses, such as a particular metabotropic glutamate receptor (Dolen et al., 2007). It may seem counterintuitive that the mental symptoms of FXS may be due to a failure to eliminate synapses during development, but we'll see in Chapter 8 that neural activity normally influences which synapses are retained, and that the coming and going of synapses continues throughout life, not just while we're growing up.

> The problems in FXS are caused by an *overabundance* of several neural proteins that would normally be in short supply, and therefore subject to competition between synapses.

But we'll put off the discussion of the role of activity in synapse survival for now so that we can talk about the second stage of synaptic development. After matching of adhesive factors brings appropriate pre- and postsynaptic partners together, there is a specific signaling system that induces the presynaptic partner to assemble mechanisms to release transmitter and induces the postsynaptic partner to assemble mechanisms to respond to the transmitter, which we take up next.

How's It Going?

1. What is the genetic mechanism at work in fragile X?
2. How does the genetic fault in fragile X affect the number of synapses in the brain?

■ 6.5 ■

PRE- AND POSTSYNAPTIC PARTNERS TIGHTLY ANCHOR ONE ANOTHER AS A SYNAPSE DEVELOPS

Once the axonal growth cone contacts a target that is "suitable" (as determined by adhesive molecules), axon and target then induce each other to construct the apparatus needed for synaptic signaling (**FIGURE 6.7A**). While many different proteins are brought into play in this process, a particular pair of membrane-embedded adhesive molecules is known to be crucial. The presynaptic axon produces a family of proteins called **neurexins** (Ushkaryov, Petrenko, Geppert, & Sudhof, 1992), which bind to postsynaptic receptors called **neuroligins** (**FIGURE 6.7B,C**; Lise & El-Husseini, 2006; Scheiffele, Fan, Choih, Fetter, & Serafini, 2000).

neurexins A family of membrane-bound proteins found in presynaptic terminals that bind to neuroligins.

neuroligins A family of membrane-bound proteins found in postsynaptic sites that bind neurexins.

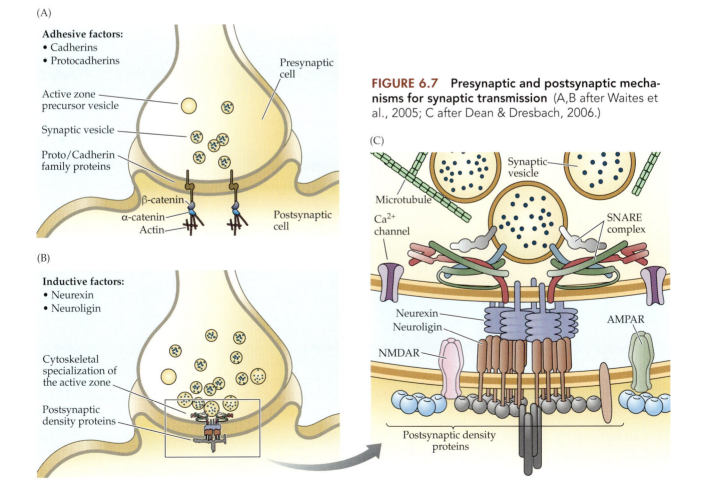

FIGURE 6.7 Presynaptic and postsynaptic mechanisms for synaptic transmission (A,B after Waites et al., 2005; C after Dean & Dresbach, 2006.)

As a mnemonic to keep track of which protein is found on which cell, you might focus on the *exi* in *neurexin* as leading to the **exi**t of transmitter, while *lig* in *neuroligin* might remind you of the **lig**and binding that happens postsynaptically.

When these two membrane proteins bind together, activation of neurexin in the presynaptic partner triggers the assembly of neurotransmitter-releasing machinery into an *active zone*, while activation of neuroligin in the target cell triggers the assembly of one of the PSDs mentioned above (Craig & Kang, 2007), including so-called scaffolding proteins (Elias et al., 2006) and receptors, to respond to that neurotransmitter (Dean & Dresbach, 2006) (see Figure 6.7C). Let's discuss how scientists demonstrated the importance of these two players in synapse development.

RESEARCHERS AT WORK

Presynaptic and Postsynaptic Receptors Trigger Synaptic Development

■ **QUESTION**: Is the interaction of neurexin and neuroligin sufficient to trigger synaptic development?

■ **HYPOTHESIS**: The binding of presynaptic neurexin and postsynaptic neuroligin is not merely something that happens during synaptic development, but actually *causes* synapse development to proceed.

■ **TEST**: Arrange to have neuroligin alone contact neuronal axons in vitro and see if that's sufficient to trigger the beginnings of presynaptic development. First transfect nonneural cells that do not normally form synapses (HEK293 cells) with neuroligin, and see if these cells can now induce formation of synapses in axonal growth cones from brainstem explants. As a further test, coat inert silica beads with neuroligin and apply them to hippocampal neurons in culture. When the beads rest against a neuron, will it begin forming presynaptic structures?

■ **RESULT**: Both the transfected nonneural cells expressing neuroligin and the silica beads coated with neuroligin successfully bound neurexin on the neurons. In both cases, there was a clustering of neurexin on the neuron's surface at the point where it contacted the provided neuroligin. In both cases the neuron also began assembling presynaptic-specific proteins such as *synapsin* and *synaptotagmin* (discussed below), which are crucial for transmitter release (**FIGURE 6.8A**). Importantly, HEK293 cells that had not been transfected to express neuroligin did not have these effects. Likewise, silica beads that had been coated with another protein (phosphatase) or with lipid alone or that had not been coated at all did not trigger clustering of synapsin at the point of contact (**FIGURE 6.8B**) (Dean et al., 2003).

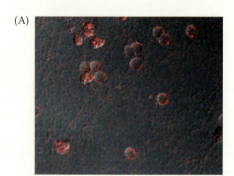
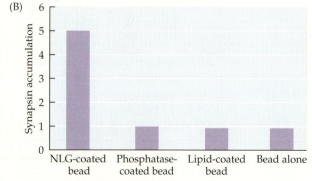

FIGURE 6.8   Neuroligin is sufficient to trigger synapse development. (A) Neuroligin-coated beads induced cultured neurons to produce the presynaptic protein synaptotagmin (red). (B) The presynaptic protein synapsin accumulates only in response to beads coated with neuroligin (NLG). (From Dean et al., 2003.)

■ **CONCLUSION**: Binding of presynaptic neurexin with postsynaptic neuroligin is sufficient to trigger the initial stages of presynaptic development. The first effect of neuroligin binding to the neuron is a clustering together of neurexin molecules in the presynaptic terminal, and this clustering of presynaptic neurexin is enough to trigger the assembly of mechanisms to release vesicular neurotransmitter.

Converse experiments, artificially providing neurexin by either transfecting cells with the gene or coating beads with the protein, were shown to trigger clustering of neuroligin and *postsynaptic* components, including neurotransmitter receptors and scaffolding proteins, in neurons (Graf, Zhang, Jin, Linhoff, & Craig, 2004). Thus while there are undoubtedly other signaling events going on between the two cells forming a synapse, the binding of presynaptic neurexin and postsynaptic neuroligin is sufficient to at least initiate assembly of both pre- and postsynaptic machinery needed for synapse formation.

To see if the neurexin-neuroligin activation is *necessary* for synapse development in vivo, researchers tried knocking out the three mammalian neuroligin genes in mice. Knocking out any one of the three neuroligins produced viable mice with apparently functional synapses, but this could be due to redundancy, where the remaining neuroligins filled in for the missing one. On the other hand, triple knockout mice, with all three neuroligin genes disabled, died at birth because of respiratory failure (perhaps due to failed neurotransmission at muscles needed for breathing). Examining the brains of the animals before birth revealed that while synapses appeared morphologically normal, synaptic transmission was dramatically reduced compared with wild-type mice, with reduced synaptic vesicle proteins presynaptically and reduced neurotransmitter receptor clustering postsynaptically (Varoqueaux et al., 2006). Thus the binding of neurexins and neuroligins appears to be both necessary and sufficient to trigger the crucial first steps of synapse formation.

Now let's consider the structures that must develop on each side of the synapse in order for it to function properly. On the postsynaptic side, activation of neuroligins induces the assembly of **postsynaptic density (PSD) proteins** (or *scaffolding proteins*, because they seem to provide a scaffold for the postsynaptic machinery), such as PSD95, PSD93, and many others (Craig & Kang, 2007). The PSD proteins are part of a family of at least 400 proteins sharing a domain (the PDZ domain) that binds the C-terminus of other proteins, making a lattice that can be built upon. Among the various PSD proteins are membrane-associated guanylate kinases (MAGUKs) that bind to the intracellular portion of neuroligin (remember this is the postsynaptic partner) and neurotransmitter receptors (Graf et al., 2004; Varoqueaux, Jamain, & Brose, 2004), as well as actin-binding proteins associated with the cell's cytoskeleton (see Figure 6.7C), so you can see how they would stabilize the postsynaptic machinery.

On the presynaptic side, a family of long-chain proteins, called **SNAREs** (for **s**oluble *N*-ethylmaleimide-sensitive **a**ttachment **re**ceptors), that were first found to be important for regulating membrane fusion in the cell's Golgi apparatus, were later found to also regulate vesicle fusion in synapses (Ungar & Hughson, 2003). One of the important SNAREs (synaptobrevin) is anchored in each vesicle and serves as a tether that other SNAREs can grip, to move the vesicle about. Two other SNAREs (syntaxin and SNAP25) are themselves anchored in the postsynaptic membrane (**FIGURE 6.9A**). Thus when these three proteins bind each other, making a **SNARE complex**, they pull the vesicle right up to the postsynaptic membrane (**FIGURE 6.9B, Panels 1 and 2**), where it's ready to fuse and release its contents. When the action potential arrives at the postsynaptic terminal, Ca^{2+} enters and binds

postsynaptic density (PSD) proteins Also called *scaffolding proteins*. A large family of proteins that bind many other proteins to anchor the components for postsynaptic responses.

SNAREs **S**oluble-*N*-ethylmaleimide-sensitive **a**ttachment **re**ceptors, a family of long-chain proteins that regulate membrane fusion.

SNARE complex A combination of SNAREs that anchor a vesicle into position for fusion to the presynaptic membrane and release of neurotransmitter.

(A)

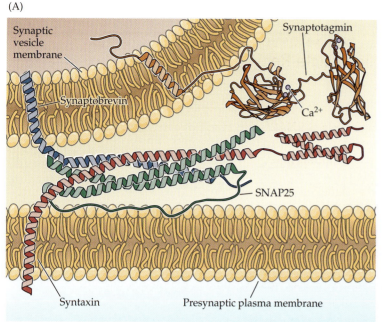

Synaptic vesicle membrane

Synaptobrevin

Synaptotagmin

Ca²⁺

SNAP25

Syntaxin

Presynaptic plasma membrane

FIGURE 6.9 **Mechanisms mediating synaptic transmitter release** (A after Chapman, 2002; B after Zhou et al., 2015.)

(B) (1) Binding of SNAREs docks the vesicle.

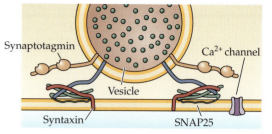

Synaptotagmin

Ca²⁺ channel

Vesicle

Syntaxin

SNAP25

(2) Synaptotagmin engages the SNAREs.

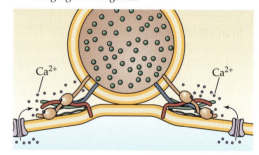

(3) When the action potential arrives, incoming Ca²⁺ binds synaptotagmin, which binds membranes, bringing them together.

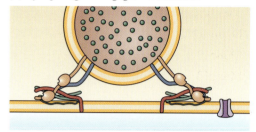

Ca²⁺

Ca²⁺

(4) The membranes fuse, releasing neurotransmitter into the cleft.

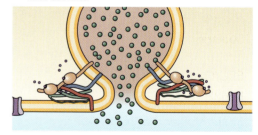

synaptotagmin An enzyme that binds to vesicle membranes and serves as a calcium sensor to trigger fusion of the membrane and release of neurotransmitter.

to another protein attached to the vesicle membrane, **synaptotagmin**, which then catalyzes the fusion of the vesicle and terminal membranes, releasing neurotransmitter from the vesicle into the synaptic cleft (**FIGURE 6.9B**, Panels 3 and 4). Thus synaptotagmin serves as a calcium sensor, triggering neurotransmitter release. These proteins represent only a fraction of those found in presynaptic terminals; a more complete list is shown in **TABLE 6.1**.

To talk aboutt the third stage of synapse development, the long-term stabilization and maintenance of a synapse, we will turn to the best-studied synapse, which is that between a motor neuron and its target muscle. Later in the chapter we'll see how the lessons learned in neuromuscular systems seem to apply to other synapses as well.

TABLE 6.1
Synaptic Proteins Involved in Active Zone Formation and Synaptic Vesicle Recycling

Protein	Function
SYNAPTIC VESICLE EXOCYTOSIS	
Synaptobrevin/Vamp, Snap25, syntaxin	Components of SNARE complex involved in synaptic vesicle docking and fusion
NSF, α- and β-SNAPs	Dissociation of SNAREs
Synaptotagmins	Calcium sensors; interact with syntaxin and Rim
N- and P/Q-type calcium channels	Calcium influx
Munc18	Binds and negatively regulates syntaxin and synaptic vesicle fusion
Munc13	Involved in synaptic vesicle priming; interacts with Rim; displaces Munc18
Rim1α	Involved in synaptic vesicle priming; interacts with Munc13, RimBPs and synaptotagmin
Rab3A	Regulates synaptic vesicle cycle; interacts with Rabphilin, Doc2, Pra1 and Rim
Complexin	Binds and regulates SNARE complex
STRUCTURAL MOLECULES OF ACTIVE ZONES	
CASK	CaMKII domain-containing MAGUK; forms a complex with MINTs, veli and calcium channels, neurexin and SynCAM
MINTs	Munc18-interacting molecules; found in complex with CASK and veli
Veli	Found in complex with CASK, MINTs and calcium channels
Bassoon and piccolo	Large structural proteins of the CAZ that interact with Pra1, profilin, Abp1 and ERC
Synapsins	Anchoring of synaptic vesicles to actin; regulation of reserve pool of synaptic vesicles
α-Liprin	Scaffold proteins that bind Rim, ERC and LAR
Spectrin	Cortical cytoskeletal protein; interacts with actin, cell-adhesion molecules and receptors
ERC/Cast	CAZ proteins that interact with piccolo, bassoon, Rim, and Liprin
RimBP	Rim binding protein; component of the CAZ
SYNAPTIC ADHESION AND SIGNALING	
β-Neurexin	Presynaptic adhesion; interacts with CASK and neuroligins; can trigger active zone formation
α-Neurexin	Presynaptic adhesion; binds and localizes N-type calcium channels to active zones
SynCAM	Homophilic cell-adhesion molecule that can induce the formation of functional active zones
CNR	Cadherin-related cell-adhesion molecules
N-cadherin	Neuronal cell-adhesion molecule; interacts with catenins and the spectrin/actin cytoskeleton
β-Catenin	Binds the C-terminal tail of N-cadherin

continued

TABLE 6.1
continued

Protein	Function
Neuroligin	Postsynaptic cell-adhesion molecule; interacts with presynaptic β-neurexins; induces active zone formation
NCAM	Member of the IgG superfamily of adhesion molecules that binds spectrin
Narp	Neural activity-regulated pentraxin; a secreted molecule that promotes clustering of AMPARs
Ephrin-B	Synaptogenic factor that promotes clustering of NMDARs after binding the EphB receptors
EphB	Tyrosine kinase receptor for ephrin-B, binds and clusters NMDAR
LAR	Receptor protein tyrosine phosphatase; binds liprin and regulates active zone assembly
SYNAPTIC VESICLE ENDOCYTOSIS	
Clathrin	Involved in synaptic vesicle endocytosis; interacts with dynamin, AP2, amphiphysin and other molecules
Dynamin	GTPase involved in pinching off synaptic vesicles during endocytosis
Amphiphysin	Binds dynamin and is involved in synaptic vesicle endocytosis
Actin	5 nm cytoskeletal filament; found surrounding the active zone as well as within dendritic spines

Source: Ziv & Garner, 2004.

Notes: AMPAR, α-amino-3-hydroxy-5-methyl-4-isoxazole propionic acid receptor; AP2, adaptor protein 2; CaMKII, calcium/calmodulin-dependent kinase II; *CASK*, calcium/calmodulin-dependent serine protein kinase; CAZ, cytoskeletal matrix assembled at active zones; IgG, immunoglobulin G; MAGUK, membrane-associated guanylate kinase-like proteins; NMDAR, *N*-methyl-D-aspartate receptor; RimBP, Rim base-pair; SNARE, soluble NSF (*N*-ethylmaleimide-sensitive fusion protein) accessory protein SNAP receptor.

How's It Going?

1. What are the presynaptic and postsynaptic signals that promote development of synaptic signaling mechanisms?
2. Describe the molecules that assemble to produce the postsynaptic responsive region.
3. Describe the presynaptic molecules that control vesicle release. Which molecule serves as a calcium sensor?

■ 6.6 ■

NEUROMUSCULAR JUNCTIONS ILLUSTRATE THAT SYNAPSE FORMATION IS A DANCE FOR TWO (OR MORE)

Before talking about the "dance for two" between developing motor neurons and their muscle targets, you need to know a bit about muscle development. In vertebrates, muscle cells are enormous cells with multiple nuclei, formed when individual cells called **myoblasts**, which have embarked on only the initial stages of differentiation into muscles, fuse together to form **myotubes** (Abmayr & Pavlath, 2012), long cylindrical structures that start making many proteins required for contraction, including actin and myosin. Once that ma-

myoblasts   An embryonic muscle cell of mesodermal origin.

myotubes   Multinucleate cells formed from the fusion of several myoblasts.

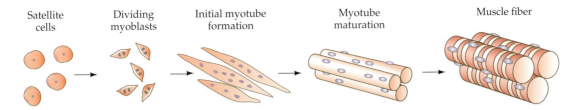

Satellite cells   Dividing myoblasts   Initial myotube formation   Myotube maturation   Muscle fiber

FIGURE 6.10 Development of muscle fibers

chinery has been organized so that the cell can contract, we call it a **myofiber** or, more commonly, a *muscle fiber* (**FIGURE 6.10**). By now you won't be surprised to learn that the arrival of motor neuron axonal growth cones facilitates the maturation of myotubes into myofibers both in vivo and in vitro. Later in life, the pattern of neural activity from innervating motor neurons can shift the biochemistry and fast- versus slow-contractile properties of muscle (Rana, Gundersen, & Buonanno, 2009; Schiaffino, Sandri, & Murgia, 2007), a theme we'll return to in Chapter 8. If there is a single overlying theme of developmental biology, it is the interaction of cells to direct one another's fate, and the interaction of motor neurons and muscle in forming a synapse is no exception. In fact, we'll see that glia also play a role. The formation of the **neuromuscular junction** (**NMJ**) turns out to be a dance for *three*.

The adult NMJ, the synapse between a motor neuron axon and its muscle target, is very large but otherwise relatively simple in structure (**FIGURE 6.11**). The motor neuron's axon terminal contains active zones, regions housing the molecular machinery for release of neurotransmitter, which in the case of vertebrates is **acetylcholine** (**ACh**). Opposite the axon terminal the muscle fiber forms deep *junctional folds*, crevasses that are filled with the enzyme

> If there is a single overlying theme of developmental biology, it is the interaction of cells to direct one another's fate.

myofibers Also called *muscle fibers*. Multinucleate cells formed from the fusion of several myoblasts that have assembled the molecular machinery for contraction.

neuromuscular junction (NMJ) The large chemical synapse found between the terminal of a motor neuron and its muscle target.

acetylcholine (ACh) The neurotransmitter released by parasympathetic fibers as well as vertebrate motor neurons.

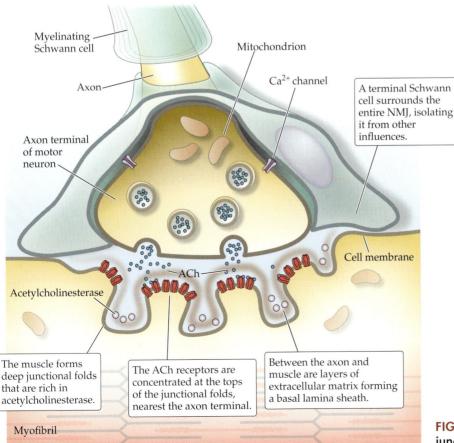

Myelinating Schwann cell
Mitochondrion
Axon
Ca²⁺ channel
A terminal Schwann cell surrounds the entire NMJ, isolating it from other influences.
Axon terminal of motor neuron
Cell membrane
ACh
Acetylcholinesterase
The muscle forms deep junctional folds that are rich in acetylcholinesterase.
The ACh receptors are concentrated at the tops of the junctional folds, nearest the axon terminal.
Between the axon and muscle are layers of extracellular matrix forming a basal lamina sheath.
Myofibril

FIGURE 6.11 The neuromuscular junction (NMJ)

acetylcholinesterase An enzyme that breaks down acetylcholine to halt neurotransmission.

acetylcholine receptors (AChRs) The integral membrane proteins, consisting of several subunits, that respond to the neurotransmitter acetylcholine.

basal lamina A particularly thick layer of extracellular matrix that surrounds mature muscle fibers and many organs.

terminal Schwann cell Specialized Schwann cells that surround and cap neuromuscular junctions, effectively isolating them from other influences.

acetylcholinesterase, which metabolizes ACh to inactivate it. The **acetylcholine receptors (AChRs)** that respond to the transmitter are concentrated at the top of the junctional folds, nearest the axon terminal. Between the axon and muscle is the extracellular matrix (ECM) we discussed earlier. Recall from our discussion of ECM in Chapter 3 that some organs, including adult muscle fibers, have multiple layers of ECM collectively known as the **basal lamina**. We'll see shortly that the portion of basal lamina at the NMJ harbors chemical cues that can play a role in regeneration of innervation. Finally, there is a glial cell that covers the motor neuron's axon terminal and is therefore called a **terminal Schwann cell** (or *perisynaptic Schwann cell*), effectively capping and isolating the NMJ from outside influences. This type of Schwann cell is distinct from those that provide myelin sheaths in the peripheral nervous system, which we'll discuss at the end of this chapter.

Because of the NMJ's large size, its isolation from other synapses, and its position in the periphery, we have been able to learn more about the formation of the NMJ than about any other synapse (Darabid, Perez-Gonzalez, & Robitaille, 2014). Fortunately, much of what has been learned about NMJ development appears to apply to other synapses, too. One important signal during NMJ development is the neurotransmitter, which in vertebrates is ACh. (In invertebrates, glutamate often serves as the neurotransmitter at NMJs, but vertebrate NMJs are cholinergic, specifically using nicotinic AChRs.) Before the motor neuron axonal growth cone and the myotube have even contacted each other, the motor neuron axonal growth cone is releasing ACh, and the myotube is expressing AChRs to respond to the transmitter.

Because growth cones do not have the obvious ultrastructural components of transmitter release, it came as something of a surprise to learn that they release ACh. The spontaneous release of ACh from motor neuron growth cones was demonstrated by using patch-clamp techniques to affix an "outside out" membrane from a muscle cell, so that ACh receptors were facing outward, on the tip of a microelectrode, then voltage-clamping the patch. By monitoring the current passing through the electrode when a molecule of ACh opened a receptor, this "sniffer-patch detector" apparatus served as a very sensitive ACh detector (Hume, Role, & Fischbach, 1983). When brought close to a motor neuronal growth cone, the ACh receptors would open, indicating that growth cones were releasing the neurotransmitter in both vertebrate systems (Young & Poo, 1983) and invertebrate systems (Yao, Rusch, Poo, & Wu, 2000). Adding ACh blockers to the medium interfered with these "blips," further supporting the idea that ACh was being released by the growth cone. This release of neurotransmitter is much more modest than will be seen in the mature motor neuron axon terminal, and indeed the growth cone has relatively little of the machinery needed for synaptic release. Nevertheless, this low-level neurotransmitter release is a first step in a series of signals between the axon terminal and its target (van Kesteren & Spencer, 2003). For example, within a few minutes of contact with a muscle target, the motor neuron growth cone increases its spontaneous release of ACh.

The initial contact also induces rapid changes in the muscle target. As we mentioned earlier, myotubes are expressing AChRs before they are contacted by any motor neuronal growth cones, indeed sometimes before the motor neuron axons have even left the CNS. Initially the AChRs are found along the entire length of the myotube, and then they clump together in a few apparently randomly placed "hotspots." These AChR hotspots that form before the motor neuron axon arrives tend to be in the middle of the myotube, so they may help guide the growth cone to the approximately correct site. But the preexisting hotspots cannot be crucial for NMJ development, because in mutants that fail to form the hotspots prior to innervation, the NMJ still forms near the middle of the muscle (S. Lin, Landmann, Ruegg, & Brenner, 2008; Zhang, Lefebvre, Zhao, & Granato,

2004). Once the growth cone gets to the myotube, the hotpots consolidate into a single large clump of receptors seen in the mature NMJ.

When the motor neuronal growth cone contacts the myotube, those preexisting AChRs embedded in the muscle membrane *migrate* to that site of contact (Liu & Westerfield, 1992). What's more, contact with a motor neuron axon also causes the muscle to increase the production of AChRs and preferentially insert them at the developing synapse (Role, Matossian, O'Brien, & Fischbach, 1985). These two changes, the aggregation of existing AChRs and the increased production of new AChRs, are triggered by two different chemical signals from motor neurons, as we'll see next.

How's It Going?

1. Describe the stages of muscle development.
2. What are the three cells that interact at a neuromuscular junction?
3. At what point does the nerve growth cone begin releasing ACh and the target muscle making receptors?
4. What two processes promote the clustering of ACh receptors in one place?

■ 6.7 ■

MOTOR NEURONAL AGRIN PROMOTES THE AGGREGATION OF ACETYLCHOLINE RECEPTORS

As we mentioned above, two different factors from motor neurons induce the newly innervated muscle to concentrate existing AChRs at the NMJ and preferentially insert newly made AChRs there. The migration of existing AChRs offers a reminder that cell membranes are really just layers of lipid (fat). Transmembrane proteins like transmitter receptors span the membrane because the portion of the protein embedded within the membrane (the *transmembrane domain*) is composed of amino acids that are especially lipophilic (more readily dissolved in lipid than in water). In the absence of anything tethering such transmembrane proteins in place, they are free to "float" along the membrane from one part of the cell surface to another. So another step that has to happen in the maturation of the NMJ, after existing AChRs have migrated to the developing NMJ and the muscle preferentially inserts additional AChRs there, is there must be some method of anchoring those AChRs to keep them in place. Keeping the clump of AChRs near the terminal is crucial for effective communication between adult motor neuron and muscle. It would do no good to form a postsynaptic site if it could simply float out from under the axon terminal.

> It would do no good to form a postsynaptic site if it could simply float out from under the axon terminal.

In addition to its modest spontaneous release of ACh, the motor neuron growth cone releases another signal, a proteoglycan called **agrin**, so named because it facilitates this *aggregation* of AChRs (Godfrey, Nitkin, Wallace, Rubin, & McMahan, 1984). In fact, agrin is normally produced by both motor neurons and muscles, as well as the Schwann cells associated with NMJs, but alternative splicing produces a distinct isoform of agrin in motor neurons that is 1,000-fold more effective at causing AChR aggregation in the muscle fiber (Hoch, Campanelli, Harrison, & Scheller, 1994). What's more, mutations that disrupt only this motor neuron–specific agrin isoform severely disrupt AChR aggregation (W. Lin et al., 2001).

When the motor neuron growth cone contacts the muscle fiber, it releases agrin to cause acetylcholine receptors to aggregate at that site. In fact, more than one motor neuron may contact the muscle fiber initially, but eventually

agrin   A large proteoglycan that serves as a signal and inductive factor at neuromuscular junctions.

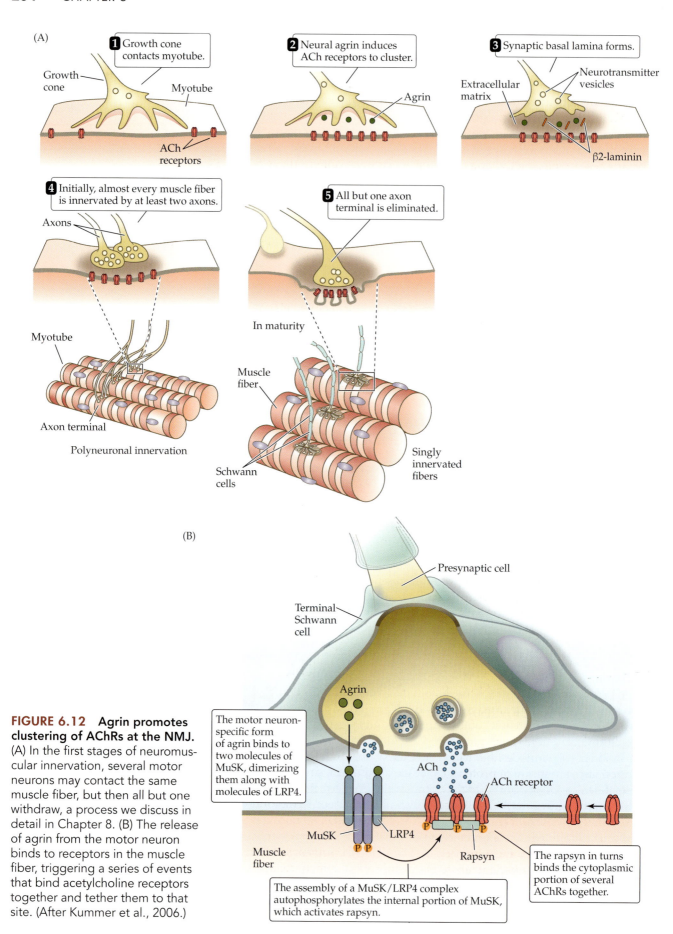

(A)

1 Growth cone contacts myotube.

Growth cone

Myotube

ACh receptors

2 Neural agrin induces ACh receptors to cluster.

Agrin

3 Synaptic basal lamina forms.

Extracellular matrix

Neurotransmitter vesicles

β2-laminin

4 Initially, almost every muscle fiber is innervated by at least two axons.

Axons

Myotube

Axon terminal

Polyneuronal innervation

5 All but one axon terminal is eliminated.

In maturity

Muscle fiber

Schwann cells

Singly innervated fibers

(B)

Presynaptic cell

Terminal Schwann cell

Agrin

ACh

ACh receptor

The motor neuron-specific form of agrin binds to two molecules of MuSK, dimerizing them along with molecules of LRP4.

MuSK

LRP4

Rapsyn

Muscle fiber

The assembly of a MuSK/LRP4 complex autophosphorylates the internal portion of MuSK, which activates rapsyn.

The rapsyn in turns binds the cytoplasmic portion of several AChRs together.

FIGURE 6.12 Agrin promotes clustering of AChRs at the NMJ. (A) In the first stages of neuromuscular innervation, several motor neurons may contact the same muscle fiber, but then all but one withdraw, a process we discuss in detail in Chapter 8. (B) The release of agrin from the motor neuron binds to receptors in the muscle fiber, triggering a series of events that bind acetylcholine receptors together and tether them to that site. (After Kummer et al., 2006.)

all but one withdraws (**FIGURE 6.12A**), a process we'll discuss in more detail in Chapter 8. When motor neuronally secreted agrin reaches the muscle, it binds to a complex receptor made of two parts: a novel **muscle-specific kinase** (**MuSK**), and a protein called **LRP4**. Released agrin binds together two MuSK molecules and two LRP4 molecules and activates the complex such that the internal domain of MuSK phosphorylates itself (**FIGURE 6.12B**). This autophosphorylation of MuSK promotes clustering with other MuSK-LRP4 complexes as well as activating a cytoplasmic protein called **rapsyn** (Gautam et al., 1995). Each rapsyn molecule in turn binds to the intracellular domains of several AChRs, effectively binding molecules of AChR together, akin to ropes lashing logs together in a raft or lashing boats together in a flotilla. In addition to lashing this "flotilla" of AChRs together, rapsyn also binds to microtubules within the muscle, anchoring this whole assembly to the muscle's cytoskeleton. These actions anchor the group of AChRs at the NMJ beneath the axon terminal that had released the agrin (Darabid et al., 2014).

Agrin also has an important role in maintaining the NMJ, namely the stabilization of AChRs so that they have a longer half-life before being replaced. The rapsyn complex holding together AChRs also phosphorylates a subunit of the AChR itself, stabilizing the overall receptor and greatly extending its half-life (Rudell & Ferns, 2013). This means the muscle doesn't have to produce as many replacement AChRs to maintain functional contact with the motor neuron controlling muscle contraction. In fact, any AChRs that fail to aggregate with the NMJ are quickly internalized by the muscle. This internalization of unaggregated receptors is hastened when the muscle is depolarized, as when actively producing action potentials for contraction. This means extrajunctional receptors are fleeting and therefore relatively rare compared with stable receptors at the NMJ.

Plus, the motor neuron releases another factor that favors production of more AChRs specifically from those myonuclei near the junction, as we discuss next.

muscle-specific kinase (MuSK) A membrane-bound kinase found exclusively in muscle fibers that serves as a receptor for agrin.

LRP4 A receptor for agrin that forms a complex with MuSK.

rapsyn A cytoplasmic protein that binds the intracellular components of several acetylcholine receptors, effectively anchoring them together.

neuregulins A family of signaling proteins, part of a much larger family of epidermal growth factors, that are membrane bound but can be cleaved to produce a diffusible signal.

ErbB A family of membrane-bound receptors that respond to neuregulins.

How's It Going?

1. What is the signal that motor neurons use to regulate the initial response of the muscle target? What are the muscle responses to that signal?
2. What are the intracellular proteins that bind ACh receptors together in the postsynaptic site?

■ 6.8 ■
NEUREGULINS BOOST LOCAL AChR EXPRESSION IN MUSCLE AND MAINTAIN TERMINAL SCHWANN CELLS

While agrin promotes aggregation and stabilization of AChRs, it does not, by itself, induce the muscle fiber to produce additional AChRs. A second motor neuron factor, a family of proteins called **neuregulins** (part of the epidermal growth factor [EGF] family), induces muscles to increase transcription of the various subunits of the AChR. Neuregulins are embedded in motor neuronal axon membranes but can also be released by cleavage, so they can act via cell-cell contact or as a released factor (Mei & Nave, 2014). In either case, neuregulins bind to **ErbB** receptor tyrosine kinases (themselves part of the EGF receptor family) on muscle fibers to boost their expression of AChR subunits (Merlie & Sanes, 1985).

Recall that muscle fibers are multinucleated cells as a result of their formation by fusion of many myoblasts. The effects of neuregulin are somewhat

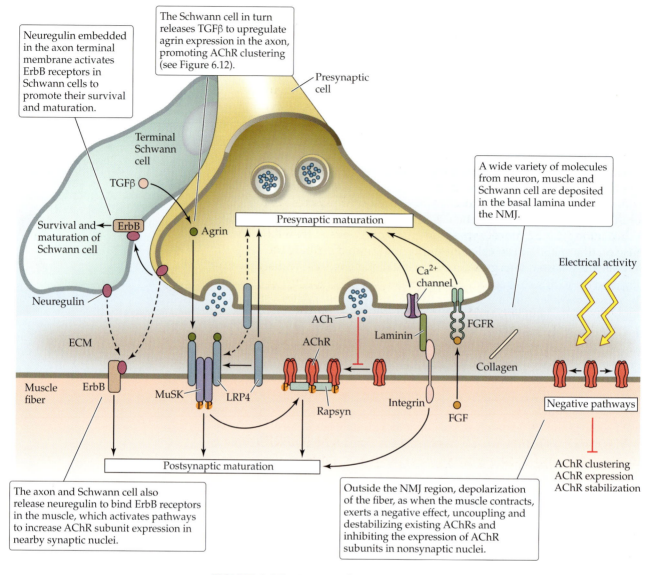

Neuregulin embedded in the axon terminal membrane activates ErbB receptors in Schwann cells to promote their survival and maturation.

The Schwann cell in turn releases TGFβ to upregulate agrin expression in the axon, promoting AChR clustering (see Figure 6.12).

A wide variety of molecules from neuron, muscle and Schwann cell are deposited in the basal lamina under the NMJ.

The axon and Schwann cell also release neuregulin to bind ErbB receptors in the muscle, which activates pathways to increase AChR subunit expression in nearby synaptic nuclei.

Outside the NMJ region, depolarization of the fiber, as when the muscle contracts, exerts a negative effect, uncoupling and destabilizing existing AChRs and inhibiting the expression of AChR subunits in nonsynaptic nuclei.

FIGURE 6.13   Neuregulin maintains terminal Schwann cells and promotes AChR expression. (After Darabid et al., 2014 and Wu et al., 2010.)

subsynaptic nuclei   The group of nuclei within a muscle fiber that lie closest to the neuromuscular junction.

local, as it is only those muscle nuclei near the NMJ, so-called **subsynaptic nuclei**, that increase AChR expression (Sanes et al., 1991). In general, electrical activity of the muscle fiber, as happens when it is activated for contraction, destabilizes the AChR, leading to internalization. In the region of the NMJ, the agrin-induced rapsyn complex phosphorylates AChRs, making them more stable and thereby counteracting this tendency. Depolarization of the muscle fiber also suppresses AChR gene expression, so only subsynaptic nuclei, under the influence of motor neuronal neuregulin, actively produce AChRs. Thus electrical activity of the muscle destabilizes expression, aggregation, and stabilization of AChRs, and the motor neuronal signals agrin and neuregulin counteract those effects to retain the receptors at the NMJ (**FIGURE 6.13**).

The ability of innervating neurons to direct the neurotransmitter responsiveness of their targets has been dramatically illustrated at the mammalian NMJ. When a rat muscle was surgically denervated and then provided with a new nerve that uses glutamate as a neurotransmitter, the muscle began expressing glutamate receptors and eventually contracted in response to glutamate transmission (Brunelli et al., 2005)! Given that glutamate is a common transmitter at NMJs in invertebrate species such as *Drosophila*, you have to wonder whether the ability of vertebrate muscle to develop glutamate responsiveness is an evolutionary holdover of mechanisms in place when one of our ancestors had glutamatergic NMJs.

In addition to the effects of motor neuronally released neuregulin on the muscle fiber, neuregulin embedded in the axon terminal also activates ErbB receptors through contact with nearby Schwann cells, inducing them to differentiate into terminal Schwann cells to cap the NMJ (see Figure 6.13) (Falls, 2003; Kummer, Misgeld, & Sanes, 2006; Rimer, 2007). This neuregulin-ErbB signaling is important for promoting the survival, proliferation, and differentiation of Schwann cells, both those that cap the NMJ (Maurel & Salzer, 2000; Syroid et al., 1996) and those that form myelin (Syed et al., 2010), as we describe at the end of this chapter.

In addition to the three cell types we've mentioned so far—the motor neuron, muscle cell, and Schwann cell—the extracellular matrix filling the synaptic cleft can also play a role in guiding synapse development, as we discuss next.

How's It Going?

1. What role do neuregulins play in neuromuscular junction maturation?
2. What is the receptor for neuregulins, and which cells in the neuromuscular junction do they act upon?

■ 6.9 ■
ONCE FORMED, THE NMJ LEAVES AN IMPRINT IN THE EXTRACELLULAR MATRIX

One of the earliest indications that motor neurons and muscles exchange signals to promote formation of the NMJ came from an experiment demonstrating the presence of such a signal in the extracellular matrix (ECM), the complex of proteins and glycoproteins that fills the space between axon terminals and the postsynaptic site. Many layers of ECM form a basal lamina around muscle fibers. It turns out that the isolated basal lamina from muscle retains an "imprint" in the synaptic region that can guide regenerating motor neuron axons back to the former site of an NMJ, and can guide regenerating muscle fibers to direct AChRs there, too, as we'll see next.

RESEARCHERS AT WORK

Neuromuscular Junctions Leave a Residue in the Basal Lamina

■ **QUESTION**: Are there any signals embedded within the basal lamina between muscles and their innervating motor neuron terminals?

■ **HYPOTHESIS**: If a signal is present in the basal lamina, then it may persist if we can somehow separate basal lamina from an existing NMJ.

■ **TEST**: The basal lamina is too flimsy to remove or manipulate surgically, so researchers instead removed the presynaptic terminals and killed the muscle fiber targets in an adult frog muscle, leaving the basal lamina behind. They removed the motor neuron terminals by cutting the motor nerve, causing the distal terminals to die back. They removed

continued

the muscle fibers by cutting them on either side of the NMJ, leaving the basal lamina there intact (**FIGURE 6.14A**). Later, the motor neuron axons sprouted to reinnervate the muscle, and myoblasts in the muscle divided and fused to regenerate new fibers. The question was whether the regenerating motor terminals would return to the former NMJ site and whether the regenerating muscles would insert AChRs at that site.

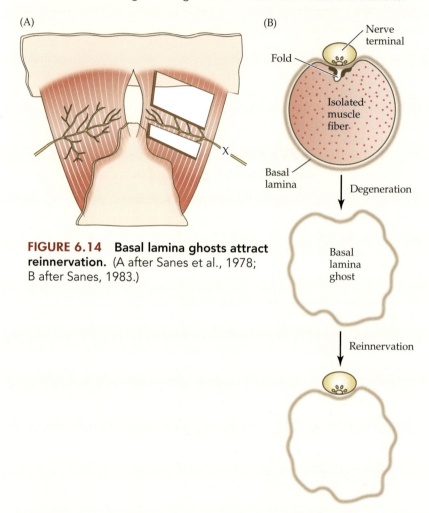

FIGURE 6.14   Basal lamina ghosts attract reinnervation. (A after Sanes et al., 1978; B after Sanes, 1983.)

■ **RESULT**: In fact, the regenerating muscle fibers formed new clusters of AChRs at the former NMJ site, even in the absence of any motor terminals (Burden, Sargent, & McMahan, 1979). Similarly, the regenerating motor terminals preferentially returned to the former NMJ site on the basal lamina even if the muscle was irradiated to keep fibers from regenerating (Sanes, Marshall, & McMahan, 1978) (**FIGURE 6.14B**). These basal lamina "ghosts" were sufficient to guide the nerves back to the original NMJ site.

■ **CONCLUSION**: One or more cues are secreted into the basal lamina to mark the site of the NMJ, and both the axon and the target muscle can use the cue(s) to re-form a synapse.

So if there is a signal embedded in the basal lamina that surrounds muscle fibers, was it the muscle or the motor neuron that secreted the signal and deposited it there? This question has been difficult to answer definitively, in part because there are so many different molecules, including a synaptic-specific isoform of laminin (β2-laminin; see Figure 6.13) (Martin, Ettinger, & Sanes,

1995), deposited in the basal lamina beneath the NMJ (Maselli, Arredondo, Ferns, & Wollmann, 2012). It seems likely that motor neuron, muscle, and even terminal Schwann cells all make contributions to the composition of synaptic basal lamina.

Now let's consider the changes in ion channels and neurotransmitter receptors that maturing neurons go through to sharpen synaptic signaling.

How's It Going?
What is the evidence that basal lamina retains signals marking the neuromuscular junction?

■ 6.10 ■
ION CHANNELS CHANGE SUBUNITS, AND THEREFORE CHARACTERISTICS, DURING DEVELOPMENT

As you may remember from previous neuroscience-related courses, the electrical activity of neurons depends on various types of **ion channels**, membrane proteins that form pores to allow particular ions to enter or leave the cell interior. The ion channels that are important for neurophysiology do not remain open all the time. Rather, they are open to allow ion flow at some times and not others. We can roughly classify ion channels into two categories depending on what types of signals control their opening:

1. **Voltage-gated ion channels** open or close depending on the cell's membrane potential, such as the voltage-gated sodium (Na^+) channels that allow Na^+ ions to enter and rapidly depolarize neurons during the action potential, and voltage-gated potassium (K^+) channels that then open to allow K^+ ions out to repolarize the neuron (**FIGURE 6.15A**).

2. **Ligand-gated ion channels** are controlled by chemical signals, called *ligands*, that bind to the ion channel. Some ligands, like neurotransmitters, bind the external surface of the ion channel, as when ACh molecules bind a nicotinic AChR, while other ligands, such as G proteins and second messengers like cyclic AMP (cAMP), bind the internal portion of the ion channel (**FIGURE 6.15B**).

ion channels   Membrane-bound proteins that, when activated, open up to allow select ions to cross the membrane.

voltage-gated ion channels   Membrane-bound proteins that open up an ion channel in response to particular membrane potentials.

ligand-gated ion channels   Membrane-bound proteins that open up an ion channel in response to a ligand such as a neurotransmitter.

(A) Voltage-gated channels

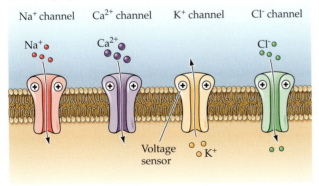

(B) Ligand-gated channels

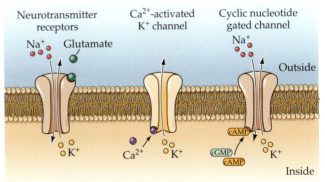

FIGURE 6.15 Types of ion channels Each channel is made up of subunits, which may change in the course of development, altering its neurophysiological properties. (A) Voltage-gated channels play a crucial role in the generation and conduction of action potentials. (B) Ligand-gated channels include ionotropic neurotransmitter receptors.

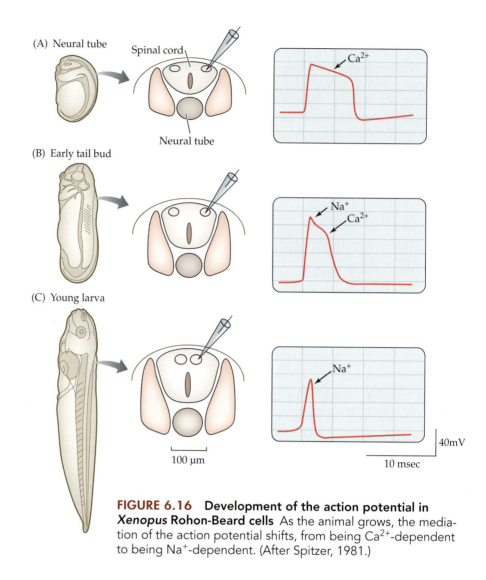

(A) Neural tube

Spinal cord

Neural tube

Ca²⁺

(B) Early tail bud

Na⁺
Ca²⁺

(C) Young larva

Na⁺

40mV

100 μm

10 msec

FIGURE 6.16   Development of the action potential in
***Xenopus* Rohon-Beard cells** As the animal grows, the media-
tion of the action potential shifts, from being Ca²⁺-dependent
to being Na⁺-dependent. (After Spitzer, 1981.)

In most cases, ion
channels are actually
made up of several
different subunits,
often derived from
separate genes.

Whether a channel is voltage or ligand gated, its opening
depends on a conformational change in the proteins to either
open or close the central pore that the ion passes through.
In most cases, the channel is actually made up of several
different subunits, often derived from separate genes. As
we'll see, the expression of ion channel subunits changes
across development, which means the electrical properties
of neurons change, too.

For example, in embryonic neurons the rapid depolarization
during the peak of the action potential is not caused by an influx of Na⁺ ions
through voltage-gated Na⁺ channels, as happens in mature neurons, but by
the influx of calcium (Ca²⁺) ions through voltage-gated Ca²⁺ channels (**FIGURE
6.16A**) (Baccaglini & Spitzer, 1977; Spitzer, 1979). The change to the mature
state is primarily due to a large increase in Na⁺ channels rather than loss of
Ca²⁺ channels (**FIGURE 6.16B**) (MacDermott & Westbrook, 1986; O'Dowd,
Ribera, & Spitzer, 1988). In both cases, voltage-gated K⁺ channels then allow
the outflow of K⁺ to restore the resting potential. However, during the phase
in which Ca²⁺ mediates the action potential, there are relatively few voltage-
gated K⁺ channels, and therefore the recovery of the resting potential takes

(A) Potassium-chloride transporter

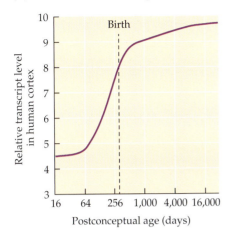

(B) Immature neuron

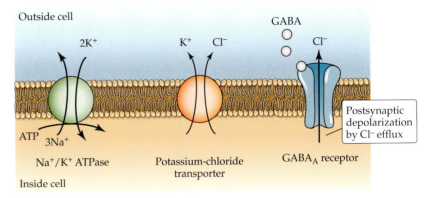

(C) Mature neuron

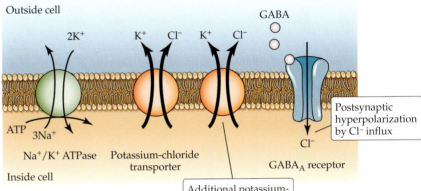

FIGURE 6.17 **Changes in chloride channels shift GABA receptors from excitatory to inhibitory.** In embryonic neurons, intracellular Cl⁻ levels are high, so opening chloride channels allows Cl⁻ to exit, depolarizing the cell. Once the cotransporter reduces intracellular Cl⁻, opening chloride channels in GABA and glycine receptors will result in Cl⁻ influx, and therefore hyperpolarization, in adult neurons. (After Kaila et al., 2014.)

much longer—as long as 100 ms rather than the 1 ms typical in mature action potentials (**FIGURE 6.16C**).

The function of ligand-gated channels also changes across development. For example, while GABA receptors in the mature nervous system are almost always inhibitory, in the developing brain they tend to be excitatory (Ben-Ari, Gaiarsa, Tyzio, & Khazipov, 2007). The GABA receptors are ion channels for chloride (Cl⁻), which is normally driven into adult neurons, hyperpolarizing them, because it is 20 times more concentrated outside than inside in mature neurons. However, that differential concentration across the neuron's membrane is due to a **potassium-chloride cotransporter** pumping Cl⁻ ions out while pumping K⁺ ions in, which is not fully expressed in the embryonic brain (**FIGURE 6.17A**). Thus immature neurons have more Cl⁻ inside than outside, so when the GABA receptor is activated, Cl⁻ exits the cell interior, depolarizing the neuron and making it more likely to fire (**FIGURE 6.17B**). But once the cotransporter has depleted the concentration of Cl⁻ inside the neuron, then opening the GABA channel allows Cl⁻ to enter the cell, hyperpolarizing it (**FIGURE 6.17C**). The same change in expression of the potassium-chloride

potassium-chloride cotransporter A specialized membrane-bound protein that pushes both potassium and chloride ions out of cells.

cotransporter is also responsible for the shift in glycine receptors from excitatory to inhibitory during postnatal development in rats (Lynch, 2004). The depolarization caused by Cl^- exit through GABA and glycine receptors in immature neurons can also activate Ca^{2+} channels (Dave & Bordey, 2009), resulting in a very long-lasting depolarization reminiscent of embryonic Ca^{2+}-mediated action potentials (see Figure 6.16A).

How's It Going?

1. What are the two main categories of ion channels, and what roles do they play in action potentials?
2. Describe the changes in the ionic basis of the action potential during development.
3. How does developmental appearance of the potassium-chloride cotransporter affect whether GABA receptors are excitatory or inhibitory?

■ 6.11 ■

EMBRYONIC SYNAPSES ARE SLUGGISH AND SLOW, THEN BECOME PROGRESSIVELY FASTER WITH DEVELOPMENT

You might think that maturation of a synapse would consist of progressively stronger postsynaptic potential (either an excitatory postsynaptic potential [EPSP] or an inhibitory postsynaptic potential [IPSP]) as development proceeds. But in fact the big change in maturing synapses is not the *amplitude* of the EPSP or IPSP, but the *time course*. Specifically, the postsynaptic potentials are very long and drawn out at first, and then as the synapse matures, they become progressively shorter. In other words, the postsynaptic machinery is somewhat sluggish in immature synapses. For example, cortical EPSPs that are about 400 ms long in newborn rats will be only 100 ms 2 weeks later (**FIGURE 6.18A,B**) (Burgard & Hablitz, 1993). At immature synapses, it also takes longer for the presynaptic machinery to release the neurotransmitter, resulting in a longer delay between arrival of the action potential to the axon terminal and the peak of an EPSP in the postsynaptic cell. It takes over 80 ms between stimulation of afferents and the peak of an EPSP in newborn rat neocortical neurons, but 2 weeks later it's more like 1 ms (compare Figure 6.18A and B). Furthermore, while there is a lot of variability in these postsynaptic responses at first, they become more uniform with maturation, as seen in the shrinking error bars with age in **FIGURE 6.18C,D**. In other words, synapses become gradually faster, sharper, and more uniform in their response as they mature.

> The big change in maturing synapses is not the *amplitude* of the EPSP or IPSP, but the *time course*.

This shortening of postsynaptic potentials is mostly due to changes in the function of the neurotransmitter receptor, which may stay open for a long time in young animals and then for a much shorter duration in adulthood. For example, hippocampal NMDA-type glutamate receptors when activated in newborn rats stay open more than twice as long as in adults (Khazipov, Ragozzino, & Bregestovski, 1995). These changes in neurotransmitter receptors during development are due to the gradual substitution of subunits, replacing a subunit of the receptor that is common in fetal or neonatal animals with another version of the subunit that is typically seen in adults. For example, glycine is a common inhibitory neurotransmitter in the spinal cord and brainstem, and

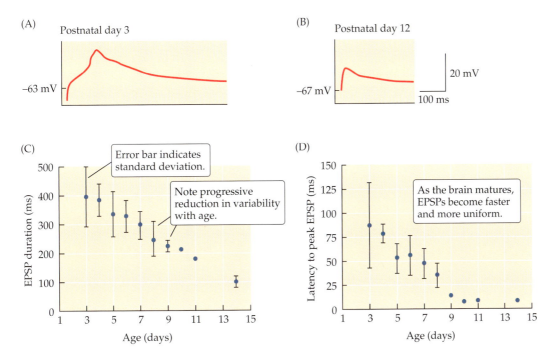

FIGURE 6.18 Maturation of excitatory synapses in rat neocortex (A,B) In addition to a progressive shortening of both synaptic delay and EPSP duration, (C,D) there is also a dramatic decline in the variability in these measures, as indicated by the shrinking vertical error bars (representing the standard deviation). Not only are the synapses becoming faster and sharper, they are also having a more consistent effect on the target neuron as the brain matures. (After Burgard & Hablitz, 1993.)

its receptor is made up of a total of five subunits of two different basic types: α and β (Lynch, 2009). There are four different versions of the α subunit (α1 through α4), so many different glycine receptors are possible, depending on which α subunits are used and whether 1 or 2 β subunits are present. In rats, the composition of glycine receptors shifts from using α2 to α1 by about postnatal day 20 (Lynch, 2004).

Once again we know the most about the molecular basis of changes in electrophysiological properties of maturing synapses at the huge NMJ. The nicotinic ACh receptors found in adult vertebrate muscle consist of five subunits: two α subunits (which provide the two ACh-binding sites found in each receptor) and three non-α subunits. Early in life, the non-α subunits are β, δ, and γ (Albuquerque, Pereira, Alkondon, & Rogers, 2009), but shortly after birth the embryonic γ subunit is replaced by an ε subunit (Missias, Chu, Klocke, Sanes, & Merlie, 1996). The γ subunit in young muscles results in the ion channel of the receptor staying open for a longer time when activated by ACh (Goldman, Brenner, & Heinemann, 1988; Missias et al., 1996). The substitution of the mature ε subunit not only sharpens the response of the ACh receptor, it also makes the receptor more resistant to degradation, further stabilizing the synapse.

As developing synapses become faster in their response, the formation of myelin means that action potentials, once generated near the axon hillock, reach the various axon terminals faster, thanks to the myelination we consider next. Taken together, these changes help the brain receive and process information sooner and send out motor commands faster, helping the maturing individual survive in a fast-paced world.

■ 6.12 ■

MYELINATION EXTENDS INTO ADULTHOOD TO HASTEN NEURONAL COMMUNICATION

myelin The fatty, whitish sheet of membrane wrapped around some axons, provided by a glial cell.

node of Ranvier The gap between successive segments of myelin in myelinated axons.

oligodendrocyte A glial cell that provides myelination for axons in the central nervous system.

Schwann cells Glial cells that provide myelin sheaths for axons in the peripheral nervous system and cap neuromuscular junctions.

myelin basic protein (MBP) An important component of myelin for both oligodendrocytes and Schwann cells.

myelin oligodendrocyte glycoprotein (MOG) An important membrane-bound glycoprotein component of myelin in the CNS.

You may recall from previous classes that **myelin** is a fatty white substance produced by glia that surrounds and electrically insulates axons, preventing current flow across the axonal membrane and thereby channeling depolarizations quickly down the axon interior to reach the next gap in myelination, called a **node of Ranvier**. The rapid sequential appearance of the action potential at each node of Ranvier is called *saltatory conduction*, and it greatly accelerates the conduction of action potentials, which may be as fast as 120 m/s (>260 mph). In contrast, nerve conduction velocity may be only 0.5 to 2 m/s in unmyelinated axons. Two types of glia provide myelin sheaths for axons: **oligodendrocytes** ("a few branches") myelinate axons in the CNS (**FIGURE 6.19A**), while **Schwann cells** myelinate peripheral axons.

Either an oligodendrocyte or a Schwann cell provides myelin by first wrapping its membranes around and around an axon (**FIGURE 6.19B**), perhaps hundreds of time. Then the glial cell produces myelin-related proteins, including **myelin basic protein** (**MBP**), various glycoproteins such as **myelin oligodendrocyte glycoprotein** (**MOG**), and cholesterol to further electrically insulate the sheath, basically by increasing the lipid content. A given oligodendrocyte may provide patches of myelin for several different neighboring axons (see Figure 6.19A), while a Schwann cell will myelinate a patch of only a single axon. Even unmyelinated axons in the periphery may be somewhat insulated from one another as Schwann cells insinuate their cytoplasm between axons, but without the multiple wrapping and production of proteins to form myelin (**FIGURE 6.19C**). We learned in Chapter 3 that oligodendrocytes arise from the proliferative zone of the ventricles from precursors they have in common with neurons and astrocytes, while Schwann cells are neural crest cells that migrate to the periphery and take on their fate upon contact with axons. Myelin is found exclusively in vertebrate species and represents an important innovation in vertebrate neural development.

For Schwann cells, the signal to wrap an axon and produce myelin is provided in the form of neuregulin (specifically, neuregulin 1 type III) protein embedded in the axon surface, which activates ErbB receptors in the Schwann cell. We saw earlier that neuregulin activation of ErbB receptors in muscle fibers promotes postsynaptic maturation, including up-regulation of AChR expression. Now the same signal from axons acts on ErbB receptors in Schwann cells to trigger myelination (Brinkmann et al., 2008; Nave & Salzer, 2006). Axons destined to be unmyelinated express little neuregulin on the axon surface, and increased expression of axonal neuregulin correlates with increased thickness of Schwann cell myelination (Feltri, Poitelon, & Previtali, 2015).

There is probably a similar axonal control of myelination from oligodendrocytes in the CNS, because oligodendrocytes in vitro respond to neuregulin by producing myelin-related proteins (Calaora et al., 2001). But the mechanism controlling myelination cannot be exactly the same for oligodendrocytes as for Schwann cells, because *neuregulin* knockout mice are severely deficient in peripheral myelin but display normal myelination in the CNS. On the other

(A)

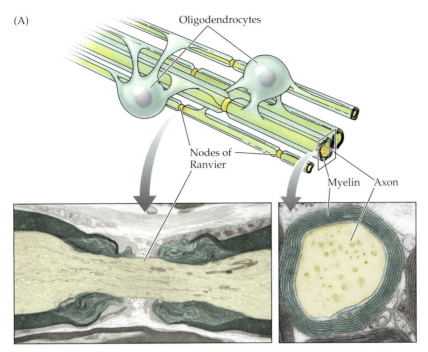

Oligodendrocytes

Nodes of Ranvier

Myelin Axon

FIGURE 6.19 **Glial cells provide myelin.** (A) Oligodendrocytes, so named because they have only a few branches, myelinate neighboring axons in the CNS. (B) Both Schwann cells and oligodendrocytes form myelin by progressively wrapping sheets of their cytoplasm around the axon. (C) Outside the CNS, Schwann cells provide layers of myelin for some axons within peripheral nerves and for individual axons outside nerves. Even unmyelinated peripheral axons are still insulated from one another to some extent by Schwann cells that insinuate their cytoplasm between axons. (Part A micrograph [left] courtesy of Dr. Mark Ellisman and the National Center for Microscopy and Imaging Research; A, micrograph [right] and C from Peters et al., 1991.)

(B) Schwann cell

Axon

(C)

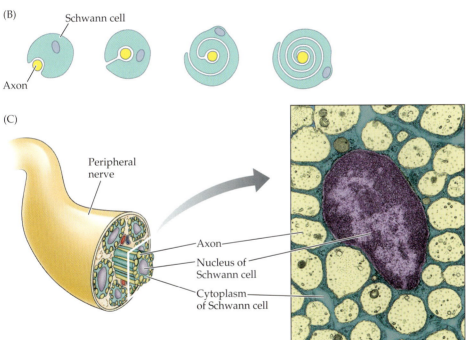

Peripheral nerve

Axon
Nucleus of Schwann cell
Cytoplasm of Schwann cell

hand, transgenic mice overexpressing either of two types of neuregulin 1 displayed hypermyelination in the CNS (Brinkmann et al., 2008). Perhaps oligodendrocytes make different ErbB receptors than Schwann cells and/or respond to different neuregulins, or even to a completely different factor from CNS axons. In any case, there's reason to think that, once formed, oligodendrocytic myelin may impede regeneration, as we see next.

How's It Going?

1. How does myelination speed the conduction of action potentials?
2. Which cells provide myelin for the CNS and the periphery?
3. What signaling system regulates the extent of myelination of a peripheral axon?

■ 6.13 ■

MYELINATING GLIA MAY PREVENT REGENERATION IN THE CENTRAL NERVOUS SYSTEM

It's been known for a long time that, while severed axons regenerate in the periphery and eventually reinnervate muscle, as happens when motor neuron axons are cut during surgery or injury, within the CNS regeneration is very limited in mammals. Severed axons in the brain or spinal cord rarely or never reinnervate their appropriate target. This is the reason there is no recovery from injuries severing the spinal cord, for example. The distinction between peripheral and CNS regeneration is driven home by dorsal root ganglion cells, which have their cell bodies in the dorsal root ganglia alongside the spinal cord and have axons reaching into the spinal cord at one end and into the periphery at the other. Despite having axonal connections in the two directions, these cells can only reestablish the connections with the axonal end in the periphery, not the spinal cord. Maybe the difference in regeneration in those two environments is due to the difference in the source of myelin.

> Severed axons in the brain or spinal cord rarely or never reinnervate their appropriate target.

There's reason to think that other glia and their reactions to injury may be at least partially responsible for the failure of CNS regeneration (Yiu & He, 2006). Keep in mind that in the embryonic brain, axonal growth cones navigate in a fairly open arena, choosing between attractive and repulsive factors in the absence of any myelin and few other cells (**FIGURE 6.20A**). After most major connections are formed, myelination commences (**FIGURE 6.20B**) which is good for the conduction of nerve impulses, but presents a much more complicated landscape for growth cones to navigate. One factor that impedes any regenerating axons in the adult CNS after injury is the presence of reactive astrocytes that form a glial scar that appears to act as a barrier to regrowing axons (Silver & Miller, 2004). But there's also ample evidence that the myelin itself actually acts as a repellent to axonal growth cones (**FIGURE 6.20C**). In vitro, growth cones that will happily grow over myelin provided by Schwann cells are repelled by myelin from oligodendrocytes (Schwab & Thoenen, 1985). Generation of antibodies to try to neutralize this repellent quality of oligodendrocytes eventually identified a membrane-bound glycoprotein, appropriately named neurite outgrowth inhibitor, or **Nogo** (Chen et al., 2000). The active isoform of Nogo is highly expressed in oligodendrocytes (but not Schwann cells) and inhibits axonal outgrowth in vitro. In vivo, mice with the gene for Nogo knocked out show significant regeneration following spinal cord injuries, which wild-type mice do not (Kim, Li, GrandPre, Qiu, & Strittmatter, 2003).

This activity of oligodendrocytes to impair CNS regeneration might seem puzzlingly maladaptive. But remember that before the advent of effective medicine and hospitals, people rarely survived brain or spinal cord damage, as opposed to a cut that might sever peripheral nerves. In other words, there might have been no selective advantage to evolving regeneration in the CNS— you'd be dead before it could happen. On the other hand, we saw in Chapter 5 that amphibians and fish can regenerate retinal connections to the brain, so perhaps evolution of some other adaptive trait in birds and mammals interfered with a previous ability of the CNS to regenerate. So it has been suggested that having oligodendrocytes provide an inhospitable environment might be adaptive in the development of complex neural circuitry. During early development axons are unmyelinated, and the major axonal pathways that we've been talking about in the last several chapters, and will continue discussing for the rest of the book, are laid down early in life. After that, as we'll see in Chapters 8–10, there are still synapses coming and going, but they tend to be relatively local changes, not changes in projection neurons (i.e., neurons that

Nogo A membrane-bound protein expressed by oligodendrocytes that inhibits neurite outgrowth. It is the product of a member of the *reticulon* gene family.

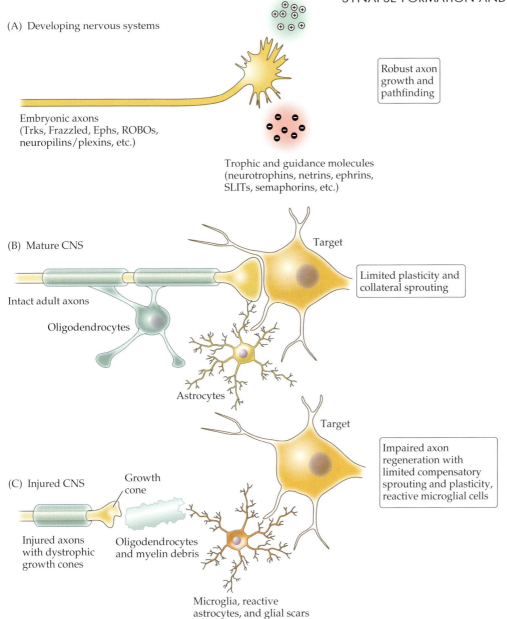

(A) Developing nervous systems

Embryonic axons
(Trks, Frazzled, Ephs, ROBOs,
neuropilins/plexins, etc.)

Trophic and guidance molecules
(neurotrophins, netrins, ephrins,
SLITs, semaphorins, etc.)

Robust axon
growth and
pathfinding

(B) Mature CNS

Target

Intact adult axons

Oligodendrocytes

Astrocytes

Limited plasticity and
collateral sprouting

Target

(C) Injured CNS

Growth
cone

Injured axons
with dystrophic
growth cones

Oligodendrocytes
and myelin debris

Microglia, reactive
astrocytes, and glial scars

Impaired axon
regeneration with
limited compensatory
sprouting and plasticity,
reactive microglial cells

FIGURE 6.20 **Oligodendrocytes may deter axonal regeneration in the CNS.**
(A) In the developing nervous system, various signals from axons and targets guide
axonal development in the relative absence of other players, including myelinating
oligodendrocytes. (B) At maturity, many signaling systems seem quiescent, promoting relatively stable connections. (C) After injury, any attempts of central neurons to
reinnervate the appropriate target must navigate around potential barriers, including myelin debris, glial scars formed by reactive astrocytes, and invading microglia
attempting to remove debris. (After Yiu & He, 2006.)

project their axons a long distance). Thus, the reasoning goes, perhaps
the spread of oligodendrocyte myelination that occurs relatively late in
development serves to "lock in" the major neural pathways, limiting
future neural plasticity to more local changes. It might be a bad idea
for extensive neural plasticity, allowing wholesale restructuring of the
CNS, to be possible in adulthood. The inability for the mature CNS to
recover following injury would be an unfortunate side effect of preserving a neural architecture that has been well tested by natural selection.
If this scenario is true, then perhaps inhibiting the signaling of factors
like Nogo would not only allow CNS regeneration, but allow for greater

Perhaps evolution of
some other adaptive
trait in birds and
mammals interfered
with a previous
ability of the CNS to
regenerate.

trembler   Mice with an autosomal dominant mutation of peripheral myelin protein-22 that results in demyelination and impaired locomotion.

peripheral myelin protein-22 (PMP22)   An important structural component of myelin in the peripheral nervous system.

jimpy   Mice with an X-linked recessive mutation of the myelin proteolipid protein that results in severe lack of CNS myelin, tremors, and convulsions.

Guillain-Barré syndrome   An autoimmune disorder attacking Schwann cells, demyelinating axons and causing varying degrees of weakness and tingling.

multiple sclerosis (MS)   An autoimmune disease attacking CNS myelin, causing highly variable degrees of impairment of motor and sensory processes.

adult neural plasticity in intact adult brains, too (J. K. Lee, Kim, Sivula, & Strittmatter, 2004; McGee, Yang, Fischer, Daw, & Strittmatter, 2005). In other words, it might be possible to reopen *sensitive periods* for neural plasticity, such as the capacity to learn a new language, in adulthood. We'll discuss sensitive periods for neural plasticity in greater depth in Chapters 9 and 10.

Several spontaneous mutations in mice offer insight into myelin-related disorders. **Trembler** mice have a mutation in the gene for **peripheral myelin protein** (**PMP22**) (Suter et al., 1992), an important component of myelin in the periphery. The animals have great difficulty walking, as they totter on the tips of their toes, and may suffer convulsions (Falconer, 1951). Discovery that the trembler mutation was in the *Pmp22* gene in mice led to the discovery that other mutations of this same gene cause a human disorder, one of the Marie-Charcot-Tooth diseases (Douglas & Popko, 2009; Timmerman et al., 1992), which begins in adolescence or early adulthood and is characterized by weakness in lower leg muscles that impairs walking, progressing to weakness in hands (Gutmann & Shy, 2015). For both the human disease and the trembler mice, the problem seems to be an *overexpression* of *PMP22*, causing abnormal myelin sheaths. While the trembler mice suffer defects in peripheral nervous system myelin, **jimpy** mice have a defect in the gene for myelin proteolipid protein, a major myelin component in oligodendrocytes, and therefore deficiencies in CNS myelin (Baumann & Pham-Dinh, 2001; Vela, Gonzalez, & Castellano, 1998).

Guillain-Barré syndrome occurs when the patient's immune system attacks Schwann cells and myelin in the periphery, an example of an *autoimmune disorder*. The attack demyelinates peripheral axons, disrupting conduction of action potentials and so leading to some loss of peripheral sensation and muscle weakness (Dimachkie & Barohn, 2013). The greatest danger is if muscles supporting breathing stop working, so the patient may be given artificial respiratory support. In most cases, the patient eventually recovers spontaneously.

A more serious autoimmune disorder is **multiple sclerosis** (**MS**), in which oligodendrocytes and the myelin they produce in the CNS are attacked. For example, the patient may produce antibodies that attack MBP and/or MOG (Berger et al., 2003). There is variability in the location of myelin damage and therefore symptoms, so the person may suffer visual loss, motor weakness or imbalance, cognitive dysfunction, or pain (Siffrin, Vogt, Radbruch, Nitsch, & Zipp, 2010). Most people with MS experience phases where the symptoms abate for a while and then recur (relapse-remitting MS) (**FIGURE 6.21A**). In animal models, acute damage to myelin or oligodendrocytes typically leads to remyelination. If remyelination fails, the bare axons appear to be especially susceptible to damage and death, meaning the axon is lost for good (Franklin & French-Constant, 2008) because, as we noted above, regeneration is extremely limited in the CNS. The remyelination process is also evident in people with MS (**FIGURE 6.21B**), but in the face of continued immune attack, progression of disease is extremely variable and difficult to predict (Miller & Mi, 2007). One hope is to learn enough about signals that maintain oligodendrocytes and encourage their remyelination to stave off symptoms. For example, although patients with MS do not seem to generate antibodies to Nogo, it may still be possible to pharmacologically inhibit Nogo signaling to facilitate remyelination in MS (J. Y. Lee & Petratos, 2013). The idea is that Nogo may normally serve as a way for competing oligodendrocytes to inhibit their rivals from myelinating a stretch of axon. Thus reducing the ability of oligodendrocytes to inhibit each other may promote remyelination (Scholze & Barres, 2012). Several animal models of MS have been developed, and they have been invaluable in screening for treatments that can slow progression of the disease (Ransohoff, 2012), but there is currently no cure.

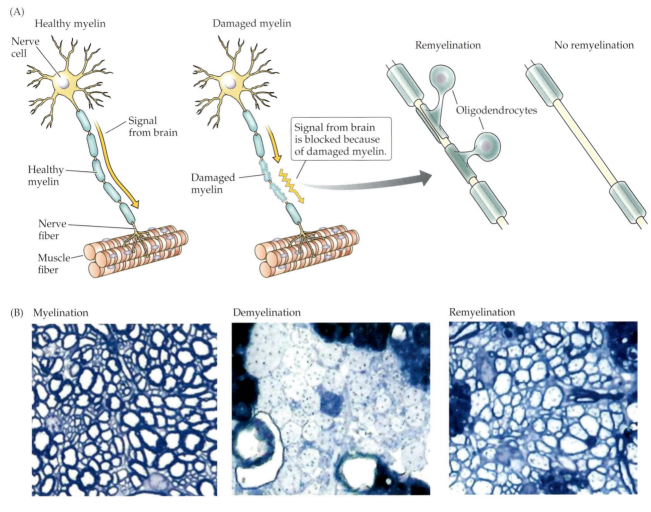

FIGURE 6.21 **The severity of multiple sclerosis may depend on the balance of demyelination and remyelination.** (B from Franklin et al., 2008.)

How's It Going?

1. How might myelination impede regeneration in the CNS?
2. What are the major disorders involving myelin?

SUMMARY

■ In unicellular organisms, calcium regulators like *calmodulin*, and transmembrane proteins such as *kinases* and *G proteins* that serve as environmental sensors, provided evolutionary precursors for components of postsynaptic responses in metazoans. Primitive metazoans like sponges use adhesive molecules such as *cadherins*, which were exploited by later-arising species for selective formation of synapses. **See Figure 6.1**

■ The basic machinery of neurotransmitter release and reception, including GABA and glutamate receptors, is found in all species with a nervous system, including jellyfish. A great leap in nervous system complexity after

our common ancestor with insects was the *duplication* of *neurotransmitter receptor* genes, followed by *divergence* that led to a greater variety of receptors for a given transmitter. **See Figure 6.2**

■ Synapses consist of (1) a *presynaptic terminal* with *synaptic vesicles* near an *active zone* to fuse vesicles with the membrane to release neurotransmitter into (2) the *synaptic cleft*, filled with *extracellular matrix* (*ECM*) of long-chain molecules to retain chemical signals, and (3) the *postsynaptic* region, including *postsynaptic densities* (*PSDs*) to detect neurotransmitter and trigger changes in the postsynaptic cell.

■ Synapses mature in stages, including (1) the initial contact where adhesive molecules favor certain connections and discourage others, (2) the assembly of synaptic machinery to increase the magnitude of synaptic transmission, and (3) the stabilization and maturation of synapses to provide more rapid, crisp synaptic transmission. **See Figure 6.3**

■ Many adhesive molecules, including cadherins, *NCAM*, and the *ephrins*, maintain some initial synaptic connections while discouraging others. There are over 100 different protocadherins, which preferentially adhere to each other (homophilic binding), enabling a very specific range of synaptic preferences. Nonpreferred connections tend to detach rather than mature further.

■ When the extracellular domain of a cadherin molecule is bound, the intracellular domain binds to β-catenin, which also binds to actin in the cell's cytoskeleton, anchoring the assembly in place. Neighboring postsynaptic regions may compete for β-catenin, and therefore for survival. **See Figures 6.4 and 6.5**

■ *Fragile X syndrome* is caused by an excess of trinucleotide repeats, which reduces the ability of the protein product to suppress transcription. Thus overall production of brain proteins is increased and an excess of synapses is seen, which seems to cause mental impairment. **See Figure 6.6**

■ Once appropriate pre- and postsynaptic partners are in contact, presynaptic *neurexins* bind to postsynaptic *neuroligins* to induce assembly of synaptic machinery in each. Activation of neurexin in the presynaptic terminal triggers construction of an active zone of vesicles and synaptic release proteins, including *SNAREs* and *synaptotagmin*. **See Figures 6.7 and 6.9**

■ Activation of neuroligin in the target cell triggers construction of a PSD of scaffolding proteins linked to neurotransmitter receptors and cytoskeletal elements, anchoring the region on the cell's surface. **See Figure 6.8**

■ *Neuromuscular junctions* (NMJs) in vertebrates form when motor neuronal growth cones releasing *acetylcholine* (ACh) contact *myotubes* that are already producing *ACh receptors* (AChRs) in random clumps. **See Figures 6.10 and 6.11**

■ Motor neuronally released *agrin* induces existing AChRs in the muscle to aggregate under the nerve terminal. Agrin binds together molecules of *muscle-specific kinase* (MuSK) and *LRP4* in the muscle fiber, resulting

in autophosphorylation of MuSK. The phosphorylated intracellular domain of MuSK activates the intracellular protein *rapsyn*, which also binds the internal portion of AChRs, lashing many receptors together. Rapsyn also binds the muscle cytoskeleton to hold the flotilla of receptors in place. **See Figure 6.12**

■ Muscle AChRs that are not stabilized by the agrin-induced complex of rapsyn are quickly internalized, a process facilitated when the muscle is electrically active, thereby reducing the half-life of extrajunctional receptors. **See Figure 6.13**

■ Motor neuronally released *neuregulins* bind to *ErbB* receptors in muscle to boost expression of AChR subunits in the target, specifically in the muscle's *subsynaptic nuclei*. Motor neuronal neuregulins also contact *terminal Schwann cells* to promote their differentiation and survival. **See Figure 6.13**

■ The *basal lamina* in the region of the NMJ retains long-chain molecules such as synapse-specific *laminin* (β2-laminin), which serve to guide regenerating axon terminals and muscle AChRs back to the original synaptic site. **See Figure 6.14**

■ Both *voltage-gated* and *ligand-gated ion channels* change subunit composition, and therefore physiological function, during development. Generally this results in slow, sluggish synaptic responses becoming more rapid and brief. **See Figures 6.15, 6.16, and 6.18**

■ GABA and glycine receptors that are almost always inhibitory in adult neurons are usually excitatory in the developing brain because the *potassium-chloride cotransporter* to reduce intracellular Cl⁻ levels is not yet fully expressed. **See Figure 6.17**

■ Axonal expression of neuregulin induces *Schwann cells* to form *myelin* sheaths in a dose-dependent manner, so axons expressing high neuregulin receive a thick sheath while those expressing less neuregulin receive a thinner sheath. Something similar is probably at work inducing *oligodendrocytes* to form myelin in the CNS. **See Figure 6.19**

■ Surviving oligodendrocytes repel growth cones, which may contribute to the limited ability for axons to regenerate in the adult CNS. *Multiple sclerosis* is caused by autoimmune attack of myelin components in the CNS. **See Figures 6.20 and 6.21**

Go to the Companion Website
sites.sinauer.com/fond
for animations, flashcards, and other review tools.

CHAPTER

7

Accepting Mortality
APOPTOSIS

BEATING THE ODDS He is perhaps the world's most famous living scientist, despite being confined to a wheelchair and able to communicate only with the help of a machine. Astrophysicist Stephen Hawking, who held the same Cambridge professorship in mathematics as Sir Isaac Newton for decades, has written several best-selling books about the universe and has even appeared on *The Simpsons* and *The Big Bang Theory*. These accomplishments are the more remarkable because at age 21 he was given a death sentence, diagnosed with **amyotrophic lateral sclerosis** (**ALS**). The name refers to apparent "starving" of muscles as they waste away (*amyotrophic*) and the degradation (*sclerosis*) of white matter in the lateral tracts of the spinal cord that is seen postmortem. It is sometimes called Lou Gehrig's disease, after the famous American baseball player who died of it at 37. There is no cure available, nor any treatment known to slow the disease. Medical care consists of postponing as long as possible respiratory failure or fatal choking as muscles weaken.

More than half the people given such a diagnosis are dead 5 years later, but Hawking has survived over 50 years (so far). It's not clear whether anyone else has ever lived so long after a diagnosis of ALS (McCoy, 2015). His debilitation has proceeded, gradually paralyzing him such that he is confined to a motorized wheelchair and can communicate only through a speech-generating device he controls with his cheek muscles, producing about one word per minute (de Lange, 2012).

The muscle atrophy in ALS was long thought to be a result of the death of motor neurons, and in England the disease is called motor neurone disease. But there was never any evidence that the disease begins in those cells or that death of motor neurons precedes symptoms; those ideas seem to have just been assumed. In fact, as we'll see in this chapter, there is now strong evidence that the disease does *not* act directly upon motor neurons, but rather acts upon some other cells, which then leads to motor neuronal death by the time the patient dies. How could a disease start in one type of cell and lead to death of other types? In this chapter, we'll find that many neurons, including motor neurons, are critically dependent on other cells for survival during fetal development. Understanding this early dependence of motor neurons may be crucial for understanding ALS.

CHAPTER PREVIEW

The previous chapters have related a beautiful unfolding of development, as progressively more complex signaling between cells results in progressively more complex structures arising as individual cells differentiate into progressively more specific shapes and functions. Since we know that a single fertilized egg, a simple sphere in shape, eventually produces the very convoluted structure called the brain (ignoring for a moment all the life-supporting organ systems that keep the brain alive), these progressive events are clearly necessary and seem entirely natural. But by the second half of the twentieth century, it became apparent that not all developmental processes involve adding cells, adding complexity, and adding structures. Initially skeptical scientists eventually had to admit that there

Astrophysicist Stephen Hawking He was diagnosed with ALS more than half a century ago. Now confined to a wheelchair, he is able to communicate only by using cheek muscles to activate the black sensor hanging from his glasses, yet he remains an active intellectual force in the world. (© Jason Bye/Alamy.)

amyotrophic lateral sclerosis (ALS) A neurodegenerative disease characterized by muscle wasting and death of motor neurons.

apoptosis Cell death that occurs as a natural process during normal development.

are also times when development consists of *regressive* processes: when synaptic connections that had formed between neurons and their targets are retracted, which will be the subject of Chapter 8, and when cells that had been born and differentiated into neurons die long before maturity, which is the subject of this chapter.

Such naturally occurring cell death is called **apoptosis**. The scientists who coined the term (Kerr, Wyllie, & Currie, 1972) noted that the second *p* is supposed to be silent, making the pronunciation "ă-puh-TOE-sis," but you will also hear people say "ā-POP-toe-sis." It derives from a Greek word meaning a "falling off" as of leaves on a tree in autumn, to emphasize that this is a natural developmental process, not a response to trauma or disease.

We'll begin this chapter by considering those cells that first hinted at neuronal death in development—spinal motor neurons. Prenatally, we each made about 50,000 more of these cells than we actually need, so they died before we were born. Then we'll learn the remarkable story of the discovery of nerve growth factor, which provided the leading model of how neuronal apoptosis is regulated in development. That perspective will provide insight into the predicament facing people like Stephen Hawking and the frustration that although we seem to understand so much, a cure eludes us.

Delving into the molecular details will make it clear that these dying neurons are not withering away, or attacked by other cells, but rather have each made an active decision to give up the ghost. They then trigger a self-destruct button, unleashing a cascade of enzymes that dismantle the cell.

Then we'll consider a domain where differences between individuals in the process of neuronal apoptosis result in individual differences in brain anatomy and behavior, specifically the development of sex differences in the brain. We'll see that in vertebrates, hormones accentuate or diminish apoptosis to make the brains of males and females differ from one another. In *Drosophila*, no hormones are involved. Instead, whether a part of the nervous system is masculine or feminine in flies is a direct result of genetic regulation where, once again, neurons that die in one sex survive to guide behavior in the other. As usual with *Drosophila*, we'll see the remarkable specificity with which genes guide behavior in these animals. We'll conclude the chapter by considering whether and how the process of sexual differentiation in other vertebrates and flies offers insight into the origins of human sexual orientation.

■ 7.1 ■

THE DEATH OF MANY CELLS IS A NORMAL PROCESS IN DEVELOPMENT

However inefficient or wasteful naturally occurring cell death may appear, natural selection hit upon this method long ago: the apoptotic process is much the same in animals and plants (Reape & McCabe, 2008), so it probably harks back to the common ancestor of both. While our interest is in the effects of apoptosis in neural development, it is also important in the formation of many body structures. For example, when our fingers and toes develop prenatally, there is initially a web of epithelium stretched between adjacent digits that is eliminated when the cells in the webbing undergo apoptosis. Natural selection may adjust when and where such apoptosis occurs. For example, ducks are

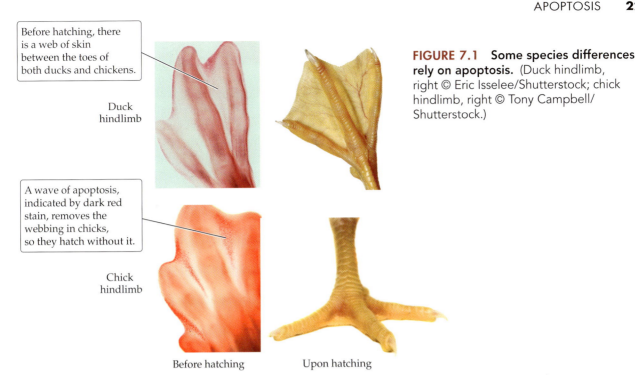

Before hatching, there is a web of skin between the toes of both ducks and chickens.

Duck hindlimb

A wave of apoptosis, indicated by dark red stain, removes the webbing in chicks, so they hatch without it.

Chick hindlimb

Before hatching   Upon hatching

FIGURE 7.1 **Some species differences rely on apoptosis.** (Duck hindlimb, right © Eric Isselee/Shutterstock; chick hindlimb, right © Tony Campbell/ Shutterstock.)

born with webbing between their toes because the apoptosis between the digits that is seen in birds such as chickens does not take place in ducks (**FIGURE 7.1**) (Merino et al., 1999). By canceling apoptosis in that webbing, natural selection improved the swimming capacity of ducks.

While cell death was documented and readily accepted for some parts of the body, as a process of sculpting particular body parts by paring away intermediate structures (Glucksmann, 1951), neuroscientists were slow to conceive that neurons might be formed and then discarded. It's one thing to produce some extra epithelium between digits and then discard it, but it seems quite another thing to produce a fully formed neuron, with dendrites and an axon reaching out for a target, only to throw it away.

You might think it would be easy to tell whether neurons have died in the developing nervous system, and in *Caenorhabditis elegans* it is—over 100 cells can be observed to die and, as in all cell fates in worms, mitotic lineage predicts which cells die. But for vertebrates, there is a basic problem in bookkeeping. Individuals get lost in the crowd, and no one has the capacity to keep track of every cell. If at one stage of development you see 20,000 neurons of a particular type, and at a later stage you see only 10,000, did the missing 10,000 die? How do you know whether instead they migrated away from that area, or de-differentiated and then differentiated into some other cell type that you don't recognize?

One person who addressed this problem was Viktor Hamburger (1900–2001), who received his doctorate working with Hans Spemann at the University of Freiburg. Hamburger worked as an instructor there until 1933 when the rector of the university, the famous philosopher Martin Heidegger, wrote to say Hamburger was dismissed because of new laws barring Jews from the faculty (Oppenheim, 2001). On a fellowship to work in the United States at the time, Hamburger remained in the States for the rest of his long career. He hit upon one way around the problem of keeping track of a population of neurons, by focusing on spinal motor neurons. As in the previous chapter about synapse development, the choice of studying motor neurons was a strategic one: motor neurons are the first neurons to appear in the developing spinal cord, and they have a distinctive appearance because they are the largest neurons in the

While cell death was documented and readily accepted for some parts of the body, neuroscientists were slow to conceive that neurons might be formed and then discarded.

Because neither Hamburger nor anyone else in the field imagined that fully formed motor neurons might normally die, he initially misinterpreted his results.

spinal cord, with multipolar cell bodies that are densely stained in Nissl stains. That means, among other things, that you at least have a chance of doing the bookkeeping, counting motor neurons. Hamburger also strategically chose to study chick embryos, where he could peek in on cell populations, and make manipulations, quite early in development. This work provided one of the earliest indications that developing nervous systems might indeed make neurons that would then simply die. But because neither Hamburger nor anyone else in the field imagined that fully formed motor neurons might normally die, he initially misinterpreted his results, as we'll see next.

How's It Going?

1. What is apoptosis, and what is the evidence that it is an ancient adaptation?

2. Why are motor neurons a good choice for studies keeping track of neuronal populations across development?

■ 7.2 ■

THE EXTENT OF DEATH AMONG DEVELOPING MOTOR NEURONS IS REGULATED BY THE SIZE OF THE TARGET

Trying to understand the regulation of spinal motor neuron number, Hamburger first replicated previous findings, showing that removing a limb bud of an early chick embryo resulted in fewer motor neurons on that side of the spinal cord, which normally would innervate that limb (Hamburger, 1934). While this result did fit the idea that the nervous system actively matches the size of the body to be innervated, Hamburger worried that the change in motor neuron number might not have been in response to the limb bud's absence, but just a general response to tissue damage on that side of the embryo. Maybe damaged limb tissue released chemicals that interfered with normal processes in the nearby spinal cord. It's relatively easy to disrupt development in such a way as to *reduce* cell number. A more convincing demonstration would be one that resulted in *more* motor neurons, so he tried that next.

RESEARCHERS AT WORK

Adding to the Periphery Prevents Apoptosis of Motor Neurons

■ **QUESTION**: Body regions with few muscles, like the chest have fewer motor neurons in the appropriate spinal cord regions (for the chest, the thoracic segments) than do body regions like the arms (innervated by cervical segments) and legs (innervated by lumbar segments). How does this matching of the number of motor neurons to muscles come about?

■ **HYPOTHESIS**: The size of the periphery controls how many motor neurons will end up in the innervating regions of the spinal cord.

■ **TEST**: Deliberately manipulate the periphery, providing either fewer or more leg muscles on one side of the body in early embryos, then when they hatch, count how many motor neurons are found in the lumbar spinal cord on the manipulated side versus the unmanipulated side of the body.

■ **RESULT**: When the leg bud was removed from one side of early chick embryos, the spinal cord at hatching had many fewer motor neurons on that side (**FIGURE 7.2A**). But this could have been due to inadvertent damage to the spinal cord, interfering with motor neuron development. So next Hamburger tried the complementary experiment, grafting an *extra* limb bud on one side of the embryo. That surgery resulted in there being *more* motor neurons on that side of the spinal cord (**FIGURE 7.2B–D**) (Hamburger, 1939; Hollyday & Hamburger, 1976).

(A) Limb bud ablation

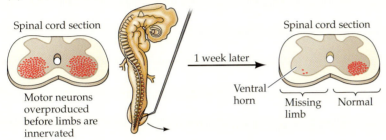

(B) Transplantation of
    supernumerary limb bud

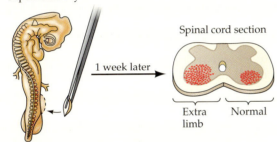

(C)   (D)

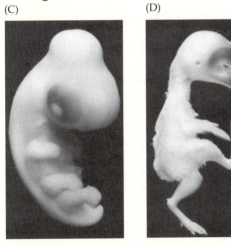

FIGURE 7.2 Target-derived trophic support regulates survival of motor neurons. (C,D from Hollyday & Hamburger, 1976.)

■ **CONCLUSION**: It seemed unlikely that any toxic substance or damage from adding another limb bud could cause *more* neurons to appear. Rather, Hamburger concluded that having a larger periphery to innervate tapped into some normal developmental process to increase the number of motor neurons on that side. But at this time Hamburger did not consider the possibility that having a larger periphery might cause fewer motor neurons to die, because he was unprepared for the idea of regressive events like apoptosis in neural development (Oppenheim, 2001). Rather, he thought the expanded periphery caused either more motor neurons to be born or more cells to differentiate into motor neurons.

FIGURE 7.3 Pyknotic cells in the embryonic mouse brain Arrows point to pyknotic cells with densely stained nuclei and, in some cases, such as that indicated at upper left, rounded droplets of a dismantled nucleus. (From Roth et al., 2000.)

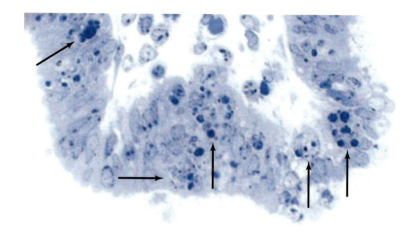

Were the cells dying, or simply de-differentiating to become some other type of neuron?

Later observations offered hints that Hamburger's manipulations had in fact affected motor neuronal survival. First, keeping track of the number of mitotic figures revealed that neurogenesis was complete quite early (Hamburger, 1948), so the limb bud manipulations could not have affected neurogenesis of motor neurons. Then keeping track of the number of cells immediately after the manipulations (Hamburger & Levi-Montalcini, 1949) made it clear that perfectly healthy-looking motor neurons and dorsal root ganglion cells were indeed disappearing, on both the manipulated side and the unmanipulated side of the body. But were the cells dying, or simply de-differentiating to become some other type of neuron?

That question was soon addressed by exploiting a characteristic appearance of apoptotic cells that had been uncovered in other organ systems (Glucksmann, 1951). It results from the nuclei collapsing upon themselves, which can sometimes be seen in regular Nissl-stained material, resulting in what is called **pyknotic** cells (from a Greek word meaning "to thicken or condense"; sometimes spelled *pycnotic*) (**FIGURE 7.3**). Thus we can also look for pyknotic cells as an indication of cells that were in the final stages of apoptosis. A disadvantage of relying on pyknotic cells to monitor apoptosis is that by the time the cell has this appearance, the process is almost over, and the remains of the destroyed cell will be cleared from sight within a few hours. That means our chance of catching a glimpse of the pyknotic cell before it disappears altogether is low. That, in turn, means that looking for pyknotic cells is generally only useful if many cells are dying, increasing our chance of catching a fraction of those "in the act" of pyknosis.

Fortunately for researchers (if not for the dying cells), sometimes there are indeed a lot of neurons undergoing apoptosis at the same time. One of the earliest demonstrations was in the frog *Xenopus laevis*, where the decline in the number of motor neurons coincided beautifully with the peak in the appearance of pyknotic cells (**FIGURE 7.4**). Using such numbers, the author was able to estimate that by the time a cell takes on a pyknotic appearance, it will only be visible for about 3 hours before disappearing (Hughes, 1961), which matched well with previous estimates in other tissues (Glucksmann, 1951). The disappearance of the pyknotic cell is facilitated by the appearance of **microglia**, which finish off the cell, dismantling its components and presumably making them available elsewhere in the body. The appearance of pyknotic cells helped persuade scientists, initially resistant to the idea that many neurons might die during normal development, that apoptosis is important for the formation of the nervous system.

pyknotic A characteristic appearance of cells undergoing apoptosis. A pyknotic cell looks as if the nucleus is collapsing upon itself.

microglia The class of glial cells that clean up debris and residue in the nervous system.

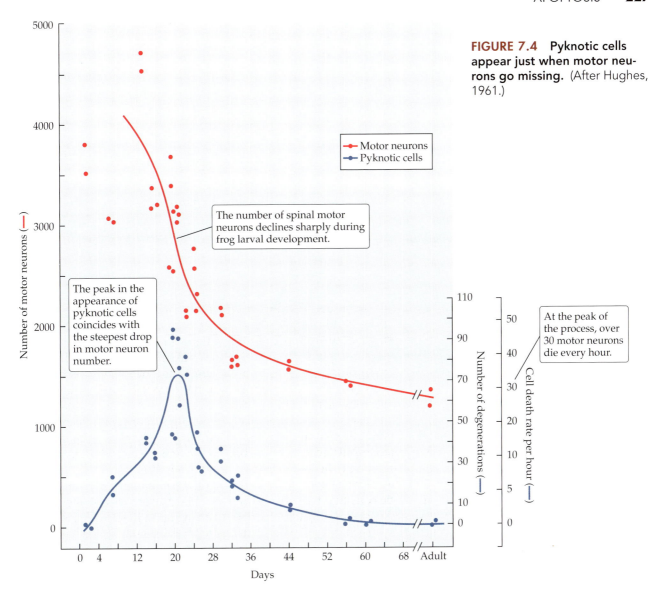

FIGURE 7.4 Pyknotic cells appear just when motor neurons go missing. (After Hughes, 1961.)

The number of spinal motor neurons declines sharply during frog larval development.

The peak in the appearance of pyknotic cells coincides with the steepest drop in motor neuron number.

At the peak of the process, over 30 motor neurons die every hour.

Hamburger found that chick embryos had about 20,000 spinal motor neurons shortly after the egg was laid, but by the second week of incubation, that number had dropped to more like 12,000, and then it remained stable (**FIGURE 7.5A**). The same loss of embryonic motor neurons was eventually documented in the human spinal cord, too (**FIGURE 7.5B**) (Forger & Breedlove, 1987a). Now it is universally accepted that apoptosis is a normal part of development throughout the nervous system (**FIGURE 7.5C,D**). For example, in the cerebral cortex there is considerable variation across regions, but on average about 30% of the neurons that arise early in life die before maturity (M. W. Miller, 1995; Nikolic, Gardner, & Tucker, 2013).

While the motor neurons' target cells have a strong effect on regulating their apoptosis, the cells that provide synapses onto the motor neurons have an effect, too. For example, deleting most afferent synapses onto developing motor neurons, by transecting the spinal cord to get rid of brain afferents, and preventing development of afferents to motor neurons from the periphery results in the loss of over a third of the motor neurons that normally survive (Okado & Oppenheim, 1984). Although the influence of afferents on apop-

Now it is universally accepted that apoptosis is a normal part of development throughout the nervous system.

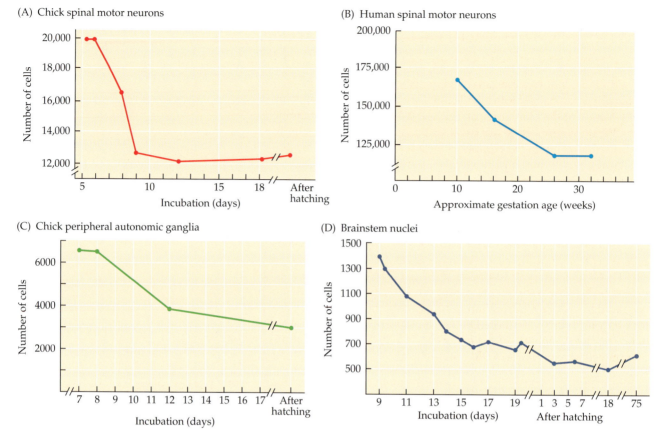

FIGURE 7.5 Many neurons die during early development. (A) The loss of motor neurons first reported in chicks was eventually reported in humans as well (B). (C) Peripheral autonomic ganglia and (D) brainstem nuclei also display a decline in the number of neurons during development. (A after Hamburger, 1975; B after Forger & Breedlove, 1987a; C after Landmesser & Pilar, 1974; D after Cowan & Wenger, 1967).

tosis is much less studied, and less understood, than the influence of target cells, in both cases there seems to be a matching of the number of neurons needed, based on either the number of cells providing synaptic input or the number of cells that require synaptic input.

Once it became clear that the size of the periphery somehow manipulated the extent of apoptosis in the nervous system in order to match the number of neurons to the number of jobs available, the question became one of mechanism. *How* does the periphery influence cell death in the nervous system? The answer to that question came about not from the study of motor neurons, but rather from the study of *sensory* neurons serving the body, as we'll see next.

How's It Going?

1. How does removing a limb bud, or grafting on an extra limb bud, affect the number of motor neurons in chicks?

2. How did monitoring pyknotic cells strengthen the case that spinal motor neurons die during development?

3. Once a cell takes on a pyknotic appearance, about how long does it take to disappear from view?

■ 7.3 ■

NERVE GROWTH FACTOR (NGF) IS DISCOVERED TO REGULATE APOPTOSIS IN SENSORY NEURONS

In 1938, after Hamburger began working in the United States, a young Italian physician named Rita Levi-Montalcini (1909–2012) lost her faculty position at the University of Turin, also because of anti-Semitic laws. Without support to work elsewhere, Levi-Montalcini simply continued her ongoing experiments on chick embryos at home. Before World War II came to a close, Levi-Montalcini's starving family was happily eating her subjects—after each experiment was over.

In her wartime experiments, Levi-Montalcini repeated Hamburger's grafts of an additional limb bud in chicks and found that these animals not only had additional motor neurons on that side of the spinal cord, they also had more **dorsal root ganglion (DRG)** cells on that side of the body (Hamburger & Levi-Montalcini, 1949). (Recall that DRGs contain the cell bodies of sensory fibers, whose branched axons extend both to the periphery to gather sensory information and into the spinal cord to deliver that information to neurons in the dorsal horn.) As with motor neurons, there were more DRG neurons early in development than upon hatching (Levi-Montalcini & Levi, 1944), indicating that the additional limb was preventing apoptosis of DRG cells. Thus having more peripheral structures reduced apoptosis of both motor and sensory cells, nicely matching the size of the nervous system to the size of the body. But how?

One clue came when a former student of Hamburger's, Elmer Bueker, implanted cells from a mouse sarcoma (malignant cancer of connective tissue) into a chick embryo. Bueker noticed that sensory fibers from the chick invaded the tumor eagerly (Bueker, 1948; Bueker & Hilderman, 1953). Not only that, the DRGs supplying those sensory fibers were noticeably enlarged (**FIGURE 7.6**). When the war ended, Levi-Montalcini got a fellowship to work

> Having more peripheral structures reduced apoptosis of both motor and sensory cells, nicely matching the size of the nervous system to the size of the body.

dorsal root ganglion (DRG)
A collection of neuron cell bodies embedded in vertebrate dorsal roots of the spinal cord that provides neurites to gather sensory information from the periphery and transmit it to the dorsal horn of the spinal cord.

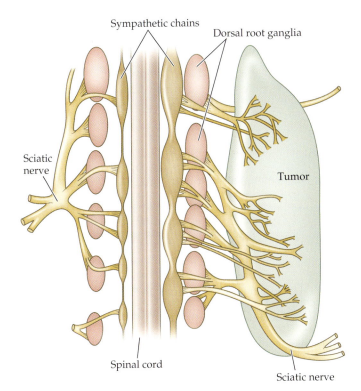

FIGURE 7.6   Chick sensory fibers invade mouse sarcoma. When a mouse tumor was implanted onto one side of chick embryos, the dorsal root ganglia and sympathetic ganglia on that side sent many axons into the tumor. The ganglia on that side also ended up being much larger than normal. (After Levi-Montalcini & Hamburger, 1951.)

in Hamburger's lab. She would stay at Washington University in St. Louis over 30 years. Together they would discover how the periphery regulates neuronal apoptosis. Hamburger encouraged Levi-Montalcini to repeat Bueker's experiments (Hamburger, 1981), and they became convinced that the sarcoma was producing some chemical substance that attracted sensory fibers and prevented their cell bodies in the DRG from dying. An unexpected finding in experiments to test this notion confirmed their ideas, as we'll see next.

RESEARCHERS AT WORK

Screening for Nerve Growth Factor

■ **QUESTION**: What chemical substance produced by mouse sarcoma cells is influencing DRG axonal growth and survival?

■ **HYPOTHESIS**: The sarcoma cells are secreting a chemical that encourages sensory cell axonal growth and prevents apoptosis.

■ **TEST 1**: If the hypothesis is correct, then fluid from the sarcoma should encourage growth of sensory fibers. To test this idea, culture explants of chick DRG in a dish, expose them to fluid from the sarcoma cells or to control fluid, and monitor the response of the DRG fibers.

■ **RESULT 1**: Within 24 hours, explants of DRG cells exposed to fluid from the sarcoma grew much more abundant "halos" of neurites than explants provided control fluid (**FIGURE 7.7A**) (Levi-Montalcini & Hamburger, 1951).

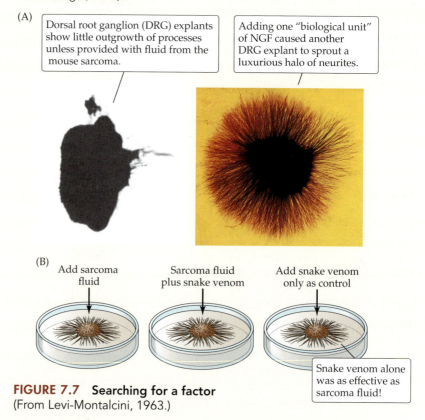

(A)

> Dorsal root ganglion (DRG) explants show little outgrowth of processes unless provided with fluid from the mouse sarcoma.

> Adding one "biological unit" of NGF caused another DRG explant to sprout a luxurious halo of neurites.

(B)

Add sarcoma fluid

Sarcoma fluid plus snake venom

Add snake venom only as control

> Snake venom alone was as effective as sarcoma fluid!

FIGURE 7.7 Searching for a factor (From Levi-Montalcini, 1963.)

■ **CONCLUSION 1**: There is a substance in the fluid from sarcomas that encourages sensory fiber growth. Is it a protein, or is it a virus that somehow triggers this response?

- **TEST 2**: Treat the fluid from sarcoma cells with an enzyme found in snake venom—phosphodiesterase—that metabolizes nucleic acids like DNA and RNA and so would neutralize any viruses in the fluid. If the fluid still encourages fiber growth from DRG explants, then by elimination, the growth-promoting substance is likely to be a protein.

- **RESULT 2**: The results were a big surprise. Not only did the snake venom not interfere with the ability of sarcoma fluid to promote DRG neurite outgrowth, but the control cultures, *exposed to snake venom alone*, without sarcoma fluid, also grew vigorously (**FIGURE 7.7B**). In fact, the snake venom was much more effective than sarcoma fluid at promoting DRG outgrowth (Cohen, Levi-Montalcini, & Hamburger, 1954; Levi-Montalcini & Cohen, 1956)!

- **CONCLUSION 2**: The chemical found in sarcoma fluid that promotes DRG neurite outgrowth is probably a protein, not a nucleic acid. More importantly, this same factor is much more abundant in snake venom than in sarcoma fluid.

Later screens would reveal an even more abundant source than snake venom for the chemical promoting DRG outgrowth: the salivary glands of male mice. By harvesting thousands of mouse salivary glands and using biochemical fractionation methods, scientists eventually identified a single protein, christened **nerve growth factor** (**NGF**), that was responsible for both promoting the outgrowth of DRG neurites and preventing apoptosis of DRG cells. In 1986 the Nobel Prize in Physiology or Medicine was awarded to Levi-Montalcini and Stanley Cohen, the biochemist she and Hamburger recruited to isolate NGF, but not to Hamburger. The decision to exclude Hamburger from the Nobel Prize was perplexing to neuroscientists of the day (Cowan, 2001; Lauder & Oppenheim, 2001; Purves & Sanes, 1987) and remains a controversial decision (**BOX 7.1**).

nerve growth factor (NGF)
The first isolated and identified neurotrophic factor, critical for the developmental survival of DRG and sympathetic cells.

BOX 7.1

THE CONTROVERSY OVER THE NOBEL PRIZE FOR NGF

Hamburger's student Elmer Bueker first implanted a mouse sarcoma in chick embryos and found that sensory fibers invaded it (Bueker, 1948). Hamburger suggested that Levi-Montalcini repeat the experiment when she came to his lab. That replication directly led to the discovery of NGF, almost all conducted in Hamburger's lab, yet Hamburger was not awarded a share of the 1986 Nobel prize, an outcome that puzzled many neuroscientists (Purves & Sanes, 1987). As one memorial tribute put it, "Although it seemed entirely appropriate to award the Nobel to Levi-Montalcini and Cohen, the absence of Hamburger appeared odd to the US neuroscience community"(Lauder & Oppenheim, 2001). Indeed, the Nobel for the work on NGF was long overdue, and because the prizes are never given posthumously, there seemed a

danger that some of the researchers might die before the committee got around to honoring their work. Once the Nobel was announced and Hamburger was omitted, you had to wonder if the committee had been waiting for him to die so they could award the prize only to Levi-Montalcini and Cohen without stirring any controversy. Hamburger was not so obliging and would live to 100 (Levi-Montalcini would live to be 103!).

On a personal note, I happened to be on sabbatical at Washington University in St. Louis in early 1988, a few years after the Nobel passed over Hamburger. Long retired from active research, he had just completed his book recounting the history of experimental embryology in Spemann's lab (Hamburger, 1988). But Viktor came to the medical

continued

THE CONTROVERSY OVER THE NOBEL PRIZE FOR NGF

school campus most Thursdays, brown bag in hand, to eat lunch and chat with the developmental neurobiologists there. Although he used a cane for support, Viktor seemed fully mobile and was intellectually engaged in the most recent research in a field he had done so much to foster. He was also keenly interested in politics, especially the civil rights movement. The only time I remember the Nobel coming up in conversation during my brief time there, it was rather indirect. Viktor reported that he was reading Levi-Montalcini's memoir (Levi-Montalcini, 1988) and that every time she related an important finding in the saga of NGF's discovery, she made a point of saying that Viktor had been out of town. It clearly hurt his feelings that Levi-Montalcini was minimizing his intellectual contributions to the work going on in his own lab. Personally, I could not fathom why the Nobel committee would deliberately and publicly hurt a distinguished elderly scientist who had not only set the intellectual stage for the discovery of NGF, but had also supported and guided the very work being honored.

Nobel Foundation rules say that permission to review the materials that led to an award, including notes about the deliberations, cannot be given until at least 50 years after the decision has been made

Viktor Hamburger 1933 (Courtesy of MBL History Project: Marine Biological Laboratory and Arizona Board of Regents.)

(Nobel Media, 2015). So in 2036 you college-aged readers are likely to be around when the Nobel committee releases records that might explain why Hamburger was snubbed. I'll try to emulate Hamburger's longevity, so that I can find out, too.

---

sympathetic ganglia The two chains of interconnected ganglia on either side of the vertebral column that receive input from the sympathetic preganglionic neurons in the spinal cord. The neurons within the sympathetic ganglia send their axons out to innervate various organs, where they usually release norepinephrine.

The scientists noted that DRGs were not the only parts of the nervous system affected by NGF. The **sympathetic ganglia** running parallel to the thoracic and lumbar segments of the spinal cord contain neuronal cell bodies that project their axons to the periphery, typically releasing norepinephrine to activate the "fight or flight" responses typical of sympathetic nervous system activation. These sympathetic ganglia in chicks also swelled in response to the sarcoma (see Figure 7.6) or snake venom, suggesting that they also respond to NGF. Indeed, NGF prevents apoptosis of sympathetic cells both in vitro (**FIGURE 7.8A**) and in vivo (**FIGURE 7.8B**).

(A)

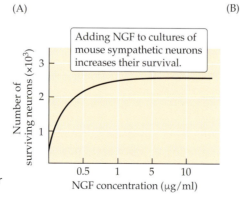

Adding NGF to cultures of mouse sympathetic neurons increases their survival.

FIGURE 7.8 **Exogenous NGF increases survival of sympathetic ganglia.** (A after Chun & Patterson, 1977; B from Levi-Montalcini, 1972.)

(B)

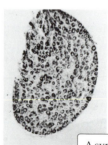

A sympathetic ganglion from a control mouse (top) is much larger than one from a mouse exposed to antibodies to NGF during development (bottom).

Having established that NGF promotes outgrowth in neurons from DRGs and sympathetic ganglia, the next question was whether NGF also prevents apoptosis in these populations and so plays a role in matching the size of the nervous system with the size of the periphery.

--- **How's It Going?** ---

1. What neurons, in addition to motor neurons, have their survival affected by manipulating the number of limb buds in chicks?

2. What was the intent of treating sarcoma fluid with snake venom in studies of DRG neurite outgrowth? What surprise was in store for scientists when they used the venom?

3. What two neuronal populations are affected by NGF?

■ 7.4 ■

NGF HAS BOTH TROPIC AND TROPHIC EFFECTS ON SELECTIVE NEURONAL POPULATIONS

Levi-Montalcini's ingenious assay for NGF activity, monitoring the process outgrowth of DRG explants, showed that NGF has a **tropic** effect, encouraging process outgrowth (**FIGURE 7.9**). The experiments implanting sarcomas in chick embryos showed NGF also has a **trophic** effect, keeping DRG and sympathetic ganglion neurons alive during the period of cell death. Since the trophic effect was on neural cells, people came to refer to NGF as a **neurotrophic** factor.

tropic   Here, referring to the capacity of a factor to attract a growing neuronal process.

trophic   Here, referring to the capacity of a factor to prevent the death of a cell, as if "feeding" it.

neurotrophic   Referring specifically to the capacity of a factor to prevent the death of neural cells.

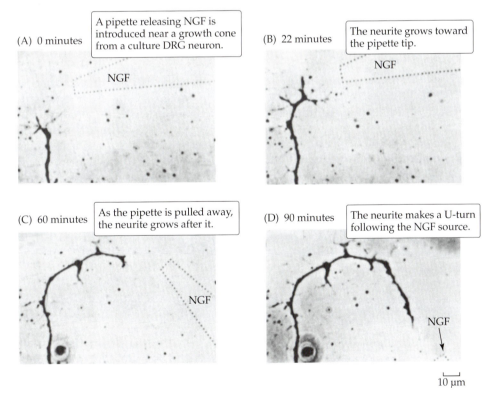

(A) 0 minutes — A pipette releasing NGF is introduced near a growth cone from a culture DRG neuron.

(B) 22 minutes — The neurite grows toward the pipette tip.

(C) 60 minutes — As the pipette is pulled away, the neurite grows after it.

(D) 90 minutes — The neurite makes a U-turn following the NGF source.

10 μm

FIGURE 7.9   NGF has a pronounced neurotropic effect on the growth cones of some neurons. This growth cone changes direction to follow a trail of NGF from an electrode (outlined in dots). The term *neurotropic* refers to such an ability to attract neurites. This is in addition to NGF's *neurotrophic* effect, preventing apoptosis in sympathetic and dorsal root ganglion neurons. (From Gundersen & Barrett, 1979, courtesy of Dr. J. Barrett.)

Once it had been shown that sarcoma cells prevented the death of DRG cells and sympathetic neurons in chick embryos, and that the substance coming from sarcoma cells that had this effect was indeed NGF, the next question was whether NGF could also prevent death of DRG cells in vivo. Indeed, injections of NGF into the limb bud on one side of a chick embryo resulted in fewer degenerating DRG cells on that side (Brunso-Bechtold & Hamburger, 1979). Still, this did not prove that the embryo *normally* uses that same substance, NGF, to regulate DRG cell apoptosis. After all, it was always possible that sarcoma-derived NGF being provided exogenously was mimicking the effect of some other, potentially unrelated protein that chicks normally produce endogenously. The most convincing demonstration that endogenous NGF is important for survival of DRG and sympathetic neurons came when scientists induced pregnant rat dams to produce antibodies to NGF. Like most antibodies, these would cross the placenta and reach the embryos. At birth, the pups from such dams indeed had drastically smaller DRGs, with many fewer neurons (Gorin & Johnson, 1979). What's more, those DRG cells that had survived development in the absence of NGF were different—they did not transport radiolabeled NGF from the periphery to their cell bodies, as most DRG neurons do in control animals. The researchers concluded that a subset of DRG neurons are NGF sensitive and need NGF to survive apoptosis and then, in adulthood, continue to respond to NGF.

The ability of neurons to retrogradely transport NGF was demonstrated in experiments using what came to be known as "Campenot chambers": a Teflon "gasket" was placed on top of cultured cells and sealed with a bit of silicon grease (Campenot, 1977), separating the fluid compartments for a neuron's cell body, dendrites, and axon (**FIGURE 7.10A**). For an NGF-sensitive neuron like a DRG cell, a sympathetic cell, or a cancer-derived cell line such as the PC12 line, some NGF must be present or the cells will die. Campenot chambers demonstrated that NGF has a local effect on growing neurites. Those neurites growing in fluid with NGF continued to grow, while other neurites, from the same cell body, growing in fluid without NGF regressed. Neurites exposed to NGF also showed a rapid influx of Ca^{2+} that spurred further growth. What's more, providing NGF to the neurites was enough to keep the cell alive, even if the fluid around the cell body was without NGF. This result suggested that NGF was being retrogradely transported from cell processes to the cell body, which was soon demonstrated when radiolabeled NGF put into the neurite compartment showed up in the cell body (Claude, Hawrot, Dunis, & Campenot, 1982). Both the NGF and the receptor that binds it, which we'll discuss next, are internalized and retrogradely transported to the cell body (**FIGURE 7.10B**), where the NGF alters gene expression to, among other things, prevent the cell from undergoing apoptosis.

> Providing NGF to the neurites was enough to keep the cell alive, even if the fluid around the cell body was without NGF.

How's It Going?

1. Compare and contrast tropic and trophic effects.

2. How did scientists manage to interfere with endogenous NGF to prove its role in DRG and sympathetic cell survival?

3. Describe a Campenot chamber. What results with Campenot chambers demonstrated that NGF at dendrites could promote survival?

4. How was it shown that NGF applied to neurites was retrogradely transported?

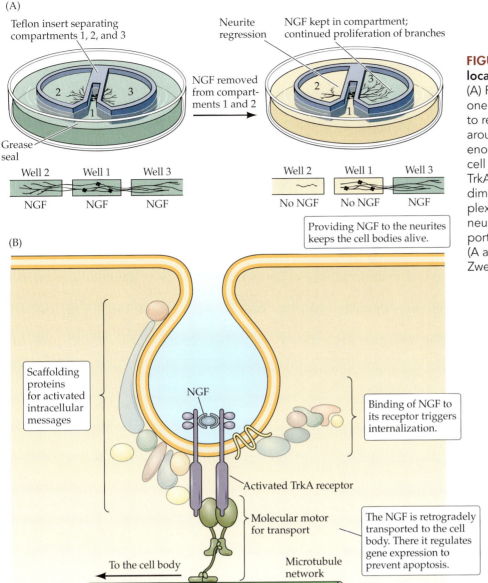

(A)

Teflon insert separating compartments 1, 2, and 3

Neurite regression

NGF kept in compartment; continued proliferation of branches

NGF removed from compartments 1 and 2

Grease seal

Well 2   Well 1   Well 3

NGF   NGF   NGF

Well 2   Well 1   Well 3

No NGF   No NGF   NGF

Providing NGF to the neurites keeps the cell bodies alive.

(B)

Scaffolding proteins for activated intracellular messages

NGF

Binding of NGF to its receptor triggers internalization.

Activated TrkA receptor

Molecular motor for transport

To the cell body

Microtubule network

The NGF is retrogradely transported to the cell body. There it regulates gene expression to prevent apoptosis.

FIGURE 7.10 NGF acts locally and in the cell body. (A) Removing NGF from around one set of neurites causes them to regress. But maintaining NGF around the other neurites is enough to prevent death of the cell bodies. (B) NGF binds two TrkA molecules, causing them to dimerize. The NGF/TrkA complex is then internalized by the neurites and retrogradely transported to the cell body. (A after Campenot, 1981; B after Zweifel et al., 2005.)

■ 7.5 ■

THE SEARCH FOR RELATIVES OF NGF REVEALS A FAMILY OF NEUROTROPHIC FACTORS AND THEIR RECEPTORS

Many scientists were inspired by the remarkable success of NGF research, but despite a name that perhaps implies it was one of a kind (*the* nerve growth factor?), it was obvious that not all neurons respond to NGF. For example, there was no effect of NGF manipulations on motor neurons, nor any visible effect on brain growth. So scientists assumed that there must be other neurotrophic factors, including some normally produced by one group of brain cells to maintain innervation from other neurons. The early attempts to find other neurotrophic factors relied on gathering large amounts of brain tissue, homogenizing it, and then using old-fashioned biochemical methods to separate the resulting fluid into "fractions." Each fraction was then tested on some assay to see whether it affected neuronal survival. Once it was determined which fraction held the factor, there were attempts to further isolate and identify the protein.

In this manner, **brain-derived neurotrophic factor** (**BDNF**) was isolated from homogenates of pig brains (hence the name) (Barde, Edgar, & Thoenen,

brain-derived neurotrophic factor (BDNF) A factor, originally isolated from pig brains, that supports the developmental survival of DRG cells, but not sympathetic cells.

FIGURE 7.11 **Members of the neurotrophin family affect overlapping classes of neurons.** (A) Different neuronal populations rely on different neurotrophic factors for survival. (B) Some of the neurotrophic factors known to support various sensory neurons from the DRG. (A from Maisonpierre et al., 1990; B after Bibel & Barde, 2000.)

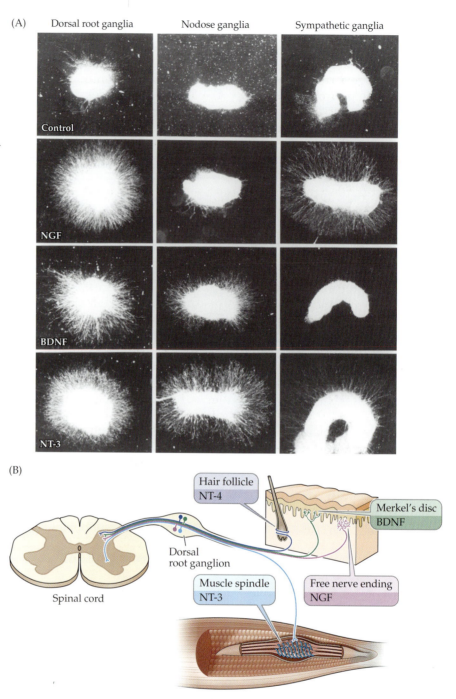

(A)

Dorsal root ganglia · Nodose ganglia · Sympathetic ganglia

Control · NGF · BDNF · NT-3

(B)

Hair follicle
NT-4

Merkel's disc
BDNF

Dorsal
root ganglion

Spinal cord

Muscle spindle
NT-3

Free nerve ending
NGF

neurotrophin-3 (NT-3)
A member of the neurotrophin family, discovered after NGF and BDNF.

neurotrophins The family of neurotrophic factors that includes NGF, BDNF, and the other neurotrophins, which are designated by number.

1982), and like NGF it supported the survival of DRG cells, but unlike NGF it had no effect on sympathetic neurons (Lindsay, Thoenen, & Barde, 1985). Once BDNF was sequenced, it was found to be similar to, but distinct from, NGF, indicating that the two are evolutionarily related.

Shortly after BDNF was isolated and found to have so many homologies with NGF, molecular biological techniques became available in which a low-stringency hybridization strategy, rather than old-fashioned biochemical fractionation techniques, could be used to look for other genes related to those two. Sure enough, a third neurotrophic factor similar to but distinct from NGF and BDNF was found, and it was named **neurotrophin-3** (**NT-3**) (Hohn, Leibrock, Bailey, & Barde, 1990), recognizing that NGF was the first and BDNF the second member of this family that had been isolated. Thus these three were seen to be members of a family, called the **neurotrophin** family. These three neurotrophins have distinct but overlapping effects on the survival and outgrowth of different classes of neurons (**FIGURE 7.11**).

Once it was clear that there was a neurotrophin family, there was a bit of a gold rush attitude in the search for other family members. Soon **neurotrophin-4** (**NT-4**) was found (Hallbook, Ibanez, & Persson, 1991; Ip et al., 1992), then *neurotrophin-5* (Berkemeier et al., 1991), but it turned out neurotrophins 4 and 5 were the same protein (Ibanez, 1996), so sometimes you'll see it called neurotrophin-4/5. There seem to be only these four neurotrophins in mammals, but teleost fishes have two more, designated neurotrophin-6 (Gotz et al., 1994) and neurotrophin-7 (Nilsson, Fainzilber, Falck, & Ibanez, 1998).

The four mammalian neurotrophins are each made from a larger propeptide, which itself appears to have no biological activity but which can be posttranslationally processed rapidly to produce one particular neurotrophin or another (Park & Poo, 2013). In some ways the name *neurotrophin* is unfortunate, because there are other proteins that are clearly *neurotrophic*, in the sense that they keep neurons from undergoing apoptosis, but are not related to the neutrophin family. For example, a factor extracted from chick embryo eyes, **ciliary neurotrophic factor** (**CNTF**), favors the survival of parasympathetic neurons that normally innervate the eye (Manthorpe, Barbin, & Varon, 1982). Another example, **g**lial cell line–**d**erived **n**eurotrophic **f**actor (GDNF), favors survival of dopaminergic neurons (Lin, Doherty, Lile, Bektesh, & Collins, 1993).

The receptor for NGF was originally studied by cancer biologists and named **tropomyosin receptor kinase**, or **Trk** (see Figure 4.3); it is also a tyrosine receptor kinase (RTK), so that name is sometimes used (Martin-Zanca, Hughes, & Barbacid, 1986). In a happy coincidence for those of us who rely on mnemonics, the receptor for the first neurotrophin, NGF, was also the first studied and so named **TrkA**. Continuing that streak, the second discovered neurotrophin, BDNF, binds the second receptor, **TrkB**, while NT-3 binds, yes, **TrkC**. However there is no TrkD receptor, so the streak breaks down. Instead, NT-4 binds to TrkB, and so do NT-3 and BDNF. In other words, the TrkB receptor binds all the neurotrophins except NGF (**FIGURE 7.12A**). The Trk receptors are said to be the high-affinity receptors for the neurotrophins, but there is also an unrelated protein, called **p75**, which binds all the neurotrophins, albeit with low affinity (**FIGURE 7.12B**).

When a neurotrophin molecule encounters a target cell, the neurotrophin binds together two of the appropriate Trk molecules, forming a dimer, and ac-

Once it was clear that there was a neurotrophin family, there was a bit of a gold rush attitude in the search for other family members.

neurotrophin-4 (NT-4) The fourth-discovered member of the neurotrophin family.

ciliary neurotrophic factor (CNTF) A trophic factor that prevents developmental death of neurons in the ciliary ganglia.

tropomyosin receptor kinase (Trk) A subfamily of the receptor tyrosine kinase (RTK) family that serve as receptors for neurotrophins.

TrkA The first-identified neurotrophin receptor, which has a high affinity for NGF.

TrkB The second-identified neurotrophin receptor, which has a high affinity for BDNF and the other neurotrophins except NGF.

TrkC The third-identified neurotrophin receptor, which has a high affinity for neurotrophin-3.

p75 A low-affinity receptor for the neurotrophins.

(A) High-affinity Trk receptors

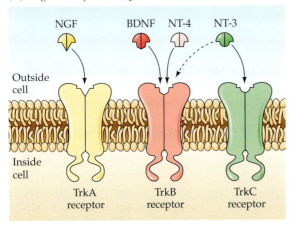

(B) Low-affinity p75 receptor

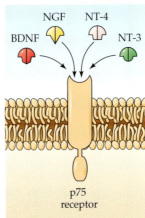

FIGURE 7.12 The neurotrophin receptors (A) The high affinity binding of various neurotrophins dimerizes different Trk molecules together to exert an effect on target neurons. (B) All the neurotrophins bind to the p75 receptor, albeit with relatively low affinity.

FIGURE 7.13   Intracellular signaling by neurotrophin receptors (A) Several different signaling pathways are initiated by neurotrophin binding to Trk receptors, but they generally result in preservation of neurons themselves, or their neurites and/or their synaptic contacts. (B) Different intracellular signaling pathways are initiated by binding the low affinity p75 receptor, and the response of the target neuron is quite variable.

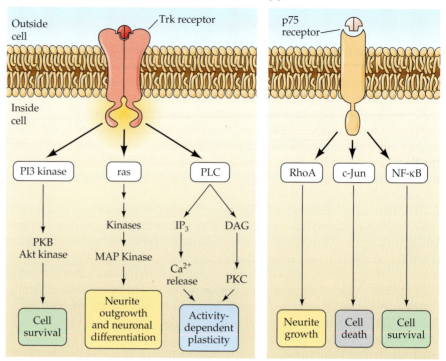

The success of research on neurotrophic factors has provided the predominant model of apoptosis in the nervous system.

tivating one or more of at least a trio of biochemical pathways inside the target cell (**FIGURE 7.13A**). Studies of mice in which one of the Trk receptors is knocked out confirmed that distinct but overlapping neuronal populations are affected by each (Snider, 1994) (**TABLE 7.1**). Neurotrophin binding of p75 activates one of several other pathways quite distinct from Trk receptors, some of which can trigger neurite outgrowth while others can control the cell's life or death (**FIGURE 7.13B**). So the response of a given cell to a neurotrophin is a function of the proportion of Trk versus p75 receptors it possesses, as well as the prior biochemical state of the cell when the neurotrophin arrives at the surface.

The success of research on neurotrophic factors has provided the predominant model of apoptosis in the nervous system—targets supply innervating neurons with a neurotrophic signal that, because it is in limited supply, causes some of those innervating neurons to die, resulting in the appropriate number of neurons surviving to adulthood. This scenario, well-established for neurons innervating targets in the periphery, may also regulate apoptosis of neurons innervating targets within the CNS, but this has been difficult to demonstrate (Southwell et al., 2012).

The term *neurotrophic factor*, derived from the notion of feeding, may conjure up the image of a neuron starving, growing progressively weaker

TABLE 7.1
Percent Neuronal Loss following Trk Receptor Knockouts

Neuronal population	TrkA	TrkB	TrkC
DRG	70–80	30–50	20
Superior cervical ganglion (SCG)	99	0	55
Trigeminal	70	25–55	65
Nodose	0	45–65	50
Vestibular	0	>80	0
Motor neurons	0	35	30

Source: After Snider, 1994.

from "malnutrition" until it succumbs to lack of resources. But while it's true that various neuronal populations do need particular neurotrophic factor(s) to survive, the neurons getting insufficient amounts of the factor don't slowly waste away. Instead, the doomed neurons actively begin the process of dismantling themselves, essentially committing suicide. Perhaps a better image of an apoptotic neuron would be a samurai warrior who, learning that his death would benefit his master, actively takes his own life. Let's discuss now the findings supporting this view of apoptosis as an active process.

How's It Going?

1. What is BDNF and how was it originally isolated?
2. What are the members of the neurotrophin family?
3. What receptors are bound by neurotrophins?

■ 7.6 ■

STUDIES IN *C. ELEGANS* PROVIDE CRUCIAL INFORMATION ABOUT THE PROCESS OF APOPTOSIS

Whenever cells die in the brain, from whatever cause, it is beneficial to get the resulting debris out of the way, and this task is accomplished by immune system cells that attack, engulf, and dismantle the material. As in *Xenopus* cells, discussed earlier, microglia accomplish this task in vertebrates, and they appear about the same time that apoptotic neurons take on a pyknotic appearance. This prompt arrival of microglia raised the question of whether they were merely cleaning up after a cell death, or were they in fact actively attacking the cell and causing it to die? While there had been several observations of pyknotic cells without apparent microglia around, suggesting microglia came after degeneration started, it was difficult to be sure the microglia hadn't been there from the beginning. Maybe a particular slice of observed tissue was in the wrong plane to detect a microglial cell, or maybe microglia had attacked the neuron and moved away before the tissue was processed. The question of which comes first, degeneration or attack by immune system cells, was first definitively settled in *C. elegans*.

Recall from Chapter 1 that in the mitotic lineage–directed differentiation of the 1,000 or so cells in the adult worm, there are cases in which a cell undergoes its final (or nearly final) mitosis, then one of the daughter cells takes on a role while the other simply dies. In fact, over 100 cells, about 12% of the total, normally die in this fashion (Hedgecock, Sulston, & Thomson, 1983). For a while, the literature referred to this mitotic lineage-determined apoptosis as programmed cell death, with the implication that it was somehow different than the apoptosis that Hamburger and Levi-Montalcini had studied. The logic of that distinction was questionable from the beginning (**BOX 7.2**), and subsequent findings confirmed that the process is much the same in worms and vertebrates.

As in vertebrates, dying cells in developing worms are quickly engulfed by immune system cells (called **phagocytes**, as they are "eating cells"), which normally clean up debris and attack invading cells. In fact, the phagocytes arrive so quickly when cells die that at first it was not clear whether the doomed cell died first and the cleanup followed, or whether phagocytes caused the cell to die.

In a program of research to understand apoptosis, many mutant worms were produced with gene mutations that affected apoptosis. These worms were called *ced* (**c**ell **d**eath abnormal) mutants, and they resulted from alterations in several different genes. Numbered in the order of their discovery, *ced-1* and *ced-2* were the first genes found, and they offered a way to test whether dying cells are murdered by phagocytes.

phagocytes Immune-related cells that attack and dismantle debris and invading microbes.

***ced* (cell death abnormal)** A collection of genes, originally isolated in *C. elegans*, that regulates apoptosis.

KERFUFFLES IN LANGUAGE: PROGRAMMED CELL DEATH

It would be hard for you to imagine the excitement when scientists working with *C. elegans* discovered that many cells produced by the cascade of mitosis normally die shortly after the final division that produced them, and that mutations in specific genes could avert those deaths. In those early years, the fact that one could predict precisely which cells would die based on mitotic lineage, and that specific genes were required for the death to occur, made the whole system seem very programmatic, like the unfolding of a meticulous and precise plan. Thus the early literature often refers to those deaths as "programmed cell death" or "pcd."

It was already clear that cell death is also a normal part of the developing nervous systems in flies and vertebrates, but of course one could never predict with certainty which particular chick motor neuron would live or die. Because apoptosis was so predictable in worms but not in other species, some researchers assumed that only worms displayed programmed cell death, and that it was quite distinct from what was happening in other species. In fact, there are still sources that apply the phrase *programmed cell death* only to apoptosis in worms.

But there's no reason to think that the death of a motor neuron in chicks is any less "programmed" or "programmatic." After all, it fulfills a systematic plan to leave the individual with an appropriate number of cells. Rather, the difference seems to be that the fate of cells, including whether they live or die, is determined almost entirely by mitotic lineage in worms, while it is determined by more complex interactions with other cells (including innervation targets) in other species.

Apoptosis is regulated by mitotic lineage in *C. elegans*, while it is regulated by neurotrophic factors from target cells in vertebrates, but the processes in both cases seem equally "programmatic." We'll see that they certainly use the same molecular pathway (see Figure 7.15). Thus you'll rarely see anyone today trying to make the case that programmed cell death is the exclusive domain of worms, and indeed some scientists refer to apoptosis in all species as programmed cell death (Buss, Sun, & Oppenheim, 2006; Purves & Sanes, 1987). In this book, I'm satisfied with simply using the term *apoptosis*, or even more simply, *death*.

RESEARCHERS AT WORK

It Was Suicide, Not Murder

■ **QUESTION**: Are cells that normally die during development in *C. elegans* actually killed by the invading immune cells called phagocytes that then clean up the debris?

■ **HYPOTHESIS**: Alternatively, apoptosis of these cells may be a **cell-autonomous response**, and the phagocytes may only come in after the cell has made that decision. In other words, maybe the phagocytes attack and clean up a cell only after it has begun dying.

■ **TEST**: Create mutant worms where the phagocytes do not show up, and see whether the cells that normally undergo apoptosis still die.

■ **RESULT**: Mutations in two different genes produced worms that appeared otherwise healthy but did not produce phagocytes to clean up debris. In both cases, almost all the cells that normally die during development still died, even with no phagocytes to clean up afterward (Hedgecock et al., 1983).

■ **CONCLUSION**: The death of cells in developing worms is a cell-autonomous decision—a case of cell suicide rather than murder by immune-related cells. CED-1 turned out to be a membrane-spanning protein produced in doomed cells that signals phagocytes to engulf them, while CED-2 functions within the phagocytes to trigger that engulfment.

cell-autonomous response The condition when a particular influence acts directly upon a cell to affect it, as opposed to acting first on another cell, which then affects the cell of interest.

Note that disabling *ced-1* means the apoptotic cell cannot signal phagocytes to engulf it and polish it off. In that case, cells that would have died persist. The apoptotic cell is still deciding whether it will live or die, either signaling for phagocytes or not, so it's definitely not murder. But because it relies on those phagocytes to complete the process, perhaps we should think of it as "assisted suicide." Similarly, in mice the death of some Purkinje cells in the cerebellum is averted if microglia (which serve as phagocytes in the mammalian brain) are disabled (Marin-Teva et al., 2004).

The next *ced* found in *C. elegans*, **ced-3**, encoded a protease (an enzyme that cleaves proteins) (Yuan, Shaham, Ledoux, Ellis, & Horvitz, 1993). The particular type of protease encoded by *ced-3* is called a **c**ysteine-dependent **asp**artate-directed prote**ase**, or **caspase**, because it acts in cysteine-rich regions where the enzyme actually cuts the protein at the site of an aspartate. This pattern of amino acids is common enough that many proteins are susceptible to caspases. Thus the protein CED-3 serves as an executioner, dicing up proteins in the cell to dismantle its metabolic machinery. The gene encodes an inert preprotein that sits waiting to be rapidly unleashed by posttranslational regulation, specifically the CED-4 protein. In worms without functional CED-3, cells that would otherwise die just sit there for a while. Without a functional CED-3 protein to act as a self-destruct button, they cannot die. Eventually these "undead" cells may differentiate to some extent, but their fates are highly variable and unlike any other cell (Avery & Horvitz, 1987).

> The protein CED-3 serves as an executioner, dicing up proteins in the cell to dismantle its metabolic machinery.

In the meantime, experiments confirmed that apoptosis is also an active process in vertebrates, as we'll discuss next.

How's It Going?

1. What are phagocytes and what role do they play?
2. What's the evidence that apoptotic cells call for phagocytes rather than die as a result of phagocyte attack?
3. What does the term "caspase" stand for, and what role do these enzymes play in apoptosis?

■ 7.7 ■

APOPTOSIS INVOLVES ACTIVE SELF-DESTRUCTION THROUGH A CASCADE OF "DEATH GENES"

The rapid disappearance of pyknotic cells offered the first hint that dying cells are not slowly withering away, but rather are quite actively dismantling themselves. This was confirmed in many ways. One of the earliest clues came from *in vitro* research on cells that are dependent on NGF—in other words, the cells will die if you don't include NGF in the fluid medium surrounding them. For example, if you dissociate sympathetic ganglia taken from a rat, the cells die quickly unless you add NGF to the medium. If you then put antibodies to NGF into the medium, effectively depriving the cells, they die quickly. Paradoxically, if when you add the antibodies to NGF, you also add antibiotic drugs that stop either gene transcription or protein synthesis, the cells do not die, even though they've been deprived of NGF (Martin et al., 1988). These results demonstrated that *apoptosis is an active process*, that the cell must be able to transcribe new genes, and make new proteins, in order to actually die. Those genes came to be known as **death genes**, whose only role is to dismantle the cell to commence apoptosis. Once again, the cell's death seems a matter of suicide, not murder.

ced-3 A gene that encodes a protease, CED-3, to dismantle cell machinery as part of the process of apoptosis.

caspases Cysteine-dependent aspartate-directed proteases, including CED-3, which cleave many different proteins.

death genes A term for those genes that are activated in cells during apoptosis, including proteases to dismantle proteins and nucleases to fragment DNA.

FIGURE 7.14 A fatal surplus of neurons
(From Kuida et al., 1998.)

(A) *caspase-9* $^{+/+}$ (Wild-type)   *caspase-9* $^{-/-}$ (Knockout)

In mice without functional *caspase-9*, neurons do not die and so the fetal brain is too large to fit in the skull.

(B) Wild-type   Knockout

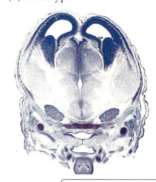 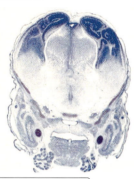

Cross sections reinforce the dense packing of cells in the brain, squeezing shut the ventricles of *caspase-9* knockout mice.

caspase-9 One of the earliest in a cascade of caspases leading to apoptosis.

caspase-3 One of the last in a cascade of caspases activated during apoptosis. It seems to act as the "executioner" that dooms the cell.

Several death genes in vertebrates are homologues of the worm *ced-3* gene, repeating the theme we've seen before, in which duplication and divergence provide vertebrates with several variants of a gene that was a singleton in invertebrate ancestors. The duplication of genes seems to afford vertebrates more options in controlling neural development. In fact, there is a cascade of caspases that complete apoptosis in mammals. The ninth caspase that was found in mammals, and so is called **caspase-9**, actually comes first in the cascade. Caspase-9 is regarded as an "initiator," cleaving only a few proteins and so doing little damage by itself. However, one of the cleavages by caspase-9 activates **caspase-3**, a powerful enzyme that cleaves so many proteins that it is called an "executioner" (Tait & Green, 2010). Among other things, caspase-3 cleaves an endonuclease inhibitor, leading to active endonucleases that fragment the DNA (Wolf, Schuler, Echeverri, & Green, 1999), dooming the cell to pyknosis. Therefore, scientists often use immunocytochemistry directed specifically at the activated caspase-3 enzyme to identify cells that are about to die.

Blocking the caspase cascade by knocking out *caspase-9* in mice results in overlarge brains that can't fit in the animals' skulls (**FIGURE 7.14**), so the mice die at birth (Kuida et al., 1998). Mice with *caspase-3* knocked out also produce

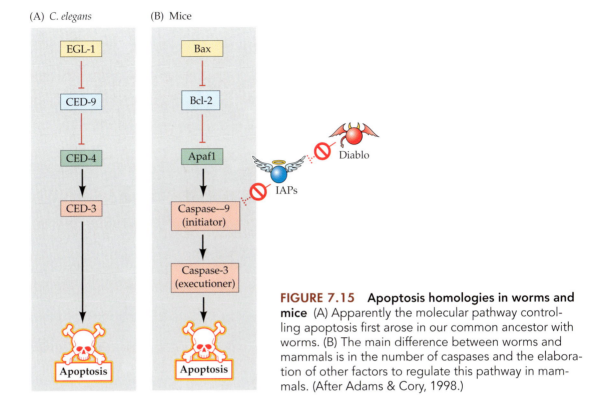

(A) *C. elegans*

EGL-1

CED-9

CED-4

CED-3

Apoptosis

(B) Mice

Bax

Bcl-2

Apaf1

Caspase-9
(initiator)

Caspase-3
(executioner)

Apoptosis

Diablo

IAPs

FIGURE 7.15 **Apoptosis homologies in worms and mice** (A) Apparently the molecular pathway controlling apoptosis first arose in our common ancestor with worms. (B) The main difference between worms and mammals is in the number of caspases and the elaboration of other factors to regulate this pathway in mammals. (After Adams & Cory, 1998.)

overly large brains (Kuida et al., 1996). Eventually the pathway of genes that execute apoptotic cells was worked out, and it turns out that homologous genes in worms and mammals act in the same sequence (**FIGURE 7.15**). The main difference between worms and mice in the sequence of gene expression underlying apoptosis is that there are additional regulatory proteins in the vertebrates (see Figure 7.15B). For example, vertebrates also produce a series of proteins, called **inhibitors of apoptosis proteins** (**IAPs**), that block caspases, including caspase-9, which initiates the cascade of caspases in vertebrates. Thus vertebrate neurons have a more complex network of intracellular signaling that determines whether a cell will undergo apoptosis.

Mitochondria also play a role in the suicide of vertebrate cells and seem to represent a "final common pathway" to apoptosis. When various influences, including the lack of neurotrophic factor at some crucial stage of development, leads a cell to apoptosis, an early event is the influx of Ca^{2+} in mitochondria, which releases a protein called **Diablo** (or Smac, for **s**econd **m**itochondria-derived **a**ctivator of **c**aspases) in response. Diablo then binds to the IAPs, releasing caspases that then execute the cell. Some proteins generally block apoptosis, such as **Bcl-2**, which works by preventing release of Diablo from mitochondria (**FIGURE 7.16**). In turn, the antiapoptotic Bcl-2 protein may itself be bound by another protein called Bax, which thereby favors apoptosis.

Another "death gene" produced by an apoptotic cell is an enzyme that is an endonuclease, which means it destroys its own DNA, irreversibly disabling the cell from functioning any longer. We can exploit this degradation of DNA to determine which cells have begun the process of apoptosis. The **T**dT-**d**UTP **n**ick **e**nd **l**abeling (**TUNEL**) method uses an enzyme (TdT: terminal deoxynucleotidyl transferase) to add labeled molecules of a thymidine analog called

Vertebrates also have proteins that generally block apoptosis, such as IAPs.

inhibitors of apoptosis proteins (IAPs) A family of proteins that block caspases and in other ways avert apoptosis.

Diablo Also known as *Smac*. A protein that promotes apoptosis, in part by blocking the action of IAPs.

Bcl-2 An anti-apoptotic protein in mammals. Its homologue in worms is CED-9.

TUNEL TdT-dUTP nick end labeling, a method to enzymatically label the various "nicks" in DNA that has been fragmented in preparation for apoptosis.

FIGURE 7.16 A common pathway to apoptosis

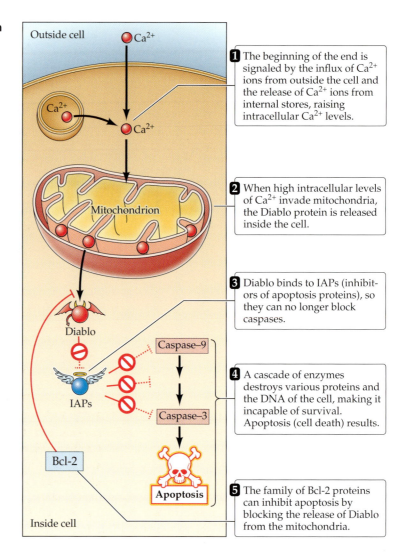

Outside cell

Ca²⁺

1 The beginning of the end is signaled by the influx of Ca²⁺ ions from outside the cell and the release of Ca²⁺ ions from internal stores, raising intracellular Ca²⁺ levels.

Ca²⁺

Ca²⁺

Mitochondrion

2 When high intracellular levels of Ca²⁺ invade mitochondria, the Diablo protein is released inside the cell.

3 Diablo binds to IAPs (inhibitors of apoptosis proteins), so they can no longer block caspases.

Diablo

Caspase–9

4 A cascade of enzymes destroys various proteins and the DNA of the cell, making it incapable of survival. Apoptosis (cell death) results.

IAPs

Caspase–3

Bcl-2

Apoptosis

5 The family of Bcl-2 proteins can inhibit apoptosis by blocking the release of Diablo from the mitochondria.

Inside cell

dUTP (deoxyuridine triphosphate) into the various "nicks" that the apoptotic cell has made in its DNA (**FIGURE 7.17A**) (Gavrieli, Sherman, & Ben-Sasson, 1992). We can then visualize the TUNEL-labeled cells by using fluorescence or immunohistochemistry, letting us know which cells were about to die before we gathered the tissue.

TUNEL-labeled cells may look otherwise normal, indicating that we have detected the earliest stages of apoptosis. But sometimes the TUNEL-labeled cell nucleus has the misshapen pyknotic appearance of rounded droplets of material, indicating the cell was in the last stages of apoptosis (**FIGURE 7.17B**). You can see the advantage of using TUNEL over looking for dying cells, as it reveals apoptotic cells that have not yet become pyknotic.

How's It Going?

1. What are death genes? How do manipulations of death genes indicate that apoptosis is an active process?

2. What are some of the additional factors that regulate apoptosis in vertebrates?

3. How does the TUNEL method label cells that are about to die?

(A)

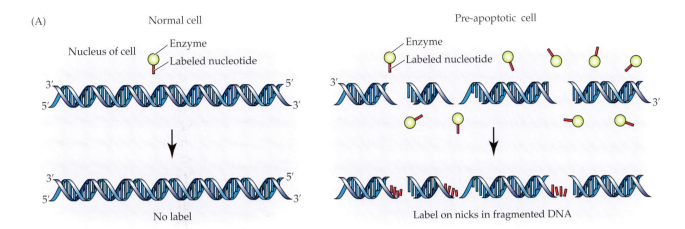

Normal cell

Pre-apoptotic cell

Nucleus of cell — Enzyme — Labeled nucleotide

No label

Label on nicks in fragmented DNA

(B)

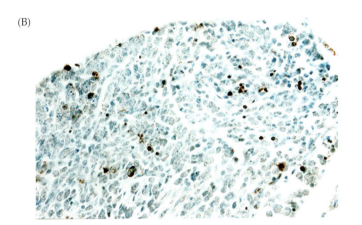

FIGURE 7.17 **TUNEL indicates apoptotic cells.**
(A) The enzyme terminal deoxynucleotidyl transferase attaches nucleotides to the free 3' ends of DNA. Such nicked ends are abundant in cells that have cleaved their DNA to begin apoptosis. If labeled nucleotides are provided along with the enzyme, they will accumulate in cells about to die. (B) The dark brown TUNEL staining indicates which cells in this dorsal root ganglion have begun digesting their own DNA at the start of apoptosis, making the DNA available for the nick end labeling that TUNEL provides. (B courtesy of Drs. Michael Vogel and Lisa Qiu.)

■ 7.8 ■

DO MOTOR NEURONS DIE IN ALS FOR LACK OF NEUROTROPHIC FACTOR(S)?

As we mentioned at the start of the chapter, amyotrophic lateral sclerosis (ALS) is characterized by a marked absence of motor neurons postmortem. We still know very little about what causes ALS, but a small minority of cases (5–10%) have a genetic cause (Boylan, 2015), including at least five different identified genes (Pasinelli & Brown, 2006). Research with animal models carrying mutations of these same genes has provided the richest insights into the disease. For example, **superoxide dismutase** (**SOD**) is an enzyme that normally acts as an antioxidant, mopping up potentially damaging free radical compounds that result from ongoing biochemical processes in the body. About a quarter of inherited ALS (which remember is a tiny minority of all ALS cases) is due to a mutation in the gene for SOD (Rosen et al., 1993). Given that SOD normally mops up free radicals, this finding at first suggested ALS might be caused by a failure to get rid of free radicals, which then damage motor neurons. But when the mutated *SOD* gene was introduced to mice, they developed symptoms of ALS (Gurney et al., 1994), *even though their endogenous* Sod *genes were intact*. In other words, the additional, mutant *SOD* was somehow causing damage, a toxic *gain of function*. Indeed,

superoxide dismutase (SOD)
A protein that normally neutralizes free radicals. Mutations of this gene can cause ALS in humans.

The additional, mutant *SOD* was somehow causing damage, a toxic *gain of function*.

mice with the endogenous *Sod* knocked out have shortened lifespans but do not show ALS-like symptoms (Reaume et al., 1996).

For some years the field widely assumed that the mutant *SOD* was acting in motor neurons to kill them, thus causing disease in the mice. But when researchers made transgenic mice that expressed mutant *SOD* only in motor neurons, no degeneration took place (Lino, Schneider, & Caroni, 2002; Pramatarova, Laganiere, Roussel, Brisebois, & Rouleau, 2001). Likewise, expressing mutant *SOD* in cell types other than motor neurons and oligodendrocytes accelerated the disease (Yamanaka et al., 2008). In mosaic mice, with cells carrying mutant *SOD* as well as cells that do not, it turned out that the wild-type motor neurons were just as likely to die as those carrying the mutant *SOD* (Clement et al., 2003). In other words, the death of motor neurons was not a cell-autonomous response to *SOD*—the gene is not acting within the motor neuron itself to kill it. Rather, *SOD* must be acting first in other cells, which then secondarily cause the motor neurons to die. And it's not clear that death of motor neurons is responsible for the symptoms of ALS anyway, because mice carrying mutant *SOD* still develop motor dysfunction and die early, even if their death gene *Bax* is knocked out so that no motor neurons die (Gould et al., 2006). Several new theories were offered to explain how *SOD* causes dysfunction, including the possibility that nearby wild-type glia could somehow prevent motor neurons from succumbing to disease. Attempts to reduce *SOD* expression in muscle fibers alone did not affect disease progression (T. M. Miller et al., 2006), so it remains unclear whether there is a single pathway for ALS. We know there are several different types of ALS, despite their commonality, and these may reflect several different cellular sites of action (Robberecht & Philips, 2013).

Now that we've reviewed the history of research on apoptosis in the nervous system, you can understand how the death of motor neurons in people, like Stephen Hawking, with ALS might be related to the loss of one or more neurotrophic factors. If so, then providing patients with the missing or reduced trophic factor might rescue motor neurons from death and halt, or perhaps slow, the disease process. Unfortunately, the exquisite sensitivity of DRG and sympathetic neurons to a single factor (NGF) may be an exception rather than the rule. In other systems, including motor neurons, it seems the cells are affected by more than one factor, so knocking out one factor, like BDNF, leads to the death of some motor neurons, but not all. If ALS is due to the loss of a single neurotrophic factor, we've yet to learn what the neurotrophic factor is or what cells normally provide it. Likewise, treatment with any given trophic factor seems to have only modest effects in animal models and, so far, no effect for people with ALS (Gould & Oppenheim, 2011; Henriques, Pitzer, & Schneider, 2010).

> If ALS is due to the loss of a single neurotrophic factor, we've yet to learn what the neurotrophic factor is.

The instances of apoptosis we've discussed so far all seem to achieve a particular goal: the approximate matching in size of neurons with their targets (**FIGURE 7.18**). Another sphere of neural development in which apoptosis appears to play a crucial role is in the development of sex differences in the nervous system, the topic we take up to close the chapter.

How's It Going?

1. Approximately what proportion of ALS cases are hereditary? Which gene is most often associated with inherited ALS?

2. What is the evidence that ALS does, or does not, commence with the demise of motor neurons?

FIGURE 7.18 Differential apoptosis matches neurons to targets.

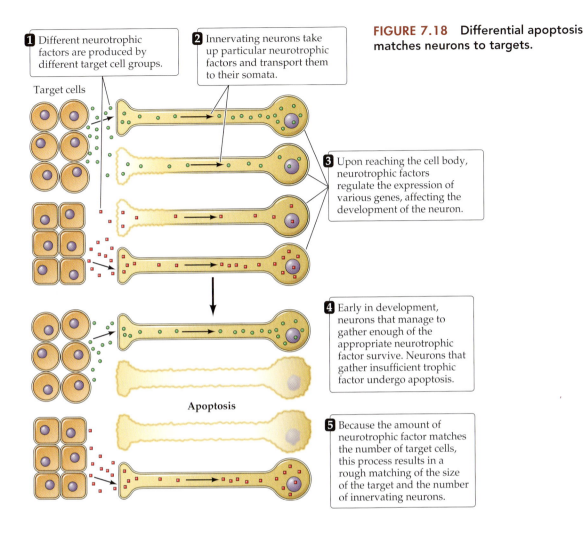

1 Different neurotrophic factors are produced by different target cell groups.

2 Innervating neurons take up particular neurotrophic factors and transport them to their somata.

3 Upon reaching the cell body, neurotrophic factors regulate the expression of various genes, affecting the development of the neuron.

Target cells

4 Early in development, neurons that manage to gather enough of the appropriate neurotrophic factor survive. Neurons that gather insufficient trophic factor undergo apoptosis.

Apoptosis

5 Because the amount of neurotrophic factor matches the number of target cells, this process results in a rough matching of the size of the target and the number of innervating neurons.

■ 7.9 ■
HORMONES DIRECT SEXUAL DIFFERENTIATION OF THE VERTEBRATE BODY AND BEHAVIOR

Almost every vertebrate species consists of two sexes, which typically display at least slightly different behaviors, including but not limited to reproductive behaviors. These sex differences in behavior are so prominent that several scientists began asking how they come about during development. They soon turned to **hormones**, chemicals released into circulation by one part of the body that reach another part of the body and affect it. The power of hormones was discovered in prehistory, when humans raising domesticated animals learned that **castration**, removing the testes, of adult males made them more docile and uninterested in mating with females. The Bible and other ancient texts also describe the castration of men, either as punishment for being on the losing side of some war, or to serve as *eunuchs*, men who could look after women in a harem without risk of impregnating them. Today we know that testes make male mammals more aggressive and increase libido through the secretion of steroid hormones called **androgens**, which include **testosterone**.

The discovery that testosterone increases interest in mating led to the idea that perhaps some males mate more readily than others because of differences in how much androgen they produce. But this idea was soon disproved. In

hormones Chemical signals released into circulation by one group of cells that affect cells elsewhere in the body.

castration Surgical removal of the gonads, typically testes.

androgens A class of steroid hormones, including testosterone, found in high concentrations in male vertebrates.

testosterone The principle androgen secreted from the vertebrate testes, found in higher concentrations in males than females.

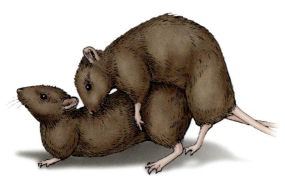

FIGURE 7.19 **Rodent mating behavior** These rats display the typical rodent mating behavior, as the male mounts the female from behind and rhythmically thrusts his pelvis against her rump. If the female is sexually receptive, she displays the lordosis posture depicted here, raising her head and rump. In the absence of lordosis, the male cannot achieve intromission and soon stops mounting the female.

There must be some androgen for mating to take place, but after that threshold amount, more androgen does not activate more behavior.

estrogen A class of steroid hormones, including estradiol, secreted by the ovaries.

receptivity Here, the willingness of a female to allow a male to mate with her.

lordosis Literally, the arching of the back that elevates the shoulders and hips; here, a posture displayed by receptive female rodents to permit copulation.

a classic study, male guinea pigs were first tested for mating vigor and classified as high, medium, or low. When the males were castrated, they eventually stopped mating altogether, as expected. Also as expected, when the males were provided with exogenous testosterone, they soon resumed mating. This influence of steroid hormones on adult behavior is sometimes referred to as an *activational effect*—the hormone activates the mating behavior. In female rodents, injecting ovarian steroids such as **estrogen** and progesterone can activate **receptivity**, meaning the females will display the **lordosis** reflex, raising the head and rump, which allows males to mate with them (**FIGURE 7.19**).

The important discovery was that the castrated males treated with testosterone soon resumed the same level of mating vigor that they had originally, such that the previously high-vigor males again showed the most vigor while the previously low-vigor males again showed less interest, even though they all got the same amount of testosterone (**FIGURE 7.20**) (Grunt & Young, 1953). Thus androgen has a *permissive effect* on mating—there must be some androgen for mating to take place, but after that threshold amount, more androgen does not activate more behavior. Later studies would show that providing castrated males with even very small doses of testosterone, one-tenth that seen in intact males, would completely restore mating behavior. Similarly, loss of libido in men who lose their testes through accident or disease is restored with testosterone treatment (Davidson, Camargo, & Smith, 1979), but individual differences in libido do not correlate with circulating levels of androgens. Apparently variation in sexual interest in male mammals is due to differences in *response* to androgens, not to differences in secretion of the hormones.

Not only do testicular hormones activate behavior in adult mammals, they also play an important role in the development of the two sexes. Early in vertebrate fetal development, the bodies of males and females look alike. Eventually they begin to diverge, with some constructing the reproductive tract appropriate for females while others form a reproductive tract appropriate for males. That process of males and females taking different developmental pathways is called **sexual differentiation**. Across mammalian species, sexual differentiation of the body is remarkably uniform. In embryos that have inher-

FIGURE 7.20 **Androgens like testosterone activate male mating behavior.** We say androgens have a permissive effect on mating behavior, because males must have some androgens to mate at all. But individual differences in the vigor with which male guinea pigs mate cannot be due to differences in androgen levels, because males provided equivalent testosterone return to their original differences in mating vigor. (After Grunt & Young, 1953.)

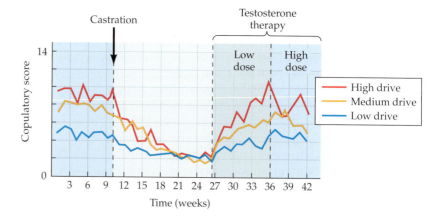

ited a Y chromosome from the father, a gene on the Y called **sex determining region of the (Sry)** starts being expressed in the gonads that until this time appeared identical in the two sexes, so-called *indifferent gonads*. Attached to the indifferent gonad are two systems of ducts, the Müllerian ducts that can develop into the female reproductive tract, and the Wolffian ducts that can develop into the male reproductive tract. At this stage, both duct systems sit side by side and look identical in the two sexes. Expression of *Sry* drives the indifferent gonad to start developing as a testis. In the absence of Sry, the indifferent gonad in XX individuals will develop as an ovary. Thus the initial stage of sexual differentiation consists of the genetic sex, presence or absence of a Y chromosome, guiding sexual differentiation of the gonad.

For tissues other than the gonads, further sexual differentiation in mammals does not depend directly on the sex chromosomes, but will instead be guided by hormonal secretions from the gonads. If testes develop, they secrete two classes of hormones that masculinize the rest of the body. A protein called *anti-Müllerian hormone* (*AMH*), suppresses development of the Müllerian ducts (by activating the caspase cascade leading to apoptosis [Allard et al., 2000]), forestalling the production of the female reproductive tract. The testes also secrete androgens, including testosterone, which drive the Wolffian ducts to develop as a male reproductive tract (epididymis, seminal vesicles, vas deferens). Androgens also direct epidermal cells to form a masculine configuration, including a penis and scrotum (the sac that will eventually contain the testes) (**FIGURE 7.21**). In the absence of a Y chro-

sexual differentiation The process by which females and males diverge from each other in structure during ontogeny.

sex-determining region of the Y (Sry) The gene on the Y chromosome of mammals that promotes the indifferent gonad to develop as a testis.

In the absence of androgens, the external genitalia form a feminine configuration.

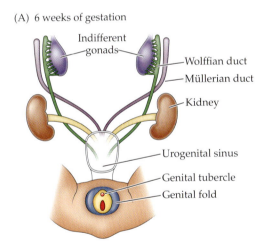
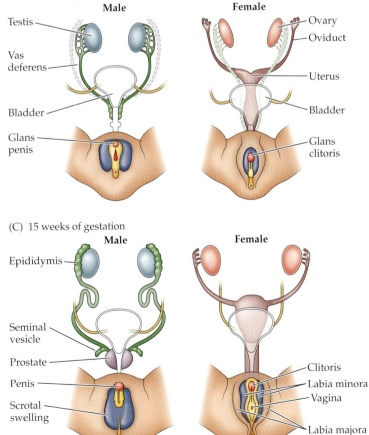

FIGURE 7.21 Sexual differentiation of the mammalian reproductive tract (A) Early in mammalian development, males and females appear alike, with indifferent gonads that resemble both ovaries and testes. (B) As development proceeds, first the gonads differentiate into recognizeable testes or ovaries. (C) Sexual differentiation of the internal reproductive tract and external genitalia is obvious by 15 weeks of gestation.

FIGURE 7.22 Women with androgen insensitivity syndrome (AIS) This photo shows a group of women with AIS and similar inter-sex conditions at a support group meeting. (Courtesy of Dr. Kimberly Saviano/AISSG-USA.)

androgen receptor (AR)
A member of the steroid receptor superfamily with a high affinity for androgens such as testosterone.

mosome, the indifferent gonads develop as ovaries, which appear to secrete little or no fetal hormone. In the absence of AMH, the Müllerian ducts develop a feminine reproductive tract, including oviducts, uterus, and inner vagina. Similarly, in the absence of androgens, the external genitalia form a feminine configuration of a clitoris, labia, and the outer vagina.

A powerful demonstration of the importance of hormones for sexual differentiation is seen in those XY individuals who inherit a dysfunctional allele for the **androgen receptor** (**AR**). Because they have a Y chromosome, *Sry* induces the indifferent gonads to develop as testes, and the testes secrete AMH to suppress development of feminine internal genitalia, as well as testosterone. However, in the absence of a functional AR, testosterone cannot masculinize the body, so the individual is born with the external phenotype of a girl. Despite being born with testes in their abdominal cavity, these individuals look just like other women (**FIGURE 7.22**). The absence of ovaries or a uterus mean these women cannot reproduce.

Starting in 1959, scientists inspired by research on development of the mammalian genitalia proposed a framework for understanding the ontogeny of behavioral sex differences that would eventually implicate apoptosis in the central nervous system. These studies explicitly asked whether the same rules governing sexual differentiation of the body might control sexual differentiation of the brain and behavior, as we'll see next.

RESEARCHERS AT WORK

Early Exposure to Androgens Organizes the Male Brain

■ **QUESTION**: How do male and female mammals come to display differences in behaviors, including mating behaviors?

■ **HYPOTHESIS**: The same testicular secretions that masculinize the external genitalia during early development may also act upon the developing brain, masculinizing it to support male behaviors rather than female behaviors in adulthood.

■ **TEST**: Inject pregnant guinea pig dams with testosterone to expose their embryos to the hormone. When the female offspring grow up, inject them with the ovarian hormones that normally activate sexual receptivity. If early exposure to androgens masculinizes the females' brains, then they should fail to display lordosis.

■ **RESULT**: Indeed, the females exposed to fetal androgens showed little or no lordosis reflex, even when given ovarian hormones that trigger lordosis in control females (**FIGURE 7.23A,B**). Later experiments would show that prenatally androgenized female rodents, if given androgens in adulthood, would mount females more readily than control females would (**FIGURE 7.23C**). What's more, male rodents deprived of fetal androgens, by either pharmacological or surgical castration early in life, would display lordosis as adults when given ovarian steroids, and they showed little mounting behavior even when treated with testosterone (**FIGURE 7.23D**) (Phoenix, Goy, Gerall, & Young, 1959).

FIGURE 7.23 Androgens act early in life to organize masculine behavior in adulthood.

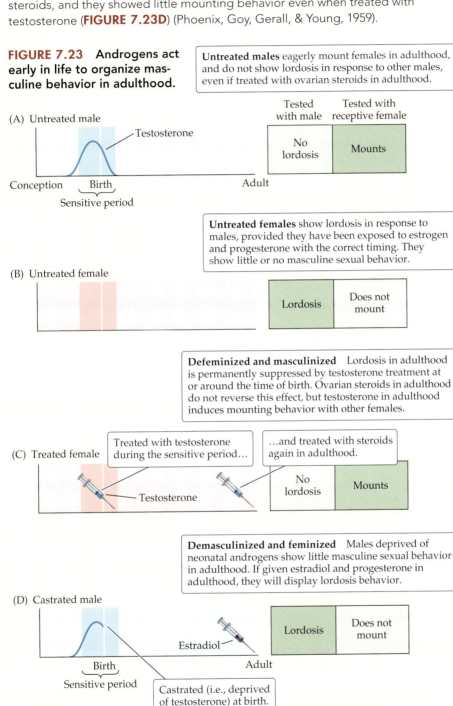

Untreated males eagerly mount females in adulthood, and do not show lordosis in response to other males, even if treated with ovarian steroids in adulthood.

(A) Untreated male

Testosterone

Conception Birth Sensitive period Adult

Tested with male: No lordosis
Tested with receptive female: Mounts

Untreated females show lordosis in response to males, provided they have been exposed to estrogen and progesterone with the correct timing. They show little or no masculine sexual behavior.

(B) Untreated female

Lordosis | Does not mount

Defeminized and masculinized Lordosis in adulthood is permanently suppressed by testosterone treatment at or around the time of birth. Ovarian steroids in adulthood do not reverse this effect, but testosterone in adulthood induces mounting behavior with other females.

(C) Treated female

Treated with testosterone during the sensitive period…

…and treated with steroids again in adulthood.

Testosterone

No lordosis | Mounts

Demasculinized and feminized Males deprived of neonatal androgens show little masculine sexual behavior in adulthood. If given estradiol and progesterone in adulthood, they will display lordosis behavior.

(D) Castrated male

Birth Sensitive period Adult

Estradiol

Castrated (i.e., deprived of testosterone) at birth.

Lordosis | Does not mount

■ **CONCLUSION**: No matter what the animal's genetic sex, those exposed to androgens during fetal development were more likely to show masculine behaviors, and less likely to show feminine behaviors, than animals that were not exposed to fetal androgens.

The same testicular secretions that masculinize the developing mammalian body also masculinize the brain during a sensitive period early in life, permanently altering the brain and therefore permanently masculinizing behavior.

organizational hypothesis The proposal that the same testicular steroids that act early in life to permanently masculinize the vertebrate body also permanently masculinize the brain and therefore behavior.

aromatization Here, the single-step reaction, catalyzed by the enzyme aromatase, that converts androgens such as testosterone into estrogens such as estradiol.

estrogen receptors Two members of the steroid receptor superfamily that have a high affinity for estrogens such as estradiol.

In some ways, the results of this experiment were not so different from those reported in earlier papers, but what made this particular study a landmark was the authors' explicit proposal of the **organizational hypothesis**: the same testicular secretions that masculinize the developing mammalian body also masculinize the brain during a sensitive period early in life, permanently altering the brain and therefore permanently masculinizing behavior. In this case, fetal exposure to testosterone rendered genetic females unable to display receptive behavior, even when provided with ovarian hormones in adulthood. Later experiments would show that, for a remarkable range of behaviors, animals exposed to androgen in development behaved like males, while animals protected from androgen exposure in development would act like females in adulthood.

There is an elegance to this idea that a single signal, androgen secreted from the testes, could act on tissues throughout the body, including the brain, to synchronize the entire body to develop in such a way that the individual could reproduce as a male. In this way, the presence of a Y chromosome triggers a chain of events that globally masculinizes the body and brain. This way of thinking also reinforces the idea that some hormones masculinize while others feminize. But other experiments soon upended that idea. It turns out that in several mammalian species, including most rodents (but not primates), testosterone masculinizes the brain not by acting upon androgen receptors, but by being first converted to estrogens (a chemical process called **aromatization**) and then acting upon **estrogen receptors** in the brain to masculinize behavior (**FIGURE 7.24**).

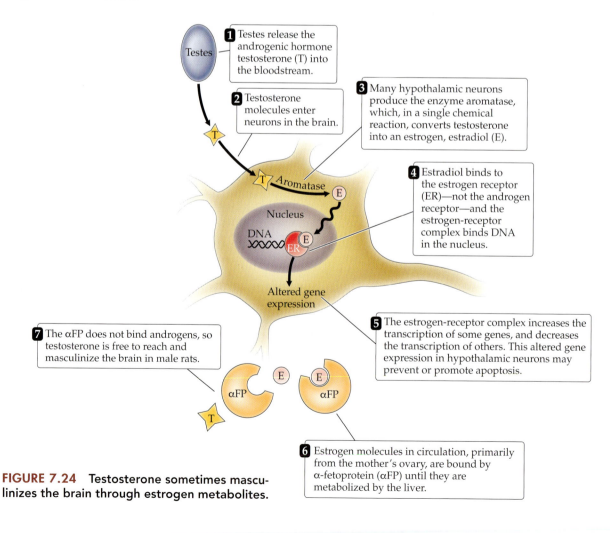

1 Testes release the androgenic hormone testosterone (T) into the bloodstream.

2 Testosterone molecules enter neurons in the brain.

3 Many hypothalamic neurons produce the enzyme aromatase, which, in a single chemical reaction, converts testosterone into an estrogen, estradiol (E).

4 Estradiol binds to the estrogen receptor (ER)—not the androgen receptor—and the estrogen-receptor complex binds DNA in the nucleus.

5 The estrogen-receptor complex increases the transcription of some genes, and decreases the transcription of others. This altered gene expression in hypothalamic neurons may prevent or promote apoptosis.

6 Estrogen molecules in circulation, primarily from the mother's ovary, are bound by α-fetoprotein (αFP) until they are metabolized by the liver.

7 The αFP does not bind androgens, so testosterone is free to reach and masculinize the brain in male rats.

Testes

Aromatase

Nucleus

DNA

ER

Altered gene expression

αFP

αFP

FIGURE 7.24 Testosterone sometimes masculinizes the brain through estrogen metabolites.

How's It Going?

1. Describe the factors that control sexual differentiation of the body in mammals.

2. Define the activational effect of hormones on behavior and give an example.

3. Describe the experiment that supported the organizational hypothesis of sex differences in behavior.

■ 7.10 ■
THE BRAIN IS ALSO SEXUALLY DIMORPHIC

The experiments we've described so far in support of the organizational hypothesis all rely on monitoring the animal's behavior, how readily it displays lordosis or mounting or some other behavior as an indication of whether the brain has a masculine or feminine configuration. At first, few people actually examined the brain to see whether it was structurally affected by early exposure to androgens, because everyone assumed the structural differences would be in things like the size and distribution of synapses, and therefore difficult to measure. But in 1976, scientists studying how the brain produces song in canaries and zebra finches made a startling discovery. The forebrain regions that were known to be important for song production were found to be dramatically larger in males (who normally sing) than in females (who normally do not). The regional volume of these nuclei was 5–6 times larger in males than females (Nottebohm & Arnold, 1976), so much larger that you didn't really need a microscope to tell whether a brain section was from a male or female. Thus not only are there sex differences in morphology, which Darwin named **sexual dimorphisms**, in the body, but there are also sexual dimorphisms in the brain.

Later research would show that steroid hormones act early in life to masculinize the brains of hatchlings, and that again it seems to be aromatized metabolites of testosterone, acting on estrogen receptors, that actually masculinize the developing brain to enlarge birdsong regions and enable the bird to sing as an adult. We'll discuss this birdsong system in more detail in Chapter 10 because, as you'll see then, the development of song also depends on social guidance, specifically the details of song syllables and patterns normally provided by the father.

For now, the importance of finding sexual dimorphism in the bird brain was that it encouraged other researchers to look for sexual dimorphisms in the brains of other species. Just 2 years later, the report came of a sexual dimorphism in the rat brain, specifically a nucleus in the hypothalamus that was 5–6 times larger in regional volume in males than in females (Gorski, Gordon, Shryne, & Southam, 1978). Again this nucleus, named the **sexually dimorphic nucleus of the preoptic area** (**SDN-POA**) was so sexually dimorphic that one could accurately sort brain sections as either male or female without the need of a microscope (**FIGURE 7.25**).

sexual dimorphism A structural difference between males and females of a species.

sexually dimorphic nucleus of the preoptic area (SDN-POA) A nucleus in the hypothalamus of many mammals that has a greater volume in males than females.

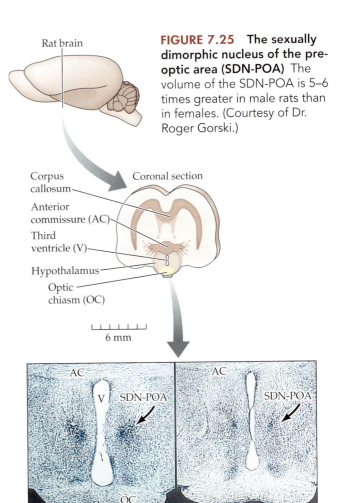

FIGURE 7.25   The sexually dimorphic nucleus of the preoptic area (SDN-POA) The volume of the SDN-POA is 5–6 times greater in male rats than in females. (Courtesy of Dr. Roger Gorski.)

Rat brain

Corpus callosum
Anterior commissure (AC)
Third ventricle (V)
Hypothalamus
Optic chiasm (OC)

Coronal section

6 mm

AC
V
SDN-POA
OC
Male

AC
SDN-POA
Female

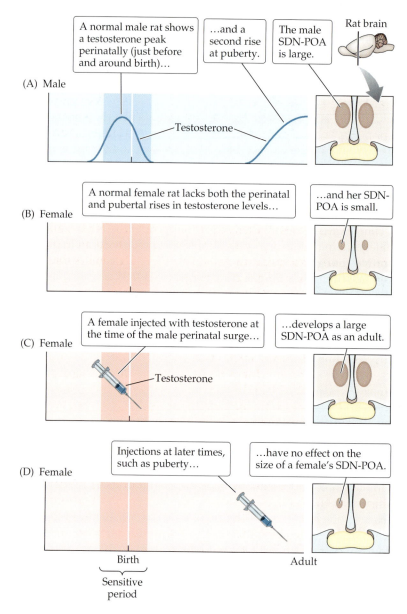

FIGURE 7.26 Exposure to androgens during a sensitive period masculinizes the adult structure of the SDN-POA.

The SDN-POA beautifully illustrated the organizational hypothesis: only exposure to androgens early in life could cause a large, masculine nucleus in adulthood. In contrast, injections of testosterone in adulthood had no effect on SDN-POA size (**FIGURE 7.26**).

How's It Going?

1. What does it mean to say a brain region is sexually dimorphic?
2. What is the significance of aromatization for the masculinization of the rodent brain?
3. What is the SDN-POA and what experimental results indicate it is organized by early hormone exposure?

■ 7.11 ■

HORMONES CAN REGULATE APOPTOSIS TO MASCULINIZE THE VERTEBRATE BRAIN

Careful analysis of the number of neurons in the SDN-POA of growing rats found evidence of apoptosis just before and just after birth. Normally there is more apoptosis in females than males, which early androgen treatment can prevent (Davis, Popper, & Gorski, 1996), indicating that androgens normally masculinize the nucleus by reducing apoptosis early in life in males. This mechanism of action, steroid hormones regulating apoptosis, is common (but not universal) in the formation of sexual dimorphism in the vertebrate brain. Greater apoptosis in one sex or the other results in sexual dimorphism (Forger, 2006). In most cases studied so far, the brain nuclei are larger in males than females, but in a brain region that is larger in females than in males, the **anteroventral periventricular nucleus** (**AVPV**), it is the males that display more apoptosis early in life (Forger, 2006; Forger et al., 2004).

Perhaps the best demonstration that sexual dimorphism in the mammalian brain depends on sex differences in apoptosis comes from studies of mice with genetic manipulations of various death genes. For example, knocking out the proapoptotic *Bax* in mice, thereby preventing apoptosis, also eliminates sex differences in several brain regions (Forger et al., 2004), and mice that overexpress the antiapoptotic gene *Bcl-2*, thereby reducing apoptosis, also show reduced sexual dimorphism in the brain (Zup et al., 2003).

> Knocking out the proapoptotic *Bax* in mice, thereby preventing apoptosis, also eliminates sex differences in several brain regions.

Another sexual dimorphism in the mammalian nervous system sculpted by a sex difference in apoptosis during development concerns spinal motor neurons. The **spinal nucleus of the bulbocavernosus** (**SNB**) consists of spinal motor neurons that innervate striated muscles (*bulbocavernosus* muscles) attached to the base of the penis in rodents (Breedlove & Arnold, 1980). In rats, both sexes possess the muscles and motor neurons before birth, and the motor neurons have established a functional synapse with the target muscles (Rand & Breedlove, 1987), but then the muscles and motor neurons die in females (Nordeen, Nordeen, Sengelaub, & Arnold, 1985). If the females are treated with androgens before the cells die, they will retain the motor neurons and muscles for life.

In genetic male rats with a spontaneous mutation that disables AR (ARKO rats), the SNB system regresses despite the presence of androgen. Thus functional AR stimulation is required to prevent its loss. Interrupting the spinal cord rostral to the SNB system, thereby disconnecting any brain afferents to the motor neurons, has no effect on survival of the cells (Fishman & Breedlove, 1985), so androgen is not acting on the brain to preserve the motor neurons. Study of mosaic rats showed that androgens are *not* acting upon SNB motor neurons themselves to spare them (**FIGURE 7.27**).

Later studies in mice used Cre-lox technology to disable AR selectively in motor neurons, but this had no effect on the number of SNB motor neurons that survived to adulthood, confirming that the motor neurons themselves are not the site of action for androgen's sparing effect. The next most likely candidate for the site of androgen action to preserve the SNB system was the target muscle, and indeed local implants of antiandrogen near the muscles in newborn rats more effectively blocked SNB survival than systemic implants (Fishman & Breedlove, 1992), indicating that some cell type within

anteroventral periventricular nucleus (AVPV) A hypothalamic region that has a greater volume in females than in males in rats and mice.

spinal nucleus of the bulbocavernosus (SNB) A sexually dimorphic collection of motor neurons in the spinal cord that innervate muscles attached to the penis in rats and mice.

RESEARCHERS AT WORK

Sometimes the Tail Wags the Dog

■ **QUESTION**: Does androgen act directly upon developing SNB motor neurons to prevent them from undergoing apoptosis?

■ **HYPOTHESIS**: If androgenic sparing of developing SNB motor neurons is a cell-autonomous response, then motor neurons without functional ARs will die even in the presence of testosterone.

■ **TEST**: Produce mosaic rats in which half the cells have a functional AR while the other half have a mutant, dysfunctional allele (ARKO). Because the *AR* gene resides on the X chromosome, XX rats carrying a dysfunctional allele on one X are normally mosaic for functional androgen receptors. This happens because random X inactivation silences one X chromosome early in embryonic development, leaving only the other X available for use by that cell and its progeny. If androgens act cell-autonomously to save SNB cells, then if we expose such female rats to androgen at birth, only those SNB motor neurons using the functional *AR* allele will survive, while motor neurons using the other, dysfunctional allele will die.

■ **RESULT**: Using immunohistochemistry to detect AR protein in the adult spinal cord, all surviving SNB motor neurons were classified as either wild type or ARKO. In fact, about half the surviving SNB motor neurons were ARKO, making no functional AR (Freeman, Watson, & Breedlove, 1996).

Androgen

FIGURE 7.27   Is androgenic sparing of SNB motor neurons a cell-autonomous response? (After Freeman et al., 1996.)

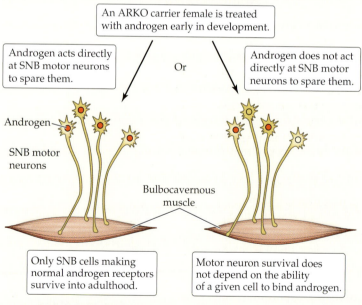

An ARKO carrier female is treated with androgen early in development.

Androgen acts directly at SNB motor neurons to spare them.

Or

Androgen does not act directly at SNB motor neurons to spare them.

Androgen

SNB motor neurons

Bulbocavernous muscle

Only SNB cells making normal androgen receptors survive into adulthood.

Motor neuron survival does not depend on the ability of a given cell to bind androgen.

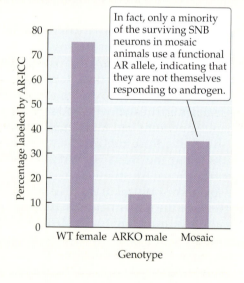

In fact, only a minority of the surviving SNB neurons in mosaic animals use a functional AR allele, indicating that they are not themselves responding to androgen.

Percentage labeled by AR-ICC (y-axis: 0–80)

WT female   ARKO male   Mosaic

Genotype

■ **CONCLUSION**: SNB motor neurons carrying a functional allele for AR were no more likely to be spared by androgen treatment than ARKO motor neurons. Since SNB motor neurons without functional AR were spared by neonatal testosterone treatment, androgen must act on some other tissue to prevent death of the motor neurons. In other words, the androgenic sparing of SNB motor neurons is not a cell-autonomous process.

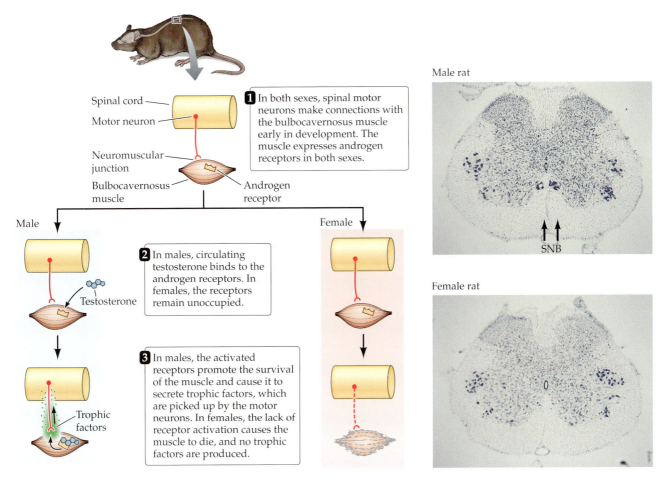

FIGURE 7.28 **Sexual differentiation of the spinal nucleus of the bulbocavernosus (SNB)**

muscle normally responded to androgen (**FIGURE 7.28**). Using Cre-lox technology in mice revealed that the muscle fibers themselves are *not* the site of androgen action, either. Rather, functional AR must be present in mesenchymal cells of the muscle for the system to be spared (Ipulan et al., 2014). Apparently androgen acts upon the mesenchymal cells, inducing survival of the muscle fibers, which in turn keep the motor neurons alive. Not only is this a case of the "tail wagging the dog," it's as if a flea is wagging the tail!

Among other things, these studies demonstrate how sexual differentiation of the body results in sexual differentiation of the nervous system. For example, the bulbocavernosus muscle is larger in men than women, and men also have more spinal motor neurons innervating these muscles (in *Onuf's nucleus* in the sacral spinal cord) than do women (Forger & Breedlove, 1986). Since the period of motor neuronal death in humans coincides with periods of androgen secretion (Forger & Breedlove, 1987a), this sexual dimorphism in humans probably comes about the same way it does in rats.

Sexual differentiation of the body results in sexual differentiation of the nervous system.

By now we've considered many examples of apoptosis in the developing nervous system, so we've seen that cell death can serve many functions. In fact, several other functions for apoptosis have been identified (Buss, Sun, & Oppenheim, 2006) and are listed in **TABLE 7.2**.

TABLE 7.2
Possible Functions of Apoptosis in Neural Development

1. Differential removal of cells in males and females (sexual dimorphisms)

2. Deletion of some of the unnecessary progeny of a specific sublineage

3. Negative selection of cells of an inappropriate phenotype

4. Pattern formation and morphogenesis (webbing in ducks versus chickens)

5. Deletion of cells that act as transient targets or that provide transient guidance cues for axon projections (guidepost cells)

6. Removal of cells and tissues that serve a transient physiological or behavioral function

7. "Systems" matching by creating optimal quantitative innervation between interconnected groups of neurons and between neurons and their targets (e.g., muscles, sensory receptors)

8. Systems matching between neurons and their glial partners by regulated glial apoptosis (e.g., Schwann cells and axons)

9. Error correction by the removal of ectopically positioned neurons or of neurons with misguided axons or inappropriate synaptic connections

10. Removal of damaged or harmful cells

11. Regulation of the size of mitotically active progenitor populations

12. The control of apoptosis as an ontogenic buffer mechanism for the accommodation of mutations involved in evolutionary change

13. Regulated survival of subpopulations of adult-generated neurons as a means of experience-dependent plasticity

Source: After Buss et al., 2006.

For now, let's turn to another species where cell death plays an important role in the development of sexual dimorphism, *Drosophila*. Hormones do not direct sexual differentiation in flies. Instead, we'll see that sexual differentiation in *Drosophila* is mostly a cell-autonomous response to sex chromosomes, including sex-specific apoptosis of neurons.

How's It Going?

1. What's the best evidence that many brain sexual dimorphisms are due to greater apoptosis in one sex than the other?
2. Describe the SNB system.
3. What experiment demonstrated that androgenic sparing of SNB motor neurons is not a cell-autonomous effect?

■ 7.12 ■

SEXUAL DIFFERENTIATION IN FLIES IS A CELL-AUTONOMOUS PROCESS

Sexual differentiation in *Drosophila* does not involve gonadal hormones and so is quite different from differentiation in mammals. In fact, in flies most cells that take on a different phenotype in the two sexes do so in a *cell-autonomous fashion*—individual cells with an XX genotype differentiate in a female fashion while cells with an XY genotype differentiate in a male fashion. Although the X and Y chromosomes of fruit flies are evolutionarily unrelated to those of mammals, the basic mechanism that individuals with XX become females and those with XY become males is the same. Unlike the case with mammals, it's not that Y chromosomes make flies male; rather, it is the absence of a second X that makes flies male. Thus an XO individual will appear like a female in mammals, but like a male in flies (in both cases they will be infertile, as the atypical chromosome complement interferes with gamete production).

> It's not that Y chromosomes make flies male; rather, it is the absence of a second X that makes flies male.

The pathways by which the dosage of genes on the X chromosomes determines sex is rather complicated but well understood (Salz, 2011). Basically, having two X chromosomes means that enough X-linked transcription factors are produced early in life to activate a gene called *Sex-lethal* (*Sxl*), the protein product of which regulates mRNA splicing to sustain its own continued production. The subsequent Sxl protein triggers a cascade of feminine gene expression in each cell. In flies with only one X, insufficient production of Sxl early in life averts production of the protein later in life, so by default cells express the genes required to take on a male phenotype. *Sxl* is the top gene in a hierarchy for sex determination—transcription factors that eventually direct male- or female-specific differentiation of a cell.

Thus insect mosaics that have different genotypes in different parts of the body may be fully masculine in some parts of the body and fully feminine in others, and so are called **gynandromorphs** (Morgan & Bridges, 1919). For example, if in an XX insect, one of the chromosomes is lost by an embryonic cell, the body parts derived from its descendants will be male while the rest of the body will be female (**FIGURE 7.29**). This cell-autonomous pattern of sex determination has been exploited to determine the site of action for behavioral mutants in *Drosophila* (Benzer, 1973; Hotta & Benzer, 1970).

Genes also affect sex-specific *behaviors* of fruit flies, including male courtship behavior. Males must engage in courtship before a female will allow mating. The male lands near a female, and if he detects favorable pheromones indicating she is a female of his own species, he will tap her abdomen with a foreleg. If she tolerates that, he will begin serenading her by extending one wing or the other in her direction and vibrating it to produce a species-specific song. If

gynandromorph A rare individual that displays both male and female features in a species in which most individuals are either uniformly male or female.

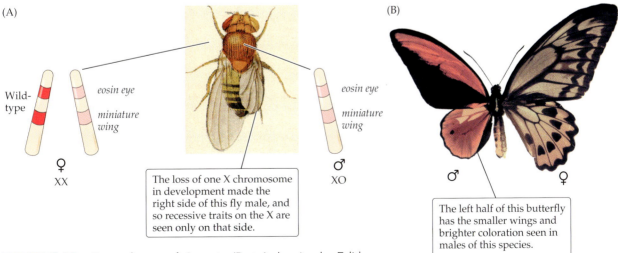

FIGURE 7.29   Gynandromorph insects (Part A drawing by Edith Wallace from Morgan & Bridges, 1919; B photo of Montreal Insectarium display by Dr. Scott F. Gilbert.)

(A)

Wild-type

eosin eye

miniature wing

♀
XX

The loss of one X chromosome in development made the right side of this fly male, and so recessive traits on the X are seen only on that side.

eosin eye

miniature wing

♂
XO

(B)

♂   ♀

The left half of this butterfly has the smaller wings and brighter coloration seen in males of this species.

Olfactory/visual cues

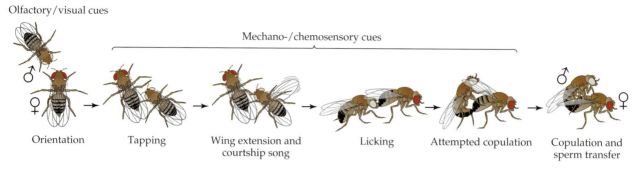

Mechano-/chemosensory cues

Orientation   Tapping   Wing extension and courtship song   Licking   Attempted copulation   Copulation and sperm transfer

FIGURE 7.30 Male courtship behavior in *Drosophila* Male flies must court females to be allowed to copulate. (After Billeter et al., 2006.)

fruitless (fru) A gene in *Drosophila* that produces functional transcripts only in males and induces the development of male-specific neural circuits that control male courtship and mating.

she stands still after this, he will lick the female's genitalia, then mount and, if she allows it, copulate with her (**FIGURE 7.30**).

Several genes downstream from *Sxl* is another transcription factor gene called **fruitless (fru)**. The gene got this name because some mutations of the gene resulted in males who did not court females, while other mutations caused males to court other males. If several of these males were caged together, they would form a chain of males, each investigating the rear end of the other! With either class of mutation, their efforts were "fruitless" in terms of reproducing (Gailey & Hall, 1989). Fascinatingly, the *fruitless* gene is present and identical in both sexes but can produce a variety of different transcripts, some of which are produced exclusively in one sex or the other. Are the male-specific transcripts actually responsible for courtship behaviors, or is some other process directing courtship behaviors and only incidentally causing male-specific transcripts to be produced?

RESEARCHERS AT WORK

Fruitless Mutants Pursue Unrequited Love

■ **QUESTION**: How does a single gene, *fruitless* (*fru*), affect so many aspects of male sexual behavior, even when it is present in both sexes?

■ **HYPOTHESIS**: Some of the many transcripts produced from *fru* are found only in males, and these male-specific transcripts control courtship behavior.

■ **TEST**: Genetically induce a female fly to produce *fru* transcripts that are normally produced only in males. Will she then display male-typical courtship behavior? Scientists replaced the normal *fru* allele with a modified allele, either one that only makes a male-typical transcript or one that only makes a female-typical transcript. Then they tested the behaviors of male and female flies expressing either a male- or female-typical transcript.

■ **RESULT**: Male flies carrying the allele that could produce only female-typical transcripts showed no courtship behavior (**FIGURE 7.31A**). More impressively, otherwise genetically normal females that could produce only male-typical transcripts of *fru* behaved like males (**FIGURE 7.31B,C**). They displayed the entire repertoire of male behaviors and directed them at females (Demir & Dickson, 2005).

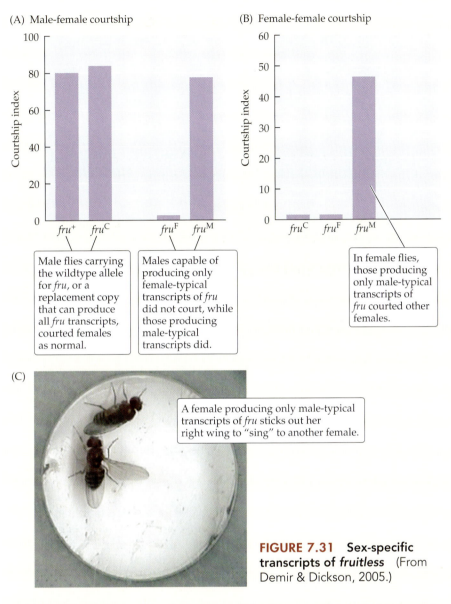

(A) Male-female courtship

Courtship index

Male flies carrying the wildtype allele for *fru*, or a replacement copy that can produce all *fru* transcripts, courted females as normal.

Males capable of producing only female-typical transcripts of *fru* did not court, while those producing male-typical transcripts did.

(B) Female-female courtship

Courtship index

In female flies, those producing only male-typical transcripts of *fru* courted other females.

(C)

A female producing only male-typical transcripts of *fru* sticks out her right wing to "sing" to another female.

FIGURE 7.31   Sex-specific transcripts of *fruitless* (From Demir & Dickson, 2005.)

■ **CONCLUSION**: Male-specific splicing of the *fru* gene is both necessary and sufficient to produce male courtship behavior, as well as a preference to direct such behavior at females.

Later research revealed that the female transcripts of fruitless (*fru*^F) appear to be nonfunctional, but there are at least three functional male transcripts (*fru*^M) (von Philipsborn et al., 2014), one of which must be present in order for males to court females. This means that the fru protein is normally translated only in males, in a widespread minority of about 2,000 neurons in the brain (**FIGURE 7.32A,B**). As in vertebrates, this network is sexually dimorphic, as a result of a sex difference in apoptosis. It is the male-specific transcript of *fru*, not hormones, that prevents apoptosis of these neurons in males (Kimura, Ote, Tazawa, & Yamamoto, 2005), producing dramatic sexual dimorphism in the brain (**FIGURE 7.32C,D**).

FIGURE 7.32 *Fruitless* **gene expression patterns** (A) A survey of the whole brain reveals a network of neurons expressing the fruitless protein in males (stained with anti-green fluorescent protein (GFP) in green; other neurons are stained [magenta]). (B) The female splice of *fru* produces no protein. (C) Genetic tools reveal neurons (in green) that express the *fru* gene in males. (D) The neurons expressing the nontranslated *fru* gene in females form a very different network from that in males. (A,B from Demir & Dickson, 2005, courtesy of Dr. B. J. Dickson; C,D from Kimura et al., 2005.)

(A) Fru protein in a male brain

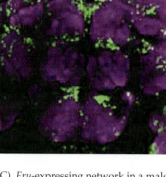

(B) Absence of Fru protein in a female brain

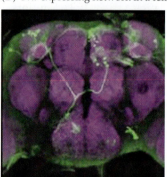

(C) *Fru*-expressing network in a male

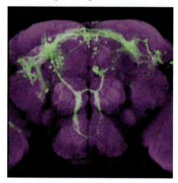

(D) *Fru*-expressing network in a female

There is one well-established instance of non-cell-autonomous sexual differentiation in flies that is reminiscent of the SNB in rodents. A pair of large abdominal muscles called the muscles of Lawrence (MOLs) is found only in male flies. However, in a reversal of the SNB where masculinization of the muscle causes masculinization of motor neurons, MOLs are masculinized by the innervating motor neurons (Currie & Bate, 1995). No matter what the genotype of the muscle fibers, as long as they are innervated by male motor neurons (Lawrence & Johnston, 1986), specifically motor neurons expressing male-specific transcripts of *fru* (Usui-Aoki, Mikawa, & Yamamoto, 2005), the muscles will grow in a masculine fashion.

The powerful effect of a single gene on *Drosophila* male reproductive behaviors, including whether to court females or males, was cited by some people as evidence that sexual orientation in humans may also be influenced by "biological" influences (Burr, 1993). But it's not at all clear that what we think of as sexual orientation in people can be modeled in flies or any other species, as we discuss next.

How's It Going?

1. How does the *Sxl* gene on the X chromosome control sexual differentiation in flies?

2. What are gynandromorphs and how do they show that sexual differentiation in flies is a cell-autonomous response?

3. What experiment demonstrated that male *fruitless* transcripts are necessary for a fly to show masculine courtship behavior?

4. What experiment demonstrated that male *fruitless* transcripts are sufficient for a fly to show masculine courtship behavior?

■ **7.13** ■

THE CONTROVERSY OVER SEXUAL ORIENTATION IN FLIES, RATS, AND PEOPLE

More than a decade ago, as the United States dealt with the political controversy over gay rights—the right of gay men and lesbians to serve in the armed forces, whether they can be protected from discrimination in the workplace and housing market, and allowed to marry—some people supporting gay rights argued that homosexuality is not "unnatural." Plenty of homosexual behavior has been documented in a wide variety of vertebrate species (Bagemihl, 2000; Poiani, 2010), so it certainly happens in nature. Likewise there's ample evidence that genes play a role in human sexual orientation (Bailey & Bell, 1993; Sanders et al., 2015), leading some authors to offer work in flies, including the dramatic ability of some *fru* alleles to cause males to court only males (Gailey & Hall, 1989), for support of the idea that human sexual orientation is fixed before birth (Burr, 1993). It's not at all clear that understanding the *fru* gene in *Drosophila* offers much insight into human sexual orientation. For one thing, while there is a single gene that determines whether male flies court females, we know that there must be many genes that influence human sexual orientation, each individually having a relatively modest influence (Sanders et al., 2015). Plus, as we've seen, the mechanism of sexual differentiation in flies is quite different from that in mammals, which rely so heavily on androgens like testosterone to masculinize structures.

> We know that there must be many genes that influence human sexual orientation, each individually having a relatively modest influence.

In fact, there is evidence that prenatal testosterone may be responsible for the fact that the vast majority of men are attracted to women. For example, two different markers for prenatal androgen—digit ratios (Grimbos, Dawood, Burriss, Zucker, & Puts, 2010; Williams et al., 2000) and acoustic emissions from the ears (McFadden & Pasanen, 1998)—indicate that lesbians, on average, were exposed to more prenatal androgen than straight women (**FIGURE 7.33**). These results support the idea that prenatal androgen programs the brain to later be attracted to women. That might then account for why the vast majority of men are attracted to women. Also, there are several cases of newborn boys who, because of severe birth defects in the abdominal wall, had to be surgically altered to resemble girls at birth and were raised as such. While this situation is rare, out of five such cases, all at puberty reported being sexually attracted to girls (Reiner, 2004; Reiner & Gearhart, 2004), as though the male-typical levels of prenatal androgen had overridden their gender of rearing.

Interestingly, body markers of prenatal androgen indicate that gay and straight men do *not* differ in prenatal androgen exposure (Grimbos et al., 2010; McFadden & Pasanen, 1998). This result suggests that boys who grow up to be gay somehow resist any influence fetal androgen may have on orientation. Perhaps there are certain genes that make the brain less sensitive to prenatal androgen in this regard.

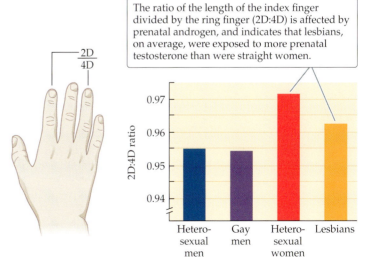

The ratio of the length of the index finger divided by the ring finger (2D:4D) is affected by prenatal androgen, and indicates that lesbians, on average, were exposed to more prenatal testosterone than were straight women.

FIGURE 7.33 **Bodily indicators of prenatal androgens** (After Williams et al., 2000.)

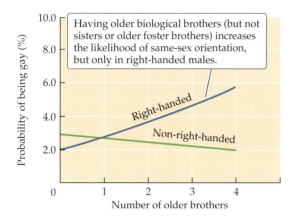

Having older biological brothers (but not sisters or older foster brothers) increases the likelihood of same-sex orientation, but only in right-handed males.

FIGURE 7.34 Fraternal birth order and sexual orientation in men (After Blanchard et al., 2006.)

There is, however, solid evidence of another prenatal influence on male orientation—the **fraternal birth order effect**: the more brothers a boy's mother carried before him, the more likely he is to be gay when he grows up (Blanchard, 2001). Each additional older brother increases the boy's chance of being gay by one-third (Cantor, Blanchard, Paterson, & Bogaert, 2002). The effect is the same whether the boys grow up together or not, and being raised with older stepbrothers has no effect (Bogaert, 2006), so the fraternal birth order effect is unlikely to be a result of social influences from the brothers. In fact, the number of older brothers only has an influence on boys who are consistently right-handed (Blanchard, Cantor, Bogaert, Breedlove, & Ellis, 2006) (**FIGURE 7.34**), as though having a brain organized in a left-handed fashion cancels out the fraternal birth order effect. This result also argues against the idea that growing up with an older brother has a mediating the effect. Alternatively, Ray Blanchard has hypothesized that a mother who has carried one or more sons may produce antibodies that cross the placenta and somehow perturb brain development in later sons (Blanchard, 2001; Puts, Jordan, & Breedlove, 2006).

> The more brothers a boy's mother carried before him, the more likely he is to be gay when he grows up.

In the meantime, there is little evidence that human sexual orientation is a matter of "choice" unless you think you chose your levels of prenatal androgen or number of older brothers. But does it really matter whether sexual orientation is chosen? As a thought exercise, imagine we discovered a switch where your brain meets the spinal cord that, when flipped, instantly reverses your sexual orientation back and forth. What right would I, or society at large, have to forbid you to ever flip that switch? And if you did flip it, why should you surrender any of your human or civil rights?

fraternal birth order effect The well-established finding that the more brothers a boy's mother carried before him, the more likely he is to grow up to be gay.

SUMMARY

- *Apoptosis* sculpts body parts by removing adjacent structures, such as the webbing between digits. **See Figure 7.1**

- Neuronal apoptosis serves to regulate the number of neurons to match the need, as the number of motor neurons in each segment of the spinal cord is appropriate for the number of muscles that need innervation. **See Figure 7.2**

- In the final hours of apoptosis, dying cells take on a characteristic *pyknotic* profile. While few pyknotic cells are visible at any single point in development, their peak matches the time when spinal motor neurons go missing, confirming that the motor neurons have indeed died rather than migrating away or differentiating into some other type of cell. **See Figures 7.3–7.5**

- Introducing mouse tumor cell lines next to chick embryos uncovered a protein that prevents apoptosis of *dorsal root ganglion* (*DRG*) cells and *sympathetic ganglia* cells. This *nerve growth factor* (*NGF*) has both *tropic* and *trophic* effects on these types of neurons, but little or no effect on others. **See Figures 7.6–7.9**

- When NGF contacts a responsive neuron, it is internalized and retrogradely transported to the cell body, where it regulates gene expression, which can forestall apoptosis and encourage process outgrowth. **See Figure 7.10**

- NGF is part of a family of *neurotrophins*, which includes *brain-derived neurotrophic factor* (*BDNF*), *neurotrophin-3* (*NT-3*), and *NT-4*, which bind with high affinity to the *Trk* family of receptors as well as the low-affinity *p75* receptor. **See Figures 7.11–7.13, Table 7.1**

- There are other neurotrophic factors in addition to the neurotrophins, including *ciliary neurotrophic factor* (*CNTF*) and *glial cell line–derived neurotrophic factor* (*GDNF*).

- Experiments in *C. elegans* proved that apoptosis was an active process, and not the result of attack from immune system cells that clean up debris. They also revealed the importance of *death genes*, including those for *caspases*, which the cell produces to dismantle itself. We can exploit the fragmentation of DNA by *TUNEL* labeling to detect dying cells even before they become pyknotic. **See Figure 7.14**

■ The main molecular pathway for apoptosis is basically the same in worms and vertebrates, but the latter species have more molecular players, such as *Diablo* and *inhibitors of apoptosis proteins* (*IAPs*). These additional regulatory pathways provide a system of checks and balances to determine the cell's survival. **See Figures 7.15 and 7.16**

■ We still don't know which cells initially die in *ALS*, but in those few cases that have a genetic basis, animal studies indicate that the motor neurons themselves are not the site of action.

■ In vertebrates, the same testicular *hormones* that masculinize the body also exert an *organizational* effect on the developing nervous system to masculinize it. *Androgens* such as *testosterone* either forestall neuronal apoptosis to provide adult males with more neurons, as in the *SDN-POA*, or augment apoptosis to leave males with fewer neurons, as in the *AVPV*. **See Figures 7.21–7.26**

■ In rodents, androgens act on developing bulbocavernosus muscles to preserve them, and thereby preserve their motor neurons in the *spinal nucleus of the bulbocavernosus* (*SNB*). **See Figures 7.27 and 7.28**

■ In flies, a cascade of transcription factors, starting with *sex-lethal* (*Sxl*) on the X chromosome, directs sexual differentiation of each part of the body in a cell-autonomous fashion. Several genes downstream from *Sxl* is *fruitless* (*fru*), a transcription factor gene that is translated only in male flies because the only transcripts of *fru* produced in females are nonfunctional. **See Figure 7.29**

■ In male flies, expression of *fru* prevents apoptosis of a network of neurons that mediate male courtship behavior. Thus expression of male-specific transcripts of *fru* is both necessary and sufficient to produce male courtship. **See Figures 7.30–7.32**

■ In humans, there is evidence that prenatal androgens may predispose individuals to be sexually attracted to females—body markers indicate that lesbians, on average, were exposed to more prenatal androgens than straight women. There is no evidence that gay men were exposed to less prenatal androgen than straight men, but the *fraternal birth order effect* demonstrates that there are prenatal influences on sexual orientation in men, too. **See Figures 7.33 and 7.34**

Go to the Companion Website
sites.sinauer.com/fond
for animations, flashcards, and other review tools.

THE EMPIRICISTS STRIKE BACK

In many ways, modern science began in Europe in the Enlightenment, roughly 1650 (the year René Descartes died) to 1750, an age when people began rejecting the traditional hierarchies of monarchs and religion. That spirit of rebellion included questioning the scientific writing of ancient authors. Of course there were many talented people conducting excellent science before this time, and all over the world, not just in Europe. Indeed up to and well into this period, probably the best physicians on Earth were in the Islamic world (Porman & Savage-Smith, 2007). But the concentration of people in Europe who were ready to throw off rulers, religious authorities, and ancient ideas in order to start thinking for themselves had an enormous impact on politics and science in the Western world. That included the so-called New World, where Western colonists expanded free thinking to shrug off European political authority (while being perfectly willing to subjugate or kill any indigenous peoples who questioned the colonists' authority).

The Enlightenment figures who would have the biggest impact on science were philosophers who insisted on the importance of experience for shaping the human mind. Because they emphasized that all knowledge derives from sensory experience, actually checking up on the world to see how things really are, they were called **empiricists** (*empirical* refers to something that can be verified, or refuted, by direct observation). The most influential were three British thinkers, John Locke (1632–1704), George Berkeley (1685–1753), and David Hume (1711–1776). Hume was the most skeptical of the three, noting that we can't really be sure the sun will rise tomorrow, and was repeatedly accused of being an atheist (which he quite ambiguously denied). But we'll zero in on Locke, who was the earliest and most influential of the three, and the one whose work is most relevant to development of the nervous system.

THE TABULA RASA AND THE IMPORTANCE OF EXPERIENCE THROUGH THE SENSES

Born in a middle-class Puritan household, John Locke nevertheless received an excellent education for his time, studying and then teaching philosophy at Oxford while also getting a medical degree. Locke was deeply immersed in

John Locke (1632–1704)

Enlightenment   The period, roughly 1650–1750, when European thinkers began rebelling against the traditional hierarchies of monarchs and religion in favor of reason and science.

empiricists   Philosophers who believed that the only way to gain knowledge of the world is through information provided by the senses.

tabula rasa   Latin for "blank slate," the idea that we enter the world with minds that are empty of innate ideas, and so we must gain information through experience.

the political rise and fall of English rulers (for his own safety he fled to the Netherlands to live for several years), and he had at least as much lasting influence in politics as in science. Almost everything you take for granted about the modern Western style of government was first proposed by Locke (*Two Treatises of Government*, 1689). He forcefully argued that sovereignty is held by the people, not a monarch; that government should follow a social contract that respects individuals' rights; and that there should be separation of church and state. As you've probably guessed, leaders of colonial rebellions in the New World were all greatly influenced by Locke's political writings.

Locke also became involved in a group of scholars at Oxford who were determined to study nature directly rather than through books and who would eventually form the Royal Society of London. These thinkers were unwilling to simply accept the teachings of earlier writers like Aristotle and Plato, insisting on testing any ideas about the natural world, including *epistemology* (the study of what we know and how we come to know it). In his *Essay Concerning Human Understanding* (1690), Locke disputed the rationalist assertion that we are born with knowledge already in our minds, what came to be called *innate ideas*. He argued that newborns bring very few ideas into the world with them, and what little understanding they have at birth, of pain versus pleasure, or sweet versus sour, for example, is probably gained through experience in the womb. Locke felt that we are born with minds that are virtually empty of knowledge, like a sheet of white paper, or what others called a **tabula rasa** (Latin for "blank slate") (**FIGURE I.1**). For Locke, the mind becomes filled with knowledge as a result of learning through experience.

Note that Locke's views on epistemology were entirely in line with his political views. England had, 150 years before, thrown off the authority of the Roman Catholic Church in favor of its own Church of England. That split had been justified by the supposed divine rights inherited by King Henry VIII, but in Locke's time, Enlightenment thinkers questioned (privately) such hereditary authority of kings, queens, and their offspring. Although it wouldn't have been healthy for Locke to say so outright, he thought that if we all begin as a blank

FIGURE I.1   A tabula rasa? (© selimaksan/Getty Images.)

slate, then anyone might gain sufficient education and wisdom to run the country as well or better than a person who happened to be born in Windsor Castle. This corollary of the idea of a blank slate was most famously stated centuries later by American behaviorist John Watson, who boasted that he could educate any healthy newborn to become a doctor, lawyer, artist, or thief (Watson, 1924). Locke's epistemological theory that we all start out with the same blank mind also supports his political belief that the sovereignty of a nation should be held collectively by the people.

In a famous passage challenging the rationalist notion that we can have ideas about things we have never experienced through our senses, Locke quotes a letter from a friend, William Molyneux, whose wife became blind from illness. The question posed in the letter has come to be known as **Molyneux's problem**:

> *Suppose a man born blind, and now adult, and taught by his touch to distinguish between a cube and a sphere of the same metal, and nighly of the same bigness, so as to tell, when he felt one and the other, which is the cube, which the sphere. Suppose then the cube and sphere placed on a table, and the blind man be made to see:* [Question:] *"whether by his sight, before he touched them, he could now distinguish and tell which is the globe, which the cube?"* To which [Molyneux] answers, *"Not. For, though he has obtained the experience of how a globe, how a cube affects his touch, yet he has not yet obtained the experience, that what affects his touch so or so, must affect his sight so or so; or that a protuberant angle in the cube, that pressed his hand unequally, shall appear to his eye as it does in the cube."*

> (Locke, Essay Concerning Human Understanding,
> Book 2, Chapter IX, par. 8).

Molyneux and Locke note that most people, when presented the problem, initially answer that of course the formerly blind man will be able to tell whether he's looking at a cube or a sphere (**FIGURE I.2**). After all, he already knows the difference in how these objects *feel*, that the cube has angles that stick out while the sphere does not. But when you ask how a blind man would know what angles *look* like, most people change their answer and say the blind man would *not* know whether he was seeing a cube or a sphere, at least not at first.

For Locke, providing vision to an adult who had been born blind was a thought experiment, something that couldn't happen in real life in his day. But since then there have been several people who gained sight for the first time in adulthood (Held et al., 2011). In general, those cases confirm Molyneux's and Locke's prediction, as these people have a hard time distinguishing objects by sight. In Chapter 9 we'll meet a man who, having learned how to ski when he was blind, and then gained vision as an adult, was unable to make much use of his new sense. In fact, he now finds skiing easier if he closes his eyes.

To give an example of how a rationalist idea might actually depend on empiricist experience, let's consider again Plato's idea of Forms, the eternal, perfect "blueprints" of objects or concepts that we recognize in their imperfect, highly variable real-life examples. As we noted in the Prologue, Plato believed that we were all born with these

Molyneux's problem A famous thought experiment that asks whether an adult enabled to see for the first time would be able to make sense of his or her vision.

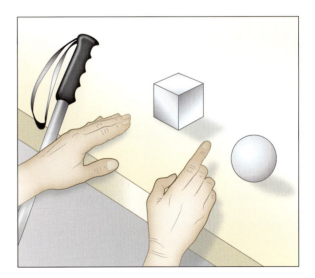

FIGURE I.2 Molyneux's problem Would a blind man who could suddenly see for the first time be able to distinguish between a cube and a sphere without touching them?

prototype   A model or concept that best typifies members of a category; a "best example."

Forms already present in our soul so that we could recognize beds, horses, and beauty. Plato's Form resembles what modern-day psychologists call a **prototype**, a model or concept that best typifies members of a category, a "best example." For instance, we might have in our minds a prototype of what "fruit" looks like. If we begin looking at lots of photos, some of which are of fruit, we will most rapidly recognize those that come closest to our prototype. That means psychologists can build up an idea of what our prototype of fruit is by manipulating images of fruit to see how quickly we recognize it.

Plato might have said we were born with the Form of fruit, but modern psychology provides evidence that we build up our mental prototype by interacting with the world (Pothos & Wills, 2011). A child has no concept of fruit until she has been exposed to lots of different foods and has been told which are fruits rather than vegetables, meat, grain, and so on. Similarly, as she's exposed to more kinds of fruit, her mental prototype of fruit is shaped by the particular examples she encounters. If Plato were to meet her and show her a fruit no one had ever seen before, say an apple with alternating purple and orange stripes, she would nevertheless know it was a fruit because it would fit closest to her prototype of fruit rather than her prototypes of other objects. Plato would attribute her recognition to the strange apple's fit to the Form she was *born* with, but psychologists would attribute it to the prototype she built up through *experience*, trying to classify objects with feedback from other people.

For an example of an abstract concept rather than an object, Plato also thought we were born with a Form of beauty so that we would recognize it when we saw it. In terms of facial beauty, this would mean we were born with a Form that lets us recognize some people's faces as very similar to that Form (beautiful) or quite different from that Form (not beautiful). It turns out that a beautiful human face is "average" in the sense that when psychologists use computers to average photographs of several women's faces, or an average of several men's faces, such morphed faces are always judged to be attractive. An average of 5 men's faces is handsome, and an average of 16 men's faces is even more handsome (**FIGURE I.3**). Perhaps when we see men's and women's faces as we grow up, we are, like the computer programs, "averaging" those faces to build up a prototype of what men look like and what women look

FIGURE I.3   The Platonic form of beauty? We all see many men's faces as we grow up. If we build up a prototype of men's faces by averaging them together as the computer program did here, we might come to see such faces as handsome. (Courtesy of the Face Research Lab, faceresearch.org.)

like, and we regard faces that come closest to that prototype as "beautiful" for women or as "handsome" for men. If so, then we weren't *born* with these Forms/prototypes as the rationalists thought. Rather, we gained them through *experience*, as the empiricists insisted.

WHAT DOES ALL THIS PHILOSOPHY STUFF HAVE TO DO WITH THIS BOOK?

The previous 7 chapters described a progressive unfolding of developing events, each resulting in an increasingly complex nervous system. Cells differentiate to become neural tissue (Chapter 1), detect their position in the nervous system and body (Chapter 2), migrate to their final positions (Chapter 3). Each cell decides on a fate as one type or other of glia or neuron (Chapter 4). Then each neuron sends out an axon that finds its way to the appropriate target (Chapter 5) and forms a synapse to transmit information (Chapter 6), after which many neurons die, to match the size of a neural region to the size of its target or to shape the individual as either male or female (Chapter 7). Recall that at each of these seven stages, proper development is critically dependent upon communication between cells, mostly very localized signals from one cell to its neighbor (the long-distance influence of hormones being an exception). But also note that this progressive unfolding of events, and the resulting progressive complexity of brain structure, is also critically dependent on *gene functioning*. Without gene products to mediate those signals between cells, and without the many transcription factors that regulate gene expression within each cell, as well as the thousands of genes encoding proteins to actually build the various types of cells we need in various regions of the brain, development will either grind to a halt or go wrong, sometimes horribly wrong as many lethal mutations attest. In all of these ways, the information we inherit from our parents exerts nearly total control over these early developmental events.

Thus in many ways, the developmental processes explored in those seven chapters are divorced from the rest of the world, all taking place within the embryo and its immediate container, either an egg or a womb. Of course that container must be in an environment that allows cells to live, can't be too hot or too cold, must have access to nutrients, and must be protected from injury and all that. But outside those sorts of *permissive* environmental factors, proper development up to this stage is actively guided by genes, not experience. Indeed, up to this point *there have been no sensory organs, and hence no sensation*. Locke's beliefs notwithstanding, how could experience have contributed *anything* to development up to this point? Rather, all this developing complexity, this unfolding program of interacting cells, is under the guidance of the genes we inherited from our truly ancient line of ancestors, stretching back millions of years. To the extent that we have a Platonic soul providing us with information before birth, it is contained in the chromosomes and maternal cytoplasm, brought together at conception, which direct the events described in the past seven chapters.

But just as there's probably a limit on how complex a nervous system can be made by having mitotic lineage alone guide cell differentiation, as in *C. elegans*, there's probably a limit on how complex a nervous system can be made relying only on information in the genome. Happily for those of us who enjoy being humans, natural selection eventually hit upon a way to bring in additional information from *outside* the genome, indeed from outside the individual, thereby enabling formation of additionally complex brains. Once the outside world begins guiding brain development, it is at least theoretically possible to produce brains nearly as complex as the universe.

This ability of the developing brain to tap the outside world, to become more complex and better adapted to that world, required the evolution of three developmental mechanisms, each dependent on those that came before, in sequence:

1. *Neuronal activity* The capacity for neuronal activity, the production of action potentials and synaptic signals, to guide developmental processes is yet another way for cell-cell interaction to direct development and mainly serves to weaken or strengthen synaptic contacts. Once such mechanisms have evolved, it's only a matter of time before a second mechanism arises to guide brain development.

2. *Sensory experience* Neuronal activity is driven by sensory information, which in turns reflects events that are occurring in the world outside the individual organism itself. In this way, sensory experience, information from outside the genome, can shape the developing brain, better adapting the individual to the way things are in the here and now. Those mechanisms in turn enable evolution of a third mechanism.

3. *Social experience* For species that provide parental care, it becomes possible for other individuals to influence the developing organisms' sensory experience, nurturing and instructing them about the world around them. In other words, social experience brings even more complexity to the developing brain, informing the young individual not only about how things are today, but even about how things were in the past (remember what Grandma said about rotary dial phones?).

Social influences are, of course, also the prerequisite for the development of culture, which has played such an important role in the evolution of our own species, as well as several other species (think of birds that migrate each year in a particular direction, each new generation learning the path by following the older birds). Natural selection led to a proliferation of such mechanisms for individuals to reproduce, as in those species that choose to mate with individuals that resemble their parents.

The remaining chapters of this book will take up these three developmental mechanisms in turn, showing how even spontaneous neuronal activity can guide neural development, primarily by adjusting synaptic connections (Chapter 8). This activity-dependent synaptic plasticity will then be used to fine tune developing sensory systems so they can accurately detect the most important stimuli in our environment (Chapter 9). That sets the stage so that social experience, especially in species like birds and mammals, can further guide neural development to maximize the chances for reproduction (Chapter 10). Following those chapters, we'll be in a position to discuss philosopher Immanuel Kant's reconciliation of rationalism and empiricism, and relate Kant's ideas to developmental neuroscience, in the Epilogue.

CHAPTER 8

Synaptic Plasticity
ACTIVITY-GUIDED NEURAL DEVELOPMENT

CRUEL AND UNUSUAL PUNISHMENT? Evan Miller was born to a poor family in rural Alabama. Neglected by his alcoholic mother, regularly beaten by his stepfather, he tried to kill himself several times, starting at age 5. At age 10, he was put in foster care for a while, but when Evan returned home, he began abusing drugs (Denniston, 2012). When he was 14, Evan and a 16-year-old friend, Colby, were smoking marijuana and drinking whiskey in the home of a 52-year-old drug dealer living in the same trailer park as Evan's family. When the man passed out, Evan stole $300 from his wallet. While Evan was trying to return the wallet, the man woke up and grabbed him by the throat. In the fight that followed, the two teenagers beat the man unconscious with a baseball bat. Then they set his trailer on fire to cover up their crime, and the man died of smoke inhalation.

When the boys were apprehended, Colby made a deal with police to testify against the younger Evan, and they were both tried as adults. Evan was convicted of two counts of murder (murder during an arson, and murder during a robbery) and received a sentence of life in prison without the possibility of parole, which at the time was mandatory under Alabama's laws. In other words, the judge had no choice but to impose that sentence.

Nine years later, the U.S. Supreme Court heard an appeal of Evan's case, contending it was "cruel and unusual punishment" to send a teenager to prison for life. Evan's lawyer noted that U.S. law does not allow 14-year-olds to drink, vote, serve on a jury, or drive a car (Totenberg, 2012), which seems to recognize that they cannot be expected to make responsible decisions because of their age. Should the horrific acts of an abused, drunken 14-year-old be sufficient reason to put him in prison for the rest of his life?

CHAPTER PREVIEW

In the first six chapters we watched the developing brain become ever more complex, with more neurons, more brain structures, and more synaptic connections, followed by a brief period in development when many neurons die, as described in Chapter 7. All seven of the involved processes—neural induction, neurogenesis, cell migration, cellular differentiation, axonal guidance, synapse maturation, and apoptosis—were orchestrated by the sequential expression of genes, which in turn was tightly controlled by cell-cell interactions as neighboring cells communicated with one another to determine their fates. So far, the communication between cells that settles their fates has consisted of four different types of chemical signals:

In prison for life Evan Miller (in white shirt) in a photo taken 3 years after the murder took place. (© Associated Press.)

synapse rearrangement The process in which some synapses are withdrawn while new synapses form.

1. Membrane-bound signaling systems between adjacent cells like the Delta-Notch system and the ephrins
2. Diffusible signals like noggin and Sonic hedgehog that spread locally
3. Retrogradely transported signals like nerve growth factor and other neurotrophins
4. Wide-ranging signals—hormones like testosterone

At this stage, once neurons develop the molecular machinery for producing action potentials and establishing synaptic connections, a new signaling mechanism becomes available to direct neural development, specifically *neuronal activity*. At some point natural selection exploited this new signal, neuronal activity, to guide further development of the brain, which is the subject of this chapter.

We'll begin by examining an instance when neuronal activity seems to have a permissive effect, namely on the death of motor neurons. That wave of apoptosis does not happen unless the motor neurons are electrically active so they can stimulate their muscle targets. The rest of the chapter will mostly be concerned with the role of neuronal activity in adjusting the strength of synapses, determining whether they grow stronger or weaker or disappear altogether. We'll see that just as an apparent overabundance of neurons is made early in development and followed by a wave of apoptosis, so there appears to be an overabundance of synapses early in development that is followed by a loss of many of those connections.

Again motor neurons offer a clear demonstration of this process. We'll see that originally each muscle fiber is innervated by several different neurons, but then most of those connections are retracted until each fiber is controlled by one and only one motor neuron. Neuronal activity is crucial for the elimination of these superfluous neuromuscular connections.

Indeed, synapses are lost throughout the developing nervous system. It turns out that during what was once described as *synapse elimination* to emphasize the loss of synapses, new synapses are also forming at the same time. So even though there may sometimes be a net loss in the total number of synapses, it still seems more appropriate to call this process **synapse rearrangement**. The bulk of this chapter will describe the crucial role of activity in guiding synapse rearrangement, deciding which existing synapses will stay or go, and which new synapses will form. We'll learn that a psychologist's speculation about the rules dictating the fates of synapses turned out to be remarkably prescient, and that a specialized neurotransmitter receptor seems to embody those rules. We'll see the important role of neuronal activity in sharpening connections between the eye and the brain before birth, prior to any visual experience. The chapter will conclude with the role of synapse elimination in the maturation of the cortex, a process that has barely even begun in most 14-year-olds such as Evan Miller. Understanding how spontaneous activity sculpts visual system development prenatally will position us to learn about the role of postnatal visual experience in creating the exquisitely sensitive human visual system, a main topic of Chapter 9.

■ **8.1** ■

MOTOR NEURONAL DEATH IS GATED BY NEURONAL ACTIVITY

Recall from Chapter 7 that nearly 50% of the motor neurons that develop in the fetal spinal cord will die as the nervous system matures. This apoptosis serves to roughly match the number of motor neurons with the size of the periphery, specifically the number of muscles that need innervation. The extent of motor neuronal apoptosis is controlled by the target muscles, which are thought to supply some mixture of neurotrophic factors that motor neurons must compete for to determine which will live and which will die. Curiously enough, that competition for neurotrophic factors appears to require neuronal activity. Silencing the communication between motor neurons and muscles seems to suspend the competition, and none of the motor neurons die until electrical activity returns.

Curiously enough, competition for neurotrophic factors appears to require neuronal activity.

For example, **curare** is a toxin, used for poison arrows by indigenous peoples of South America, that paralyzes prey by blocking nicotinic acetylcholine receptors, including those at the neuromuscular junction. When chick embryos are exposed to curare during the period when motor neurons normally undergo apoptosis, motor neuronal death is postponed (**FIGURE 8.1**) (Pittman & Oppenheim, 1979). When the acetylcholine blockade is lifted, motor neuron death resumes, leaving the chick with about the same number of motor neurons as in untreated embryos. One hypothesis to explain this result revolves around the fact that the curare treatment increases the number and size of neuromuscular junctions, which might provide more neurotropic factor(s), with the result that no motor neuron dies. In support of that explanation, when several different drugs to block neuromuscular junctions were tried at different doses, only those treatments that resulted in expanded neuromuscular junctions also postponed motor neuron death (Oppenheim, Bursztajn, & Prevette, 1989). Thus it appears that when muscles are electrically silenced, which would normally indicate an absence of motor neuron innervation, the fibers release signals to entice more nerve inputs, thereby making more neurotrophic factors available to the motor neurons. Without a scarcity of neurotrophic support, motor neuronal apoptosis is postponed. In this way, electrical activity normally "starts the race" among motor neurons, which then compete for trophic factors and therefore for survival.

curare A toxin found in various South American plants that blocks nicotinic acetylcholine receptors.

Even after the competition for motor neuronal survival is over, each muscle fiber is still innervated by more than one motor neuron.

Interestingly, even after the competition for motor neuronal survival is over, each muscle fiber is still innervated by more than one motor neuron. Then later in development the surviving motor neurons will compete again, this time for which muscle fibers they will control in adulthood, as we'll see next.

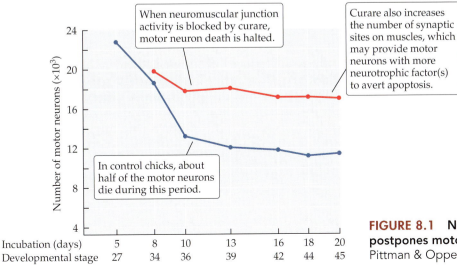

When neuromuscular junction activity is blocked by curare, motor neuron death is halted.

Curare also increases the number of synaptic sites on muscles, which may provide motor neurons with more neurotrophic factor(s) to avert apoptosis.

In control chicks, about half of the motor neurons die during this period.

| Incubation (days) | 5 | 8 | 10 | 13 | 16 | 18 | 20 |
| Developmental stage | 27 | 34 | 36 | 39 | 42 | 44 | 45 |

FIGURE 8.1 **Neuromuscular blockade postpones motor neuron apoptosis.** (After Pittman & Oppenheim, 1979.)

How's It Going?

1. What is the effect of altering the electrical activity of muscle fibers on motor neuronal apoptosis?
2. What is the evidence that such neuromuscular blockade works by affecting availability of neurotrophic factors from the muscle target?

■ 8.2 ■
DEVELOPING MUSCLE FIBERS START OFF WITH POLYNEURONAL INNERVATION

motor unit   A motor neuron and all the muscle fibers that it innervates.

polyneuronal innervation   The condition of having more than one neuron innervate a target, such as a muscle fiber in newborn rats.

A **motor unit** is a single motor neuron and all the muscle fibers that it innervates. For any given muscle, some motor units will be large, as when a single motor neuron stimulates hundreds or even thousands of individual muscle fibers. Other motor units will be small, as when the motor neuron innervates perhaps 50 or 100 muscle fibers. In general, motor neurons with larger cell bodies and axons tend to control many fibers (i.e., constitute a large motor unit), while smaller motor neurons tend to control fewer fibers. This variety in motor unit size makes it possible to adjust muscle tension, allowing us to fine-tune our motor performance. In general, we recruit smaller motor units first, and larger motor units later (Conwit et al., 1999; Henneman, Somjen, & Carpenter, 1965) as muscle tension increases, so we can get the muscle force about right from the start and then add or subtract units to get the force we want.

Note that an important adjunct for this recruitment of different motor units to fine-tune force is that *each muscle fiber is controlled by one and only one motor neuron.* If this were not true, then to the extent that there was overlap between motor units, say several different motor neurons controlling the same fibers, then we would lose the ability to recruit a new increment in force by recruiting a new motor unit. Indeed, in adulthood, each muscle fiber is singly innervated—controlled by only one motor neuron.

However, in newborn rats, almost every muscle fiber is innervated by at least two different motor neurons, so they receive **polyneuronal innervation**. This was first demonstrated physiologically by electrically stimulating the nerve containing axons that innervate a given muscle, then measuring the excitatory postsynaptic potential (EPSP) in individual muscle fibers. If the nerve is provided a very low level of electrical stimulation, no action potential will be generated in any motor neuron axons, and so of course no EPSP will be seen in the fiber. Gradually increasing the strength of the electrical stimulation will eventually generate an action potential in an axon innervating that particular muscle fiber, and an EPSP will be generated. In adults, stronger stimulation of the nerve will continue to elicit an EPSP, but it will never be any larger, because an action potential has already been produced in the one and only motor neuron axon innervating that fiber. In contrast, when this experiment was done in *newborn* rats, increasing stimulation of the nerve did result in larger EPSPs (Redfern, 1970). Tellingly, the increased EPSPs were not continuously gradual. Rather, they always happened in increments (**FIGURE 8.2A**), indicating that there were several distinct inputs, each recruited at a particular level of stimulation, which then combined their EPSPs (**FIGURE 8.2B**).

Thus after birth each muscle fiber loses several motor neuron inputs until, by adulthood, each fiber receives innervation from one and only one motor neuron. Please note that this elimination of many terminal branches of motor neuron axons has nothing to do with apoptosis, because the elimination of polyneuronal innervation happens several days *after* motor neuron apoptosis

(A)

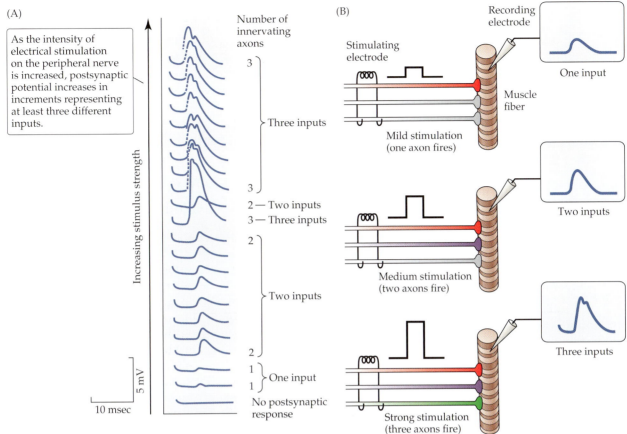

As the intensity of electrical stimulation on the peripheral nerve is increased, postsynaptic potential increases in increments representing at least three different inputs.

Increasing stimulus strength

Number of innervating axons

3 } Three inputs

3

2 — Two inputs
3 — Three inputs

2

Two inputs

2

1 } One input
1

No postsynaptic response

5 mV

10 msec

(B)

Recording electrode

Stimulating electrode

One input

Muscle fiber

Mild stimulation (one axon fires)

Two inputs

Medium stimulation (two axons fire)

Three inputs

Strong stimulation (three axons fire)

FIGURE 8.2 **Physiological evidence of polyneuronal innervation of newborn rat muscle fiber** (A after Redfern, 1970.)

is complete. In this later process, all the motor neurons survive; they differ only in terms of which particular muscle fibers they innervate. Also note that while the number of axons innervating each neuromuscular junction is reduced until there is only one input, that remaining neuromuscular junction is growing considerably larger (**FIGURE 8.3**). This expansion of the junction innervated by the sole remaining motor neuron axon is probably important for making sure the muscle fiber fires every time the axon fires.

Surveying lots of rat muscle fibers in this way revealed that at 5 days of life, every muscle fiber was polyneuronally innervated, but by the time the pups were weaned at 21 days of life, virtually no fiber received more than one input (**FIGURE 8.4A**) (Brown, Jansen, & Van Essen, 1976). Examining rats prenatally made it clear this process of muscle fibers eliminating excess innervation, going from being polyneuronally innervated to singly innervated, starts before birth. In fact, the peak of polyneuronal innervation happens just before birth, at which point muscle fibers have an average of three inputs (**FIGURE 8.4B**) (Dennis, Ziskind-Conhaim, & Harris, 1981). Repeated viewings of developing neuro-

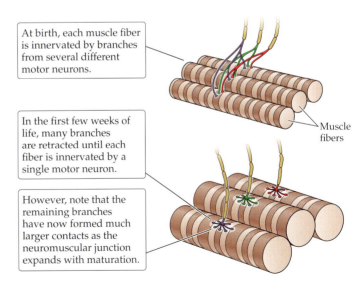

At birth, each muscle fiber is innervated by branches from several different motor neurons.

In the first few weeks of life, many branches are retracted until each fiber is innervated by a single motor neuron.

Muscle fibers

However, note that the remaining branches have now formed much larger contacts as the neuromuscular junction expands with maturation.

FIGURE 8.3 **Neuromuscular synapse elimination** At birth, virtually every muscle fiber in rats is polyneuronally innervated, but by adulthood, each fiber will receive input from only a single motor neuron.

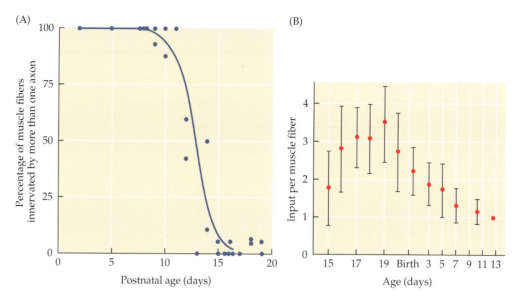

(A)

Percentage of muscle fibers innervated by more than one axon

Postnatal age (days)

(B)

Input per muscle fiber

Age (days)

FIGURE 8.4 **Loss of polyneuronal innervation in rat muscle fibers** (A after Brown et al., 1976; B after Dennis et al., 1981.)

> The elimination of many terminal branches of motor neuron axons has nothing to do with apoptosis.

muscular junctions allows us to watch one motor neuron's axon retract while the remaining motor neuron's axon takes its place (**FIGURE 8.5**).

A similar loss of polyneuronal input is seen in the cerebellum, where neurons in the brainstem's inferior olive nucleus provide climbing fibers to extensively innervate Purkinje cells in the cerebellar cortex (**FIGURE 8.6A**). In adulthood, each Purkinje cell receives climbing fiber input from only a single brainstem neuron. Thus gradually increasing electrical stimulation of the inferior olive will elicit an all-or-none response in the adult Purkinje cell—once you've stimulated the single olivary neuron innervating it, the EPSP won't get any larger (**FIGURE 8.6B**). But in newborn rats, varying the strength of electrical stimulation of the olive can provide varying sizes of EPSPs, indicating more than one input per Purkinje (**FIGURE 8.6C**) (Mariani & Changeux, 1981). As development proceeds, the average number of climbing fiber inputs per Purkinje declines until, after 2 weeks of life, each receives only a single such input (**FIGURE 8.6D**). Thus each olive cell innervates a single Purkinje, and each Purkinje gets a climbing fiber from a single olive cell. Then further synaptic refinement takes place as the climbing

The axon terminal from a motor neuron expressing a blue marker occupies most of the junction on the 11th day of life.

Two days later the blue neuron contributes little to the junction.

The blue terminal forms a retraction bulb as it pulls away from the junction.

Now the junction is occupied by the green terminal alone.

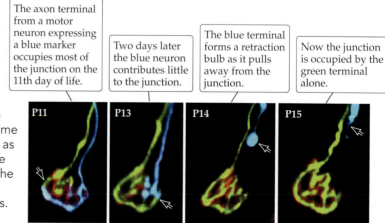

P11

P13

P14

P15

FIGURE 8.5 **Competing motor neuron axon branches** Repeated views of the same neuromuscular junction allow us to watch as innervation from one axon is lost while the remaining axon takes over the junction. The red-labeled bungarotoxin marks the acetylcholine receptors beneath the terminals. (From Walsh & Lichtman, 2003.)

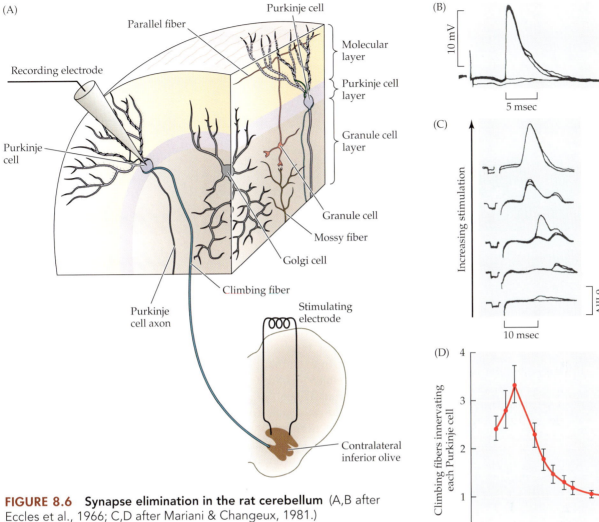

FIGURE 8.6 **Synapse elimination in the rat cerebellum** (A,B after Eccles et al., 1966; C,D after Mariani & Changeux, 1981.)

fiber, which initially innervated both the soma and the dendrites of the Purkinje cell, retracts its input to the soma, resulting in the adult pattern of innervation only of the dendrites. This retraction of inputs can also be monitored from the point of view of the inferior olive cell—initially it may innervate several Purkinje cells, but then it retracts from all but one (**FIGURE 8.7**).

We still don't understand how the loss of polyneuronal innervation onto muscle fibers and Purkinje cells is controlled, but it must be a competitive process because each target ends up with exactly one input. If the motor neurons were simply retracting a random sample of, say, 60% of their terminals, then by chance alone some fibers would retain more than one input and some would have none! Because both those outcomes are exceedingly rare in mature muscles, there must be some *communication* between the target and the several neurons innervating it, to avoid loss of all input. And there must be some *competition* between the two inputs, for every target to end up with one and only one input.

> The loss of polyneural innervation onto muscle fibers and Purkinje cells must be a competitive process because each target ends up with exactly one input.

There is also strong evidence that electrical activity is important for this competitive process in both the cerebellum (Andjus, Zhu, Cesa, Carulli, & Strata, 2003) and the neuromuscular junction. For example, electrically stimulating the motor nerve accelerates the pace of synapse elimi-

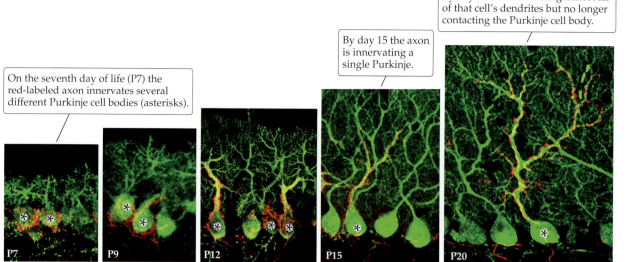

On the seventh day of life (P7) the red-labeled axon innervates several different Purkinje cell bodies (asterisks).

By day 15 the axon is innervating a single Purkinje.

By day 20 it is innervating almost all of that cell's dendrites but no longer contacting the Purkinje cell body.

FIGURE 8.7 Climbing fiber retraction from cerebellar Purkinje cells The climbing fiber from a single neuron in the brainstem's inferior olive possesses a red fluorescent marker. (From Hashimoto et al., 2009.)

> It is the *asynchronous* arrival of action potentials down the axons of various motor neurons that drives elimination of polyneuronal innervation.

tetrodotoxin (TTX) A toxin found in the ovaries of pufferfish that blocks voltage-dependent sodium channels and thus prevents conduction of action potentials.

bungarotoxin A toxin found in the venom of certain snakes that irreversibly binds to and blocks nicotinic acetylcholine receptors.

> The motor axons that are more successful at driving the muscle fiber are retained at the expense of less effective motor axons.

nation in neuromuscular junctions, and chronically blocking action potentials in the motor nerves by treating them with **tetrodotoxin** (**TTX**), which blocks voltage-dependent Na+ channels and so stops action potentials, forestalls the process (Thompson, 1983). Likewise, preventing activation of the neuromuscular junction with nicotinic cholinergic blockers like curare causes the polyneuronal innervation to persist. Interestingly, if care is taken to stimulate all the motor axons at the same time so that their action potentials arrive synchronously, then neuromuscular synapse elimination halts (Busetto, Buffelli, Tognana, Bellico, & Cangiano, 2000). It is the *asynchronous* arrival of action potentials down the axons of various motor neurons that drives elimination of polyneuronal innervation. In fact, recordings of action potentials in developing motor neuron axons reveal that they are synchronously active before birth, when polyneuronal innervation of fibers is prominent, and start firing asynchronously just as synapse elimination begins (Busetto, Buffelli, Cangiano, & Cangiano, 2003).

How might activity gate the elimination of polyneuronal innervation at neuromuscular junctions? One possibility is that the various inputs to each muscle fiber are competing for which is most effective at driving the fiber to fire (and therefore contract). In support of this idea, applying another toxin that blocks nicotinic ACh receptors, **bungarotoxin**, to just one part of a neuromuscular junction results in the loss of ACh receptors in that region and a withdrawal of the nerve terminal branch that had been over it (Balice-Gordon & Lichtman, 1994). The loss of the terminal branch is accompanied by loss of rapsyn (see Chapter 6) in the postsynaptic region opposite the terminal (Culican, Nelson, & Lichtman, 1998), suggesting that the forces binding together the nerve and target are lost. This loss of the silenced terminal branch doesn't happen just because the receptors are inactive, because using the drug to block the *entire* neuromuscular junction results in no loss (Balice-Gordon & Lichtman, 1994). Rather, it appears that *within a junction*, those terminal branches that fail to excite their underlying ACh receptors are lost if and only if neighboring branches are successful at stimulating the muscle fiber. In other words, the motor axons that are more successful at driving the muscle fiber are retained at the expense of less effective motor axons (Lichtman & Colman, 2000). In the end of such competition, each fiber ends up with input from one, and only one, motor neuron.

All of these findings suggest that young neuromuscular junctions behave like *Hebbian synapses*, where activity determines whether synapses will become stronger or weaker, a topic we'll discuss in detail later in this chapter.

How's It Going?

1. What is a motor unit, and what's the pattern of innervation like within a motor unit?

2. How can scientists use gradual stimulation of motor nerves to determine the number of inputs to an individual muscle fiber?

3. What is the course of synapse elimination at neuromuscular junctions, and what is the evidence that neuronal activity plays a role?

■ **8.3** ■

AUTONOMIC NEURONS REFINE THEIR INPUTS AND OUTPUTS

A similar loss of polyneuronal innervation is seen in the autonomic nervous system. You may recall from previous classes that the **autonomic nervous system** consists of two parts: the *sympathetic* and *parasympathethic* systems. The **sympathetic nervous system** consists of neurons in the spinal cord that send their axons out the ventral roots to synapse upon neurons in the chain of **sympathetic ganglia** that lie along either side of the vertebral column. The neurons in the sympathetic ganglia in turn send their axons to provide noradrenergic innervation to the various organ systems. In the other half of the autonomic nervous system, the **parasympathetic nervous system**, neurons from either the brainstem or sacral spinal cord innervate **parasympathetic ganglia**, which are scattered around the body, each close to the organ that it innervates (see Figure 3.20). While sympathetic neurons release norepinephrine and have a noradrenergic effect on the target organ, parasympathetic neurons usually release acetylcholine, which typically has the opposite effect. For example, sympathetic norepinephrine speeds up the heart while parasympathetic acetylcholine slows it down.

In some autonomic ganglia, most of the neurons are innervated by a single axon in adults. But in a situation reminiscent of that seen in developing muscle fibers, these autonomic ganglion neurons in newborn mammals are almost always polyneuronally innervated. Then, as the animals grow, almost all of the polyneuronal innervation is lost, and neurons that received an average of five inputs at birth will, at maturity, receive only a single input on average (**FIGURE 8.8**) (Lichtman, 1977).

autonomic nervous system
The neural system that regulates activity of many organ systems and consists of the sympathetic nervous system and the parasympathetic nervous system. It is largely outside of our conscious control.

sympathetic nervous system
The portion of the autonomic nervous system that generally prepares the body for action.

sympathetic ganglia The two chains of interconnected ganglia alongside the vertebral column that receive input from the sympathetic preganglionic neurons in the spinal cord. The neurons within the sympathetic ganglia send their axons out to innervate various organs, where they usually release norepinephrine.

parasympathetic nervous system The portion of the autonomic nervous system that generally facilitates relaxation and recuperation of the body.

parasympathetic ganglia
The ganglia scattered throughout the body that receive input from either the brainstem or sacral spinal cord. The neurons within the parasympathetic ganglia send their cholinergic axons to innervate various organs.

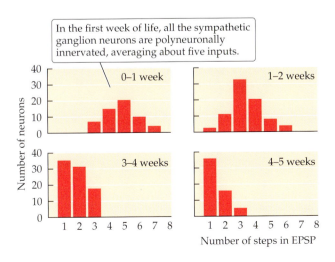

In the first week of life, all the sympathetic ganglion neurons are polyneuronally innervated, averaging about five inputs.

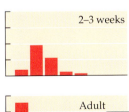

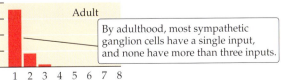

By adulthood, most sympathetic ganglion cells have a single input, and none have more than three inputs.

FIGURE 8.8 Reduction in the number of inputs to rat autonomic ganglion cells during development Each ganglionic neuron initially has several inputs, but eventually most of them are retracted. (After Lichtman, 1977.)

FIGURE 8.9 **The pattern of synapse rearrangement in autonomic ganglia** (After Purves & Lichtman, 1978.)

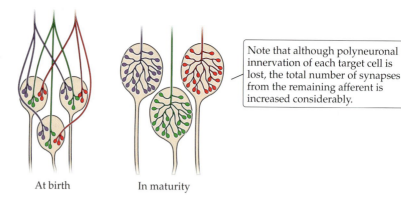

Note that although polyneuronal innervation of each target cell is lost, the total number of synapses from the remaining afferent is increased considerably.

At birth   In maturity

This loss of polyneuronal innervation was monitored physiologically, as with developing muscle fibers, by gradually increasing electrical stimulation of afferents and noting the number of increments in the EPSP elicited in the ganglionic neuron (see Figure 8.2). When developing autonomic cells were examined anatomically, it turned out that while the number of inputs per neuron was declining, the size of the remaining input(s) was growing considerably (**FIGURE 8.9**). In other words, while fewer afferent neurons are innervating the neuron in adulthood, those remaining afferent(s) make a lot of synapses on that neuron (Purves & Lichtman, 1978). This means that while some synapses are indeed being eliminated, there is an overall net gain in the total number of synaptic inputs.

The competition between autonomic afferents is also evident when some of the axons are cut during synapse elimination. The remaining afferents sprout and restore the pattern of innervation that would have resulted without the manipulation (Purves, 1975). One interesting outcome of this competitive process is that the afferents seem to be competing for synaptic space. For example, in some autonomic ganglia there is a great deal of variability in how many inputs each neuron receives. In those cases, there is a strong positive correlation between the number of dendrites the autonomic neuron has and the number of inputs it receives (Purves & Hume, 1981) (**FIGURE 8.10**). This

(A)

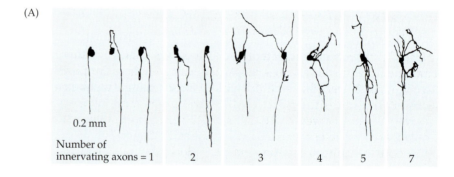

0.2 mm

Number of innervating axons = 1   2   3   4   5   7

FIGURE 8.10 **Some neurons offer more synaptic space than others.** (A) Neurons with more extensive dendrites tend to have more neuronal afferents. (B) Plotting the data reveals a remarkably constant relationship between synaptic space and number of inputs. (From Purves & Hume, 1981.)

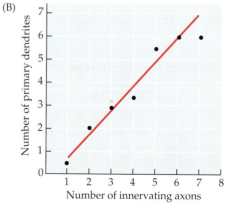

outcome indicates that the afferents are indeed competing for synaptic space on their targets, and that targets with larger dendrites will retain more afferents (Purves, 1983). Thus the overall picture of synapse elimination in neuromuscular junctions, the cerebellum, and the autonomic nervous system is that inputs to a neuron compete with one another and that powerful afferents become more powerful, while weaker afferents grow weaker. Long before these data were gathered, a psychologist predicted that synapses would compete in this manner, as we discuss next.

How's It Going?

1. What are the two divisions of the autonomic nervous system, and how do they have opposite effects on target organs?
2. What is the normal course of development of innervation patterns in autonomic ganglia, and what's the evidence that this represents a competition between inputs?

■ 8.4 ■

DONALD HEBB SPECULATED ABOUT NEURAL PLASTICITY

In 1949 psychologist Donald Hebb published a book, *The Organization of Behavior*, that proposed a mechanism of synaptic plasticity that could, in theory, account for many behavioral phenomena, including habituation, classical conditioning, and memory. While Hebb regarded his proposal for how the effectiveness of synapses might change with activity as a mere reworking of ideas from previous writers, his presentation was so effective that today such a synapse is usually called a **Hebbian synapse**:

> *"When an axon of cell A is near enough to excite a cell B and repeatedly or persistently takes part in firing it, some growth process or metabolic change takes place in one or both cells such that A's efficiency, as one of the cells firing B, is increased"* (Hebb, 1949, p. 62).

Note that the proposal is not simply that active synapses will be strengthened, but rather that *successful* synapses, those that manage to drive the postsynaptic cell to fire, will be strengthened (**FIGURE 8.11A**). A corollary of this rule is that unsuccessful synapses, those that, when active, rarely cause the postsynaptic cell to fire, will become weaker. The proposal has sometimes been summarized as "cells that fire together wire together" (Shatz, 1992, p. 64).

Also note that inherent to the concept of Hebbian synapses is *competition* between inputs to a given neuron. If the input that is most successful at causing the neuron to fire will be strengthened, its efficiency as one of the cells making the neuron fire will be increased, which means that other synapses will become relatively less effective. In other words, synaptic inputs that are rarely active just before the target neuron fires will become less important in driving that neuron. Thus Hebbian synapses represent a sort of Darwinian selection process, whereby effective synapses become more effective, and ineffective synapses become even less effective.

When Hebb made his proposal, he was primarily thinking about neural plasticity in adulthood, and at that time most scientists believed that synapses became stable by adulthood. So he envisioned existing synapses becoming stronger or weaker, not that synapses would come and go in the mature brain. In the years since, it has become clear that indeed new synapses are formed, and existing synapses disappear, throughout life. In other words, the brain has

Hebbian synapse A synapse that grows stronger when it repeatedly succeeds in driving the postsynaptic cell to fire and grows weaker when it repeatedly fails to drive the postsynaptic cell.

FIGURE 8.11   Hebbian synapses at work "When an axon of cell A is near enough to excite a cell B and repeatedly or persistently takes part in firing it, some growth process or metabolic change takes place in one or both cells such that A's efficiency, as one of the cells firing B, is increased." (Hebb, 1949, p. 62.) (A) Hebb envisioned existing synapses getting stronger or weaker. (B) But in addition, strong inputs may sprout to form additional synapses, and weak inputs may be retracted.

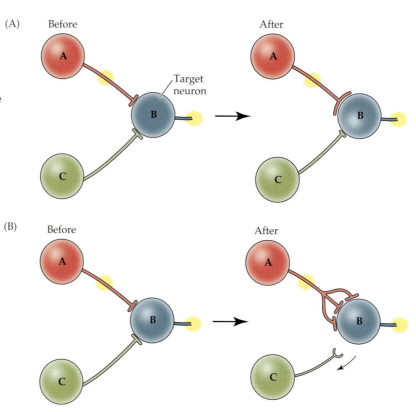

It has become clear that indeed new synapses are formed, and existing synapses disappear, throughout life.

turned out to be even more plastic than expected, changing not just in terms of functional strength of synapses, but also in terms of structural properties, such as synapses and dendrites. As we noted in Chapter 3 when we talked about adult neurogenesis, it turns out that whole neurons come and go in the adult brain, too. I like to say that every year when I attend the Society for Neuroscience meetings, the brain turns out to be more plastic than it was the year before.

In fact, as we'll see, Hebbian synapses do exist in the brain, and the competition between inputs mediated by Hebbian processes does sometimes result in weak synapses being retracted and in strong inputs sometimes growing larger, or sprouting additional terminals to produce new synapses (**FIGURE 8.11B**). Note that this additional wrinkle to Hebbian synapses means there *must be communication between the pre- and postsynaptic neurons.* If Hebbian synapses always stayed put, merely became weaker or stronger, it would have been enough for either the postsynaptic neuron to vary how many receptors were present or for the presynaptic neuron to vary how much neurotransmitter to release. But if synapses subjected to Hebbian competition may disappear, then there must be communication between the two neurons, if only to coordinate the loosening of binding between them (see Chapter 6). We'll see that indeed there is communication between pre- and postsynaptic neurons in Hebbian synapses, and that change in synaptic strength involves changes on both sides of the synapse.

■ 8.5 ■

LONG-TERM POTENTIATION (LTP) CONFIRMS THE EXISTENCE OF HEBBIAN SYNAPSES

hippocampus   A portion of the limbic system that is known to be important for learning and memory

In the 1970s, people studying neurophysiology in rabbits discovered a form of synaptic plasticity that seemed to embody the synapses Hebb had proposed. They were examining the **hippocampus** (Latin for "seahorse," which its three-dimensional shape resembles), a part of the limbic system that was already known to be involved in learning. The two main parts of the hippocampal formation are the hippocampus itself and the **dentate gyrus** (so called because in cross section it appears to have *denta*—"teeth").

The three-dimensional shape of the hippocampal formation is rather complex, which is one reason it long fascinated neuroanatomists. Imagine two lengths of rain gutter, one turned upside down, with one lip of each gutter sitting in the trough of the other. One length of rain gutter represents the layer of neuronal cell bodies that make up the hippocampus proper, while the other represents the neuronal cell bodies of the dentate gyrus. Now take the two nested gutters and bend them in the shape of a *C*, then shrink them and place them beneath the back of the cerebral cortex, with both ends of the *C* facing rostrally. One end of the *C* sits under the parietal cortex, and the other is embedded in the temporal cortex (**FIGURE 8.12A**). These neurons also have highly stereotyped synaptic connections with one another and the rest of the brain. For example, neurons in nearby entorhinal cortex send axons in a bundle that seems to perforate the hippocampal formation and so is called the *perforant pathway* (**FIGURE 8.12B**). These perforant fibers synapse upon all the hippocampal neurons, including dendrites of neurons in the dentate gyrus. Eventually it was found that at least three different pathways of afferents in the hippocampal formation display Hebbian plasticity, but it was first shown in this perforant pathway onto dentate gyrus neurons in rabbits, as we'll see next. The crucial insight was to electrically stimulate some excitatory afferents very rapidly in a burst, say 15 times per second for 15 seconds, which is called a **tetanus**. This was intended to make sure the postsynaptic target neurons would fire—in other words, that the synapses would be very successful at driving the postsynaptic neuron.

dentate gyrus A portion of the hippocampal formation. It is known for continued neurogenesis in adulthood.

tetanus A sustained burst of rapid neuronal firing.

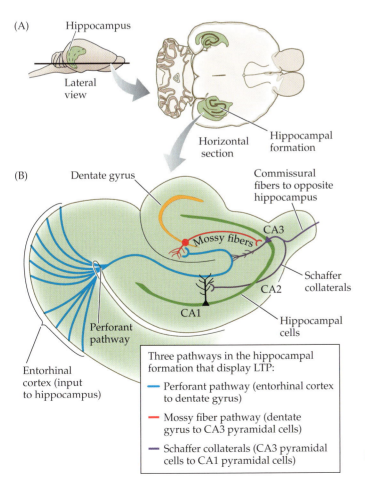

(A) Hippocampus
Lateral view
Horizontal section
Hippocampal formation

(B) Dentate gyrus
Commissural fibers to opposite hippocampus
CA3
Mossy fibers
Schaffer collaterals
CA2
CA1
Hippocampal cells
Perforant pathway
Entorhinal cortex (input to hippocampus)

Three pathways in the hippocampal formation that display LTP:

— Perforant pathway (entorhinal cortex to dentate gyrus)

— Mossy fiber pathway (dentate gyrus to CA3 pyramidal cells)

— Schaffer collaterals (CA3 pyramidal cells to CA1 pyramidal cells)

FIGURE 8.12 **The hippocampal formation and various synaptic pathways**

■

RESEARCHERS AT WORK

Cells That Fire Together Wire Together

■ **QUESTION**: Do some synapses in the brain actually behave like Hebbian synapses?

■ **HYPOTHESIS**: If many excitatory afferents are electrically stimulated at the same time by being subjected to a tetanus, then they will succeed in making the postsynaptic neurons fire. If the synapses behave in a Hebbian fashion, then they should be made stronger after the tetanus.

■ **TEST**: Stimulate the perforant pathway in anesthetized rabbits, and record excitatory postsynaptic potentials (EPSPs) in neurons of the dentate gyrus. If the afferents are stimulated at a low rate, every half second, the resulting EPSPs are very stable in size. Now rapidly stimulate the afferents, producing a tetanus, for 15 seconds. After that, return to the low rate of stimulation to see if EPSPs are different after the tetanus.

■ **RESULT**: Low rates of stimulation indeed produced a very stable amplitude of EPSPs in the dentate gyrus. During the tetanus, EPSPs were much higher, as expected given temporal summation of excitatory synapses. Following the tetanus, EPSPs were initially smaller for a few seconds, but then they gradually increased in size such that 20 minutes later the EPSPs were again very stable in size but more than 50% larger than they had been before the tetanus (**FIGURE 8.13A**). In other rabbits, providing additional tetanus episodes continued to increase the amplitude of the EPSPs, an effect that lasted for hours (**FIGURE 8.13B**) (Bliss & Lomo, 1973).

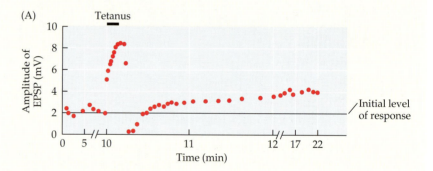

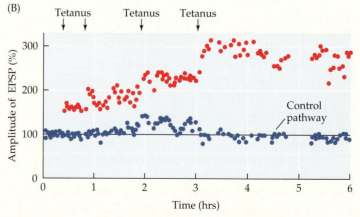

FIGURE 8.13 Long-term potentiation (LTP) in the rabbit perforant pathway to the dentate gyrus (After Bliss & Lomo, 1973.)

■ **CONCLUSION**: Perforant path afferents to the dentate gyrus behave like Hebbian synapses: when the afferents have successfully driven the postsynaptic neurons to fire, as during a tetanus, the synapses become stronger, an effect that is stable and relatively long lasting. This long-lasting strengthening of synapses after a tetanus came to be called **long-term potentiation** (**LTP**; *potentiation* is an English word for "enhancement" or "augmentation"). In a companion paper, the authors were also able to demonstrate LTP in awake, unanesthetized rabbits using electrodes that had been surgically implanted (under anesthesia, of course) beforehand and that this LTP could last for *weeks* (Bliss & Gardner-Medwin, 1973).

Later researchers were able to see LTP in slices of tissue taken from the hippocampal formation and in all three of the pathways marked in Figure 8.12B. LTP would be seen in many different vertebrate and invertebrate species and may be a characteristic of all excitatory synapses (Malenka & Bear, 2004). But the hippocampus remains the best-studied locus of examples of LTP. Most of the hippocampal pathways use excitatory synapses that are glutamatergic, and it would be in these synapses that we would learn how various classes of glutamate receptors underlie Hebbian plasticity, as we'll see next.

How's It Going?

1. How do Hebbian synapses change in response to neuronal activity?
2. What is LTP, and how does it conform to Hebbian rules of neuronal plasticity?
3. How might LTP relate to the process of memory formation?

■ 8.6 ■
A CLASS OF GLUTAMATE RECEPTORS ENFORCES HEBBIAN RULES

The finding of synaptic circuits like those in the hippocampus that display the Hebbian-like properties of LTP excited many neuroscientists, especially those studying learning and memory. Hebb had already explained how synapses of that sort could underlie simple learning like classical conditioning.

The excitement was heightened when neuroscientists discovered a neurotransmitter receptor system responsible for the Hebbian-like behavior of hippocampal LTP. Recall that there are several classes of receptors that normally respond to the amino acid neurotransmitter glutamate. Some glutamate receptors are **metabotropic**, meaning that they use second messenger systems to affect function in the receiving cell. Other classes of glutamate receptors are **ionotropic**, meaning that the receptors include an ion channel that opens in response to the neurotransmitter, and it is these types of receptors that underlie Hebbian properties at synapses in the hippocampus and many other brain regions. There are at least three classes of ionotropic glutamate receptors, each named after the synthetic drugs that bind them best—the **AMPA** (α-amino-3-hydroxy-5-methyl-4-isoxazolepropionic acid), **NMDA** (N-methyl-D-aspartate), and **kainate** receptors. These glutamate receptors all share considerable homology (Traynelis et al., 2010), indicating that they evolved from a common precursor. (A closely related subfamily, called delta (δ) receptors, share

long-term potentiation (LTP) Long-lasting strengthening of synaptic strength seen after the induction of a tetanus in presynaptic afferents. It is observable in several neuronal pathways of the hippocampal formation.

metabotropic Referring to a class of neurotransmitter receptor that uses a second messenger system to biochemically alter the postsynaptic target.

ionotropic Referring to a class of neurotransmitter receptor that includes an ion channel to affect the electrical potential across the postsynaptic cell's membrane.

AMPA α-amino-3-hydroxy-5-methyl-4-isoxazolepropionic acid, a synthetic compound that binds one class of ionotropic glutamate receptors with high affinity.

NMDA N-methyl-D-aspartate, a synthetic compound that binds with high affinity a class of ionotropic glutamate receptors that are known for imparting a Hebbian-like plasticity.

kainate A synthetic compound that binds a class of ionotropic glutamate receptors with high affinity.

(A) Normal synaptic transmission

(B) Induction of LTP

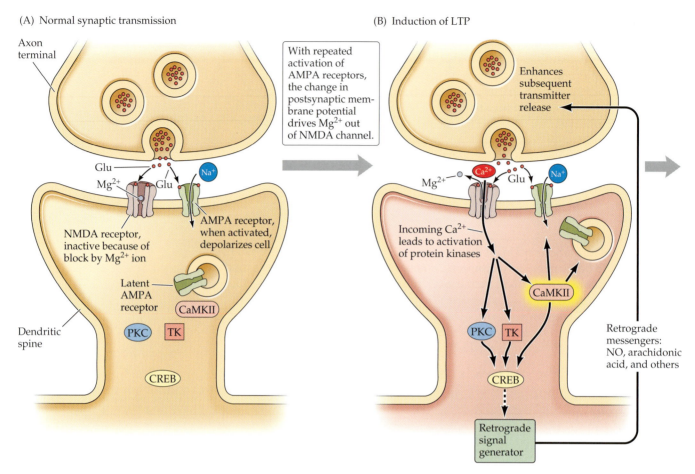

With repeated activation of AMPA receptors, the change in postsynaptic membrane potential drives Mg^{2+} out of NMDA channel.

Axon terminal

Glu
Mg^{2+} — Glu

Na^+

NMDA receptor, inactive because of block by Mg^{2+} ion

AMPA receptor, when activated, depolarizes cell

Latent AMPA receptor

CaMKII

Dendritic spine

PKC   TK

CREB

Enhances subsequent transmitter release

Mg^{2+} Ca^{2+} Glu Na^+

Incoming Ca^{2+} leads to activation of protein kinases

CaMKII

PKC   TK

CREB

Retrograde signal generator

Retrograde messengers: NO, arachidonic acid, and others

FIGURE 8.14 Synaptic plasticity mediated by NMDA-type glutamate receptors

much of this homology but do not bind glutamate, so their function remains unknown [Schmid, Kott, Sager, Huelsken, & Hollmann, 2009]).

LTP in several hippocampal circuits is caused by the actions of two of these classes of glutamate receptor. AMPA receptors act in the relatively simple, classical fashion you are probably familiar with: glutamate binds the receptor, an ion channel opens, allowing positively charged ions like K^+ and Na^+ to enter, depolarizing and hence exciting the postsynaptic neuron. Such AMPA receptors mediate the excitatory response to a single activation of input to hippocampal circuits. The fact that slowly repeated stimulation of afferents every half second or so produces a very stable excitatory response indicates that these AMPA receptors are very stable and reliable under these conditions (**FIGURE 8.14A**).

However, the other class of ionotropic glutamate receptors, the NMDA receptors, behave quite differently. When the afferents are stimulated at a slow rate, these NMDA receptors bind the released glutamate and change their configuration to open up an ion channel, but under these conditions no ions pass though the channel, so the NMDA receptor does not contribute to the steady excitatory response of the postsynaptic neurons. Why don't ions pass through the activated NMDA receptors under these conditions? Because when the postsynaptic neurons are only mildly activated, they remain sufficiently polarized that the NMDA receptor's ion channels are quickly blocked by Mg^{2+} ions that, like Na^+ ions, are attracted to the target neuron's negatively charged

(C) Enhanced synapse, after induction of LTP

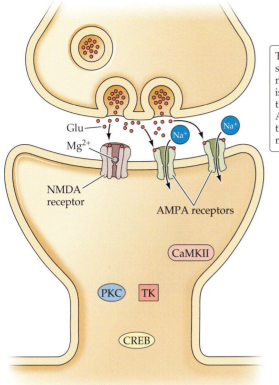

The synapse is now stronger because more transmitter is released and there are more AMPA receptors in the postsynaptic membrane.

Glu

Mg^{2+}

Na^+ Na^+

NMDA receptor

AMPA receptors

CaMKII

PKC   TK

CREB

interior. But the NMDA receptor's channel is too small to allow these Mg^{2+} ions to pass through and so they block the channel up (see Figure 8.14A).

Things get more interesting when the postsynaptic neurons are more depolarized, as can happen when many excitatory signals arrive together, including when the afferents are stimulated with a tetanus. When the postsynaptic neurons are depolarized a bit more, then the Mg^{2+} block is lifted as the interior negative charge no longer outweighs the ion's concentration gradient (Mg^{2+} is more plentiful inside neurons than outside). Now the NMDA receptor's ion channel is open to Ca^{2+} ions, which enter the target neurons and initiate several changes (**FIGURE 8.14B**). First, the Ca^{2+} activates Ca^{2+}/calmodulin-dependent protein kinase (CaMKII), an enzyme that phosphorylates proteins, including several transcription factors. This activation of CaMKII results in the rapid insertion of new AMPA receptors in the postsynaptic membrane. That means that when the steady, slow activation of afferents resumes after the tetanus, a greater EPSP is seen.

The entry of Ca^{2+} ions through postsynaptic NMDA receptors also triggers a cascade of events that will affect presynaptic function to strengthen LTP. Eventually the postsynaptic region releases a **retrograde messenger**, so called because it sends information back across the synapse in the opposite direction of regular synaptic communication. It is unclear what the retrograde signal is, exactly, but it may be arachidonic acid and/or a soluble gas like nitric oxide. Whatever its particular identity, the retrograde signal induces changes in the presynaptic terminal, with the result that in the future more glutamate will

When the postsynaptic neurons are depolarized a bit more, the Mg^{2+} block is lifted.

retrograde messenger Here, a chemical signal emanating from the postsynaptic cell that affects the presynaptic cell.

BOX
8.1

DOES HIPPOCAMPAL LTP MEDIATE LEARNING?

Much of the initial excitement about the discovery of LTP in the 1970s was the possibility that this might be a mechanism for storing memories in the brain. A long-lasting change in synaptic strength might somehow encode a particular memory. In the years since, considerable evidence has accumulated to support the idea. First, there are suggestive correlations, such as the similar time courses of LTP induction and memory formation (Lynch & Baudry, 1991; Staubli, 1995). More convincingly, pharmacological manipulations that interfere with LTP also interfere with memory formation. For example, drugs that block NMDA receptors (Morris, 1989) or inhibit the CaMKII (Serrano et al., 1994) that mediates NMDA-induced changes in the postsynaptic cell also interfere with memory formation. Also, knocking out the gene for the NMDA receptor subunit in only the hippocampus of mice impairs both the generation of LTP there and the animals' memory (Shimizu, Tang, Rampon, & Tsien, 2000). Conversely, mice engineered to overexpress NMDA receptors in the hippocampus have enhanced LTP *and* better than normal memory (Cao et al., 2007; Tang et al., 1999).

Of course, what you'd really like is to see LTP induced not by a tetanus, but by a learning experience, and then figure out how the resulting change in synaptic strength is related to the animal's memory at some later time. The problem is figuring out exactly which synapses might be affected by the learning experience in order to see if an LTP-like change has taken place. Perhaps the closest anyone's

How to build a smarter mouse   Presented with two objects, a mouse will spend less time investigating the one it remembers having seen before. Mice overexpressing NMDA receptors in the hippocampus have enhanced LTP compared with control mice, and they also display a better memory in such tasks. (Courtesy of Princeton University.)

come to that demonstration is the finding that a single trial avoidance learning task in rats caused a change in glutamate receptors in one part of the hippocampus and that this learning-induced change made the synapses insensitive to an additional, tetanus-induced LTP (Whitlock, Heynen, Shuler, & Bear, 2006).

be released when an action potential arrives at the terminal (**FIGURE 8.14C**). Thus the greater response seen after the tetanus in LTP is due to two changes: more AMPA receptors in the postsynaptic membrane, and more glutamate release from the presynaptic terminal. Because NMDA receptors are found to mediate LTP in invertebrates such as *Aplysia* (Tarabeux et al., 2011), it appears that this mechanism underlying synaptic plasticity has been around a long time. There was never any debate about whether LTP *can* underlie learning, only debate about whether LTP *actually does* underlie learning, as we discuss in more detail in **BOX 8.1**.

NMDA receptor activation can also lead to longer-lasting changes in the synapse. Usually this involves a cascade of changes triggered by the inrush of Ca^{2+} ions at the NMDA receptor, activating several kinases, which in turn activates CREB (**c**yclic AMP **r**esponse **e**lement-**b**inding protein), which regulates gene expression (**FIGURE 8.15**). In this way the synapse may be further strengthened, and this may also trigger spine formation to entice the successful presynaptic neuron to sprout and form additional synapses between the two neurons.

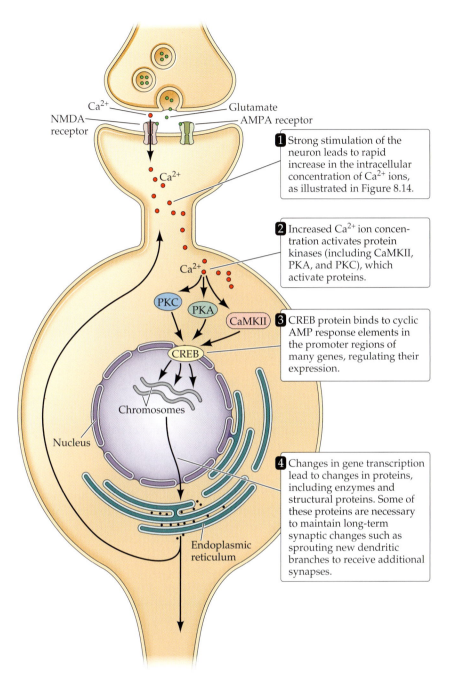

Ca²⁺ — Glutamate

NMDA receptor

AMPA receptor

1 Strong stimulation of the neuron leads to rapid increase in the intracellular concentration of Ca²⁺ ions, as illustrated in Figure 8.14.

Ca²⁺

2 Increased Ca²⁺ ion concentration activates protein kinases (including CaMKII, PKA, and PKC), which activate proteins.

Ca²⁺

PKC

PKA

CaMKII

3 CREB protein binds to cyclic AMP response elements in the promoter regions of many genes, regulating their expression.

CREB

Chromosomes

Nucleus

4 Changes in gene transcription lead to changes in proteins, including enzymes and structural proteins. Some of these proteins are necessary to maintain long-term synaptic changes such as sprouting new dendritic branches to receive additional synapses.

Endoplasmic reticulum

FIGURE 8.15 Long-term consequences of activation of NMDA receptors

Note that the change in synaptic strength mediated by NMDA receptors is gated by neuronal activity per se; there's no rule that says the activity must in some way relate to any sensory process. Indeed, the NMDA receptors have no mechanism to distinguish between neural stimulation originating in sensory experience versus some other trigger, such as a tetanus provided by a scientist. Nevertheless, in the following chapter we'll see how sensory experience can sometimes drive the activity that strengthens or weakens Hebbian connections. But before visual experience fine-tunes brain connections, neural activity that is independent of sensory experience will direct synaptic rearrangement, as we'll see next.

■ **8.7** ■

THE BRAIN MUST INTEGRATE INPUT FROM THE TWO EYES

retinal ganglion cells (RGCs) The class of retinal neurons that send their axons out the optic nerve to transmit visual information to the brain.

optic nerve The bundle of axons of retinal ganglion cells that exit the eye and project to the brain.

nasal Referring to the nose. For example, the nasal retina is the medial portion of the retina in vertebrates.

contralateral Referring to the opposite side of the body or brain.

temporal Here, referring to the side of the head. For example, the temporal retina is the lateral portion of the retina in vertebrates.

ipsilateral Referring to the same side of the body or brain.

lateral geniculate nucleus (LGN) A nucleus of the thalamus that receives projections from the eyes.

primary visual cortex (V1) Also called *striate cortex*. The region of the occipital lobe where most visual information first arrives in the cortex.

layer IV The fourth layer down from the pial surface in vertebrate neocortex. It is predominated by input from the thalamus and other cortical regions.

monocular Referring to one eye.

binocular Referring to two eyes.

transsynaptic transport The transfer of a chemical marker across a synapse.

To understand the role of prenatal neuronal activity in the developing visual system, we'll first need to review some neuroanatomy, specifically how information from the eyes reaches the brain. It is not true that the left eye sends information to the right cortex (and right eye to the left cortex). Rather, it is the left *visual field* that sends information to the right cortex (and vice versa), but *both* eyes get input from the left visual field, so both eyes provide information to the right cortex. Likewise, the right visual field reaches both eyes, so they both send axons to the left cortex.

How does that happen? Recall that in the eye, **retinal ganglion cells (RGCs)** send their axons out of the eye in a bundle forming the **optic nerve**, which is called the *optic tract* once it enters the brainstem. In mammals, some of the axons in the right optic nerve will remain on the right side of the brain while the rest will cross the midline; the same split occurs in the left optic nerve. The RGC axons arising from the *medial* portion of each retina, or the **nasal** ("close to the nose") retina, will send their axons across the midline to the **contralateral** ("opposite-sided") cerebral hemisphere. Those retinal axons coming from the *lateral* portion of each retina, or **temporal** ("next to your temples") retina, will stay on the **ipsilateral** ("same-sided") hemisphere. In primates, cats, and other mammals with front-facing eyes, about half the axons from each eye cross the midline while the other half stay on the same side.

Given the way light projects onto the retinas, this pattern of connections means that information from the right visual field reaches the left hemisphere, while information from the left visual field reaches the right hemisphere (**FIGURE 8.16**). But note that both eyes receive information from both halves of the visual field, so both eyes send axons to both sides of the brain. Light from the right visual field reaches the temporal retina of the left eye and the nasal retina of the right eye, and they both project to the left side of the brain.

In mammals, most of those retinal axons will innervate neurons in a portion of the thalamus called the **lateral geniculate nucleus** (**LGN**; *geniculate* means "knee," which refers to the cross-sectional resemblance of the human LGN to a bent knee) (**FIGURE 8.17**). In adults, input from the two eyes remains segregated, as neurons in some layers of the LGN get input from RGCs from the ipsilateral eye, while other LGN layers get input from the contralateral eye.

LGN neurons send their axons to **primary visual cortex**, also known as **V1** or *striate cortex* (see Figure 8.22). Like all outside input to cerebral cortex, these axons synapse upon neurons in **layer IV** of the cortex. In adults, the input to layer IV of visual cortex is still segregated by eye, by which I mean each cortical neuron receives input exclusively from either the ipsilateral eye

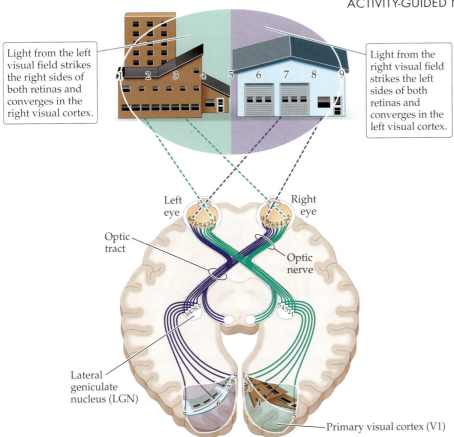

Light from the left visual field strikes the right sides of both retinas and converges in the right visual cortex.

Light from the right visual field strikes the left sides of both retinas and converges in the left visual cortex.

FIGURE 8.16 **Bringing information from the two eyes together** (After Frisby, 1980.)

Left eye

Right eye

Optic tract

Optic nerve

Lateral geniculate nucleus (LGN)

Primary visual cortex (V1)

Thus both eyes project to both sides of the brain, but left visual field goes to the right side of the brain, while the right visual field goes to the left hemisphere.

or the contralateral eye. Some patches of layer IV get input from the left eye; other patches of layer IV get input from the right eye. In other words, each cortical neuron in adult layer IV receives **monocular** ("one-eyed") input. Layer IV neurons send their axons to higher and lower layers of cortex, and it is these neurons outside layer IV that receive visual information from both eyes. In other words, visual cortical cells in all the layers *except* IV receive **binocular** ("two-eyed") information.

We can visualize these patterns of connections by injecting radioactively labeled amino acids into one eye. They will be taken up by RGCs and incorporated into various proteins, some of which will be transported down the RGC axons. Some of those proteins will be released by the axon terminals onto LGN neurons and subsequently transported down *their* axons projecting to the cortex. Because the radiolabeled amino acids are carried across synapses, we call this **transsynaptic transport**. If we wait the right amount of time after injecting the radioac-

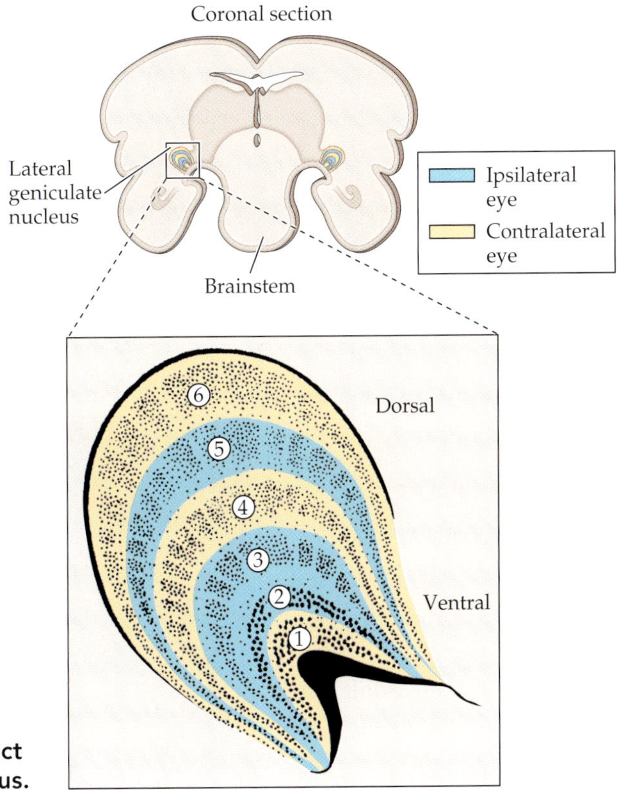

Coronal section

Lateral geniculate nucleus

Brainstem

Ipsilateral eye

Contralateral eye

Dorsal

Ventral

FIGURE 8.17   **In adults, the two eyes each project axons to distinct layers of the LGN of the thalamus.**

FIGURE 8.18
Transsynaptic transport of label from the eye to visual cortex By injecting radiolabeled amino acids into one eye, we can trace the distribution of information from that eye in the LGN and cortex.

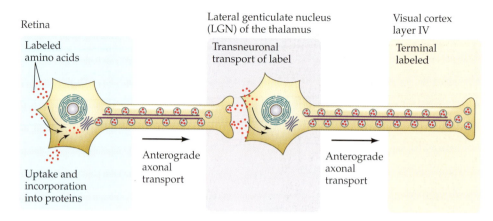

autoradiogram A preparation where tissue "takes its own picture" by exposing chemical film to radioactively labeled markers.

tive amino acids into one eye, we can remove the visual cortex, section it so that the input layer, layer IV, is revealed, and place it next to light-sensitive photographic film. As radioactive particles in the labeled amino acids are released, they will expose the film (**FIGURE 8.18**). Later we can treat the film to produce an **autoradiogram** (so called because the tissue "took its own picture"). We'll see silver grains wherever the labeled proteins ended up and see that, within the adult layer IV, input from the injected eye is found exclusively in bands of cortex about 0.5 mm wide (**FIGURE 8.19**). As each band of

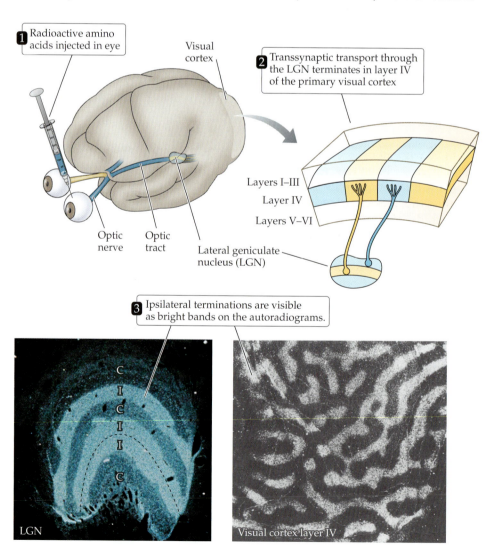

FIGURE 8.19 **Monocular innervation in the adult visual system** (LGN micrograph courtesy of P. Rakic; ocular dominance columns from LeVay et al., 1980.)

Day 15

Fifteen days after birth the radioactive label from one eye is evenly distributed across layer IV of visual cortex.

Day 22

In a 22-day-old kitten, input from the injected eye has begun to segregate into particular patches of layer IV.

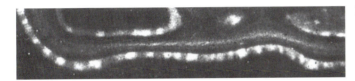

Day 39

By postnatal day 39, ocular dominance bands are sharply defined in alternating bands of light (from injected eye) and dark (from the other eye).

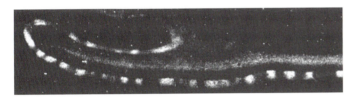

Day 92

At 92 days, the bands are even sharper, indicating completely monocular input from either the injected eye (white) or the other eye (black)."

FIGURE 8.20 **Ocular inputs to visual cortex segregate during development** (From LeVay et al., 1978.)

layer IV receives input exclusively from one eye or the other, they are called **ocular dominance bands**.

Now that we've reviewed connections in the *adult* visual system, we can address the question of how they come about during development. In fact, early in development, as in a 15-day-old kitten, label from the injected eye is distributed more or less evenly across layer IV of visual cortex, indicating that the eye is projecting throughout the layer. As we see the result when injecting either eye, this means that both eyes are projecting throughout layer IV. But as development proceeds, the ocular dominance bands in layer IV become progressively more distinct, as input from the two eyes becomes segregated (**FIGURE 8.20**). In kittens, this emergence of ocular dominance bands becomes more distinct after their eyes open (about 10–14 days after birth), so you might suspect that visual experience drives segregation of the inputs. But a similar segregation of input from the two eyes happens in primates before birth (Crowley & Katz, 2002), so that cannot be driven by visual experience. What's more, in kittens the segregation in visual cortex is preceded by segregation of ocular input to the LGN before the eyes have opened (Shatz, 1983). Also, the segregation of eye input to the LGN and cortex in mice happens even before rods and cones in the retina can respond to light (Godement, Salaun, & Imbert, 1984), so *visual experience* cannot play a role in the segregation of input from the two eyes. However, that still leaves room for *neuronal activity* to play a role, as we'll see next.

ocular dominance band
A region of neural tissue that is predominated by input from one eye or the other, such as in layer IV of primary visual cortex.

As development proceeds, the ocular dominance bands in layer IV become progressively more distinct, as input from the two eyes becomes segregated.

■ 8.8 ■

EVEN SPONTANEOUS, APPARENTLY RANDOM ACTIVITY CAN PROVIDE ORDER

There are at least two stages in development when neuronal activity is crucial for development of the visual system. The earliest known role for activity occurs before birth and therefore before there has been any chance for visual experience. After birth, it is activity arising from visual experience that plays an important role, and we'll discuss those studies at the start of the next chapter.

Early in fetal development, many LGN neurons receive input from both eyes. But by the time fetal development comes to an end, the adult pattern, where each layer of the LGN receives input from either the ipsilateral eye or the contralateral eye, is firmly established (see Figure 8.17). In fact, the segregation of input in the LGN begins even before photoreceptors in the retina can react to light (Huberman, Feller, & Chapman, 2008), so this segregation cannot depend on *visual experience*. Nevertheless, there is excellent evidence that *neuronal activity* is required for the segregation to take place. Before birth there is spontaneous activity of retinal ganglion cells, including waves of activity that roll across the retina before birth. These waves of activity seem to begin at random spots in the retina, but once begun, they spread to neighboring neurons in waves, back and forth across the retina. Over time, the entire retina will be active, as various patches are active at different times, such that there is locally correlated activity (Feller, Butts, Aaron, Rokhsar, & Shatz, 1997). Several different mechanisms are responsible for these spreading waves of activation—for example, at first there are electrical synapses, *gap junctions*, between retinal neurons, such that spontaneous activity in one readily spreads to neighbors (Syed, Lee, Zheng, & Zhour, 2004). Later, just before birth, it is the spontaneous release of glutamate from retinal bipolar cells that causes nearby RGCs to fire (Bansal et al., 2000). The spread of these waves of activity from cell to cell means that there is correlated activity between ganglion cells on the retina—whenever a given RGC fires, its neighbors are likely to fire, too. Thus, in general, neighboring RGCs tend to fire at about the same time, not because they are activated by visual stimuli, but simply because they are neighbors and spontaneous activity spreads across adjacent neurons. The spontaneous activity in the retina comes to an end just about the time the eyes open up in kittens and ferrets, and about the time of birth in monkeys. In other words, spontaneous waves of retinal activity normally end about the time visual experience begins.

In general, neighboring RGCs tend to fire at about the same time simply because they are neighbors and spontaneous activity spreads across adjacent neurons.

Considerable evidence indicated that this spontaneous activity of neighboring RGCs is important for segregation of input from the two eyes, first in the LGN and then in layer IV of visual cortex. For example, infusing TTX to block all activity in the LGN causes the eye-specific layers of input to fail to form. Instead, both eyes continue to innervate all the LGN layers. But that is a rather drastic manipulation, and it may have nothing to do with the spontaneous activity arising from the eyes. What was needed was a way to keep the LGN functional while manipulating spontaneous activity in the eye, as we see next.

Spontaneous Waves of Retinal Activity Form Ocular Dominance Bands in the LGN

■ **QUESTION**: Does the spontaneous activity in the retina play a role in the development of eye-specific innervation in the LGN?

■ **HYPOTHESIS**: Blocking the waves of spontaneous activity that cause neighboring RGCs to fire at similar times will delay or prevent segregation of ocular inputs to LGN layers.

■ **TEST**: Inject into the eye a drug (epibatidine, a neurotoxin secreted from the skin of certain South American frogs) that blocks the waves of spontaneous activity in young ferrets, in which segregation of ocular inputs to the LGN normally occurs postnatally, in the first 9 days after birth. Inject the drug into one or both eyes, to see if segregation of ocular input to the LGN is affected.

■ **RESULT**: When saline was injected into one eye, the normal segregation of ocular input between postnatal days 1 and 9 proceeded as usual (**FIGURE 8.21A,B**), so injection by itself does not affect the process. But when the drug was injected into both eyes, blocking spontaneous activity in both retinas, further segregation of ocular input to LGN layers was completely blocked (**FIGURE 8.21C**). Ocular innervation of the LGN of treated ferrets, 9 days later, was no different than in 1-day-olds, so the maturation process had been completely halted. Inputs from both eyes were completely intermingled in the heart of the LGN. Injection of the drug into only one eye resulted in a different pattern: segregation of ocular input to LGN layers still took place, but projections from the silenced eye shrank dramatically while the other eye was enlarged (**FIGURE 8.21D**). In other words, the LGN regions innervated by the silenced eye were smaller than normal, while those from the unmanipulated eye expanded (Penn, Riquelme, Feller, & Shatz, 1998).

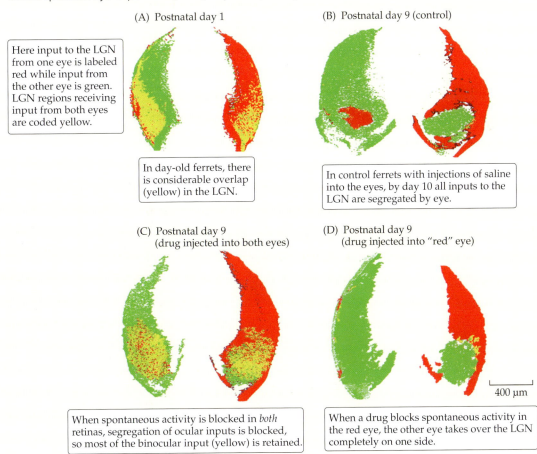

(A) Postnatal day 1

Here input to the LGN from one eye is labeled red while input from the other eye is green. LGN regions receiving input from both eyes are coded yellow.

In day-old ferrets, there is considerable overlap (yellow) in the LGN.

(B) Postnatal day 9 (control)

In control ferrets with injections of saline into the eyes, by day 10 all inputs to the LGN are segregated by eye.

(C) Postnatal day 9 (drug injected into both eyes)

When spontaneous activity is blocked in *both* retinas, segregation of ocular inputs is blocked, so most of the binocular input (yellow) is retained.

(D) Postnatal day 9 (drug injected into "red" eye)

When a drug blocks spontaneous activity in the red eye, the other eye takes over the LGN completely on one side.

400 μm

FIGURE 8.21 Ocular dominance innervation of ferret LGN
(From Penn et al., 1998.)

continued

■ **CONCLUSION**: Spontaneous activity in the retina is indeed important for segregation of ocular inputs to the LGN. The two eyes seem to compete for innervating the LGN, in such a manner that the more active afferents to the LGN are maintained while less active afferent inputs to the LGN are lost. In this experiment, the more active afferents would also be more successful in driving postsynaptic activity in the LGN. In normal animals, where both eyes are on an equal footing in this competition, each displaying an equal level of spontaneous activity, the various ocular bands in the LGN are more or less equal in extent.

Later studies using other toxins that block spontaneous retinal activity generally confirmed these findings. Mice with a knockout of the gene required for spontaneous activity in the retina also failed to segregate inputs to the LGN (Muir-Robinson, Hwang, & Feller, 2002). Blocking spontaneous retinal activity also disrupts development of ocular dominance columns in visual cortex (Huberman, Speer, & Chapman, 2006).

> In any LGN neuron that initially receives input from axons from two different eyes, those two inputs will almost never be active at the same time.

Note that these results fit well with what one would expect from a Hebbian process. There is a correlation in the electrical activity of neighboring retinal ganglion cells, but that means this spontaneous activity *between* the two eyes is completely uncorrelated. In other words, when axons from one eye are conducting action potentials, those from the other eye are not likely to be active (or, are no more or less likely to be active than at any other time). This asynchrony between the two eyes means that any time several action potentials from several different axons arrive at the LGN at about the same time, they are likely to be from the same eye. And it is only when several afferents are active at about the same time that the postsynaptic LGN neurons are likely to reach threshold and fire, so those synapses would be strengthened (they were successful at driving the LGN neuron). In contrast, in any LGN neuron that initially receives input from axons from two different eyes, those two inputs will almost never be active at the same time. That means that those synapses cannot *both* be successful at driving the LGN neuron. Only groups of synapses that come from the same eye are able to "gang up" on the LGN neuron to fire it (**FIGURE 8.22**). Thus each LGN neuron will come to be innervated by either one eye or the other.

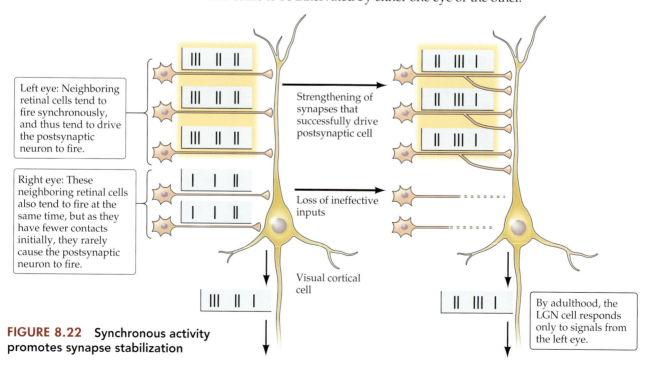

Left eye: Neighboring retinal cells tend to fire synchronously, and thus tend to drive the postsynaptic neuron to fire.

Right eye: These neighboring retinal cells also tend to fire at the same time, but as they have fewer contacts initially, they rarely cause the postsynaptic neuron to fire.

Strengthening of synapses that successfully drive postsynaptic cell

Loss of ineffective inputs

Visual cortical cell

By adulthood, the LGN cell responds only to signals from the left eye.

FIGURE 8.22 Synchronous activity promotes synapse stabilization

Why do the inputs from each eye tend to be congregated together in a particular region (in ferrets) or particular layers (in cats and primates)? One answer is that there is already some segregation of inputs from the two eyes at the very start of the competition, and this initial segregation must be due to axonal pathfinding of the sort we described in Chapter 5, where adhesive molecules start axons on the approximately correct trajectory, and recognition systems and synaptic preferences of the sort described in Chapter 6 correct minor errors. In other words, activity-independent mechanisms must get the axons to the LGN, and to the *approximately* correct portion of the LGN for each eye, and then spontaneous activity-dependent mechanisms fine-tune the connections to segregate input from the two eyes. Then yet another period of further fine-tuning will occur when not just spontaneous activity, but visual experience, puts the final touches on the visual system wiring, a topic we'll take up in the next chapter. Before that, we'll close this chapter by examining the consequences of synapse rearrangement throughout cortex, not just in V1.

How's It Going?

1. Describe the pattern of spontaneous activity in the retinas of fetal mammals.

2. What's the evidence that this spontaneous activity affects synapse rearrangement between the eye and the brain before birth?

3. How does the pattern of monocular or binocular innervation of layer IV in visual cortex change during development?

■ 8.9 ■

THE GRAY MATTER OF HUMAN CORTEX THINS AS WE MATURE

As we noted in Chapter 4, expansion of brain size often involves prolonging the later stages of development rather than the earlier stages, and so it is that the greater size of the human brain over other mammals is primarily in an expansion of the late-developing cerebral cortex. In fact, as we noted before, the human cerebral cortex at birth is still, in many ways, fetal-like, supporting the notion that our childhood represents prolonged fetal development outside the womb (see Figure 4.16A). In the first year of life, there is a dramatic expansion of dendrites from neurons in the cerebral cortex (**FIGURE 8.23**). Of course that means there's also an expansion of synapses onto those dendrites (Conel, 1939, 1947, 1959), which means there's also a proliferation of axon terminals that we cannot see in preparations like the one in Figure 8.16, which reveals only dendrites.

To count synapses directly, scientists have to resort to electron microscopy, because synapses are too small to detect readily with light microscopes. Electron microscopy to count synapses is rather tedious work, as each photomicrograph represents a tiny fraction of the whole

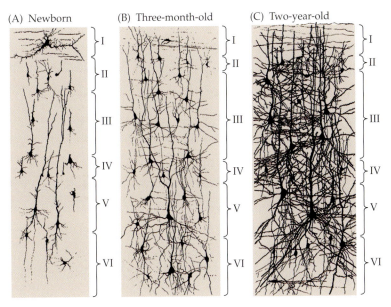

FIGURE 8.23 **Proliferation of dendrites in human cerebral cortex** (From Conel, 1939, 1947, 1959.)

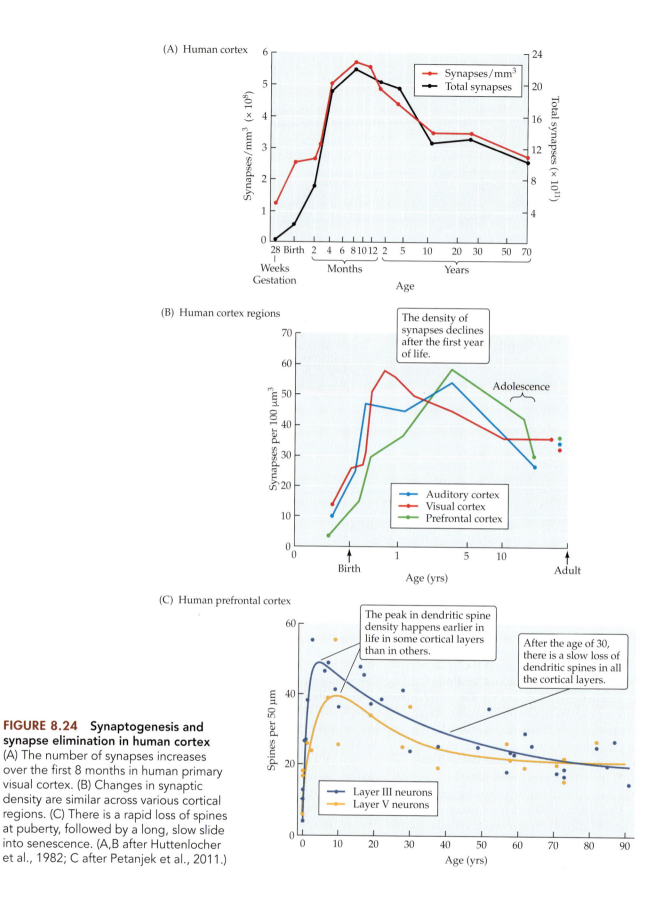

FIGURE 8.24 Synaptogenesis and synapse elimination in human cortex (A) The number of synapses increases over the first 8 months in human primary visual cortex. (B) Changes in synaptic density are similar across various cortical regions. (C) There is a rapid loss of spines at puberty, followed by a long, slow slide into senescence. (A,B after Huttenlocher et al., 1982; C after Petanjek et al., 2011.)

volume, so scientists must also resort to sampling to generate reliable numbers. Happily for us, several neuroscientists undaunted by these challenges showed that, as the views of dendrites indicated, there is tremendous postnatal growth in the numbers of cortical synapses, in terms of both density of synapses per cubic millimeter and total number of synapses (**FIGURE 8.24A**).

(A) Human cortex

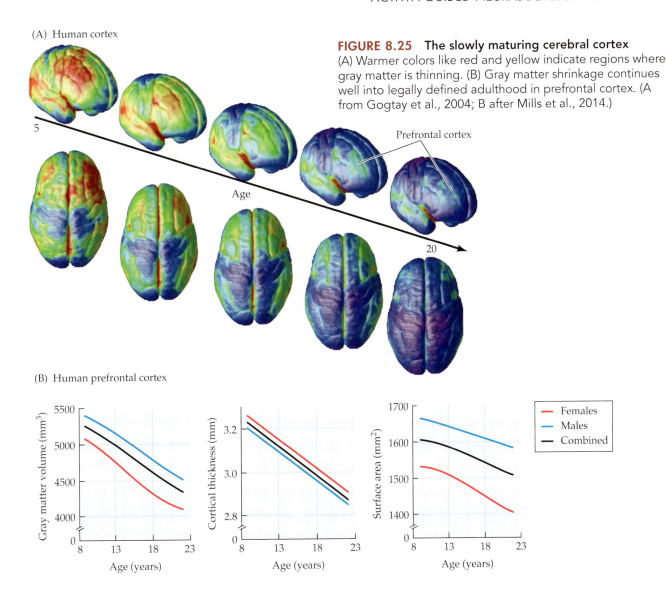

5

Age

Prefrontal cortex

20

FIGURE 8.25 **The slowly maturing cerebral cortex**
(A) Warmer colors like red and yellow indicate regions where
gray matter is thinning. (B) Gray matter shrinkage continues
well into legally defined adulthood in prefrontal cortex. (A
from Gogtay et al., 2004; B after Mills et al., 2014.)

(B) Human prefrontal cortex

Females
Males
Combined

But also note that the peak in the number of synapses in primary visual
cortex is reached just before the first birthday, after which there is a prominent
decline in total synapses over the next 10 years or so. In other regions of cortex,
the peak in synaptic density may occur a little later (**FIGURE 8.24B**), but the
overall pattern is the same—we have our peak number of synapses in childhood, followed by a sharp decline until adulthood. Note that the decline in
synaptic density in the prefrontal cortex (green graph in Figure 8.24B) occurs
during adolescence, well after the decline in visual cortex. Some scientists
have asked even more refined questions about the number of synapses, such
as the density of dendritic spines in various layers of prefrontal cortex. These
also indicate a sharp rise in synapses until ages 4–10 or so, followed by a
pronounced decline until adulthood. These studies also give insight into the
aging process after adulthood, because there is also a much more gradual loss
of dendritic spines from age 30 to 90 (**FIGURE 8.24C**).

The net loss of synapses after childhood has one consequence that can be
visualized by noninvasive methods like MRI—namely a thinning of the overall
layer of **gray matter** on the outer surface of the cortex. As **FIGURE 8.25** shows,
there is rapid shrinkage of gray matter in pretty much all the cortical regions
in the first 12 years of life. Note that after that, most cortical regions are stable
with the exception of the frontal lobe, which continues to mature from age 16

gray matter The outer portion
of the vertebrate cortex predominated by cell bodies, rather than
myelin, hence dark in color in
postmortem preparations.

FIGURE 8.26 **Activity-dependent myelination** (A after Fields, 2015; B after Mensch et al., 2015.)

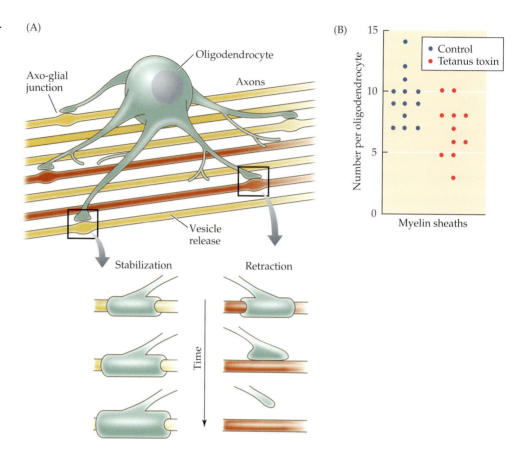

(A)

Oligodendrocyte

Axo-glial junction

Axons

Vesicle release

Stabilization

Retraction

Time

(B)

Number per oligodendrocyte

- Control
- Tetanus toxin

Myelin sheaths

Most cortical regions are stable with the exception of the frontal lobe, which continues to mature from age 16 to 20.

white matter The inner portion of the vertebrate brain, consisting primarily of myelinated axons coursing to or from the cerebral cortex to other brain regions, hence light in color in postmortem preparations.

prefrontal cortex The portion of the vertebrate cortex found at the extreme anterior pole.

to 20. Even at age 20, there is continued loss of synapses and shrinkage of gray matter in the frontal lobe, including prefrontal cortex. While overall cerebral gray matter is declining from 10 years of age to adulthood, the volume of **white matter**, the myelinated axons coursing to and from the cortical gray matter, is increasing, as you'd expect since overall brain size is pretty stable after age 15 or so.

This change in white matter with age is also affected by neuronal activity. Several studies have documented changes in myelination correlating with specific behaviors, such as learning to read or play the piano, or making executive decisions (Fields, 2015). The best evidence that these changes in myelination are regulated by neuronal activity come from studies of zebrafish, where scientists were able to observe the initial contact between an oligodendrocyte and an axon in vivo (**FIGURE 8.26A**). Interestingly, often when an oligodendrocyte process contacts an axon, it is soon withdrawn again. But if action potentials are blocked in the axon using tetanus toxin, such withdrawals are even more common (**FIGURE 8.26B**) (Mensch et al., 2015). In other words, the extent of myelination is regulated by the electrical activity of the axon such that active axons are more likely to become myelinated than inactive axons.

The relatively late maturation of frontal cortex in particular has been offered as an explanation of why teenagers, whose bodies may appear relatively mature, nevertheless sometimes show poor judgment in governing their own behavior. The frontal lobe, especially the rostral-most **prefrontal cortex**, has repeatedly been associated with impulse control (Kane & Engle, 2002; Sebastian et al., 2014). Early support came from the famous nineteenth-century case of Phineas Gage, who despite exaggerations that have been written about his behavior after an iron rod damaged his frontal lobes, clearly had a more impulsive personality following the accident (Macmillan, 2000). In modern clinical studies, damage to prefrontal cortex impairs measures of attention and executive function, such as being able to shift strategies in the Wisconsin card sort task (Kane & Engle, 2002), where you are asked first to sort cards by color, then by shape, etc. People with damage to prefrontal cortex have a hard time preventing themselves from sorting by the previous category.

Data such as these have been cited in questions of whether children should be judged in the same way as adults in weighing legal responsibility and punishment. Can we blame a person for having an immature prefrontal cortex? Even before the U.S. Supreme Court took up the case of Evan Miller, whom we discussed at the start of the chapter, they had noted that "developments in psychology and brain science continue to show fundamental differences between juvenile and adult minds"—for example, in "parts of the brain involved in behavior control" (Steinberg, 2013).

In the ruling for *Miller v. Alabama* (2012), the Supreme Court did not rule that giving a life sentence for a crime that someone committed as a 14-year-old was necessarily "cruel and unusual punishment," but it did take issue with the fact that the judge, in setting the sentence, was not legally allowed to consider the offender's age, underprivileged background, or chance for rehabilitation. Thus they struck down mandatory sentences of life in prison for crimes committed by juveniles.

Justice Elena Kagan, writing for the 5-4 majority, said, "Mandatory life without parole for a juvenile precludes consideration of his chronological age and its hallmark features—among them, immaturity, impetuosity, and failure to appreciate risks and consequences…. It prevents taking into account the family and home environment that surrounds him—and from which he cannot usually extricate himself—no matter how brutal or dysfunctional" (*Miller v. Alabama*, 2012, paragraph II).

In Evan Miller's case, the Supreme Court ruled that the original judge for the case must set a new hearing to determine the appropriate sentence, and that the judge is free to consider these various factors when determining how long Evan, now 26 years old, must remain in prison. As of this writing, 4 years after the Supreme Court demanded Evan receive a new sentencing hearing, it has not yet happened.

Of course, thinning of our cortex is not the only thing changing as we age from 14 to 21—so do our circumstances in terms of social relationships, schools, jobs, and the inevitable accidents of life. One change that can be a matter of life and death is that, despite the fact that our reaction time declines as we age (Olson, 1991), we become more careful drivers once we leave the teenage years behind us (**FIGURE 8.27**). For most of us, the teenage years are a time of intense learning about people and how the world works, and surely those experiences play an important role in determining which synapses are lost or retained, weakened or strengthened, as our gray matter gets thinner and the white matter grows.

In the following chapter we'll see many examples where we know experience, sensory stimulation, has a profound effect on the strength and retention of synapses. Much of that discussion will again center on the visual system, where visual experience will prove crucial for proper construction of neural circuits that let us see.

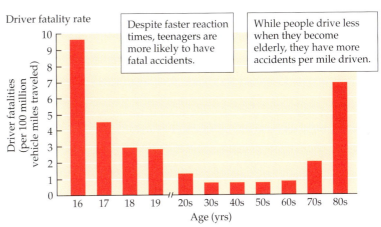

FIGURE 8.27 Evidence of impulsiveness? (After Fatality Analysis Reporting System of the National Highway Traffic Safety Administration, n.d.)

SUMMARY

■ Following neuronal apoptosis, another regressive process is the loss of many synapses during *synapse rearrangement*. The loss of some synapses is usually accompanied by the production of new synapses, so depending on the system and stage of development, there may be a net gain or loss of total synapses.

■ Neuronal activity plays a role in early developmental events such as neuronal apoptosis, as when motor neuron death occurs only if the target muscles are electrically active. **See Figure 8.1**

■ In newborn muscles, every fiber is *polyneuronally innervated*, but most of the initial contacts from each motor neuron are retracted over time until each fiber is innervated by one and only one neuron. This is a competitive process that occurs only if the junctions are electrically active. Presynaptic terminals that activate the muscle fiber are more likely to remain than those that are ineffective. **See Figures 8.2–8.5**

■ In autonomic ganglia, an initially exuberant pattern of polyneuronal innervation is followed by a loss of most inputs until each neuron receives extensive synaptic input from a relatively small number of afferents, sometimes a single afferent. **See Figures 8.8–8.10**

■ *Hebbian synapses* are strengthened or duplicated when they effectively drive the postsynaptic cell to fire, and they are weakened or withdrawn when they are ineffective. **See Figure 8.11**

■ Several pathways in the *hippocampal formation* function in a Hebbian fashion, as when driving excitatory afferents though a *tetanus* results in a long-lasting enhancement of the synapses, called *long-term potentiation* (*LTP*). **See Figures 8.12 and 8.13**

■ In several hippocampal pathways, LTP is mediated by *NMDA*-type glutamate receptors. When the postsynaptic neuron is depolarized, as happens after a tetanus, a magnesium block on the NMDA receptor's channel is lifted, such that glutamate triggers an inrush of Ca^{2+} ions, which leads to insertion of additional *AMPA*-type glutamate receptors in the postsynaptic site and eventually an increase in glutamate release from the presynaptic terminal. **See Figures 8.14 and 8.15**

■ In the mammalian fetus, both eyes innervate wide regions of the *lateral geniculate nucleus* (*LGN*) of the thalamus, but as development proceeds, each eye comes to innervate separate, *monocular* domains of the LGN. The process of segregation of inputs to the LGN depends on waves of spontaneous activity that cross the retinas, which drive synaptic competition, following Hebbian rules, between the eyes. **See Figures 8.16, 8.17, and 8.21**

■ Similarly, information from both eyes is broadly distributed across all of *layer IV* in *primary visual cortex* (*V1*) at first but then becomes segregated such that ocular dominance bands result, with each cortical neuron in layer IV receiving monocular input. The segregation of ocular dominance bands in layer IV may also depend on a Hebbian-type synaptic competition between the two eyes, but if so, it would be in response to spontaneous activity, not visual experience, because it occurs before birth in primates. In the next chapter we'll see that almost all neurons in visual cortex outside layer IV receive *binocular* input, a process that depends on Hebbian-like competition driven by visual experience. **See Figures 8.18–8.20, and 8.22**

■ The number of synapses in the human cortex increases rapidly until the first few years of life, then declines sharply until adulthood. The loss of synapses from adulthood to old age is much more gradual. **See Figures 8.23 and 8.24**

■ The net loss of synapses from childhood to adulthood can be gauged by the thickness of the *gray matter*, which becomes progressively thinner until about 20 years of age. Among cortical regions, the last to complete the thinning of gray matter is the *prefrontal cortex*. The prefrontal cortex has been implicated in executive control and inhibition, so its late development may account for the relatively impulsive behavior of adolescents. Myelination of the cortex increases during adolescence, and there is evidence that neuronal activity regulates this process. **See Figures 8.25 and 8.26**

Go to the Companion Website
sites.sinauer.com/fond
for animations, flashcards, and other review tools.

9

Fine-Tuning Sensory Systems

EXPERIENCE-GUIDED NEURAL DEVELOPMENT

LEARNING TO SEE A chemical explosion destroyed 3-year-old Michael May's left eye and damaged the surface of his right eye so badly that he was blinded. Mike could tell whether it was day or night, but otherwise he couldn't see anything. An early attempt to restore his sight with corneal transplants failed, but Mike was undaunted. He learned to play Ping-Pong using his hearing alone by learning to interpret the sound cues. Mike also enjoyed riding a bicycle, but his parents made him stop after he crashed first his brother's bike, then his sister's.

As an adult, Mike became a champion skier, marrying his instructor and raising two sons. He also started his own company, making equipment to help other blind people navigate on their own. When Mike was 46, technical advances made it possible to restore vision in his right eye. As soon as the bandages were removed, he could see his wife's blue eyes and blond hair. But even years later, he cannot recognize her face unless she speaks to him, or recognize three-dimensional objects like cubes or spheres unless they are moving. Mike can still ski, but he finds he has to close his eyes to avoid falling over. On the slopes, seeing is more distracting than helpful. Why does Mike have a hard time using his restored vision?

CHAPTER PREVIEW

Mike May seems to provide a modern answer to a seventeenth-century query that came to be known as **Molyneux's problem**: whether a person born blind could, upon being made to see as an adult, recognize objects by use of his newfound sight alone. As we saw in the Interlude, empiricist philosopher John Locke agreed with Molyneux that such an adult, suddenly made to see, would not be able to distinguish a cube from a sphere based on sight alone, at least at first, until he had enough visual experience to learn what cubes and spheres look like. What was originally a thought experiment soon became, with the advent of techniques to surgically remove lenses of the eye clouded by cataracts, an empirical question that was addressed several times over the centuries (Cheselden, 1728; Gregory & Wallace, 1963;

Michael May with his child
(© Florence Low.)

Molyneux's problem A famous thought experiment that asks whether an adult enabled to see for the first time would be able to make sense of his or her vision.

Sikl et al., 2013). In general, the reports confirmed Molyneux and Locke's prediction that the person would not be able to distinguish objects based on sight alone at first. But we'll see that modern examples, like Mike May, show that even after years of visual experience in adulthood, these individuals still have a hard time distinguishing objects and have almost no ability to distinguish faces.

In this chapter we'll come to understand why restoring the visual image to the retina in adulthood is not enough to ensure normal vision, even after years of visual experience. We'll see that visual experience early in life exerts a powerful influence on the development of synaptic connections between the eye and the brain. Without this experience-driven guidance, the visual system cannot function properly. In frogs, cats, monkeys, and people, visual experience is crucial for the development of proper synaptic connections in the brain and, therefore, sight. And for the mammals, at least, the older we get, the less able the brain is to make use of that instructive experience to provide us with the marvel of sight. Nor is vision the only sense that relies on experience for proper development. We'll see that hearing and odor detection also rely on experience to guide synaptic connectivity and that even in adulthood experience normally maintains brain connections underlying the sense of touch.

In earlier chapters we learned about signaling systems that guide axons to the approximately correct target, independent of activity, and such processes are at work in the developing visual system (Hagihara, Murakami, Yoshida, Tagawa, & Ohki, 2015). But then in Chapter 8 we saw instances when natural selection produced mechanisms by which neural activity, including spontaneous activity, could affect the fate of synapses, serving to fine-tune those approximate connections, for example by segregating eye inputs to the lateral geniculate nucleus (LGN) and layer IV of visual cortex. The development of Hebbian synapses, which are strengthened when they successfully drive the postsynaptic neuron and weakened if they rarely drive the postsynaptic neuron, made such activity-guided development possible. Once that mechanism arose, natural selection could exploit it to guide sensory development, as we'll discuss in this chapter. We'll see that those same Hebbian synapses, now driven by sensory experience rather than spontaneous activity, are crucial for the development of sensory systems. In Chapter 10, we'll see how other individuals can then tap into those sensory systems to further guide neural development through social influences such as language.

■ 9.1 ■

HUMANS CAN ADAPT TO SEEING THE WORLD IN A NEW WAY

As an undergraduate in philosophy at the University of California in Berkeley, George Stratton (1865–1957) became intrigued by the new subfield of psychology. He went on to get his doctorate at the University of Leipzig, where Wilhelm Wundt had established the first psychology research laboratory. Stratton returned to Berkeley to teach in the philosophy department, eventually helping to found a new department of psychology.

For his PhD thesis in Leipzig, Stratton had built a pair of glasses with prisms that reversed the view left to right and made objects appear upside down. It was too hard to fuse the images from both eyes with the glasses, so he covered one eye entirely and tried to move about the house with vision in his open eye reversed. To avoid any chance of normal vision during the experiment, Stratton kept both eyes closed when he removed the glasses at night, sleeping with a blindfold until putting the glasses back on the next morning. His movements were clumsy at first, but after 3 days he was able to move about his home without bumping into things (Stratton, 1896). Initially, he was constantly aware that his vision had been inverted, but toward the end of the experiment, Stratton noticed that when he attended to his vision, his experience was that his vision was normal, but he felt as if his *head* was in an odd position, as if he were looking between his legs (Stratton, 1896)! When he removed the glasses after 3 days, vision immediately seemed normal and his motor capacity was unaffected.

On his return to Berkeley, Stratton extended the experiment to 8 days, venturing outside to walk around his garden and the "village." Again his movements were clumsy at first—pouring a glass of milk was a matter of trial and error. A few days later Stratton noted strange experiences—looking at a fireplace, he felt he was seeing it out of the back of his head (Stratton, 1897)! By the seventh day, Stratton felt like his hands and feet were properly oriented in this reversed visual scene, and he had no problem moving about, reaching for objects, etc. What changed in the course of the experiment was his *experience of his body*—his experience of where his hands and feet were and in which direction he was moving his eyes. This new experience of his body meshed well with the (inverted) visual scene, which is why movement was easy. Whenever he forced himself to recall where his various limbs really were, he felt like he was looking at a scene with his body upside down (Stratton, 1896). Removing the lenses 8 days later, Stratton found the scene bewildering for several hours, although he did not feel that things were upside down. For a while, he was at a loss about which hand to use to grasp a door handle.

Later researchers continued experiments such as these, some managing to reverse vision in both eyes at once (**FIGURE 9.1**). Eventually one even rode a bicycle about town with prisms on (Ewert, 1930)! Thus it was established that we can learn to use visual information even after it has been reoriented drastically. For one thing, this explained why we normally don't experience our vision as reversed from reality, even though the image of the world on the retina is reversed and flipped by the lens at the front of the eye. Having grown up with this visual input, we learn to move and experience our bodies in congruence with that input, and it's interesting that we can unlearn that relationship, and learn a new one, after just a week of wearing prisms. From a neuroscience perspective, this adaptation cannot be due to any change in the connections from the eye to the brain, because the adaptation taking place in just a few days, and the even faster return to normal vision when the prisms are removed, is much too fast to allow wholesale changes in existing connections. Rather, the adaptation must take place at higher levels of the brain, regions that *interpret* the information from the eyes differently, probably by rapid changes in the strengths of existing synapses that take part in analyzing visual input.

Can other animals learn to reinterpret such information from the eyes?

FIGURE 9.1 **A dashing young scientist about town** It's not clear whether he also wore a suit while riding a bicycle. (Photo from Ewert, 1930.)

This new experience of his body meshed well with the (inverted) visual scene, which is why movement was easy.

How's It Going?

What were the conclusions of experiments with reversing prisms conducted by George Stratton and others?

■ 9.2 ■

RETINAL GANGLION CELLS IN ADULT AMPHIBIANS AND FISH CAN REESTABLISH CONNECTIONS TO THE TECTUM

The ability of people to compensate for reversed vision led to the question of whether other animals could do this as well. Rather than trying to affix lenses in front of the eyes of animals, Roger Sperry instead used surgery. By severing the connections around the perimeter of a newt's eye, but carefully keeping the optic nerve intact, then rotating the eye 180° and suturing it back in place in this reversed position, he effectively accomplished the same sort of reversal. Now visual stimuli above the newt, which had previously fallen upon the ventral retina, instead fell on what had been the dorsal portion of the retina. Immediately after recovering from the anesthetic, the animal's behavior made it clear that a visual reversal had been accomplished. Offered a mealworm overhead, the hungry animal sent its tongue down instead of up (see Figure 5.18A). Likewise, visual stimuli on the newt's left fell on the part of the retina that had formerly been stimulated by objects on the right, and so on. Sperry would later do similar experiments in frogs and fish and see the same result: the animal's behavior made it clear that the visual reversal had been accomplished. Would these animals eventually compensate for this reversed vision and effectively use their visual system? No. Even after four months, amphibians with this reversal still could not use the eye properly, and if the eye was surgically reversed back into the original position, they were able to use the eye properly as soon as they recovered from anesthesia (Sperry, 1943). Thus, unlike people who experience reversed vision, newts, frogs, and fish never seemed to learn to compensate for the reversed visual input.

> Unlike people who experience reversed vision, newts, frogs, and fish never seemed to learn to compensate for the reversed visual input.

This result alone is interesting. Note that in the preparations described so far, the connections from the animal's eye to the brain were left intact, just as in the humans wearing prisms. But whereas humans appear to use higher brain centers to readjust behavior to fit the reversed visual input, the amphibians do not. In fact, when Stratton took off his prisms after 8 days, he was able to function pretty well within just a few hours, too fast for any rewiring of connections. But amphibians were unable to adapt to visual reversal even after months.

Recall from Chapter 5 that Sperry went much further in his experiments with fish and amphibians, because they, unlike mammals, show regeneration after severing of the optic nerve. If the optic nerve is cut in a fish or frog, the animal is of course blind in that eye at first, but in the course of a few weeks, its behavior demonstrates that it can again use that eye to navigate its world. As we noted in Chapter 5, when the optic nerve is cut and the eye is reversed, the retinal ganglion cells that reinnervate the tectum do so by reestablishing their former pattern of connection. This meant, among other things, that the animal's behavior was altered by the reversal, just as if the eye had been reversed without cutting the optic nerve. And as when the eye was reversed without cutting the optic nerve, the animals with regenerated connections from the reversed eye never learned to compensate for that reversal.

You might ask why this would be important if it hinged on a regeneration ability that humans don't have. But of course one reason to study regeneration that is possible only in other species is to figure out why it's not possible for us. If you were to understand that, you might be able to come up with an intervention that would permit regeneration in humans, too. Furthermore,

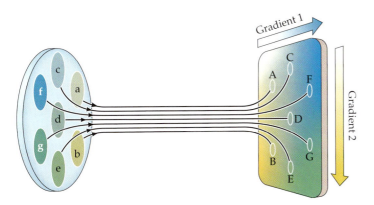

FIGURE 9.2 Possible versions of chemoaffinity

there was the hope that the processes at work in guiding the regenerating optic system to connect properly to the *adult* brain might be a recapitulation of whatever forces guide retinal axons to find their way in the first place, during *development*. It was relatively easy to use physiological recordings of tectal neurons' responses to light stimuli to map out the projections of the visual field (and therefore of the retina) in a single afternoon's recording session. Such mappings were nearly impossible in embryos.

Also recall from Chapter 5 that to explain the ability of retinal ganglion cells to reestablish their original pattern of connections to the tectum, Sperry proposed the **chemoaffinity hypothesis**, that each retinal neuron had a specific chemical identity that caused it to synapse only with a neuron in the appropriate portion of the tectum. The retinal cell would seek out the tectal cell with the correct "address." The chemical signatures to match retinal cells with tectum could consist of many different chemical cues or could be as simple as consisting of perpendicular gradients of just two cues that provide each tectal cell with a unique combination of the two factors (**FIGURE 9.2**). In Chapter 5, we saw that at least one cue to guide reinnervating retinal axons is the distribution of ephrins that interact with ephrin receptors found in the retinal growth cones. A gradient of ephrin expression in the tectum is matched by a gradient of ephrin receptor expression in the retina, and the advance of growth cones is stopped when they match (see Figure 5.20).

These findings about ephrins certainly vindicate Sperry's chemoaffinity hypothesis about regeneration of retinotectal innervation. After regeneration, the original two-dimensional map of visual space falling upon retinal ganglion cells (**FIGURE 9.3A**) is once again topographically projected by their axons onto the tectum, but because the eye has been rotated, the projection of visual space onto the tectum has been rotated, too (**FIGURE 9.3B**).

The ability of fish and amphibian to regenerate optic nerves captured the imagination of a generation of neuroscientists who tried to push the boundaries of what chemoaffinity could accomplish. But as scientists tried ever more elaborate manipulations to test the chemoaffinity hypothesis, it soon became clear that there were some instances where a strict chemoaffinity alone could not explain the pattern of reinnervation, as we discuss next.

chemoaffinity hypothesis
Roger Sperry's proposal that axonal growth cones seek out a particular target cell based on chemical signals that mark both. It explains how regenerating retinal cells manage to reestablish their synaptic contacts with tectal cells.

How's It Going?

What is the chemoaffinity hypothesis, and how does it relate to optic nerve regeneration?

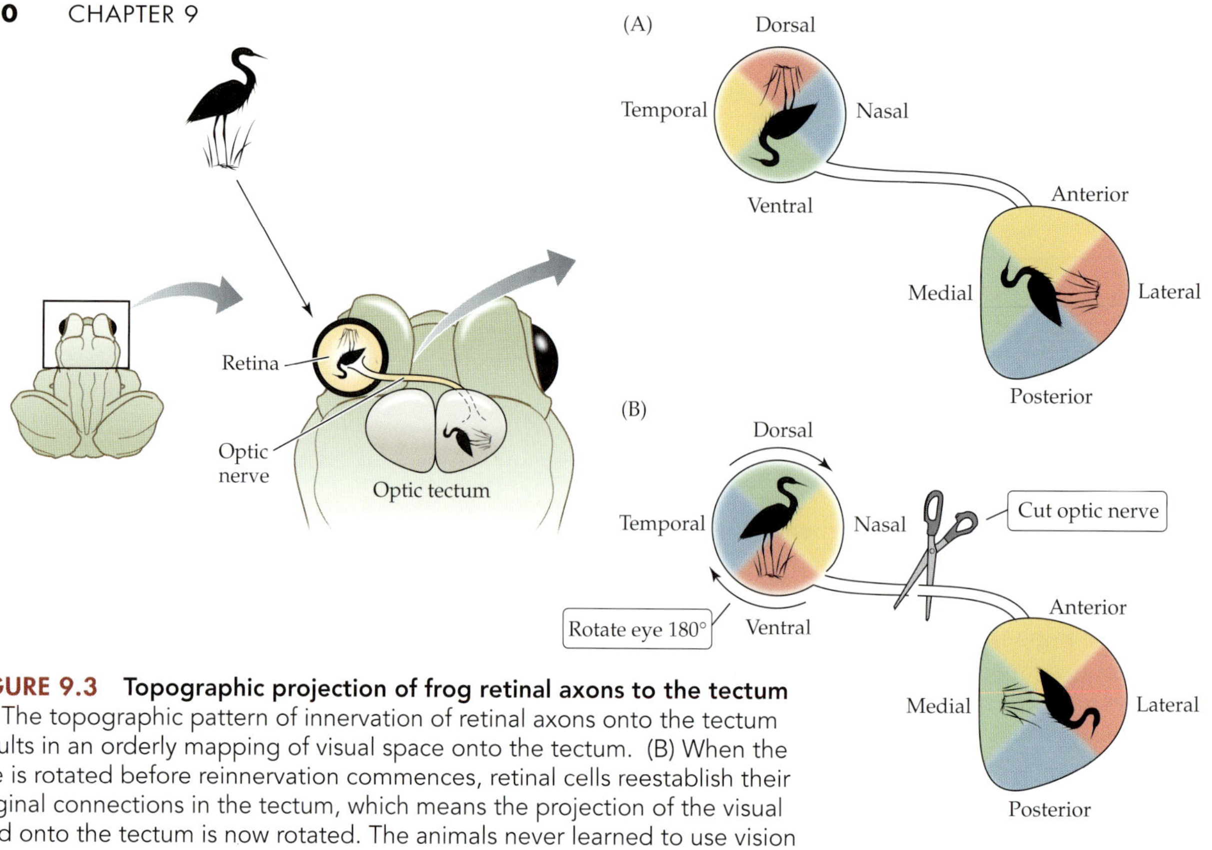

FIGURE 9.3 **Topographic projection of frog retinal axons to the tectum** (A) The topographic pattern of innervation of retinal axons onto the tectum results in an orderly mapping of visual space onto the tectum. (B) When the eye is rotated before reinnervation commences, retinal cells reestablish their original connections in the tectum, which means the projection of the visual field onto the tectum is now rotated. The animals never learned to use vision in that eye to localize visual stimuli.

■ 9.3 ■

VARIOUS PERMUTATIONS OF RETINOTECTAL REGENERATION REFUTE A STRICT VERSION OF CHEMOAFFINITY

To test the limits of the chemoaffinity hypothesis, ever more elaborate permutations on Sperry's original experiments of cutting the optic nerve were conducted. Scientists tried grafting two ventral halves of eyes where the original eye was, or grafting two temporal halves, or lesioning half the retina, or lesioning half the tectum, and so on. These experiments quickly disproved a strict interpretation of chemoaffinity, because if you challenged the system, you could in fact persuade retinal ganglion cells to innervate tectal neurons that they would never normally innervate. Note that this state of affairs resembles the issue of "commitment" of cell fate in Chapter 1. To say that regenerating retinal cells are committed to reinnervate their original tectal targets has no meaning except in the context of what challenge(s) you present. If you simply cut the optic nerve, or cut the optic nerve and reverse the eye, then retinal cells reinnervate their original tectal targets. But if you present a greater challenge, then the retinal neurons will innervate new tectal targets, as we'll see.

In one variant of retinotectal regeneration, the optic nerve was sectioned and then half the retina was lesioned or surgically removed. It was thought that if the remaining retinal neurons followed a strict chemoaffinity pattern, then they should innervate the half of the tectum they originally innervated, leaving the other half of the tectum uninnervated. But what actually happened was that the remaining retinal cells spread out their

> If you challenged the system, you could in fact persuade retinal ganglion cells to innervate tectal neurons that they would never normally innervate.

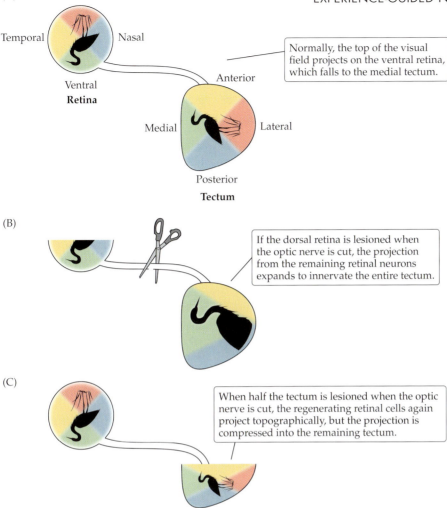

Normally, the top of the visual field projects on the ventral retina, which falls to the medial tectum.

(B)

If the dorsal retina is lesioned when the optic nerve is cut, the projection from the remaining retinal neurons expands to innervate the entire tectum.

(C)

When half the tectum is lesioned when the optic nerve is cut, the regenerating retinal cells again project topographically, but the projection is compressed into the remaining tectum.

FIGURE 9.4 **Tectal expansion and compression indicate a flexible type of chemoaffinity.**

projection pattern, each cell innervating about twice as many tectal neurons. Importantly, the topographic pattern of retinal innervation was preserved (nasal retina projected to anterior tectum, dorsal retina to medial tectum, etc.), but now the projection was expanded (Schmidt, 1978) (**FIGURE 9.4A,B**). This restoration of the pattern of innervation certainly conformed to the chemoaffinity hypothesis, but the fact that some retinal cells were innervating tectal cells that they normally would not indicates a certain degree of flexibility in that pattern of innervation.

In a converse class of experiments, the optic nerve was cut and then half the tectum was lesioned. If the retinal ganglion cells strictly reconstructed their original innervation pattern, the half of the retinal cells that normally innervated the remaining half of the tectum should reestablish their innervation, leaving the other half of the retinal cells without a target. But in fact, in these experiments all the retinal neurons eventually innervated the remaining tectum, and again the topographic pattern of the original innervation was reestablished, but that projection was now compressed to fit within the remaining tectum (Meyer & Wolcott, 1987; Schmidt, 1983) (**FIGURE 9.4C**). Together, these types of experiments, where the retinal topography was altered, were referred to as "tectal expansion" and "tectal compression" experiments.

At least some of the plasticity in projections seen in tectal expansion and compression experiments is independent of neural activity in the eye, because topographic retinotectal projections form even if the retinal neurons are silenced with tetrodotoxin (TTX; recall that this toxin blocks voltage-gated sodium channels and so stops the conduction of action potentials) (Meyer & Wolcott, 1987). Presumably gradients such as the ephrins, which we discussed in Chapter 5 (see Figure 5.20), guide retinal axons to the generally correct portion of the tectum to reestablish the original pattern of innervation. However, it soon became clear that the initial projections of regenerating retinal axons, and those established during activity blockade such as with TTX, were less refined than those established when the retinal neurons were active (Schmidt & Edwards, 1983). Likewise, keeping the recovering animals in darkness to reduce retinal activity, or keeping them in darkness with intermittent exposure to strobe lights, which stimulates all retinal cells at the same time, resulted in roughly correct but unrefined retinotectal mapping (Schmidt & Eisele, 1985).

> Chemoaffinity can account only for the initial, roughly correct topographic map of retina onto the tectum.

Thus it appears that chemoaffinity can account only for the initial, roughly correct topographic map of retina onto the tectum. After that, electrical activity of the retina, in which neighboring ganglion cells are likely to fire at about the same time in response to visual stimuli, sharpens the grain of the map (Olson & Meyer, 1994). As in Chapter 8, when we learned about the impact of spontaneous activity in the retina before birth, these results indicate a Hebbian synaptic process, whereby neighboring retinal cells, firing more or less synchronously, sum together to fire tectal targets, thereby solidifying those connections at the expense of other synapses. With guidance factors getting retinal axons to the approximately correct portion of the tectum, this later, activity-driven competition between retinal inputs sorts them out, refining projections to fine-tune that map.

This scenario explains why tectal expansion and compression happen: if there are fewer retinal cells competing, then the remaining neurons expand their projection to the tectum. When fewer tectal targets are available, then the retinal neurons must compete over less space and so form a compressed map onto the tectum.

The ability of regenerating retinotectal connections to adapt, reestablishing a topographic projection but altering the precise pattern of which particular retinal cell innervates which particular tectal cell, might have been dismissed as an odd result of the rather bizarre preparations (how often in nature do frogs suffer a severed optic nerve and live long enough for it to regenerate?). But then workers realized that the natural development of the animals *requires exactly this sort of flexibility in the innervation pattern*. In fish and amphibians, unlike mammals and birds, animals can continue to grow long after they reach sexual maturity. Thus each year the animal's eyes grow larger, which includes adding more photoreceptors and related neurons to the retina (Kaslin, Ganz, & Brand, 2008). For example, a sexually mature goldfish may quadruple body length as it grows further, which results in a sixfold increase in retinal area (Johns & Easter, 1977).

These additional patches of retina are always added to the *periphery* of the retina, representing an outermost ring of new neurons (Stenkamp, Barthel, & Raymond, 1997). That means, of course, that new neurons are added to all quadrants of the eye—dorsal, ventral, nasal, temporal (Otteson & Hitchcock, 2003). There is also an ongoing addition of neurons to the tectum as the animal grows, but those additional neurons are added *exclusively to the posterior tectum* (Raymond & Easter, 1983; Straznicky & Gaze, 1972) (**FIGURE 9.5**). No matter what the age of the animal, that topographic projection of retina to tectum is preserved, but if year by year more retina is added to the periphery, while only posterior tectum is added, then there must be a continual shifting of individual retina-to-tectal neuron

> There must be a continual shifting of individual retina-to-tectal neuron innervation to preserve the overall projection as the animal grows.

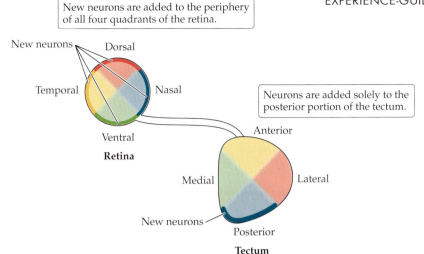

New neurons are added to the periphery of all four quadrants of the retina.

New neurons

Dorsal

Temporal

Nasal

Ventral

Retina

Neurons are added solely to the posterior portion of the tectum.

Anterior

Medial

Lateral

New neurons

Posterior

Tectum

FIGURE 9.5 Normal growth of the retinotectal system requires a constant shifting of the map. Retaining the topographic map of retina onto the tectum requires a subtle shifting of connections, so old retinal ganglion neurons must be innervating different tectal neurons than before.

innervation to preserve the overall projection as the animal grows (Gaze, Keating, Ostberg, & Chung, 1979).

How can the topographic projection be maintained while the individual connections shift as the animal grows? The answer to that question was first hinted at when scientists managed to create an even more bizarre frog, the first one in history that had two eyes innervating a single tectum, as we'll see next.

How's It Going?

How do tectal expansion and compression experiments, as well as the normal growth of retinotectal projections, force an amendment to the chemoaffinity hypothesis?

■ 9.4 ■

VISUAL EXPERIENCE FINE-TUNES FROG RETINOTECTAL CONNECTIONS

In frogs, as in many species subject to predation, the eyes are directed to the sides in order to provide as wide a visual field as possible (**FIGURE 9.6A**), presumably to make it easier to detect any predators that might approach. This arrangement means there is very little overlap in the visual fields of the two eyes, and almost all the optic nerve fibers will cross the midline at the optic chiasm. Thus each tectum gets input almost exclusively from a single source, the contralateral eye. As we'll see a bit later in this chapter, predatory species like cats and humans tend to have their eyes directed forward, so there is ample overlap in the visual fields of the two eyes, and about half the axons in each optic nerve, specifically those from the temporal retina, do not cross the midline (**FIGURE 9.6B**). Having our eyes directed forward in this manner provides important binocular cues to judge how far away objects are. We sacrifice a wider field of view to get better depth perception. In both prey and predator species, the net result is the same: the left visual field projects to the right brain while the right visual field projects to the left side of the brain.

In yet another variation of a test of Sperry's chemoaffinity hypothesis, researchers asked what would happen in a situation that would be truly bizarre in frogs—when two eyes innervate the same tectum. Would retinal axons still form a topographic map of visual space on the tectum, and would the homologous portions of the two retinas innervate the same tectal targets? As with tectal expansion and compression experiments, the outcome of having two retinas innervate a single tectum again argued against a strict version of chemoaffinity. What's more, the outcome reinforced the notion that visual experience matters, as we show next.

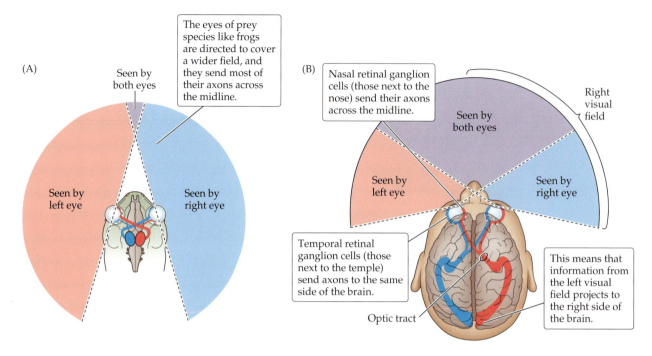

FIGURE 9.6 Placement of the eyes varies across species. Note that in both cases, the left visual field projects to the right side of the brain, and vice versa.

RESEARCHERS AT WORK

Three-Eyed Frogs Show Us the Way

■ **QUESTION**: What happens when two eyes are directed to innervate a single tectum in frogs?

■ **HYPOTHESIS**: If both eyes follow a strict version of chemoaffinity, then homologous regions of the two retinas will innervate the same tectal targets, and both eyes will reproduce the typical topographic map of visual space, overlapping each other.

■ **TEST**: In embryonic frogs, before the stage when retinal axons normally reach the brain, an eye primordium from a donor frog was grafted onto the head of a recipient frog. In ten such cases, the third eye seemed to be incorporated onto the head (**FIGURE 9.7A**). To map the innervation of the third eye into the tectum, radiolabeled amino acids were injected into one of the eyes 2–5 days before sacrifice. The tectum was then processed for autoradiography to reveal the termination pattern of axons from the injected eye.

■ **RESULT**: In several cases the supernumerary eye innervated a single tectum. Whether the normal eye or the supernumerary eye was injected, the pattern of innervation in the tectum was the same: rather than a uniform mapping of retina onto the tectum that is normally seen between eye and brain (**FIGURE 9.7B**), the eye innervated only isolated bands of tectum, about 0.5 mm wide, which ran in a rostral-caudal orientation (**FIGURE 9.7C**) (Constantine-Paton & Law, 1978).

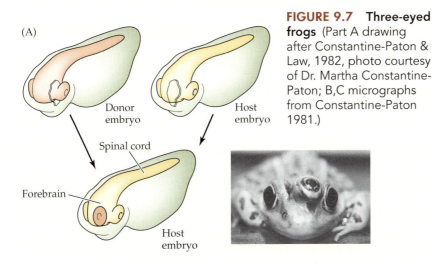

FIGURE 9.7 Three-eyed frogs (Part A drawing after Constantine-Paton & Law, 1982, photo courtesy of Dr. Martha Constantine-Paton; B,C micrographs from Constantine-Paton 1981.)

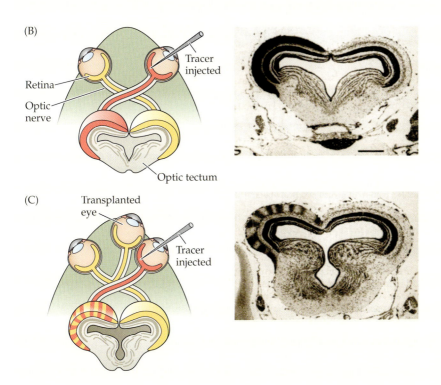

■ **CONCLUSION**: Later experiments injecting an anterograde tracer (horseradish peroxidase [HRP]) into one eye and radiolabeled amino acids into the other would confirm that the stripes of tectum that were not innervated by one eye were exclusively innervated by the other eye (Law & Constantine-Paton, 1981). What's more, physiological mapping of visual receptive fields revealed that each eye also projected in a topographically correct fashion, providing a map of retina onto the tectum, but each was interrupted by the intervening stripes innervated by the other eye. In other words, the two eyes projected in a topographic manner, as the chemoaffinity hypothesis would predict, but each innervated mutually exclusive stripes of tectum, as if they were competing for space, and once they established connections, they excluded inputs from the rival eye.

Follow-up experiments demonstrated that the segregation of input from the two eyes into the tectum is an activity-dependent process. Blocking electrical activity by injecting all three eyes with TTX to block action potentials early in the process prevented segregation from ever occurring—instead both eyes projected evenly over the tectum (Reh & Constantine-Paton, 1985). Electrical activity was also important for *maintaining* segregation of input to the tectum—if TTX injections were made *after* the segregation would normally take place, the segregation became progressively more blurred such that 4 weeks later there was no segregation at all. Importantly, the size of the retinal axonal projection arbors was greater during TTX blockade and became smaller when electrical activity was allowed. These results indicate that activity promotes segregation of inputs by shrinking the projections from individual retinal neurons, thereby sharpening the projection of visual space onto the tectum.

> Activity promotes segregation of inputs by shrinking the projections from individual retinal neurons, thereby sharpening the projection of visual space onto the tectum.

We saw in the previous chapter that spontaneous activity in the retina can accomplish such mapping as long as neighboring retinal neurons tend to fire together, as happens in mammals before birth. But in the frogs, it turned out that not just neural activity, but *visual experience-driven activity*, is responsible for fine-tuning of regenerating retinal projections to the tectum. Raising frogs in the dark blocked fine-tuning of inputs to the tectum in the first place, but if the animals were then exposed to a lighted environment as adults, the fine-tuning finally happened (Keating, Dawes, & Grant, 1992). Likewise, raising frogs in conditions such that their only exposure to light consisted of strobe flashes, where every retinal neuron was stimulated at the same time (thus putting them all on an equal footing for any activity-dependent competition), also prevented the fine-tuning of the retinal map onto the tectum (Brickley, Dawes, Keating, & Grant, 1998).

These results indicated that a Hebbian synapse type of competition was going on between the two eyes, and as NMDA-type glutamate receptors had been implicated in such competition in other systems, scientists asked whether they might be involved in three-eyed frogs as well. To test that idea, they exposed the tectum to an NMDA receptor–specific blocker, APV, after the segregation would normally take place. They found that 2 weeks of APV treatment completely abolished the segregation of ocular input to the tectum without any apparent effect on the electrical activity of the eyes (Cline, Debski, & Constantine-Paton, 1987). If instead of sacrificing the frogs at this stage, the APV treatment was halted, within 2 weeks the segregation of input to the tectum was reinstated. These results demonstrated that even in the stable three-eyed frogs, the established segregation of eye input to the tectum is maintained by the eyes' continuing competition with one another for target neurons.

> Even in the stable three-eyed frogs, the established segregation of eye input to the tectum is maintained by the eyes' continuing competition with one another for target neurons.

This competition between the two eyes for synaptic space in the brain, which must be artificially induced in frogs, is an essential component of visual development in predatory species like monkeys, cats, and humans, where input from the two eyes normally converges in each cerebral hemisphere (see Figure 9.6). And in these cases, as in three-eyed frogs, Hebbian-type competition seems to mediate the fine-tuning of synaptic connections, as we'll see next.

How's It Going?

1. What happens when two eyes innervate the same tectum in frogs, and what is the role of visual experience in this process?
2. Describe the results of an experiment that implicated a particular neurotransmitter receptor in the retinotectal projections of three-eyed frogs.

■ 9.5 ■

MAMMALS REQUIRE VISUAL EXPERIENCE DURING A SENSITIVE PERIOD TO DEVELOP FUNCTIONAL VISION

Sometimes people do not see well with one of their eyes, even though it's intact and a sharp image is focused on the retina. Such impairments of vision are known as **amblyopia** (from the Greek *amblys*, "dull" or "blunt," and *ops*, "eye") or "lazy eye." Sometimes there may be no apparent cause of the amblyopia and only separate vision testing of the eyes will reveal the problem. In those cases, a patch over the "good" eye some of the time forces the person to use the "lazy" eye, and this can improve vision in that eye (**FIGURE 9.8**).

One cause of amblyopia that is readily apparent is when an eye is turned inward (the person is cross-eyed) or outward, conditions known as **strabismus**. Children born with such a misalignment see a double image rather than a single fused image, which apparently leads them to ignore input from the deviated eye. By the time an untreated child reaches the age of 7 or 8, pattern vision in the deviated eye may be completely suppressed. This cause of amblyopia is typically addressed by surgery to correct the strabismus.

Our understanding of amblyopia and strabismus advanced tremendously thanks to visual-deprivation experiments with animals. These experiments revealed startling changes related to disuse of the visual system in early life. **Monocular deprivation** (depriving one eye of light) can be readily accomplished in animals by suturing the eyelids of one eye, or putting in a contact lens with frosted glass. Both manipulations will still admit some light to the retina but will prevent any form vision, because no visual image is focused on the retina. Such monocular deprivation in an infant cat or monkey produces profound structural and functional changes in the thalamus and visual cortex. If such deprivation is maintained for several weeks during development, when the eye is opened the animal will be functionally blind in that eye. Although light enters the eye and the cells of the retina respond, the brain seems to ignore the messages and the animal is unable to detect visual stimuli. If the deprivation lasts long enough, the animal is *never* able to recover eyesight. Thus, early visual experience is crucial for the proper development of vision, and there is a **sensitive period** during which these manipulations of experience can exert long-lasting effects on the system. These effects are most extensive during the early period of synaptic development in the visual cortex (**FIGURE 9.9**). After the sensitive period, monocular deprivation has little or no lasting effect on visual ability.

FIGURE 9.8 Hey there, you with the stars in your eye (Courtesy of Patch Pals, www.PatchPals.com.)

amblyopia Reduced visual acuity of one eye, that is not caused by optical or retinal impairments.

strabismus A deviation of one or both eyes, such that they do not converge on the same region of visual space.

monocular deprivation A procedure of depriving one eye of light.

sensitive period Here, the period during ontogeny when a particular manipulation must be made to affect neural development.

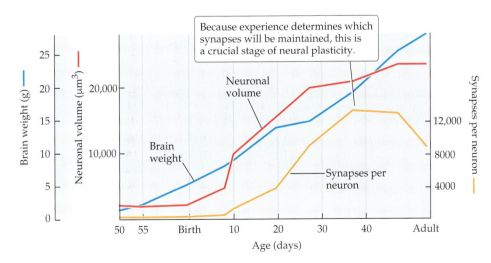

FIGURE 9.9 Development of visual cortex in cats The sensitive period for visual deprivation in cats to affect vision extends from day 10 until nearly adulthood, when synapses per neuron (yellow) first increase rapidly and then decline significantly in the pubertal period before adulthood. (After Cragg, 1975.)

These twentieth-century experiments with animals can be related to Molyneux's problem from the seventeenth century, discussed at the start of the chapter, of whether a person who first experienced vision as an adult would be able to make sense of the world. Molyneux and Locke predicted the person would be unable to recognize objects at first but would be able to learn how to distinguish objects by sight alone. In the centuries in between, Molyneux's problem was addressed several times when cataracts clouding the lenses at birth were surgically removed in people.

For example, when eighteenth-century surgeon William Cheselden removed cataracts from a teenage boy, "He knew not the shape of any thing, nor any one thing from another, however different in shape or magnitude." Cheselden goes on to imply that the boy learned to distinguish objects: "but upon being told what things were, whose form he knew before from feeling, he would carefully observe, that he might know them again" (Cheselden, 1728, p. 448). This report, and others that came after, implied that such people did learn to see eventually, which seems to contradict Mike May's experience. But if you carefully read the reports, you can find hints that the patients' recovery of sight, including the ability to distinguish objects, was far from complete. For example, note the portions I've emphasized in Cheselden's report on his patient: "He learned to know [by sight], *and again forgot* a thousand things in a day…. Having forgot which was the cat and which the dog, *he was ashamed to ask*; but catching the cat (which he knew by feeling) he was observed to look at her steadfastly, and setting her down, said, 'So Puss! I shall know you another time.'"

The phrases I've emphasized suggest that recovery of sight, or at least the ability to distinguish a cat from a dog, was very poor even after both feeling and seeing the animals. Cheselden attributed the boy's mistakes to a faulty memory rather than faulty vision.

In fact, more recent investigations indicate that recovery of sight depends on the age at which the cataracts are removed. Babies born with cataracts in industrialized countries usually have them removed a few months after birth and will have good vision. When the cataracts are removed later in life, when the individuals are teenagers, they are poor at recognizing objects by sight alone (**FIGURE 9.10**), at least at first. However, with more visual experience their ability to distinguish objects does improve (Held et al., 2011; Ostrovsky, Andalman, & Sinha, 2006; Ostrovsky, Meyers, Ganesh, Mathur, & Sinha, 2009).

In contrast, when cataracts are removed in adulthood, people are generally disappointed at how little use they can *ever* make of vision. Facial recognition seems the hardest aspect of vision to gain among people who have been blind for a long time. Individuals who had cataracts occluding their vision for just the first 6 months of their lives are impaired at recognizing faces even 9 years later (Le Grand et al., 2001).

> When cataracts are removed in adulthood, people are generally disappointed at how little use they can *ever* make of vision.

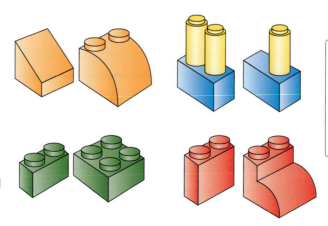

FIGURE 9.10 **Testing visual object recognition** (From Held et al., 2011.)

> Newly sighted individuals allowed only to *touch* these various toy blocks could distinguish them easily. However, when allowed only to *see* them, they were poor at predicting which objects corresponded to the blocks they'd identified earlier.

We've seen that Mike May could not recognize his wife's face after his surgery (Fine et al., 2003). Even after more than a decade of visual experience, Mike still has significant deficits in distinguishing three-dimensional objects (**FIGURE 9.11**) or faces (Huber et al., 2015). Another man, blinded as an infant and given vision 50 years later, came to recognize his friends from about 15 feet away, but based on their *clothes*, not their faces (Gregory & Wallace, 1963). In yet another case, a man who became blind rather late in life, at 17 years of age, had his sight restored 53 years later. While he learned to distinguish objects and judge the distance of objects, he had poor face recognition even eight months after his surgery (Sikl et al., 2013). His difficulty is all the more surprising given that he presumably had reasonable facial recognition abilities until he was 17. Apparently the 50-year absence of seeing faces robbed him of a skill he had acquired as a child. We have to wonder whether synaptic connections crucial for facial recognition had withered for lack of experience to maintain them.

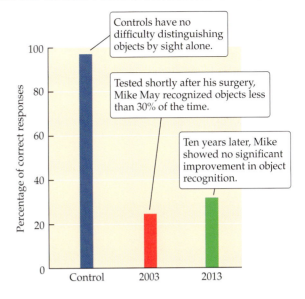

FIGURE 9.11 graph annotations:
- Controls have no difficulty distinguishing objects by sight alone.
- Tested shortly after his surgery, Mike May recognized objects less than 30% of the time.
- Ten years later, Mike showed no significant improvement in object recognition.

FIGURE 9.11   Object recognition after visual recovery in adulthood (After Huber et al., 2015.)

Why does Michael May have such poor vision despite the clear images entering his eye? Had the accident happened to him as an adult, the surgery to let light back into his eye would have restored normal vision. But, like a kitten fitted with opaque contact lenses, Mike was deprived of form vision—in his case, for over 40 years. Because this deprivation began when he was a child, synaptic connections within his visual cortex were not strengthened by the patterns of light moving across the retina. In the absence of patterned stimulation, synapses between the eye and the brain presumably languished and disappeared.

In one sense, Mike was lucky that his blindness came as late as it did. He had normal form vision for the first 3½ years of his life, and that stimulation may have been sufficient to maintain some synapses that would otherwise have been lost. These residual synapses are probably what allow him to make any sense whatsoever of his vision today. Michael loves having sight, but as he himself says, usually he has to "guess" what he's seeing.

Thus modern investigation of Molyneux's problem in humans seems to conform to the outcome of animal experiments, in which young animals that are deprived of vision long enough during development seem unable to see as adults, even with the opportunity for visual experience.

How's It Going?

1. What are the effects of monocular deprivation on the vision of monkeys, cats, and humans?

2. Contrast the effects of monocular deprivation in development versus adulthood. How do these findings relate to the modern answer to Molyneux's problem?

■ 9.6 ■

PHYSIOLOGICAL RECORDINGS REVEAL HOW VISUAL DEPRIVATION IMPAIRS SIGHT

Electrophysiological recordings of neurons in visual cortex have revealed how visual deprivation during the sensitive period can lead to blindness in the deprived eye. Using a microelectrode introduced into visual cortex, you can present light to one eye or the other to find what type of stimuli affect each neuron's firing rate. In cats, monkeys, and ferrets, most cortical neurons

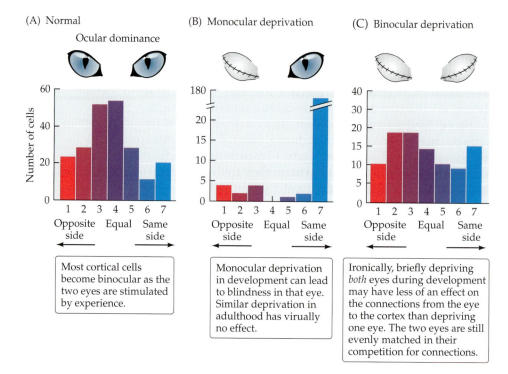

FIGURE 9.12 Ocular dominance histograms are affected by visual experience. (After Hubel & Wiesel, 1965, and Wiesel & Hubel, 1965.)

(A) Normal

Ocular dominance

Most cortical cells become binocular as the two eyes are stimulated by experience.

(B) Monocular deprivation

Monocular deprivation in development can lead to blindness in that eye. Similar deprivation in adulthood has virually no effect.

(C) Binocular deprivation

Ironically, briefly depriving *both* eyes during development may have less of an effect on the connections from the eye to the cortex than depriving one eye. The two eyes are still evenly matched in their competition for connections.

ocular dominance histograms Plots depicting the distribution of visual cortical neurons in terms of the extent to which they respond to stimuli in both eyes.

binocular deprivation A procedure of depriving both eyes of light.

can be excited equally by light presented to either eye. In fact, the neuron typically responds not just to light at any two spots on the two retinas, but to stimulation of *corresponding portions* of the two retinas, that is, the portions of the two retinas that are normally stimulated by a particular part of the visual field. An exception to this rule is the group of neurons in layer IV where monocular input from the lateral geniculate nucleus forms the ocular dominance bands we discussed in the Chapter 8. For the other layers of visual cortex, Nobel Prize–winning neuroscientists David Hubel and Torsten Weisel came up with a scale to rate each neuron in terms of how well it responds to light in the two eyes. If a neuron in the left visual cortex responded only to light in the eye on the opposite side (the contralateral side, i.e., the right eye), then they gave that a rating of 1. Neurons that responded only to light in the eye on the same side (the ipsilateral eye) were rated as 7. If the neuron was equally affected by both eyes, it could be rated as 4 and so on, each neuron graded for how monocular or binocular it was in response. By surveying lots of visual cortical neurons and rating each for whether it responded to one or both eyes, they constructed **ocular dominance histograms**, which portray the strength of response of cortical neurons to stimuli presented to either the left or the right eye (**FIGURE 9.12A**).

Monocular deprivation during development results in a striking shift from the normal graph; most cortical neurons respond only to input from the nondeprived eye (**FIGURE 9.12B**). In cats the susceptible period for this effect is the first four months of life. In rhesus monkeys the sensitive period extends to six months of age. After these ages, visual deprivation has relatively little effect.

Further experiments revealed that during early development, synapses are rearranged in the visual cortex, and axons representing input from each eye compete for synaptic places, much as retinal axons compete for synaptic space in the tectum of three-eyed frogs. Like the Hebbian synapses we introduced in Chapter 8, active, effective synapses predominate over inactive synapses. Thus, if one eye is "silenced," synapses carrying information from that eye are retracted while synapses driven by the other eye are maintained. The competitive nature of this process can be seen in a finding that would otherwise appear paradoxical—depriving animals of light to *both* eyes (**binocular deprivation**) has relatively little effect on ocular dominance in visual cortex (Wiesel & Hubel, 1965), as if with both eyes being equally

If one eye is "silenced," synapses carrying information from that eye are retracted while synapses driven by the other eye are maintained.

deprived, neither is able to out-compete and displace the other for cortical projections (**FIGURE 9.12C**). (But, if both eyes are deprived for long enough, the animal will be functionally blind in adulthood.)

Visual deprivation has these effects by reducing neural activity in the eye, since injections of TTX to block production of action potentials of retinal ganglion cells have similar effects. If one eye is silenced with TTX, the other eye comes to predominate in influencing cortical neurons (Chapman, Jacobson, Reiter, & Stryker, 1986). Just as binocular visual deprivation has relatively modest effects on ocular dominance in visual cortex, as if neither eye is able to outcompete the other, so too does binocular TTX treatment have only modest effects on ocular dominance distributions (Stryker & Harris, 1986).

The effects of competition between the two eyes were also seen in experiments designed to model what happens in children with strabismus, as we see next.

RESEARCHERS AT WORK

Strabismus in Kittens Drastically Alters Visual System Connections

- **QUESTION**: What is the effect of strabismus early in life on the pattern of synaptic connections in the visual system?

- **HYPOTHESIS**: Strabismus will alter which portions of the two retinas are stimulated by any given object in visual space. Visual stimuli that would otherwise fall upon corresponding portions of the two retinas will now fall on disparate regions. This means visual stimuli will no longer activate convergent pathways from the two eyes onto neurons in the brain.

- **TEST**: The scientists produced an animal replica of strabismus (or "squint") in kittens by surgically cutting an extraocular muscle in one eye, causing the eyes to diverge.

- **RESULT**: The ocular dominance histogram of these animals reveals that the normal binocular sensitivity of visual cortical cells is greatly reduced (**FIGURE 9.13**). A much larger proportion of visual cortical cells is excited by stimulation of either the right or the left eye in these animals than in control animals (Hubel & Wiesel, 1965).

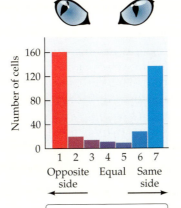

If one eye is deviated, each cortical cell will respond to only one eye or the other, resulting in poor depth perception.

FIGURE 9.13 Induced strabismus reduces binocular innervation in visual cortex. (After Hubel & Wiesel, 1965.)

- **CONCLUSION**: The induced strabismus caused a dramatic shift in the synaptic connections between the eyes and the visual cortex: most cortical neurons received information exclusively from one eye or the other, with only a minority receiving binocular input. Because comparison of input to the two eyes is a crucial component of visual depth perception, such animals would be expected to have poor depth perception, despite the fact that the brain received input from both eyes.

The explanation for this effect of strabismus is that before any visual experience was possible, axonal guidance cues, like the ephrins, and neural activity-guided competition through spontaneous waves of activity in the retina provided a rough mapping of visual space onto the cortex. Thus both eyes were already providing information to the roughly correct portions of the cortex. That means that information from the corresponding portions of the two retinas was already converging upon the same cortical regions. In that case, visual experience, such as an object moving in the world, would simultaneously stimulate the corresponding regions of the two retinas, which would then fire together to eventually stimulate some overlapping target neurons in the cortex. That simultaneous stimulation from the two eyes would likely cause the cortical neurons to fire and thus strengthen those synapses in a Hebbian fashion (see Figure 8.22). Thus as the topographic map of visual space onto the cortex is being sharpened by experience, there is also a strengthening of binocular inputs from corresponding regions of the two retinas.

But after surgery to induce strabismus, the two eyes are no longer aligned, so visual stimuli falling on the misaligned eyes no longer stimulate corresponding regions of the two retinas. Apparently the initial rough mapping of retina to cortex is not broad enough that the impulses provided by these two disparate parts of the retinas ever arrive at a single cortical neuron. Without simultaneous, convergent input from the two eyes onto cortical neurons, no binocular connections are cemented. Only if binocular inputs onto cortical neurons are already provided, as by nonexperiential factors in normally aligned eyes, can the synapses be strengthened by experience. This is how strabismus results in cortical neurons that respond to one eye or the other but rarely to both (see Figure 9.13).

> Without simultaneous, convergent input from the two eyes onto cortical neurons, no binocular connections are cemented.

A demonstration that activity from the two eyes is competing for input to cortex led to a similar outcome. Chronic electrodes were implanted in kittens in order to directly stimulate the two optic nerves for several hours per day. If the two nerves were stimulated simultaneously, most of the cortical neurons responded to binocular visual stimuli, as if the equal activity kept either eye from dominating in the competition. But if the nerves were stimulated *alternately*, most cortical neurons came to respond exclusively to one eye or the other (Stryker, Chapman, Miller, & Zahs, 1990), resembling the situation seen in strabismus (see Figure 9.13).

What this understanding of visual development means for children with strabismus is that it is important to surgically correct the misalignment of the two eyes relatively early in life, generally before the age of 2, if they are to enjoy full binocular vision (Banks, Aslin, & Letson, 1975; Fawcett, Wang, & Birch, 2005) (**FIGURE 9.14**). Presumably, this age is before binocular inputs to cortical neurons have withered entirely. If the surgery is performed in adulthood, only a minority of patients benefit from increased binocularity (Mets, Beauchamp, & Haldi, 2003).

> For children with strabismus it is important to surgically correct the misalignment of the two eyes relatively early in life if they are to enjoy full binocular vision.

This line of research altered the typical medical response to childhood strabismus. Formerly, the surgery was thought to be merely cosmetic, and so the philosophy was to wait until the children were older, when the surgery would be easier and the risks of general anesthesia lower. Nowadays, the idea is to correct the strabismus as early in life as is safe for the child. When the eyes are realigned during childhood, the child learns to fuse the two images and has good depth perception as an adult. But if

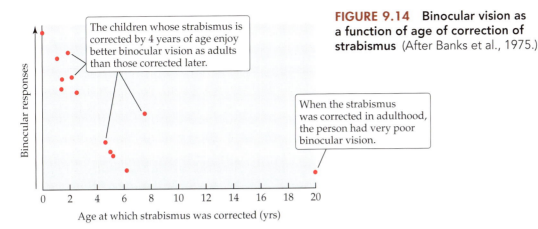

FIGURE 9.14 Binocular vision as a function of age of correction of strabismus (After Banks et al., 1975.)

Binocular responses / Age at which strabismus was corrected (yrs)

The children whose strabismus is corrected by 4 years of age enjoy better binocular vision as adults than those corrected later.

When the strabismus was corrected in adulthood, the person had very poor binocular vision.

the realignment is not done until adulthood, it's too late to restore acute vision to the turned eye.

Like the competition between the eyes in three-eyed frogs, the effects of neural activity on ocular dominance patterns in mammals appear to be mediated by NMDA-type glutamate receptors enforcing Hebbian-type competition. Infusions of the NMDA receptor–specific blocker APV into the visual cortex of kittens prevented a shift in ocular dominance patterns in response to monocular deprivation (Bear, 1996; Kleinschmidt, Bear, & Singer, 1987). Likewise, systemic treatment with an NMDA receptor blocker also affects ocular dominance histograms in kittens (Daw et al., 1999).

Neurotrophic factors may also be playing a role in experience-driven synapse rearrangement. For example, if the postsynaptic cells are making a limited supply of a neurotrophic factor, and if active synapses take up more of the factor than inactive synapses do, then perhaps the inactive axons retract for lack of neurotrophic factor. Brain-derived neurotrophic factor (BDNF) has been implicated as the neurotrophic factor being competed for in the kitten visual cortex (McAllister, Katz, & Lo, 1997) and in the frog retinotectal system (Du & Poo, 2004). So perhaps ineffective synapses wither for lack of neurotrophic support from their targets. Scientists are also zeroing in on activity-regulated expression of genes such as those for insulin-like growth factor, which can regulate the strength of synapses on V1 neurons (Mardinly et al., 2016).

One demonstration suggests visual experience in everyday life may affect our perception. In **FIGURE 9.15**, the numbers and letters along the bottom line appear more slanted than those above, but in fact the slant is the same (Whitaker & McGraw, 2000). One theory of why we see a difference here that doesn't exist is that our experience reading digital clock readouts and italic fonts may tune synapses in our brain to perceive them as more upright than they really are—an effect lost when the figures are presented backward.

Visual experience is not only important for proper development of vision; it can also be important for fine-tuning hearing. Rather than shifting vision in one eye by inducing strabismus, scientists shifted hearing in one ear in young owls, as we'll see next.

The effects of neural activity on ocular dominance patterns in mammals appear to be mediated by NMDA-type glutamate receptors enforcing Hebbian-type competition.

FIGURE 9.15 Which lines are more slanted? Most people perceive the mirror-reversed instance to be more slanted, but in fact the angles are identical. Has our experience reading clocks and italic print affected visual system connections to reduce our perception of the slant?

How's It Going?

1. What is the effect of induced strabismus on ocular dominance histograms in kittens?

2. What do these results suggest about the best course of treatment for children born with strabismus?

■ 9.7 ■

OWLS CAN USE VISUAL EXPERIENCE TO FINE-TUNE THEIR AUDITORY MAPS

Hearing is very important for owls to find their prey in the dark, and the responsiveness of neurons in the owl tectum reflects the importance of auditory cues for these animals. The tectal neurons respond to sounds, not according to their pitch or volume, but according to *where in space they came from*. Thus there is a roughly spherical topographic map of auditory space across the surface of the owl tectum. In fact, these neurons are multimodal, because they will also respond to visual stimuli, and again they respond primarily to where the stimulus falls in the visual field. These two maps of space, one of where visual stimuli arise, and the other of where auditory stimuli arise, are normally aligned, superimposed upon one another, with individual neurons responding to either modality. Presumably this alignment of the two maps helps guide the owl's behavior toward stimuli, such as a mouse, either visible at dusk or scurrying on the barn floor in complete darkness.

A series of elegant experiments established that this alignment of visual and auditory space in the owl tectum is guided by experience, especially experience early in life. If a plug was placed in one ear in adult owls, they made large errors in localizing sounds, with responses shifted in the direction of the open ear—presumably because the sound was louder in that ear, which would normally mean the sound had come from that side. Adult animals never seemed to adjust to the ear plug, but owls younger than 8 weeks at the time of plugging eventually compensated and could locate the sources of sounds.

Vision is crucial for young owls to adapt to the ear plug: if the owlets were deprived of vision, they never adjusted their behavior to make proper use of sound localization. What's more, when other owlets were outfitted with prisms to shift the visual field by 10° (**FIGURE 9.16**), the adjustment of the auditory localization map on the tectum was matched to the visual shift (Knudsen, 1985). The induced mismatch between the auditory and visual tectal maps provoked a remapping of auditory space in the owlets that was no longer possible in the brains of adults (Bergan & Knudsen, 2009). This alignment of visual and auditory space appears to be another instance of Hebbian synapses at work. As long as auditory and visual space are aligned, any stimulus that can be both seen and heard will activate converging input onto particular tectal target neurons, likely making them fire and so constituting a successful input that would be preserved at the expense of ineffective synapses. Presumably that convergence of visual and auditory input is normally responsible for aligning the two maps in the tectum. But inducing a shift in either visual stimuli (with prisms) or auditory stimuli (with an ear plug) induces a mismatch, so a given stimulus will activate one

FIGURE 9.16 Shifting the visual field in an owlet If an owlet is fitted with prisms that shift the visual field 10° to the right, they can learn to make use of that information to retrieve prey. This experience will also shift the mapping of visual space onto the tectum. Adult owls fitted with such prisms never adapt to the shift. (Courtesy of Dr. Eric Knudsen.)

region of the tectum through visual channels and another region of tectum through auditory channels (Witten, Knudsen, & Sompolinsky, 2008). By now you might not be surprised to learn that this experience-dependent shifting of tectal maps in owls is dependent on the activity of NMDA receptors (Knudsen, 1999).

Next we'll see how experience plays a role in refining connections for a quite different sensory modality—olfaction.

How's It Going?

1. What sort of stimuli drive activity of neurons in the owl tectum?
2. What is the role of experience in establishing this pattern?

■ 9.8 ■
OLFACTORY RECEPTOR MAPS ARE ALSO SCULPTED BY EXPERIENCE

Mammalian odor detection relies on a sheet of **olfactory sensory neurons (OSNs)** in the olfactory epithelium within the nasal cavity (see Figure 3.15). Humans possess about 10 million OSNs, while bloodhounds have over 200 million, which is one reason they can discriminate more odors than we can (Quignon, Rimbault, Robin, & Galibert, 2012). The OSNs project their axons though tiny openings in the skull to synapse in the **olfactory bulb**, specifically upon tiny, roughly spherical clusters of cells known as **olfactory glomeruli**. Each glomerulus relies on mitral cell neurons to relay information to the prepyriform cortex (also known as *primary olfactory cortex*; note olfaction is the only sense to report directly to cortex without passing through the thalamus) as well as the amygdala.

Each OSN possesses a membrane-bound **olfactory receptor protein**, the extracellular domain of which detects a particular odorant, while the intracellular domain activates a specific G protein, named G_{olf} in recognition of its importance in olfaction. G_{olf} knockout mice are unable to detect odors (Belluscio, Gold, Nemes, & Axel, 1998), a condition known as **anosmia**. G_{olf} in turn activates an olfactory-specific cAMP-gated channel that allows cations in to depolarize the cell, triggering an action potential. Mice with the gene for this channel knocked out are also anosmic, without other obvious behavioral deficits (Brunet, Gold, & Ngai, 1996).

Another reason other mammals can detect more odors than we can is because they have more functional receptor proteins than we do. While both humans and mice have about 1,000 different olfactory receptor genes, most of ours have accumulated mutations, so only about 350 are fully functional (Crasto, Singer, & Shepherd, 2001; Glusman, Yanai, Rubin, & Lancet, 2001). Since we can discriminate at least 5,000 different odors, at least some odors must activate a distinctive combination of olfactory receptors and we detect them by pattern coding. In mice, nearly all 1,000 olfactory receptor proteins seem to be at work, and they can be divided into four receptor subfamilies. Each receptor subfamily is expressed by a separate band of OSNs in the mouse olfactory epithelium (**FIGURE 9.17**) (Vassar, Ngai, & Axel, 1993). As we don't know which odors are detected by which olfactory receptor proteins, or combination of receptor proteins, we don't know what this map reflects about odors (are there four basic dimensions of odor?). Presumably this is a topographic map of some quality of odors that we don't recognize and therefore have no words to describe.

olfactory sensory neurons (OSNs) The primary sensory neurons of the olfactory system, which contact and recognize particular odorants.

olfactory bulb An anterior projection of the brain that terminates in the upper nasal passages and, through small openings in the skull, receives axons from olfactory receptor neurons.

olfactory glomeruli Spherical clusters of cells in the vertebrate olfactory bulb that process information from a particular class of olfactory sensory neurons and therefore a particular odor.

olfactory receptor protein A protein embedded in the membrane of olfactory sensory neurons that binds odorants on the extracellular surface and triggers second messenger systems within the neurons.

anosmia The condition of being unable to detect odors.

Other mammals can detect more odors than we can because they have more functional receptor proteins than we do.

(A)

Olfactory bulb

Brain

Olfactory
epithelium

(B)

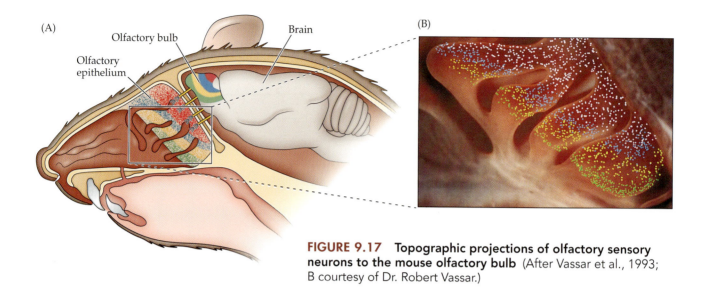

FIGURE 9.17 **Topographic projections of olfactory sensory neurons to the mouse olfactory bulb** (After Vassar et al., 1993; B courtesy of Dr. Robert Vassar.)

neuropilins A family of membrane receptor proteins that govern growth cone guidance. They bind and respond to semaphorins.

semaphorins A family of secreted and membrane-bound molecules that often serve to repulse growth cones, including those that express neuropilin or plexins, and so direct them away from a boundary.

Importantly, in adulthood almost every OSN expresses only one of these 1,000 different olfactory receptor genes, but in newborn mice there are many more OSNs that co-express two or more. The loss of neurons expressing more than one receptor protein appears to be due to apoptosis, since treatments to block apoptosis prevent the loss of such OSNs (Tian & Ma, 2008). The dying of co-expressing OSNs appears to be gated and/or guided by sensory experience, because surgically closing the nostril prevents their loss. Even in young adult mice, closing the nostril results, a month later, in an increase of OSNs expressing more than one receptor protein. This result also suggests that apoptosis to eliminate co-expressing receptor neurons is an ongoing, perhaps lifelong, process (recall from Chapter 3 that olfactory receptor neurons are continually replaced throughout life). Thus olfactory experience seems to sharpen the tuning of OSNs to specific stimuli by eliminating those that might respond to more than one odorant.

Olfactory experience also sharpens the pattern of innervation of the OSNs onto the glomeruli in the bulb. Remarkably, the axons of all OSNs expressing any particular receptor gene converge onto just two of the thousands of glomeruli in the bulb (Imai, Sakano, & Vosshall, 2010). As we've seen in other models, initial projections are directed by gene expression alone. For example, guidance of OSN projections begins as their axons sort themselves by gradients of expression: those expressing higher levels of the guidance receptor **neuropilin** are repulsed by their neighbors that express higher levels of the ligand, **semaphorin** (Imai et al., 2009) (recall we discussed semaphorin's action as a repulsive cue in Chapter 5). This presorting directs axons to the approximately correct anterior-posterior position in the bulb.

Expression of various adhesive molecules, including ephrins, that guide further axonal sorting is affected by the electrical activity of OSNs (Serizawa

Maintenance of the topographic mapping of OSNs onto the bulb also depends on neuronal activity.

et al., 2006). Maintenance of the topographic mapping of OSNs onto the bulb also depends on neuronal activity. In a mosaic analysis, the OSNs that expressed a dysfunctional allele for the olfaction-specific cAMP-gated channel, and therefore did not fire action potentials in response to odorants, lost out in competition with OSNs expressing the wild-type allele (Zhao & Reed, 2001). Blocking the nostril to prevent exposure to

(A) Non-occluded

When the nostril is open, only a few of the olfactory sensory neurons that are unable to respond to odors (blue) survive in the olfactory epithelium to project into glomeruli (blue circles) in the bulb.

Occluded

When the nostril is occluded, many more unresponsive olfactory sensory cells survived and project to the bulb.

(B)

A dorsal view of the two bulbs shows many more of the olfactory sensory neurons that cannot respond to odors remain to project to glomeruli on the side with an occluded nostril.

FIGURE 9.18 **Experience-dependent competition between olfactory sensory neurons** (From Zhao & Reed, 2001.)

odors also blocked the competition, such that even the disabled OSNs survived and retained projections to particular glomeruli within the bulb (**FIGURE 9.18**). In other words, inputs from silent receptor neurons are normally eliminated when facing competition from neighboring active neurons but will survive if competition is suspended by blocking olfactory experience.

The further refinement of OSN connections to specific glomeruli appears to be dependent on olfactory experience per se. For example, in young mice OSNs expressing a particular olfactory receptor protein sometimes project to more than just the two glomeruli per bulb that is seen in adults. In other words, they are projecting to a glomerulus that is also innervated by OSNs expressing a different olfactory receptor. As the animals grow, these misplaced projections to inappropriate glomeruli tend to disappear, probably because the offending OSNs have died in the ongoing apoptosis seen in OSNs. But if one nostril is closed in young mice, preventing any olfactory stimuli reaching the OSNs, the misprojecting neurons persist into adulthood (Zou et al., 2004).

We'll conclude this chapter by considering yet another sensory modality where experience influences the development and maintenance of neuronal connections. In the case of touch, we even have good evidence of experience-guided development of synaptic connections in humans, as well.

If one nostril is closed in young mice, preventing any olfactory stimuli reaching the OSNs, the misprojecting neurons persist into adulthood.

FIGURE 9.19 **Effects of early whisker manipulations on mouse somatosensory cortex** (Part A and photo courtesy of Dr. T. A. Woolsey; B–E after Cowan, 1979.)

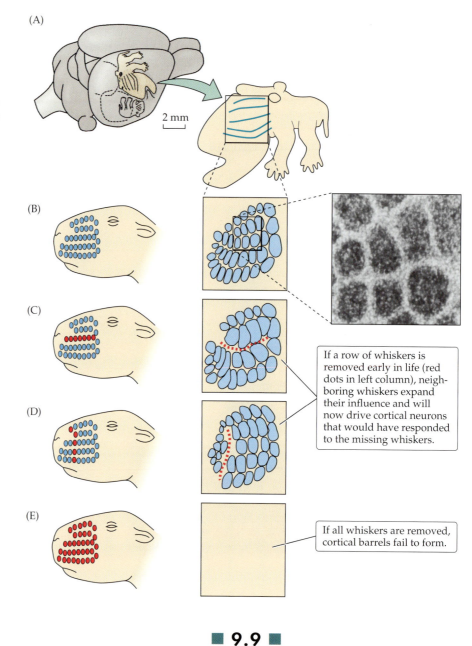

2 mm

If a row of whiskers is removed early in life (red dots in left column), neighboring whiskers expand their influence and will now drive cortical neurons that would have responded to the missing whiskers.

If all whiskers are removed, cortical barrels fail to form.

■ **9.9** ■

TACTILE EXPERIENCE GUIDES THE FORMATION OF TOPOGRAPHIC MAPS IN SOMATOSENSORY CORTEX

Sensory experience also plays an important role in touch, as has been very convincingly demonstrated in studies of the whiskers on the faces of mice. The whiskers provide tactile information to the animals when navigating narrow passages, often in darkness. If you've ever had a pet mouse, you may have noticed that the whiskers are not distributed randomly on the muzzle; rather, they form an orderly array of five rows on each side of the snout. Sensory receptor cells at the base of each whisker detect any deflection and report to somatosensory cortex. Remarkably, if you examine that portion of somatosensory cortex, you will see clusters of cells called **whisker barrels** because their arrangement makes them look like barrels squeezed together. The layout of these cortical barrels corresponds to a map of the whiskers (**FIGURE 9.19A**).

The sensory information from each whisker competes for space in cortex. If a row or column of whiskers is surgically removed early in life, the portion of somatosensory cortex that would have received information from those

whisker barrel A column of somatosensory cortex that receives sensory information from a single whisker.

(A) Representation of the left
    hand in primary somatosensory
    cortex in right hemisphere
    of monkey brain

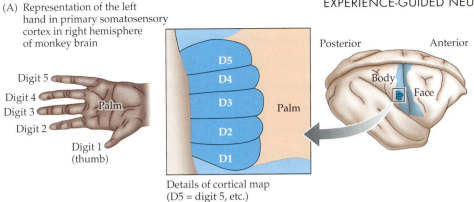

Digit 5
Digit 4
Digit 3
Digit 2
Palm
Digit 1
(thumb)

Posterior   Anterior
Body
Face

Palm

Details of cortical map
(D5 = digit 5, etc.)

FIGURE 9.20
Experience alters the topographic mapping of body afferents to somatosensory cortex. (After Merzenich & Jenkins, 1993.)

(B) Experiment 1

D3 surgically removed.

D5
D4
D2
D1

D5
↓ D4
↑ D2
D1

Several weeks later, areas representing D2 and D4 have expanded, replacing the representation of D3.

(C) Experiment 2

Monkey trained to keep two fingers in contact with rotating stimulus disk.

D5
D4
D1
Stimulus disk

D5
D4
↑ D3
↓ D2
D1

Areas representing stimulated digits expand and replace part of the areas that formerly represented adjacent digits.

whiskers is invaded by afferents reporting from the remaining whiskers on either side (**FIGURE 9.19B–E**). The maximum effect is achieved if the whiskers are removed at birth; removal on the fifth day of life results in a shrinkage, rather than disappearance, of the whisker barrel (Durham & Woolsey, 1984). The dendrites of cortical neurons within the barrel are also affected by the loss of tactile input (Brown et al., 1995). Touch information from the whiskers, like all somatosensory information in vertebrates, is relayed through the thalamus, and whisker manipulations also affect maps there (Mosconi, Woolsey, & Jacquin, 2010; Zantua, Wasserstrom, Arends, Jacquin, & Woolsey, 1996).

Somatosensory information competes for cortical space in primates, too, and this competition can be observed even in adults. By recording from cells in primary somatosensory cortex (S1) while gently stimulating various portions of skin, researchers can construct a topographic map of the fingers and hand in monkeys (**FIGURE 9.20A**). This pattern of representation in S1 is consistent across individual monkeys. When one digit is surgically removed in adulthood and S1 mapped several weeks later, the cortical region that would have responded to tactile stimuli from that digit now responds to either of the neighboring digits (Merzenich et al., 1984). In other words, information from the remaining digits has invaded the cortical region formerly representing the missing digit (**FIGURE 9.20B**). Conversely, if we arrange for two digits to be

FIGURE 9.21 Adult experience affects somatosensory cortex in rat dams. (A) The normal mapping of somatosensory information from the rat ventrum. (B) In female rats that are not lactating, information from the ventrum (blue) reaches a limited region of somatosensory cortex. (C) In lactating rats, which receive a great deal of tactile stimulation from nursing pups, a much greater proportion of somatosensory cortex is devoted to the ventrum. (B,C after Xerri et al., 1994.)

(A) Female rat

Location of nipples on ventrum

Primary somatic sensory cortex

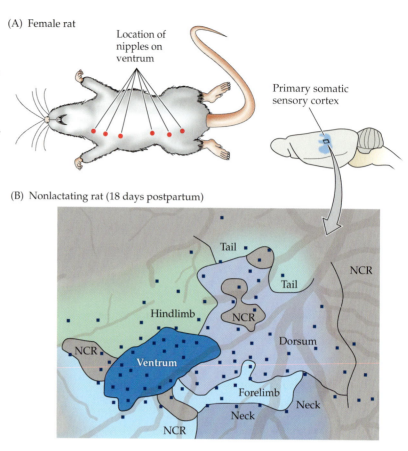

(B) Nonlactating rat (18 days postpartum)

(C) Lactating rat (19 days postpartum)

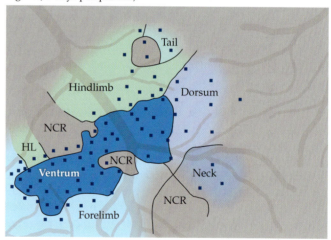

exposed to *more* sensory information than the others by training the monkey to make fine tactile discriminations with those particular digits, then the cortical representations of those digits expand (**FIGURE 9.20C**) (Wang, Merzenich, Sameshima, & Jenkins, 1995).

Somatosensory maps are also plastic in adult rats. As female rats approach the end of pregnancy, they spend a lot of time self-grooming, particularly licking and cleaning the nipples in preparation for the arrival of pups. Once the pups arrive, there is continued stimulation of the nipples as they nurse. That additional stimulation from pups alters somatosensory maps of the region, because a few weeks after the pups are born, the maps from mothers that are actively nursing show more representation of the ventral surface of

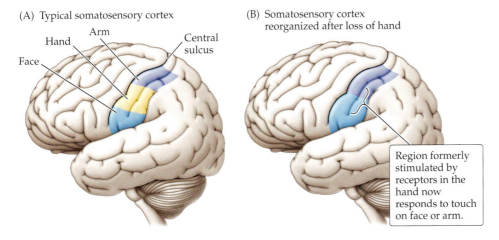

(A) Typical somatosensory cortex

Face
Hand
Arm
Central sulcus

(B) Somatosensory cortex reorganized after loss of hand

Region formerly stimulated by receptors in the hand now responds to touch on face or arm.

FIGURE 9.22 **Plasticity in somatosensory maps in humans** (B after Yang et al., 1994.)

the body than do maps from control dams whose litters were removed after birth (**FIGURE 9.21**) (Xerri, Stern, & Merzenich, 1994).

Plasticity of cortical somatosensory maps has also been demonstrated in humans. For example, musicians who play stringed instruments have expanded cortical representations of the left fingers (Munte, Altenmuller, & Jancke, 2002; Ripolles et al., 2015), presumably because they have been using those fingers to depress strings for precisely the right note. People who suffer accidental amputations of the hand reinforce the notion of adult plasticity. Normally the map of the body on human S1 (the so-called homunculus) is such that information from the hand is sandwiched between those regions of S1 representing the face and the upper arm (**FIGURE 9.22A**). However, in a man who lost his hand as an adult, fMRI imaging showed that the cortical region that would have received information from the hand was now receiving information from the face and upper arm (**FIGURE 9.22B**) (Yang et al., 1994). In another case, a man who had lost his hand at age 19 received a transplanted hand 35 years later. A few months after the transplant, when the man reported he could detect touch from the transplanted hand, fMRI imaging confirmed that cortical representation of that hand was in the typical part of S1, between the face and upper arm (Frey, Bogdanov, Smith, Watrous, & Breidenbach, 2008).

> Plasticity of cortical somatosensory maps has also been demonstrated in humans.

Taken together, these studies indicate that neuroplasticity driven by sensory experience is widespread in the somatosensory system and that such plasticity is present, to some extent, well into adulthood and probably for the entire life span.

Thus for every sensory system we've explored, and probably for those we haven't, experience plays a crucial role in sharpening the connections between sensory receptor neurons and the brain, and this sharpening of connections aids our discrimination of various stimuli, which is clearly important for survival and reproduction. Once natural selection hit upon the use of sensory experience to guide neural development, it became possible to tap yet another source of information to guide development, namely information provided by other individuals. In Chapter 10 we'll examine the many cases where social stimuli guide neural development. In our own species, where language plays such a huge role in our cognition and behavior, the social acquisition of language must have been a pivotal point in our evolution.

┌─ **How's It Going?** ─┐

1. What are whisker barrels, and what is the role of activity in forming them?

2. What is the evidence that tactile experience affects the projection of afferents from the body onto primary somatosensory cortex in rats, monkeys, and humans?

─ SUMMARY ─

■ Humans fitted with reversing prisms can learn to use the inverted vision to guide movement, presumably through higher-level processing, but amphibians with reversed vision cannot. However, adult amphibians and fish can regenerate connections from the eye to the brain after a severing of the optic nerve. **See Figure 9.1**

■ The *chemoaffinity hypothesis* proposed that the regenerating retinal axons recognized some unique identity of tectal neurons to reestablish the original projections. At least one factor guiding this reinnervation is expression of ephrin receptors by retinal ganglion cells and ephrins by tectal neurons. **See Figures 9.2 and 9.3**

■ However, experiments resulting in tectal expansion and compression demonstrated some flexibility in the regeneration of retinotectal connections. Likewise, normal growth of the periphery of the retina and the posterior tectum requires continuous shifting of the topographic projection from the retina. **See Figures 9.4 and 9.5**

■ In three-eyed frog preparations, two eyes innervating the same tectum segregate, each eye projecting to a separate rostral-caudal strip of tectum. This segregation is activity dependent, as visual experience allows neighboring retinal neurons to dominate particular tectal targets, stabilizing those synapses. In normal optic nerve regeneration, this experience-dependent process fine-tunes topographic mapping of visual space onto the tectum. **See Figure 9.7**

■ In monkeys and cats, an eye silenced by visual deprivation or TTX during a *sensitive period* in development will show little or no evidence of vision in that eye. Such visual deprivation in adulthood has relatively little effect on visual ability. **See Figure 9.9**

■ Humans deprived of vision early in life have difficulty identifying objects and faces even after years of visual experience in adulthood. **See Figures 9.10 and 9.11**

■ Ocular dominance patterns in visual cortex are established and maintained by visual experience, such that the active eye comes to predominate activity of cortical neurons. The competitive nature of this process explains how *binocular deprivation* may have relatively little effect on ocular dominance. **See Figure 9.12**

■ *Strabismus* shifts one eye enough that objects in visual space no longer strike corresponding portions of the two retinas. If this happens early in life, cortical neurons retain connections from one eye or the other, but very few receive binocular input. **See Figure 9.13**

■ Surgical correction of strabismus must occur in childhood for the person to develop full binocular visual capacity, presumably by preserving binocular innervation of cortical neurons. **See Figure 9.14**

■ Owlets use experience to align maps of visual space and auditory space in the tectum, in a Hebbian-like process mediated by NMDA receptors. **See Figure 9.16**

■ Each *olfactory sensory neuron* (*OSN*) expresses only one of the thousands of genes for *olfactory receptor proteins*. Each OSN expressing a particular receptor projects to only two glomeruli in the *olfactory bulb*. Olfactory experience causes misprojecting OSNs to die, sharpening the map of olfaction onto the bulb. **See Figures 9.17 and 9.18**

■ The mapping of tactile information from the body to somatosensory cortex is maintained by experience-guided activity of afferents. Removing whiskers in young mice results in expanded input to the cortex from the surviving neighboring whiskers, at the expense of the loss of input from the missing whiskers. **See Figure 9.19**

■ In monkeys, reducing afferent activity from a digit results in less representation in somatosensory cortex (S1). Increasing activity of a digit causes its representation in S1 to expand. Similarly, in humans who lose a hand in adulthood, the representation of body parts in neighboring S1 expands, a process that can be reversed if a new hand is transplanted. These results indicate that mapping of tactile afferents to S1 is an experience-dependent competitive process going on throughout life. **See Figures 9.20–9.22**

Go to the Companion Website
sites.sinauer.com/fond
for animations, flashcards, and other review tools.

CHAPTER

10

Maximizing Fitness
SOCIALLY GUIDED
NEURAL DEVELOPMENT

GENIE: GROWING UP ALONE The middle-aged woman had come to the Los Angeles welfare agency seeking financial assistance, but the social worker was more interested in the woman's daughter, "Genie." The girl appeared to be 6 or 7 years old and walked in such a curious fashion, holding her hands up and hopping like a bunny. When she wasn't hopping, Genie rocked her body back and forth. Plus, the girl seemed completely unaware of any of the other people around her. Then the social worker was shocked to learn that this tiny little girl was 13 years old, twice the age she appeared to be! Suspecting severe malnutrition, the social welfare officers investigated the girl's home and discovered the horrible truth. Genie had been living alone, locked in a room in the back of the house, eating the small amount of food that was tossed in to her. She had spent her days strapped, naked, to a potty-chair. At night Genie was diapered and placed in a cage with a chicken wire lid. Her father, who had enforced Genie's abuse, committed suicide before facing criminal charges.

Genie showed almost no evidence of language—the only words she said were "Stop it" and "No more." She would sometimes erupt into silent rages, hitting and tearing at her own body. She drooled and spat constantly and masturbated openly and excessively. A team of therapists and scientists rallied around Genie to provide her every opportunity to recover from her horrific childhood. With proper nutrition, she quickly gained weight and grew taller, but could her mind recover?

CHAPTER PREVIEW

As we've followed the developing individual in previous chapters, the nervous system has gotten progressively more complex in structure and functional capacity. Of course this is the story of biological development, going from simple structures to ever more complex structures until old age breaks the whole organism down again. Because one of the functions of the brain is to store information, learning about the world and using that learning in the future, we can think of that increasingly complex nervous system as gaining information. At first, the only source of information available to the embryo comes from genes, which regulate one another's expression in an unfolding pattern so that a rather complex nervous system is formed without need for any other source of information. Later,

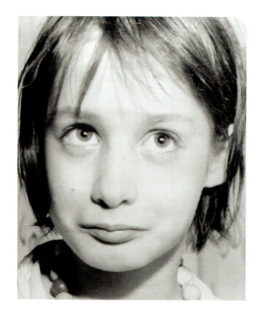

Genie Due to her severe malnutrition growing up, Genie appeared younger than she was. Upon removal from her abusive home, she showed little interest in other people and did not seem to be able to tell them apart. (© Bettmann/ Getty Images.)

conspecific An individual of the same species.

neural activity can determine the fate of synaptic connections (Chapter 8), so once sensory systems are online, the developing brain has a new source of information to guide development—any detectable event in the environment offers a chance for an even more complex and more adaptive brain (Chapter 9).

For those organisms that are cared for by their parents and interact extensively with other **conspecifics** (members of the same species) before they reach maturity, the ability of sensory experience to guide neural development offers another rich source of information to bring further complexity to the brain—social experience. By interacting with parents, developing individuals can gather information, not only about the environment today, but about past events as well. They can learn where to find food, how to travel safe routes for seasonal migration, and how to recognize and attract a suitable mate. This final chapter of the book concerns that rich source of nearly infinite amounts of information to guide neural development, interactions with other individuals.

We'll begin by discussing a false dichotomy that has plagued the study of neural development, the idea of instinctive or innate behavior. Close examination of that vague attempt to separate simple from complex behavior leads us to a much more useful, and more clearly defined, distinction—namely, those behaviors that properly develop only through interactions with conspecifics, that is, social experience. Then we'll consider the social interaction that is crucial for survival in many species, including all the birds and mammals— parental care. We'll see that parental behavior also provides the young animal with information about the species, including what sort of mate will be appropriate when it grows up. In many bird species, parental behavior also affords the young chicks the opportunity to learn what sort of song the male offspring should sing in adulthood, and what sort of song the female offspring should find attractive when they grow up.

Perhaps no species benefits more from parental behavior than our own. As we noted in Chapter 4, the human brain is about as big as it can be at birth to fit through the mother's pelvic girdle safely. Then after birth, brain growth continues at such a rapid pace, a rapidity seen only in fetal development in other species, that we can think of the first few years of life as fetal development outside the womb (see Figure 4.17). Another indication of how crucial this stage of development is for proper neural development can be seen when we consider primates that are reared with ample food, water, and shelter, but without parents. As we'll see, missing out on social guidance during development produces a thoroughly dysfunctional monkey, one that would have no chance of reproducing in the wild. We'll examine these same issues in humans, where social interactions are crucial for development of language, which shapes so much of our thinking. The absence of social experience severely stunted Genie's mental development. After parental care is complete, peer interactions among juveniles provides stimulation critical for brain development, including myelination. In considering the role of social influences on the human brain, we'll discuss IQ scores to conclude the chapter. It's clear that IQ tests measure an important human trait and that they are influenced by genes. However, we'll also find that cultural developments have, in just one century, greatly boosted humanity's intelligence, at least as measured by IQ tests. Thus in considering the controversial issue of group differences in average IQ scores, we'll find plenty of reason to think that those differences may have nothing to do with genetic differences between the groups and could easily be due to environmental influences alone.

■ 10.1 ■
THE TERMS *INSTINCT* AND *INNATE* ARE SO VAGUE THAT THEY ARE WORTHLESS

Ancient Greek philosophers like Heraclitus and the Stoics insisted that humans are the only animals possessing "reason," what we today regard as logical thinking, in contrast with purely emotional behavior of animals. Incorporating this viewpoint into Christian theology, Albertus Magnus (1200–1280) and his student Thomas Aquinas (1225–1274) proposed the concept of **instinct** to explain how animals could display complex behavior, despite having no soul (Beach, 1955; Büchner, 1880/2013). The idea was that our soul guides our behavior, earning us either eternal reward or damnation after death, while animals rely strictly on instinct. Thus animals could not be held accountable for their behavior (or go to heaven), a view embraced by seventeenth-century philosopher René Descartes, who insisted that animals have no soul and therefore are mere "machines" (Descartes, 1637, part V).

> Thomas Aquinas proposed the concept of instinct to explain how animals could display complex behavior, despite having no soul.

In contrast, Darwin insisted that our behavior was part of a continuum that had evolved over time. He used the term *instinct* to mean simply an inherited behavioral tendency acquired by natural selection, without any consideration of how an individual comes to display the behavior (Darwin, 1859). In other words, he did not concern himself with how the behavior arises *during ontogeny*, which is understandable given the limited understanding of genetics in his day. Use of the term *instinct* grew steadily in the study of animal behavior until the early 1920s, when several critiques pointed out that the term was difficult to define and therefore hard to test (Z. Y. Kuo, 1921, 1922), as we'll see shortly. Use of the term then dropped precipitously, declining further still after similar critiques a generation later (Beach, 1955; Lehrman, 1953) (**FIGURE 10.1**). Perhaps we're finally rid of the term *instinct*, but over the past 30 years another term applied to behavior, **innate**, has arisen like a zombie to fill in for *instinct* in scientific reports, with the very same problems as the older term.

There are two problems with applying the term *instinctive* or *innate* to behavior. First, if it's not actually defined, the reader can't know what the writer means by the term. This is a big problem, because one review found at least 26 different definitions of the term *innate* applied to behavior (Mameli & Bateson, 2006)! Second, no matter which of the several meanings is intended, it's difficult, often impossible, to test whether a particular behavior actually meets each criterion.

instinct   A vaguely defined term said to typify some behaviors of some nonhuman animals.

innate   A vaguely defined term, roughly equivalent to *instinctive*, said to typify some behaviors of some nonhuman animals.

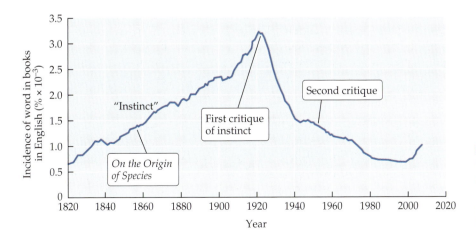

FIGURE 10.1   Popularity of the term *instinct* in books in English This Google Ngram analysis is based on incidence of the word in over 5 million books in English. (After Michel et al., 2011.)

Four common ideas have been associated with the terms *innate* and *instinctive* when applied to behavior. Critical evaluation of each criterion would result in applying the term to either every behavior or none:

1. **Behavior that has a genetic component** We've seen in the first seven chapters of this book that genes play a crucial role in the developing nervous system, so of course *genes play a role in the appearance of every behavior of every organism*. For example, behavior geneticists find that genes influence not just psychiatric disorders (Geschwind & Flint, 2015), but even apparently arbitrary behaviors like television watching (Plomin, Corley, DeFries, & Fulker, 1990) and political attitudes (Alford, Funk, & Hibbing, 2005). Are those innate behaviors? An adjective that can be applied to every behavior no longer serves any purpose. Sometimes writers might talk about behavior that is *"largely* genetic" or has a *"strong* genetic influence," but unless there's an objective definition of when genetic influence graduates from being a mere "influence" to a *"strong* influence," there's no way to test whether a behavior meets that criterion.

2. **Behavior that the nervous system is organized to support** Like the previous criterion, this definition, which crops up occasionally (Haidt & Joseph, 2007; Liedtke et al., 2011), would apply to *every* behavior of *any* organism that has a nervous system. If the nervous system was not "pre-wired" to permit the individual to display the behavior, then *by definition* the animal could not possibly display it!

3. **Behavior that does not rely on experience** The problem with this definition is that neural activity (Chapter 8) and experience (Chapter 9) play a crucial role in normal brain development. Every sensory system that has been examined in detail provides evidence of experience sharpening synaptic connections. It's difficult to think of a behavior that is not a response to some sensory input (a behavior that appears at random, in response to nothing, is unlikely to be adaptive). If experience was crucial for development of the sensory system that detects cues triggering the behavior, then experience was important for the animal to display that behavior. In that case, no behavior would be innate.

There is also a practical problem with describing instinctive/innate behavior as independent of experience—it is impossible to try an infinite number of sensory manipulations to prove that none of them affect the behavior. Older experiments tried to do this by raising animals in isolation, but the problem is deciding which experiences, exactly, to deprive the animal of (Lehrman, 1953). Unless you isolate the animal from *all experiences*, it's always possible that there was a particular experience that played an important role in the development of the behavior (Hailman, 1969). Furthermore, given the results reviewed in Chapter 9, we would expect that an individual that had grown up isolated from *all* sensory experience would develop a very abnormal nervous system, including degraded sensory systems.

4. **Behavior that is unlearned** Like the previous meaning, this criterion is difficult to test, because it is hard to prove that no learning has taken place. Even while you're observing an animal, it may be learning something without your knowledge. And how do you know, or prove, that the individual didn't learn some component of the behavior *before* you began observing? Even in the egg (Gottlieb & Kuo, 1965) or uterus (DeCasper & Fifer, 1980; Locke, 1993), the animal may be learning something.

Lest you think I'm being unfair in criticizing the term *innate*, let's consider a specific example. With a few sporadic exceptions, the term *innate* was rarely applied to behavior in the neuroscience literature until an influential article applied the term to egg-laying behavior in the sea slug *Aplysia* (Scheller & Axel, 1984).

Subsequently, more and more papers, primarily dealing with invertebrates such as *Drosophila* and *C. elegans*, but also vertebrates (Axel & Carniol, 2014), applied the term *innate* to behavior, usually without defining what, exactly, *innate* meant. At least the 1984 article defines the concept, so I've added letters to break down that definition for analysis: "[innate behaviors] are [A] shaped by evolution and are inherited by successive generations; [B] largely unmodified by experience or learning, they are [C] displayed by all individuals of a species and [D] not by other species" (Scheller & Axel, 1984, p. 54).

Examining each of the four components of this definition reveals that they can be applied to either all behaviors or no behaviors or they are untestable. The first two parts of the definition are simply restating the criteria we listed above and have all those problems identified over half a century ago (Beach, 1955; Lehrman, 1953): *all* behaviors are shaped by evolution and influenced by genes, and it is impossible to exclude any experiential effects or learning. What's the threshold for being "largely" unmodified by experience or learning? Added to those old problems, this definition adds two additional, less common, criteria. As for part C—the behaviors are "displayed by all individuals of a species"— that is untestable, requiring us to observe every individual. And does this mean a behavior shown by only half of a species, say females only, could not possibly be innate? As for part D—that the behavior is displayed by no other species—quite apart from again setting up the impossible requirement that we now must observe *all the other species on Earth*, is clearly nonsense. Really, if two different species from the genus *Aplysia* displayed egg-laying behavior, would the behavior then no longer be considered innate in either species? How then could behaviors evolve over time to appear in more than one descendant species?

By the time you actually ran all these tests to see if a particular behavior met the criteria to be considered innate, you might ask yourself *why you even care*. Will you conduct experiments to probe the role of genes or the environment upon the behavior differently, depending on whether the behavior is innate? Will you interpret the results of those experiments differently, depending on whether the behavior is innate? If you stopped using the word at all, what would you lose?

You might wonder why, if the term is as problematic as I claim, it keeps popping up, first as *instinct* and then zombified as *innate*? Philosophers studying the history of the term *innate* suggest it is really a vague "folkbiology" concept, mixing up several ideas such as whether the behavior is species-typical, important for that species' ecological niche, insensitive to the environment, and so on (Linquist, Machery, Griffiths, & Stotz, 2011). In the end, it's hard to shake the feeling that *instinct/innate* is used primarily, and probably unconsciously, to emphasize that ancient notion that animals are simple ("just a machine" [Descartes, 1637, part V]) while we humans are complex, having a "soul." But if the goal is to separate "simple" from complex nervous systems or behaviors, a more useful distinction might be whether the behavior is shaped by social influences. Of course, for mammals and birds, the first and arguably most important social stimulation is provided by parents, our next topic.

> By the time you actually ran all these tests to see if a particular behavior met the criteria to be considered innate, you might ask yourself *why you even care.*

■ 10.2 ■
SPECIES WITH PARENTAL BEHAVIOR DEVELOP THE MOST COMPLEX BRAINS AND BEHAVIOR

We've see that the important dividing line between simple and complex behaviors cannot be between those that are instinctive/innate/genetic versus those that are acquired/learned/experience dependent. That is a false dichotomy, as all behaviors are influenced by genes, and experience has at least the potential to influence any behavior. But an important gradation in the complexity of behavior can be made: between those behaviors that can be displayed by individuals without any social interactions during development and those behaviors that are properly displayed only with the guidance of parents and/ or social interactions with other conspecifics.

In many species, including nearly all invertebrates and most fish, amphibians, and reptiles, individuals never meet their parents. For example, female sand wasps display a very specific sequence of behaviors: dig a nest, fly off to paralyze a caterpillar, bring the caterpillar to the bottom of the nest, lay an egg on the prey, close the opening to the nest, then periodically bring more caterpillars to each of several nests to provide for the growing larvae (**FIGURE 10.2A**). Each female sand wasp displays this sequence of behaviors *even if it never observes its mother doing them* (Tinbergen, 1958). In other words, the behavior is displayed without any explicit instruction or an opportunity to observe another individual. Of course individuals of these species eventually interact with each other (otherwise they would never reproduce), and their brains are surely affected by those encounters, but they enter that social arena without any parental guidance and are successful.

You can probably think of several vertebrate species in which individuals never meet their parents and therefore cannot rely on their social guidance to develop properly. Consider sea turtles hatching on the beach weeks after their mother departed, frogs hatching from a clump of eggs deposited in a pond, or those fish that reproduce by releasing eggs and sperm on a beach at

(A)

(B)

(C)

FIGURE 10.2 Many animals do not require parental guidance to develop. (A) Sand wasp females display a very specific sequence of behaviors to build nests without ever seeing others do it. (B) Sea turtle hatchlings are on their own as they scramble to the sea. (C) In most fish and amphibian species, individuals never meet their parents. (A © Itsik Marom/Alamy; B © Konrad Wothe/Minden Pictures/Getty Images; C © BiosPhoto/Alamy.)

some point in the lunar cycle (**FIGURE 10.2B,C**). If we hatch a fish egg alone in an aquarium, experience or learning may play a role in the development of the behaviors it displays as an adult, but at least we can be sure that *social interactions* weren't required for them to appear.

While there are many species of fish, amphibians, and reptiles that display parental behavior, they are the exceptions, not the rule. On the other hand, *all* birds and mammals *require* parental behavior to survive, and it's probably no coincidence that those species have the largest brains and display the most complex behaviors. Among mammals, humans have the most extensive parental behavior and therefore the most extensive social influences on the brain as it goes through a remarkably long, fetal-like development outside the womb, as we discussed in Chapter 4 (see Figure 4.16A). It seems likely that this extensive social experience normally drives that extensive brain development and complex behaviors, especially social behaviors like language. An adjunct to that idea would be that absence of social stimulation in these species would result in abnormal brain development. In the remainder of this chapter, we'll see that both are true: social experience directs brain development in species with parental care, and deprivation from social exposure profoundly stunts brain development.

How's It Going?

1. What are some of the meanings attached to the terms *instinct* and *innate*, and why are they problematic?
2. How could we test whether social experience is required for a fish to display particular behaviors?

◼ 10.3 ◼
MATERNAL BEHAVIOR CAN REGULATE THE STRESS RESPONSE OF OFFSPRING

The evolution of avian and mammalian development has made parental behavior obligatory—offspring cannot survive without an extended period of parental assistance immediately after hatching or birth. Parents provide shelter and food, which are of course required, but they provide a lot more, too. Even in rats, maternal attention makes a difference. Some rat mothers are more attentive than others. The attentive mothers spend more time nursing their pups and more time licking them clean. For example, rat mothers that lick their pups' anogenital region more cause more motor neurons to survive in the spinal nucleus of the bulbocavernosus (SNB) (Moore et al., 1992).

Another demonstration of maternal influence in mice can be seen in inbred strains, where every individual has virtually the same genotype. These strains display consistent differences in behavior. For example, mice of the B6 strain tend to spend more time exploring the center of an open field than BALB mice. We may be tempted to think this and other differences in behavior between these strains are due to differences in their genes alone (we may even be tempted to think of them as innate differences). But when B6 embryos were transferred into the uteri of BALB dams, and especially when they were also reared by BALB dams, the B6 mice in adulthood acted like BALB mice in many ways (**FIGURE 10.3**).

Another effect of maternal behavior can be seen in rats, where pups raised by more attentive dams are less fearful in adulthood and show less dramatic responses to stress, including releasing smaller amounts of *corticosterone*, a **glucocorticoid** steroid hormone released from the adrenal glands in response to stress (in humans, the principle glucocorticoid hormone is *cortisol*). Con-

glucocorticoid   A class of steroid hormones, primarily released from the adrenal cortex, named for their role in regulating glucose metabolism.

FIGURE 10.3 Maternal factors and postnatal care can affect the behavior of genetically identical mice. (After Francis et al., 2003.)

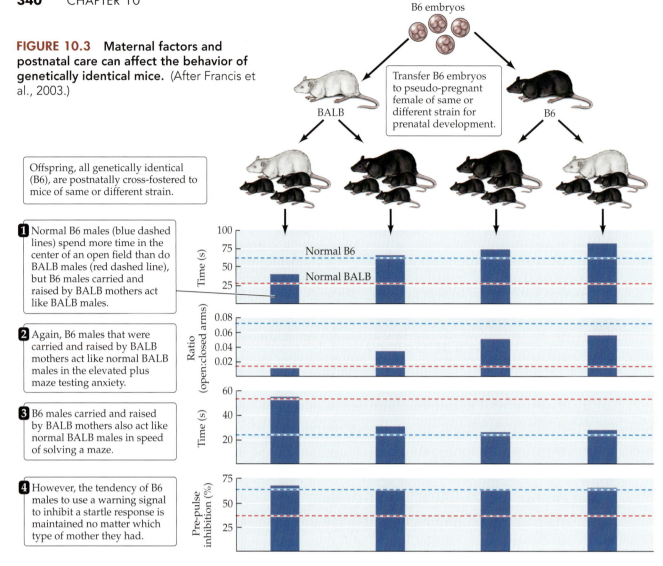

Offspring, all genetically identical (B6), are postnatally cross-fostered to mice of same or different strain.

1 Normal B6 males (blue dashed lines) spend more time in the center of an open field than do BALB males (red dashed line), but B6 males carried and raised by BALB mothers act like BALB males.

2 Again, B6 males that were carried and raised by BALB mothers act like normal BALB males in the elevated plus maze testing anxiety.

3 B6 males carried and raised by BALB mothers also act like normal BALB males in speed of solving a maze.

4 However, the tendency of B6 males to use a warning signal to inhibit a startle response is maintained no matter which type of mother they had.

glucocorticoid receptor
A member of the steroid receptor superfamily that normally binds and responds to glucocorticoids.

epigenetic A variably defined term that often refers to a change in the genome, other than the sequence of nucleotides, that has a lasting effect on expression of a gene.

An inattentive mother can alter the brains of her daughters to make them inattentive mothers, too.

sequently, stress doesn't upset the health of these adults as much as it does those raised by inattentive mothers. No matter what genes the rat pups carry, if they are raised by attentive mothers, they grow up to be less fearful and release less corticosterone in adulthood. If raised by less attentive mothers, they are more fearful and release more corticosterone (Weaver et al., 2005).

How does maternal behavior have this effect? In the brain, neurons possessing the **glucocorticoid receptor** detect adrenal stress hormones like corticosterone, regulating any further release of the hormone, as part of a negative feedback loop. In pups that are raised by an inattentive mother, the gene for the glucocorticoid receptor gets modified in hypothalamic neurons; specifically, the promoter is methylated, reducing glucocorticoid receptor expression (**FIGURE 10.4A**). When these pups grow up, because of reduced glucocorticoid receptors in the hypothalamus, they are less responsive to negative feedback, so more glucocorticoids are released and they are more anxious. What's more, when the daughters of these inattentive mothers grow up and have pups of their own, they are less likely to groom their own pups! Thus an inattentive mother can alter the brains of her daughters to make them inattentive mothers, too.

This mechanism, altering a gene's expression without affecting the sequence of nucleotides, is often referred to as an **epigenetic** mechanism. Having an

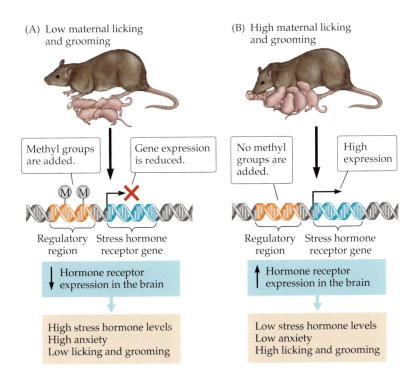

(A) Low maternal licking and grooming

(B) High maternal licking and grooming

Methyl groups are added.

Gene expression is reduced.

No methyl groups are added.

High expression

Regulatory region

Stress hormone receptor gene

Regulatory region

Stress hormone receptor gene

↓ Hormone receptor expression in the brain

↑ Hormone receptor expression in the brain

High stress hormone levels
High anxiety
Low licking and grooming

Low stress hormone levels
Low anxiety
High licking and grooming

FIGURE 10.4 **Maternal care can have epigenetic effects that can be transmitted across generations.** (After Hackman et al., 2010.)

attentive mother epigenetically modifies the glucocorticoid receptor gene in the pups, which grow up to epigenetically modify the genome of their own pups, so this epigenetic mechanism is acting across generations. Note that this more modern usage of the term *epigenetic* is rather different from the older term epigenesis that we discussed in Chapter 1. In fact, there is some confusion about even this modern usage of the term *epigenetic*, as we discuss in **BOX 10.1**.

Conversely, when rat mothers groom their pups extensively, that stimulation of the pups causes a protein cofactor to bind to the receptor gene, which prevents methylation of the gene in the brain (**FIGURE 10.4B**). These pups

BOX 10.1

KERFUFFLES IN LANGUAGE: *EPIGENETIC*

Recall from Chapter 1 that *epigenesis* was the term Aristotle used to describe the process he observed in embryonic chicks—the gradually changing shape of the body as it acquires new structures and grows more complex with time. This perspective contrasted with *preformationism*, the proposal that the structures in all their complexity were there all along, in really, really tiny form, and that development was a matter of simple growth—a change in size but not in complexity. The demise of preformationism, with its obvious limitations, meant that both terms fell into disuse. Thus the term *epigenesis* was available for a second career, with a new meaning for the adjective form, *epigenetic*.

Unfortunately, what exactly is meant by this term is something of a moving target. Initially, the modern term *epigenetic* referred to a change that one could

inherit, other than a mutation in the sequence of nucleotides in DNA, that altered the expression of a gene (Ledford, 2008). Thus a gene that becomes methylated in the parent may be passed on as a methylated gene in an offspring.

But in some circles, the definition has been expanded so that epigenetic change is any change in chromosomal structure other than the sequence of nucleotides that affects gene expression, *whether it can be inherited or not*. Thus any influence, even in adulthood, that affected the acetylation or methylation of the DNA molecule, making the gene more or less likely to be expressed, would be considered to have had an epigenetic effect. Note that this definition, unlike the previous one, does not require that the change be passed onto offspring.

continued

BOX
10.1

continued

KERFUFFLES IN LANGUAGE: *EPIGENETIC*

This usage is especially problematic, as probably any change that affects an organism is also changing gene expression in cells somewhere in the body (including the brain). For example, the classical action of steroid hormones is to alter gene expression—should that be considered an epigenetic influence? Are adolescents who are exposed to gonadal steroid experiencing epigenetic changes? When a traumatic event and subsequent glucocorticoid release alters gene expression in the brain, is that an epigenetic effect? Or when a transcription factor is expressed in a cell and alters expression of other genes, is that an epigenetic effect? If so, then anything that affected an organism would fit the definition, at which point the term *epigenetic* would no longer serve any purpose.

To make matters even more confusing, sometimes a gene may not be methylated in the parent but is nevertheless more likely to be methylated in the offspring if it is inherited from the father rather than the mother (Perez et al., 2015). Or to mix up matters even more, what about the case we describe in the main text, when rat dams grooming their pups alter the methylation of the glucocorticoid receptor gene in the pups' brains, which causes those offspring to groom their own pups in the same manner? Here methylation patterns are repeated across generations, not because they were *inherited*, but because they were passed on through social experience. Unfortunately, there is no widely adopted consensus on what should or should not be considered epigenetic. This ambiguity of meaning in the field is problematic because, as the figure shows, the term is becoming increasingly more common in the literature. Thus in reading this book and other sources, you'll have to rely on context to determine what the authors mean by *epigenetic*.

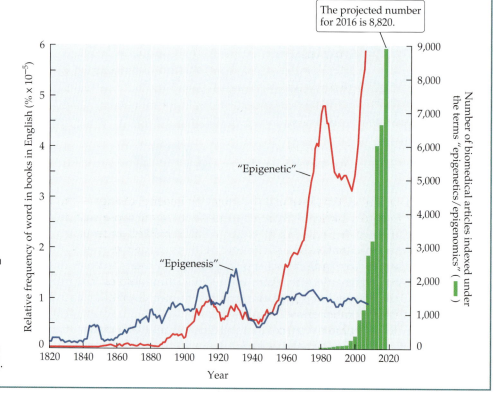

The projected number for 2016 is 8,820.

***Epigenetic* in books and PubMed articles in English** Data from 2016 are projected from the first 3 months of the year. Search terms: ("epigenomics" [MeSH Terms] OR "epigenom- ics" [All Fields] OR "epigenetic"[All Fields]) AND English[Language].

display greater expression of the glucocorticoid receptor gene in the brain, making them more sensitive to negative feedback, which reduces their adrenal steroid responses as adults (Hackman, Farah, & Meaney, 2010). These pups are also less anxious, and the female pups are more likely to be attentive mothers when they grow up (D. Liu et al., 1997).

Does this fascinating mechanism, where early neglect imprints a gene and thereby decreases its expression, have any relevance for humans? Examination

(A)

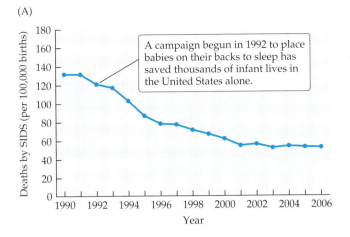

A campaign begun in 1992 to place babies on their backs to sleep has saved thousands of infant lives in the United States alone.

FIGURE 10.5 Back to sleep (A) The Back to Sleep campaign saved the lives of thousands of infants. (B) However, infants placed on their bellies to sleep reach several motor development milestones sooner than babies placed on their backs. (A from Task Force on Sudden Infant Death Syndrome & Moon, 2011; B after Davis et al., 1998.)

(B)

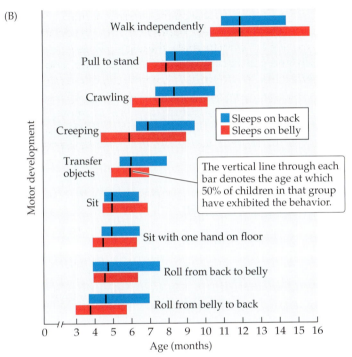

The vertical line through each bar denotes the age at which 50% of children in that group have exhibited the behavior.

of the brains of suicide victims revealed the same "epigenetic" changes—the methylation of the glucocorticoid receptor gene—but only in those suicide victims who had been abused or neglected as children (McGowan et al., 2009). The implication is that early neglect modified the gene in these people, making them less able to handle stress and therefore more likely to become depressed and commit suicide. The brains of suicide victims with no history of early abuse did not show the epigenetic changes, so their depression may have been a response to other, more recent, influences.

Of course, abusive parents represent an extreme example of social influence on behavior that is relatively rare. But a more modest difference in parental behavior affecting human development was demonstrated by an unintended experiment conducted in the United States. Scientists discovered that **sudden infant death syndrome** (**SIDS**, sometimes called crib death), the unexplained failure to breathe that sometimes strikes infants before the age of 1, is more common when babies are placed on their stomachs to sleep. An NIH campaign started in 1992, Back to Sleep, encouraged parents to place babies on their backs to sleep. This campaign was remarkably successful, reducing the incidence of SIDS by over 50% in 10 years (**FIGURE 10.5A**) (Task Force on Sudden Infant Death Syndrome & Moon, 2011), saving the lives of tens of thousands of infants. But there turns out to be a slight disadvantage for babies who are put on their backs to sleep—fewer chances to crawl on all

sudden infant death syndrome (SIDS) Also called *crib death*. The unexplained failure to breathe that sometimes kills infants in the first year of life.

fours. When scientists compared babies who had been put on their backs or stomachs to sleep, they found that those placed on their backs were slightly delayed in reaching various milestones in motor behavior (**FIGURE 10.5B**) (Davis, Moon, Sachs, & Ottolini, 1998). There is no evidence that this delay in motor behaviors has a detrimental effect on the infants (Pin, Eldridge, & Galea, 2007), and of course a slight delay in motor development is a small price to pay for reducing the risk of death. But this episode illustrates how a seemingly small difference in parental behavior can have a measurable effect on the behavioral development of their offspring.

Another important consequence of parents' behavior is that the young have a chance to learn what adults of their species look like, which can be important for later mate choice, as we see next.

How's It Going?

1. Describe an experiment that demonstrates that not all differences in the behavior of inbred mouse strains are due to differences in genes.
2. How do differences in maternal behavior affect gene expression in the rat brain?
3. How do differences in laying babies down to sleep affect motor development?

■ 10.4 ■

MANY SPECIES LOOK TO THEIR PARENTS TO RECOGNIZE MATING PARTNERS

In some bird species, especially those that spend a lot of time on the ground, the hatchlings avidly follow their mother in the search for food. The young birds learn to recognize their mother in a process called **imprinting**. Chicks will learn to follow any individual, even a human or an inanimate object, if they see it moving at a relatively slow rate in the first few hours after hatching (Spalding, 1872). Growing up on a farm, I watched two rescued eggs hatching and was surprised to find that those two chicks eagerly followed me wherever I went for weeks, and avoided other people or even hens. The famous ethologist Konrad Lorenz once had a batch of graylag geese hatchlings imprint on him (**FIGURE 10.6A**). Imprinting illustrates how natural selection can favor experience-dependent plasticity to shape behavior. Rather than genetically specifying a complicated neural program to recognize "hen," natural selection hit upon a simple program by which hatchlings look for a slow-moving object and memorize that as the thing to follow. Except in very rare, artificial conditions (as when Lorenz or I supervised egg hatching), that object is the mother that will protect the hatchlings and help them find food. In other words, the chicks' predisposition to seek out a social relationship improves their chances for survival.

Even in the egg, ducklings are influenced by auditory stimulation, such as their mother's calls, which helps them identify her at hatching. If the egg is incubated without this auditory stimulation, the duckling is much less likely to follow her call. Interestingly, the ducklings also vocalize before hatching, and exposing an egg to those sounds alone also helps the hatchling recognize the maternal call (Gottlieb, 1981, 2007). In zebra finches, auditory communica-

> Rather than genetically specifying a complicated neural program to recognize "hen," natural selection hit upon a simple program by which hatchlings look for a slow-moving object and memorize that as the thing to follow.

imprinting   Here, the tendency of young animals, primarily birds, to visually note the characteristics of a parent and then follow that parent.

(A)

(B)

FIGURE 10.6 **Imprinting for fun and conservation** (A) These graylag geese had imprinted on Konrad Lorenz as chicks. (B) These whooping cranes imprinted on this ultralight aircraft as chicks and are now following it along the traditional migratory route of this species that nearly became extinct. (A © Nina Leen/The LIFE Picture Collection/Getty Images; B © Jeffrey Phelps/Aurora Photos/Getty Images.)

tion before hatching can serve another purpose—when the weather is hot, the parents make a particular call while incubating the eggs. The embryos exposed to this call develop more slowly, which somehow helps the offspring better survive the warmer weather (Mariette & Buchanan, 2016) (a good thing in the face of global warming).

Imprinting was exploited by conservationists hoping to revive whooping crane populations in the wild. Young whooping cranes were imprinted on an ultralight aircraft, which guided them to their traditional winter feeding grounds (**FIGURE 10.6B**). Before the species became endangered, adults who had previously made the migration showed younger animals the route (www.operationmigration.org). Imagine how difficult it might be for natural selection to rely on genes alone to guide the migration.

You might think, Well how hard would it be for genes to make the animal feel like flying south in the winter and north in the summer? Indeed, there are demonstrations of such directed migratory restlessness (Helm & Gwinner, 2006), even in birds raised apart from conspecifics. For example, nest parasites, such as the Eurasian cuckoo, lay their eggs in the nests of other species, letting the foster parents raise the chicks. In adulthood, the fostered cuckoos will migrate to winter in central Africa, even if the foster parents are a nonmigratory species that stays put. But not all bird migration flight plans are as simple as "fly south." In Europe, storks fly south to Africa for the winter, but to avoid long paths over water, they must either fly along the eastern edge of the continent, over Spain and the short gap of water at Gibraltar, or fly along the western edge, over Turkey and the Middle East (Bobek et al., 2008). Storks in western Europe follow the western path, while those in eastern Europe follow the eastern path (**FIGURE 10.7**). When fledgling eastern storks about to migrate for the first time were displaced to western Europe after the local birds there had left, they took off to the southeast (and perhaps perished in the Mediterranean). But eastern storks displaced to

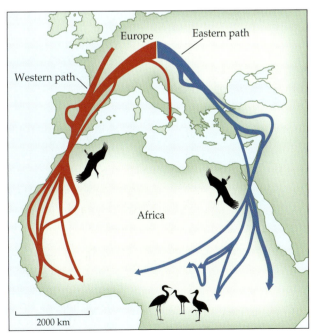

FIGURE 10.7 **Migratory pathways of European storks that winter in Africa** (After Bobek et al., 2008.)

(A)

(B)

FIGURE 10.8   **Not all finches look alike.** (A) Zebra finches and (B) Bengalese finches look quite different. Zebra finch hatchlings raised by Bengalese foster parents will, in adulthood, court and attempt to mate with Bengalese finches. (A courtesy of David McIntyre; B © Arco Images GmbH/Alamy.)

sexual imprinting   The tendency of many birds and mammals to prefer mating with individuals that resemble their parents.

the west *before* the local birds had departed altered their course, migrating southwest along with the local storks (Gwinner, 1990). So whatever direction the eastern storks may be inclined to follow, if social cues are available, they will use them to learn the migratory path.

Later in life, another type of imprinting, **sexual imprinting**, improves the hatchlings' chance for reproduction. When the birds reach sexual maturity and begin seeking mates, the individuals they find most attractive are those that resemble their parents. This preference is not built into their nervous system before hatching, but is again a result of learning. If birds are raised by foster parents, as when zebra finch hatchlings are raised by Bengalese finches (**FIGURE 10.8**), then the newly matured birds seek to mate with individuals like their foster parents, not their biological parents (Immelmann, 1972). In nature, where animals are almost always reared by their biological parents, sexual imprinting allows the hatchlings to recognize their own species so that (normally) they seek out a partner that has at least a chance of producing fertile offspring. A little later in this chapter, we'll see another important reason for songbirds to attend to their parents, as males learn to produce a song like that of the father that raised them (whether a biological father or foster father), and females grow up to prefer that type of song in choosing their own mates.

> Sexual imprinting allows the hatchlings to recognize their own species so that they seek out a partner that has at least a chance of producing fertile offspring.

In a similar fashion, birds tend to prefer foods, home areas, and habitats that resemble those they experienced shortly after hatching (Immelmann, 1975; Mettke-Hofmann & Gwinner, 2003). You can see how this would be a more likely strategy for natural selection to stumble upon—if the food, home area, and habitat were successful in getting this individual to sexual maturity, it may be a successful reproduction formula for it, too. Because your parents were successful in reproducing, following their example may work for you, too. (Conversely, "if your parents didn't have any children, you probably won't either.") Of course, parents are not the only conspecifics available as models for survival, as we see next.

How's It Going?

1. What are imprinting and sexual imprinting?
2. What's the evidence that young birds learn their migratory pathways from adults?

■ 10.5 ■
OBSERVATIONAL LEARNING CAN TRANSMIT BEHAVIORS ACROSS GENERATIONS

"You can observe a lot by watching." Like many aphorisms from the great baseball player Yogi Berra, this quote works because beneath the double-talk lies an important truth: you can learn a lot by watching. (Another example: "Always go to other people's funerals, otherwise they won't go to yours.") Of course you can learn a lot by observing inanimate objects, but **observational learning** refers to things one individual learns by observing another individual. We've seen one example of observational learning already, when storks that had been displaced took a new migratory route by accompanying the older, local birds.

Our dog, Dipsy, taught me about observational learning. She never showed any interest when we loaded the dishwasher until my in-laws visited with their dog. Their dog wasn't just interested in the process of loading the dishwasher; she got right in there and licked off every soiled plate as soon as it was put in. Dipsy clearly observed this behavior, because once the visiting dog left, Dipsy became an avid participant in loading our dishwasher for the rest of her days. (Don't worry, she had no interest in licking the dishes after they were washed.) This is also an example of modeling, when one individual imitates another.

Modeling behavior has been documented many times in nonhuman animals. One famous example was among Japanese macaques, also called snow monkeys. Researchers studying a wild troop of the monkeys left yams on the beach where a river emptied into the ocean. They hoped to tempt the monkeys into the open to be observed. One day a female monkey took her yam and dipped it into the river, apparently to wash the sand off. The researchers eventually named her Imo ("yam" in Japanese), and the other monkeys were watching her, too. At first only Imo's siblings and mother imitated her, but gradually other monkeys took up the practice until the whole troop was doing it. Later Imo began dipping her yam into seawater, presumably because she liked the seasoning from the salt. Again the other monkeys observed Imo's behavior, and soon they started dipping yams into seawater, too (Kawai, 1965). Imo died long ago, but this behavior has been passed on for several generations (**FIGURE 10.9A**).

In England, learning spread in a similar way when birds known as blue tits, related to the chickadees of North America, learned to open the tinfoil tops of milk bottles to steal cream (**FIGURE 10.9B**). As the bottles had been in use many years before this began, and the number of birds stealing cream increased rapidly, an individual bird must have learned the trick first, and others must have imitated it. More recently

observational learning The process by which an individual first learns how to accomplish some task by watching another individual do it.

(A)

(B)

(C)

FIGURE 10.9 Instances of observational learning (A) A Japanese macaque dips her yam into seawater before eating it. (B) Blue tits in England learned to break through the foil tops of milk bottles to skim off the cream at the top. (C) Gulls have learned to prey upon surfacing whales, grabbing chunks of fat. The gulls often zero in on those whales that have already been attacked and so are missing patches of skin on their backs. (A ©Cyril Ruoso/JH Editorial/Minden Pictures/Corbis; B ©Roger Wilmshurst/Alamy; C © Ralph Lee Hopkins/Getty Images.)

another new behavior rapidly spread among the closely related great tits of Hungary: when food becomes scarce in winter, they've started flying into caves to kill and eat hibernating bats! Again, this behavioral innovation seems to be passing down from generation to generation (Estok, Zsebok, & Siemers, 2010). Another sad foraging innovation is spreading among gulls. Some learned to peck on the backs of whales surfacing to breathe, gouging out chunks of food (**FIGURE 10.9C**). Once enough birds have done this, removing the tough skin over the patch, its even easier to get at fat under the surface, so other gulls zero in on the lesions, to the detriment of the whales. The gulls have learned that baby whales are especially vulnerable. The number of right whales with such lesions on their backs in one population increased from 2% in the 1970s to 99% by 2011 (Maron et al., 2015).

Once you become aware of these modern examples of innovations in foraging behavior that spread across individuals, and are passed down across generations by observation, you start to wonder how many other foraging behaviors of species in the wild had their origins in this manner. If the innovation took place before our ancestors began leaving written reports, we would regard them as part of the animals' natural history. We might even assume that the foraging behavior arose because genes favoring the behavior had been selected for (we might even call them instinctive, or innate), not realizing the role played by observational learning. In the cases we've reviewed, the behavior spread much too fast to result from any change in the gene pool.

In contrast, the most famous example of observational learning in birds, singing, is clearly an important part of their evolved life history, as we'll see next.

■ 10.6 ■

BIRDSONG IS A LEARNED BEHAVIOR WHERE YOUNG MALES MODEL THEIR FATHER'S SONG

In many bird species, especially among the Passerines, males sing to attract mates. The males of different species sing very different songs. Although the songs of males from a single species are similar to one another, there are noticeable individual differences in the songs produced by different males. What's more, if you study the songs of, say, white-crowned sparrows in different regions, you find there are "dialects": males in Berkeley, California, sing slightly different songs than the white-crowned sparrows north or south of there (Marler, 1997) (**FIGURE 10.10**). These individual differences come about because males learn their song by listening to their father's singing. If you take nestling white-crowned sparrows from Berkeley and place them into a nest of sparrows from across the bay, the males will learn the song of their foster father and grow up singing that dialect rather than the Berkeley dialect that their biological father sings.

Sparrows reared in isolation, without the opportunity to hear an older male, produce a very abnormal, unrecognizable song (**FIGURE 10.11A,B**). Normally, when young males exposed to their father's song first begin singing, they produce disjointed syllables in no particular order and produce a wide variety of vocalizations, which is why this stage is referred to as **plastic song**. During plastic song, the male must be comparing the vocalizations it makes to a remembrance of the father's song, because if the young male is deafened, the song is more or less arrested at this stage (**FIGURE 10.11C**). As long as the young male can hear his own vocalizations, he continues developing his song, which eventually becomes a recognizable variant of the song he'd heard earlier. At this stage, the song is said to have been *crystallized* and, in many species, is more or less stereotyped from this point on. If the male is deafened after song

plastic song An early stage in birdsong learning in which the song is highly variable.

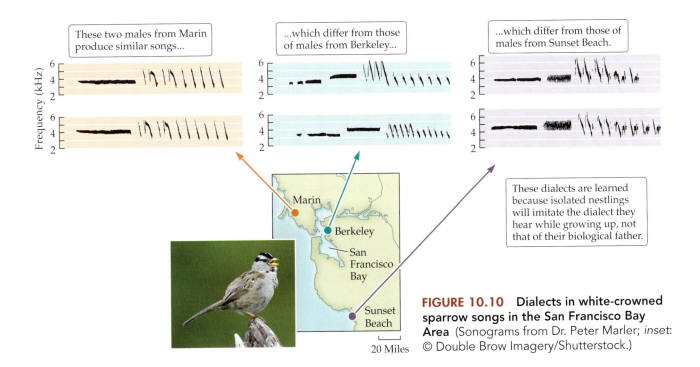

These two males from Marin produce similar songs...

...which differ from those of males from Berkeley...

...which differ from those of males from Sunset Beach.

Frequency (kHz)

Marin

Berkeley

San Francisco Bay

Sunset Beach

20 Miles

These dialects are learned because isolated nestlings will imitate the dialect they hear while growing up, not that of their biological father.

FIGURE 10.10 Dialects in white-crowned sparrow songs in the San Francisco Bay Area (Sonograms from Dr. Peter Marler; *inset:* © Double Brow Imagery/Shutterstock.)

crystallization, the song is relatively unaffected, indicating that the auditory feedback required for acquiring the song is not needed to maintain it. Because there can be a considerable delay between when the young male hears the adult song and when he begins producing the song himself, he must keep a neural representation of the song, said to be a *template*. In one study, swamp sparrow nestlings were reared in isolation and exposed to recordings of swamp sparrow song. Eight months later, without any rehearsal in between, they began plastic song and eventually produced recognizable copies of the song they had heard months earlier (Marler & Peters, 1981).

Interestingly, white-crowned sparrows reared in "borderlands," where they may hear two different dialects, may learn both dialects and exploit that ability later. Nestlings raised in isolation and hearing recordings of two different dialects will later incorporate elements of both into their plastic song. But if during that plastic song phase they are again exposed to recordings, this time of only one dialect or the other, they end up settling into that dialect for their own song (Nelson & Marler, 1994). In the wild, this ability to tweak the song to settle into the local dialect of a region may help their quest to attract a female from that region.

So how far can young males go in acquiring whatever song they hear? In nature, they must be exposed to songs of different species, so what prevents a young male from memorizing a song from some other species, which would be a disaster for attracting a female of his own species? Despite the importance of learning in the process, natural selection results in young songbirds with a predisposition to attend to some songs more than others.

crystallization Here, the process by which a young male bird makes the transition from highly variable plastic song to an adult song, which tends to be rather stereotyped.

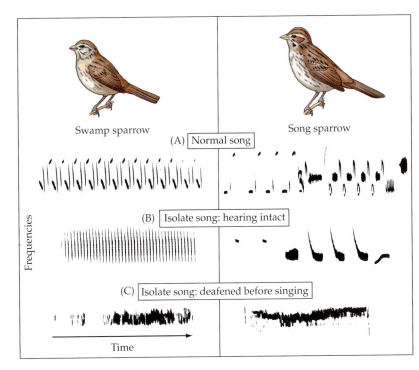

Swamp sparrow

Song sparrow

(A) Normal song

Frequencies

(B) Isolate song: hearing intact

(C) Isolate song: deafened before singing

Time

FIGURE 10.11 Birds raised in isolation from singing males develop an abnormal song (After Marler & Sherman, 1983, 1985.)

RESEARCHERS AT WORK

Sparrows Are Predisposed to Learn Species-Specific Song Elements

■ **QUESTION**: Young male birds are liable to hear the songs of other species. For example, swamp sparrows and song sparrows live within earshot of one another in much of the United States. How do they know which song to memorize and later reproduce?

■ **HYPOTHESES**: Males may have a predisposition to attend to only the song of the male that feeds them (typically the father). Alternatively, males may have a predisposition to attend to songs only of their own species, not those of any species. Or, perhaps male songbirds can learn to produce any song they are exposed to in development.

■ **TEST**: Raise swamp sparrows in isolation, and expose them to recordings of song. Splice together elements of song from swamp sparrows and from song sparrows in various patterns. In some cases, put song sparrow syllables together in patterns usually heard in swamp sparrows. Let the males grow up, and analyze what type of song they produce.

■ **RESULT**: The swamp sparrows were extremely selective in which elements they used. They only incorporated syllables from swamp sparrows into their own song, even if they'd been exposed to song sparrow syllables in patterns usually heard from swamp sparrows (**FIGURE 10.12**) (Marler & Peters, 1977).

FIGURE 10.12   Selecting the proper syllables (From Marler & Peters, 1977.)

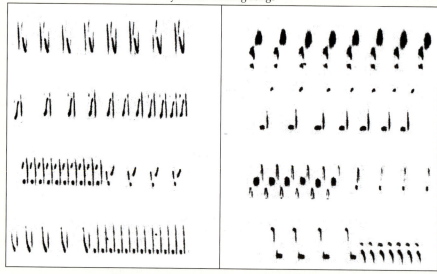

Songs produced by two swamp sparrows
exposed only to training songs

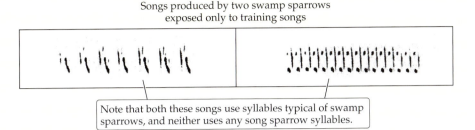

Note that both these songs use syllables typical of swamp
sparrows, and neither uses any song sparrow syllables.

■ **CONCLUSION**: While songbirds clearly attend to songs they hear in
development to produce their own song, they are predisposed to accept certain
syllables and to reject others. Thus young swamp sparrows raised in earshot of
song sparrows do not incorporate song sparrow syllables into their song. Note
in Figure 10.11B that *isolated* swamp sparrows and song sparrows each produce
syllables typical of their own species and not the other. What they lack without
hearing a tutor is the *pattern* of syllables typical of their species.

As we noted in Chapter 7, the brain regions underlying birdsong are sexually
dimorphic in those species, such as white-crowned sparrows, zebra finches, and
canaries, where males sing and females do not. (Fascinatingly, in songbird species
where both sexes sing "duets" as part of their
courtship, these same brain regions are much
less dimorphic [Brenowitz, Arnold, & Levin,
1985]). These sexually dimorphic brain regions
include a forebrain nucleus called the higher vocal center (HVC), which sends axons to another
sexually dimorphic nucleus called the robustus
archistriatum (RA) (Nottebohm & Arnold, 1976),
which in turn innervates the motor neurons of
the nucleus of the twelfth cranial nerve, which
controls vocalization (**FIGURE 10.13A**). In adult
birds, lesions anywhere along this so-called short
pathway severely disrupt song.

A fascinating aspect of birdsong learning
involves another nucleus, the LMAN (lateral
magnocellular nucleus of the anterior neostria-
tum), that is part of a "long pathway" between
the HVC and the brainstem (see Figure 10.13A).
If LMAN is lesioned before zebra finch males
have started producing song, their song is severely disrupted. On the other hand, lesions of
LMAN after the period of sensory learning is
over, when the male has started practicing the
song he learned earlier, have virtually no effect
(Bottjer, Miesner, & Arnold, 1984). Thus LMAN

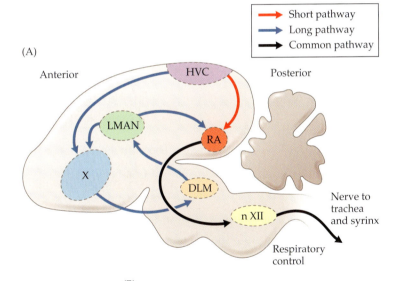

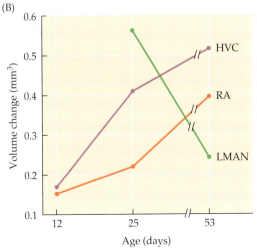

FIGURE 10.13 Birdsong nuclei in the brain (A) In adults, lesions
anywhere along the "short pathway" severely disrupt song. The
"long pathway" is important for acquiring song. Lesions of LMAN
in adulthood have no effect on song, but if the nucleus is lesioned
before the male has begun singing, the song is disrupted. (B)
After males have learned the song to be copied, LMAN shrinks in
absolute size, even as overall brain size continues to grow. (A after
Arnold, 1980; B after Bottjer et al., 1985.)

seems to play a transient role in the learning of the song to be imitated later, after which this nucleus is unneeded. Fascinatingly, the volume of LMAN reflects its transient role: the nucleus actually gets smaller in size, and loses about half its synapses, after the male begins practicing his song (**FIGURE 10.13B**) (Bottjer, Glaessner, & Arnold, 1985). This is the only example I know of in which a postfetal vertebrate brain region permanently shrinks after it has fulfilled its function. The closest approximation is in seasonally breeding vertebrates, where the sizes of neural regions important for mating behavior wax and wane with the seasons (Cooke, Hegstrom, & Breedlove, 2002; Forger & Breedlove, 1987b; Nottebohm, Nottebohm, & Crane, 1986). This role of LMAN in learning the song seems to depend on N-methyl-D-aspartate (NMDA) type glutamate receptors, as their pharmacological blockade disrupts the young male's ability to reproduce the tutor's song (Basham, Nordeen, & Nordeen, 1996).

Of course, humans learn to vocalize, too, as we see next.

How's It Going?

1. What is the evidence that young male songbirds learn the song they will sing as adults?
2. What is the role of auditory feedback in the acquisition and maintenance of birdsong?
3. What is the evidence that birds have a predisposition to learn certain elements of song?
4. Which brain region is important in song learning, but not maintenance, and how do we know that?

■ 10.7 ■

HUMANS ARE PREDISPOSED TO LEARN LANGUAGE WITHOUT ANY FORMAL TRAINING

The number of English words is now estimated to be one million and growing (Michel et al., 2011)—no one could define them all without consulting a dictionary. Still, even the average high school graduate is thought to know 50,000–60,000 words. Knowing so many words means that, in practical terms, a speaker might construct an infinite number of different sentences. Because language has this vast capacity to produce so many different sentences, it is said to be *generative*, capable of producing lots of "offspring." Most linguists believe that humans are born with the capacity to learn and use language (Chomsky, 1957).

Genetic studies support this idea. Analysis of a British family with a rare heritable language disorder led to the identification of a gene, *forkhead box P2* (*FOXP2*), that appears to be important for human language. Children with a specific mutation of the *FOXP2* gene take a long time to learn to speak (Lai, Fisher, Hurst, Vargha-Khadem, & Monaco, 2001), and they display long-lasting difficulties with some specific language tasks, such as learning verb tenses (Nudel & Newbury, 2013). The pattern of brain activation in these individuals during performance of a language task is different from that seen in typical speakers—they show underactivation of Broca's area (**FIGURE 10.14**), a brain region important in language (Fisher & Marcus, 2006). Given its powerful effects on behavior, you may not be surprised to learn that the FOXP2 protein is a transcription factor, so it probably regulates the expression of many genes. That gene regulation appears to have been related to communication for quite some time, because interfering with the *FOXP2* gene disrupts ultrasonic vocalizations in rats and mice (French & Fisher, 2014) and song learning in birds (Haesler et al., 2007).

Interfering with the *FOXP2* gene disrupts ultrasonic vocalizations in rats and mice, and song learning in birds.

forkhead box P2 (FOXP2) A gene encoding the transcription factor FOXP2, which is implicated in language in humans, song in birds, and ultrasonic vocalizations in mice.

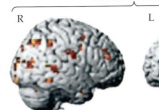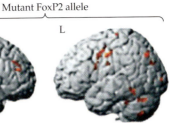

Controls   Mutant FoxP2 allele

R   L   R   L

Broca's area

In a contrast to the language deficits caused by mutations in *FOXP2*, **Williams syndrome** is caused by a loss of about 30 genes from chromosome 7 (de Luis, Valero, & Jurado, 2000). People with Williams syndrome speak freely and fluently with a large vocabulary, yet they may be unable to draw simple images, arrange colored blocks to match a sample, or tie shoelaces. Their language proficiency, which is accompanied by a highly social nature and music skill, contrasts with their otherwise pronounced intellectual deficiency, suggesting language (and social interest) may be distinct from other cognitive skills, as you might expect if language is a very specific adaptation.

All of us learn this remarkable tool, language, without explicit training, but our mastery of it nevertheless totally depends upon interaction with other people. Although there are considerable differences across individuals in the age at which various language milestones are reached, the sequence of language acquisition is remarkably consistent across children (**TABLE 10.1**).

FIGURE 10.14 **An inherited language disorder** The pattern of brain activation during a language task is different in control subjects (left) than in those with a mutant *FOXP2* allele. (From Fisher & Marcus, 2006.)

Williams syndrome A disorder characterized by fluent linguistic function but poor performance on standard IQ tests and great difficulty with spatial processing.

TABLE 10.1
Typical Stages of Childhood Language Development

Age	Receptive language	Expressive language
Birth–5 months	Reacts to loud sounds Turns head toward sounds Watches faces that speak	Vocalizes pleasure and displeasure (laugh, cry, giggle) Makes noises when talked to
6–11 months	Understands "no-no" Tries to repeat sounds	Babbles ("ba-ba-ba, da-da-da") Gestures
12–17 months	Attends to book for about 2 minutes Follows simple gestures Tries to imitate simple words	Points to objects, people Says 2–3 words to label objects
18–23 months	Enjoys being read to Follows simple commands Points to body parts Understands simple verbs	Says 8–10 words (maybe with unclear pronunciation) Asks for foods by name Starts combining words ("more milk")
2–3 years	Understands about 50 words Understands pronouns Knows spatial concepts ("in," "out")	Says about 40 words Uses pronouns such as "you," "I" Uses 2- to 3-word phrases
3–4 years	Understands colors Understands groupings of objects (foods, clothes, toys, etc.)	Is mostly understandable by strangers Expresses ideas, feelings
4–5 years	Understands complex questions Understands "behind," "next to"	Says about 200–300 words Uses some irregular verb past tenses ("ran," "fell") Engages in conversation
5 years	Understands >2,000 words Understands sentences >8 words long Can follow a series of three directions Understands time sequences (what happened first, second, last)	Uses complex and compound sentences

Sources: Adapted from American Speech-Language-Hearing Association, n.d., National Institutes of Health, 2014, and PRO-ED Inc., 1999.

parentese   The lilting, singsong style of speaking often directed at infants across cultures.

phoneme   A particular speech sound used in language. Different human languages may use different subsets of phonemes.

Across cultures (Boysson-Bardies, 2001), people speak to babies in a high-pitched, singsong fashion with slow, exaggerated pronunciation (Falk, 2004) that is called **parentese**. The lilting qualities of parentese convey emotional tone and reward, helping the baby to attend to speech and use developing memory skills to attach meaning to previously arbitrary speech sounds. Babies will work hard to hear this sort of speech. What sort of work? No, they are too young to learn to press a bar to hear sounds, but they will suck vigorously on a pacifier if that exposes them to speech sounds.

One reason babies attend to speech is that they spend their first year of life learning all the particular speech sounds, called **phonemes**, that are in use around them. Across all known human languages, over 140 different phonemes have been reported, but most languages use only a subset of these. English, for example, uses about 44 different phonemes. Some phoneme differences that we easily distinguish are physically very similar. For example, the syllables *ba* and *pa* are a lot alike and differ only in terms of how soon we vocalize (make a hum in the back of the throat) after we pop our lips apart.

You might think newborns would be poor at discriminating phonemes and must learn to tell them apart. But in fact, babies are born with the ability to distinguish *every* phoneme that is known. How do we know what newborns can distinguish when they cannot talk to us? A pioneering study exploited habituation, the tendency to ignore a repeated stimulus, to help us "read the minds" of infants, as we see next.

RESEARCHERS AT WORK

The Habituation Response Allows Us to Read Babies' Minds

■ **QUESTION**: Can babies distinguish between phonemes that are physically similar?

■ **HYPOTHESIS**: Babies who have habituated to one phoneme will notice the difference in the other, slightly different phoneme.

■ **TEST**: Have babies suck on a pacifier for a chance to hear sounds. If they are given the same phoneme repeatedly, they will habituate and suckle less. If they are then given a new sound, they will renew their sucking if they can actually tell that the phoneme is new.

■ **RESULT**: The babies did indeed increase sucking when presented with a new phoneme, even if it was only subtly different from the first phoneme (**FIGURE 10.15**) (Eimas, Siqueland, Jusczyk, & Vigorito, 1971).

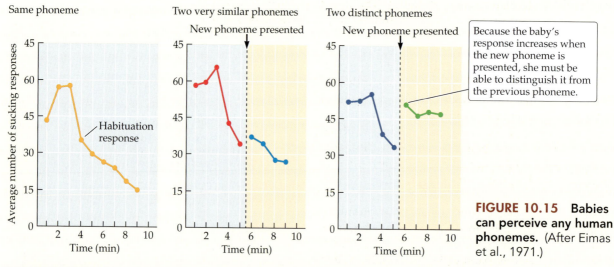

FIGURE 10.15 **Babies can perceive any human phonemes.** (After Eimas et al., 1971.)

■ **CONCLUSION**: Young babies can distinguish phonemes, even if they are only subtly different from one another. Later research would use similar methods to show that young babies can distinguish all the phonemes that have been found in *any* language.

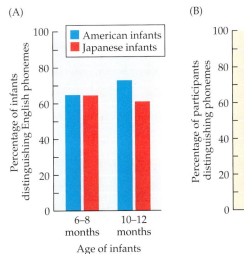

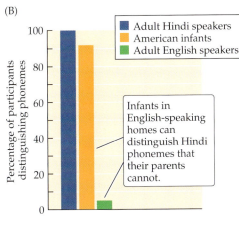

Infants in English-speaking homes can distinguish Hindi phonemes that their parents cannot.

FIGURE 10.16 Sharpening phoneme detection (A after Kuhl et al., 2006; B after Werker et al., 1981.)

Adult monkeys can also discriminate between human phonemes (Ramus, Hauser, Miller, Morris, & Mehler, 2000), so this ability may reflect a basic property of the primate auditory system. But there's more to the story about babies. By attending to the phonemes in the language spoken around them, human babies, who begin life babbling nearly all the phonemes known in all human languages, soon come to use only the subset of phonemes in use around them. Not only that, babies also get better and better at distinguishing the phonemes they're exposed to. But as they get more and more exposure to the phonemes in use around them, they slowly *lose* the ability to distinguish other phonemes. For example, Japanese newborns can distinguish between the sounds for *r* versus *l*, but if they hear only Japanese while growing up, they will find it hard to tell those sounds apart as adults. The divergence in the ability of American versus Japanese babies can be seen in the first year of life (**FIGURE 10.16A**) (Kuhl et al., 2006). As another example, native English-speaking adults have a very difficult time distinguishing some of the phonemes in Hindi, one of the official languages of India, yet 6- to 8-month-old babies from English-speaking households can detect these different Hindi phonemes just fine (**FIGURE 10.16B**) (Werker, Gilbert, Humphrey, & Tees, 1981). Babies begin losing the ability to distinguish phonemes they have not been exposed to at about the age they themselves start making halting language-like sounds, at 6–8 months of age.

> Babies, who begin life babbling nearly all the phonemes known in all human languages, soon come to use only the subset of phonemes in use around them.

Thus just as there is a sensitive period during which songbirds must hear a father's song to be able to duplicate it in adulthood, there is a sensitive period during which humans must hear phonemes to be able to detect and produce them readily. This is one reason why it is much more difficult to acquire a new language, without having an accent, if one begins as an adult. Similarly, we have a much easier time learning the grammar of a language if we begin learning it as children (**FIGURE 10.17**).

Note that, because of the difficulty of phoneme discrimination after childhood, if a phoneme becomes rare in a language for some reason, eventually there will be more and more babies that fail to hear it, and so fail to detect it. Once a phoneme falls out of the language entirely, then the next generation of babies will not be able to discriminate it, and at that point it

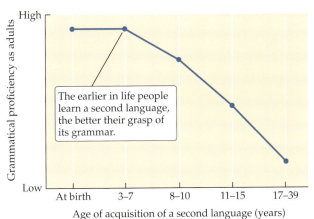

The earlier in life people learn a second language, the better their grasp of its grammar.

FIGURE 10.17 The critical period for acquiring fluency in a second language (After Johnson & Newport, 1989.)

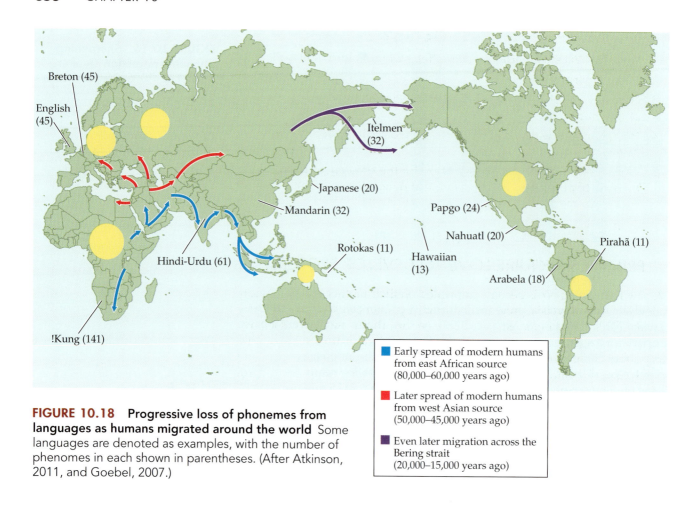

FIGURE 10.18 **Progressive loss of phonemes from languages as humans migrated around the world** Some languages are denoted as examples, with the number of phenomes in each shown in parentheses. (After Atkinson, 2011, and Goebel, 2007.)

Legend from figure:
- Early spread of modern humans from east African source (80,000–60,000 years ago)
- Later spread of modern humans from west Asian source (50,000–45,000 years ago)
- Even later migration across the Bering strait (20,000–15,000 years ago)

> Once a phoneme falls out of the language entirely, then the next generation of babies will not be able to discriminate it.

would be very difficult for the language to ever incorporate the phoneme again. In other words, there may be a "ratchet" enforcing phoneme loss: once it drops from a language, it is lost forever.

This possible mechanism for the loss of phonemes in a population may have affected the changes in languages in the course of human migration during prehistory. In a controversial theory, one linguist has suggested that human languages have shed more and more phonemes as people have migrated farther and farther from our origin in Africa (Atkinson, 2011). In a survey of over 500 languages from around the world, those with the most phonemes are found in southern Africa, which includes the so-called click languages, like the one spoken by the !Kung people of Kalahari, which has 141 phonemes. In contrast, the languages with the fewest phonemes are the Oceania languages like Hawaiian (13 phonemes) and the Amazonian languages such as Pirahã (only 11 phonemes) (**FIGURE 10.18**). The fossil record indicates that humans arose in southern Africa, where the most phonemic diversity is seen in modern languages, and anthropological studies make it clear that Oceania and South America were the last places to be colonized by humans. Fascinatingly, a study of genetic diversity also finds the greatest diversity in southern Africa, with increasingly restricted genetic diversity as one follows the presumed path of human migration out of Africa (Creanza et al., 2015; Tishkoff et al., 2009). As with phonemes, once an allele was lost from a migrating population, it would be unlikely to return.

Of course, developing humans need more than just language exposure from other people. We need love, too, as we'll see next.

───── **How's It Going?** ─────

1. Contrast Williams syndrome with effects of mutations in *FOXP2*.
2. What is parentese?
3. How can we tell whether infants can distinguish between two phonemes?
4. What happens with the ability of infants to distinguish phonemes as they grow?
5. What is the controversial explanation for the differences in the number of phonemes across human languages?

■ 10.8 ■
PRIMATES REQUIRE LOVE TO DEVELOP PROPERLY

We saw that young birds become imprinted on their mother and will avidly follow her. A controversial program of research by American psychologist Harry Harlow (1905–1981) demonstrated that newborn rhesus monkeys also have a drive to recognize and stay close to a mother figure (Harlow, Dodsworth, & Harlow, 1965). Harlow separated young monkeys from their mothers and raised them in clean, warm, safe environments with their nourishment provided by bottles placed in cloth-covered models of a monkey mother (**FIGURE 10.19A**). He found that the young monkeys would cling to the model, not just when eating, but between meals as well. Several observations indicated that the monkeys drew comfort from the models. For one thing, if a monkey happened to be away from the model when an experimenter approached, it would run and cling to the model, in what resembled stranger anxiety in human infants. If a monkey was separated from the model, it would show behavioral signs of stress, such as sucking its thumb and rocking back and forth. When given the opportunity, it would quickly run back to the model.

What does the isolated infant monkey look for in a mother? Like human newborns that prefer looking at faces rather than abstract shapes, infant monkeys prefer mother models with faces, even if they are crude models such as that in Figure 10.19A. They also prefer models that offer nourishment in the form of a baby bottle, but what young monkeys really love is a soft mother

Young monkeys love the feel of soft things, which would normally be their mother's belly.

(A)

(B)

FIGURE 10.19 **The science of love** (A) Harry Harlow found that infant rhesus monkeys separated from their mothers would form strong attachments to mother models. When stressed, the monkeys would cling to the models and seemed to draw comfort from them. Soft, cuddly models were preferred over hard models, no matter which model provided milk. (B) Motherless monkeys cling to each other. Peer interactions in monkeys can go a long way to preventing the behavioral abnormalities that result from growing up without a mother. (A,B © Nina Leen/The LIFE Picture Collection/Getty Images.)

model. When given the choice between a bare-wire model mother that gave milk or a soft, cloth-covered model that did not, Harlow's young monkeys chose the soft, cloth-covered model. Young monkeys love the feel of soft things, which would normally include their mother's belly.

Despite the comforts of the model, these motherless monkeys were permanently damaged by the absence of a real, interactive mother. As adults they could not get along with other monkeys, alternately running away from them and attacking them. This erratic behavior upset other, normal monkeys, so they often attacked the motherless monkeys. This abnormal social behavior meant, among other things, that the motherless monkeys were hopeless at having sexual relationships, never figuring out how to interact with a willing partner. More dramatically, females that had been raised without a real mother were themselves horrible mothers. They were artificially inseminated, and when their babies were born, the motherless mothers would neglect, attack, or even kill the babies. Just as these motherless mothers, when young, yearned to hold onto their screen-and-cloth models, their own unfortunate offspring kept trying to return to their abusive mothers, in heartbreaking scenes.

Fascinatingly, much of the damage caused by being raised without a real mother could be averted if the young monkeys were given regular exposure to similar-aged peers (Suomi, Delizio, & Harlow, 1976; Suomi, Harlow, & Kimball, 1971) (**FIGURE 10.19B**). Motherless monkeys that regularly played with their peers were much more socially adept as adults. These results suggest that one therapy for abused children might be interaction with other, more supportively parented children.

These experiments were and remain controversial because of the obvious distress the young monkeys experienced. In today's world it may seem that the experiments were pointless, showing what we already know. But what we know today is very much a result of these experiments. It's important to note that when these studies began, many "experts" in child rearing urged parents not to get too close to their children. Spending too much time with children was thought to make them too emotionally dependent on their parents. Well into the 1950s and 1960s, many child-rearing books urged parents not to get too physically or emotionally close to children to avoid "spoiling" them (Blum, 2002). In this context, Harlow's results, including wrenching films of young motherless monkeys rocking with anxiety and curling into balls in withdrawal, shocked the world into realizing how important cuddling and emotional support are for proper social development. Bucking the trend of psychologists of his day who felt science could only study behavior you could see and who were unwilling to use terms such as *mind* or *desire*, Harlow insisted that he was studying *love*. Those isolated monkeys loved their cloth-model mothers, and normal monkeys loved their real mothers and benefited enormously from the love those real mothers gave back to them. Harlow felt his experiments proved that young individuals have a critical need for love, and his graphic results convinced almost everyone (even if they didn't want to think of the social experience as love).

> When these studies began, many "experts" in child rearing urged parents not to get too close to their children.

The studies also boosted efforts to study **attachment**, the strong emotional bond or tie between infants and their caregivers (often the mother). Those studies would reveal the importance of secure attachment between infant and caregivers for the child's future mental adjustment (Ainsworth et al., 1978). As with Harlow's monkeys, children with insecure attachment to caregivers, typically a result of negligent parenting, benefit from having positive interactions with other children (Simpson, Collins, Tran, & Haydon, 2007).

Another heart-wrenching scene of motherless monkeys was the rhythmic body rocking they displayed whenever deprived of their cloth mothers. Genie,

attachment Here, the strong emotional bond between an infant and one or more caregivers.

whom we met at the start of the chapter, raised with very little human contact with either caregivers or peers, also rocked rhythmically. And like the mother-less monkeys, Genie never fully recovered from her isolated beginnings. She clearly formed social bonds with the adults who tried to look after her, and with intensive efforts her vocabulary expanded somewhat. But Genie never learned more than a few hundred words, rarely stringing more than a few words together for sentences, and had no understanding of grammar (Curtiss, 1977).

Was Genie mentally disabled from the beginning, as her abusive father believed? Did early malnutrition stunt a brain that would otherwise have been fine? We'll never know for sure. The rapid growth in her vocabulary after her release makes it clear that she could have learned a lot more than she did while in captivity. Also, Genie's performance on IQ tests, while well below that of other children her age, continued to improve—every year her mental age increased by about a year. This is not what we would expect to see in someone who was mentally disabled. Yet in the end, Genie never achieved a normal IQ and, with her limited language, was never able to lead a self-sufficient life. Eventually the funding to study Genie faded away, and she became just another ward of the state. Her mother has since died and, now in her 60s, Genie lives in an adult home for the mentally disabled.

As our discussion of babies' perceptual development makes clear, and Genie's severely stunted language ability emphasizes, social interaction is absolutely crucial for the development of human language, a higher-order, symbolic system of communication. Furthermore, an individual's skill with language makes up a large part of what we generally consider human intelligence and is also an important component of what we measure in IQ tests, the final topic of this chapter. But first let's consider the importance of social interactions after parental care has ended.

> In the end, Genie never achieved a normal IQ and, with her limited language, was never able to lead a self-sufficient life.

How's It Going?

1. Describe the results of Harry Harlow's experiments with monkeys.
2. Do you think monkey mothers normally provide their offspring with love?
3. What manipulation appears to at least partially counteract the problems of monkeys that grow up without a mother?

■ 10.9 ■
POSTNATAL SOCIAL STIMULATION CONTINUES TO AFFECT BRAIN DEVELOPMENT

Just as the sensory experience that is crucial for fine-tuning synaptic connections in the developing brain continues to have at least some effects in the mature, adult brain, so too does social experience continue shaping the mammalian brain long after parental care is complete. A long series of studies examined the effects of **environmental enrichment** on the development of the brain in rats following weaning (when the rats are taken away from their mother so they can no longer nurse). These studies typically assigned weanling rats to one of three conditions: (1) *isolated condition*, where each animal resided in a cage alone; (2) *standard condition*, where two or three animals shared a cage; or (3) *enriched condition*, where many rats shared a much larger cage filled with various objects (toys, basically) that were periodically replaced for variety (**FIGURE 10.20A**). The earliest studies reported differences in whole-brain levels of the enzyme that normally breaks down acetylcholine, acetylcholines-

environmental enrichment   The stimulation provided to domesticated animals intended to boost brain development, including social stimulation and access to physical objects.

FIGURE 10.20 **Experimental manipulation of housing conditions.** (B,C after Greenough, 1976.)

(A)

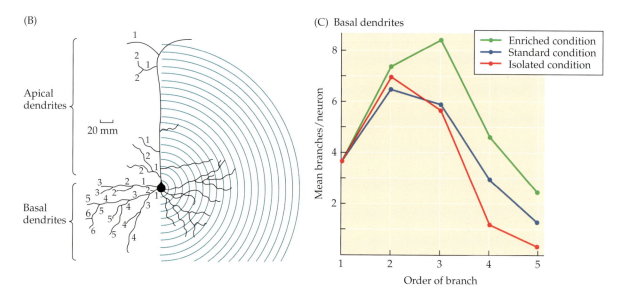

(B)

(C) Basal dendrites

terase, which was examined primarily because a sensitive assay was available at the time (Krech, Rosenzweig, & Bennett, 1962; Rosenzweig, Krech, Bennett, & Diamond, 1962).

Later studies found that animals in enriched conditions had thicker layers of cortex (Bennett, Diamond, Krech, & Rosenzweig, 1964; Diamond, Krech, & Rosenzweig, 1964), with neurons there sporting more complex dendrites (Greenough, Volkmar, & Juraska, 1973). As you might expect, the more complex the environment the rats were exposed to, the larger and more complex their brains became (**FIGURE 10.20B**). Usually the largest differences were between rats kept in isolated versus enriched conditions, but often there were differences between rats in isolated and standard conditions (**FIGURE 10.20C**), which differed primarily in whether the animals had the opportunity to interact with conspecifics. Thus, one lesson of this research program is that social stimulation continues after parental care is over, and this social stimulation

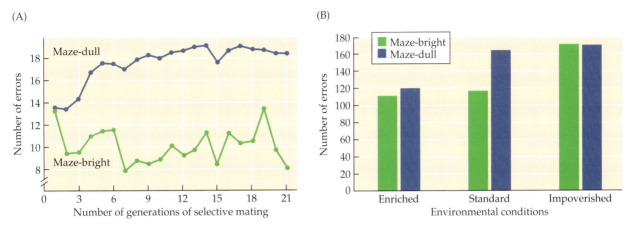

FIGURE 10.21   Selective breeding for maze running ability

spurs brain development. Many behavioral differences between rats raised in these conditions correlated with the brain changes—generally animals raised in more complex conditions performed better in tests of learning and memory.

The importance of postweaning rearing conditions was revealed in a follow-up to a classic demonstration of the role of genes in learning. Robert Tryon selectively bred laboratory rats for maze learning, pairing males and females who made few errors ("maze bright") or pairing males and females that made many errors ("maze dull"). After seven generations of such selective breeding, the offspring of the maze bright line almost all performed better than any offspring of the maze dull line (**FIGURE 10.21A**) (Tryon, 1940, 1942). This was an early demonstration of the importance of inheritance for learning ability, although interestingly, the strains did not differ in learning ability generally (Searle, 1949), only in the type of maze learning for which they were bred. Even for that narrow ability, postweaning environment made a difference, too. The behavioral differences that Tryon reported could be replicated only when the animals were kept in standard conditions. If the rats were isolated after weaning, both strains performed equally poorly (Cooper & Zubek, 1958). If they were kept in enriched conditions, both strains performed equally well (**FIGURE 10.21B**).

Another domain in which postweaning social experience makes a difference is in sexual behavior. After weaning, rat pups, like the young of many mammalian species, engage in juvenile play. In many species, including rats (Meaney, Stewart, Poulin, & McEwen, 1983), mice, monkeys (Alexander & Hines, 2002), and humans (Pasterski et al., 2005), males typically engage in more rough-and-tumble play than females. During this play, animals spend a lot of time investigating and smelling each others' butts, which may be important for later sexual functioning. For example, compared with male rats caged with other males after weaning, male rats raised in isolation are much less likely to become sexually aroused in response to odors from receptive females (Cooke, Chowanadisai, & Breedlove, 2000). What's more, the brain region known to receive and process information about mating odors, the posterodorsal aspect of the medial amygdala, is significantly smaller in isolated males than in males raised with cage mates (**FIGURE 10.22**).

One lesson of this research program is that social stimulation continues after parental care is over, and this social stimulation spurs brain development.

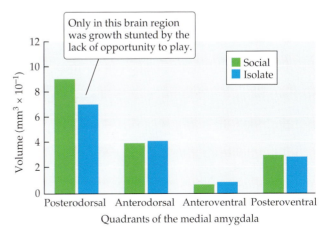

FIGURE 10.22   **Social isolation affects brain structure** (After Cooke et al., 2000.)

Social stimulation after weaning has also been shown to be crucial for the myelination of the brain, which continues well into adulthood and which may be important for learning, as we discuss next.

RESEARCHERS AT WORK

Social Stimulation Alters Neuregulin Signaling to Promote Myelination

■ **QUESTION**: What is the mechanism by which social isolation affects brain myelination in mice?

■ **HYPOTHESIS**: Normally, axons express neuregulin, which acts on ErbB receptors in oligodendrocytes, inducing the glial cells to wrap the axon and form myelin. Perhaps social isolation impairs myelination by affecting this signaling system between axons and glia.

■ **TEST 1**: Mice reared in isolation perform poorly on a test of working memory, compared with mice reared in standard conditions with a cage mate or mice reared in enriched conditions. Isolated mice also had reduced myelination in the prefrontal cortex (**FIGURE 10.23A**). Later tests would show that only isolation during the first 2 weeks postweaning, from postnatal day 21 (P21) to P35, had these effects. Isolation from P35 to P65 had no measurable effects. To test whether interference with the neuregulin-ErbB signaling system could mimic these effects of social isolation, the experimenters used Cre-lox technology (see Appendix) to knock down ErbB expression either during or after this sensitive period for social effects.

■ **RESULT 1**: Knockdown of ErbB expression in the Cre+ mice starting at P19 reduced expression of two myelin-related proteins—myelin basic protein (MBP) and myelin-associated glycoprotein (MAG)—and also impaired working memory in a manner similar to that seen after social isolation (**FIGURE 10.23B**). Knocking down ErbB expression beginning at P36 had no such effects (data not shown) (Makinodan, Rosen, Ito, & Corfas, 2012).

FIGURE 10.23 Social isolation stunts myelination and working memory. MBP is myelin basic protein and MAG is myelin-associated glycoprotein. (After Makinodan et al., 2012.)

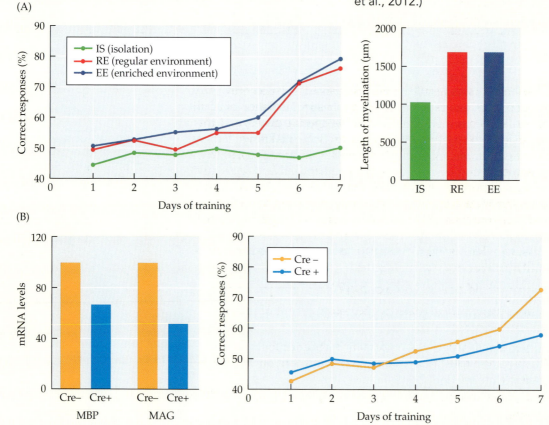

(A)

(B)

■ **TEST 2**: To ask whether the neuregulin-ErbB signaling system mediates these effects of social isolation, the researchers measured expression of several genes in mice isolated from P21 to P35 or housed in standard conditions with a cage mate.

■ **RESULT 2**: Isolated and control mice did not differ in expression levels of the three ErbB genes measured (data not shown), but there was less neuregulin 1 (NRG1) expression in the prefrontal cortex of isolated mice than of controls. What's more, this reduced expression was fully accounted for by type III of neuregulin 1 (**FIGURE 10.23C**) (Makinodan, Rosen, Ito, & Corfas, 2012).

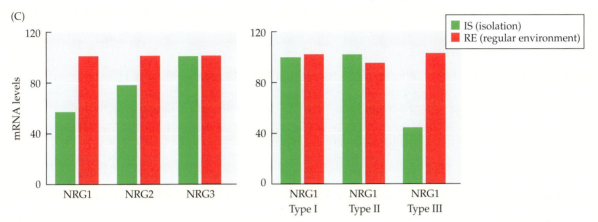

(C)

■ **CONCLUSION**: These findings are consistent with the hypothesis that social isolation immediately after weaning reduces the expression of neuregulin 1 (type III) by neurons in the prefrontal cortex of mice, which results in reduced myelination of their axons. This reduced myelination may be responsible for the deficits in working memory that isolated mice display, because artificially interfering with neuregulin signaling (by knocking down ErbB receptor expression) during that period also impairs prefrontal cortex myelination and working memory.

While these authors found significant effects of social isolation only early in the postweaning period, others have reported that social isolation of adult mice also decreases expression of oligodendrocyte-specific genes, and that results in thinner myelin sheaths in prefrontal cortex, which is known to play a role in social behavior. They saw no such changes in other regions, such as the cerebellum. Importantly, returning the animals to social housing could reverse these deficits accumulated in adulthood (J. Liu et al., 2012). Thus it appears that social stimulation in juvenile mammals continues to boost brain development and therefore contribute to adaptive behavior long after parental care is complete.

Now let's consider what such developmental responses to environmental enrichment might mean for human intelligence.

How's It Going?

1. How does manipulating the postweaning environment of rats affect brain development?

2. How did Tryon demonstrate the importance of genes for learning? What's the role of postweaning environment on this demonstration?

3. What is the mechanism by which social stimulation after weaning aids myelination?

■ 10.10 ■
INTELLIGENCE TESTS DEMONSTRATE THE PERVASIVE EFFECTS OF CULTURE

intelligence quotient (IQ)
A measure of intelligence based on performance on certain tests of abstract reasoning.

Flynn effect The increase in average human performance on IQ tests that has been observed worldwide over the past century.

Tests designed to measure human intelligence were originally intended to help place school children in the most appropriate classrooms. Later they were used in World War I to screen millions of American men to decide who should be officers, and whether some men were too unintelligent to even serve as ordinary soldiers (there was a special version of the test for those men who could not read). These tests would eventually give rise to **intelligence quotient (IQ)** tests that today measure where an individual ranks among the thousands of people who have taken the test before him. Even if you've never taken an official IQ test, such as the Stanford-Binet or Wechsler, you have probably taken "achievement tests" such as the SAT or ACT, and they essentially measure the same thing. An important component of all these tests is verbal ability, especially the use of abstract reasoning using language. They have been so fined-tuned over the years that they are very *reliable* in the technical sense that a given individual will earn about the same score no matter which version she takes or when she takes it. The scores are calculated so that the average IQ score will be 100, and the variability of scores will be such that the standard deviation (the average distance of scores from the mean) will be 15 IQ points. This means that 68% of the population will have an IQ score of 100 ± 15 points, in other words, a score between 85 and 115, while 95% of scores fall between 70 and 130 (**FIGURE 10.24A**).

One can certainly argue whether every important aspect of human intelligence is measured by such tests—they don't assess musical talent, athletic ability, or social facility. But there's no denying that IQ tests measure something real, that IQ matters. In other words, IQ tests are also *valid*, because knowing a person's IQ allows you to predict, in a statistical sense, important things like performance in school, lifetime earnings, and lifetime health. In the workplace, the correlation between IQ and wages is pretty impressive—people with IQ scores in the top 20% make nearly half again as much as people in the bottom 20% (**FIGURE 10.24B**). Of course, education can boost earnings, too, but even if you consider only people who earn a college degree, those with a higher IQ still make more money, on average, than those with a lower IQ (**FIGURE 10.24C**).

> There's no denying that IQ tests measure something real, that IQ matters.

IQ scores also correlate with good health. One of the most compelling demonstrations of this correlation is that children who received high IQ scores when they were 11 were *more likely to be alive* 65 years later than were low-scoring children (Whalley & Deary, 2001). Children with high IQs were also less likely to be hospitalized for accidents over the following 50 years (Lawlor, Clark, & Leon, 2007). The consistent correlations between IQ and health suggest that people who inherit good genes and grow up in a favorable environment are able to maximize their brain development and therefore score well on IQ tests.

Now keep in mind that these are correlations and we are comparing averages of *groups* of people, which means that if you know only the IQ scores of two people, you can't be sure which person will make more money. Similarly, while there is a correlation between IQ scores and various professions, you can find people with IQs between 95 and 110 in almost any profession (**FIGURE 10.25**). So I'm not saying IQ is destiny; I'm just saying IQ is one important factor in aspects of life that matter—health and prosperity. Still, if you set aside people with very low IQ scores (who often have serious health issues as well), there is little evidence that IQ correlates with happiness or life satisfaction (Veenhoven, 2010). So obviously other factors matter in life, too.

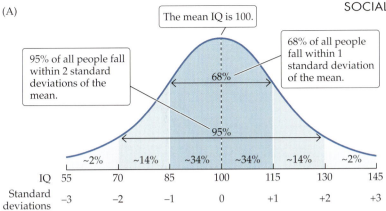

(A)

The mean IQ is 100.

95% of all people fall within 2 standard deviations of the mean.

68% of all people fall within 1 standard deviation of the mean.

68%

95%

~2%   ~14%   ~34%   ~34%   ~14%   ~2%

IQ	55	70	85	100	115	130	145
Standard deviations	−3	−2	−1	0	+1	+2	+3

FIGURE 10.24 IQ matters. (A) The average IQ is arbitrarily set to 100, and the standard deviation to 15, by adjusting the performance required to get a particular score. (B) IQ scores correlate with income. (C) Education improves earnings, too, but even when one controls for level of education, income still varies with IQ. (B,C after Ceci & Williams, 1997.)

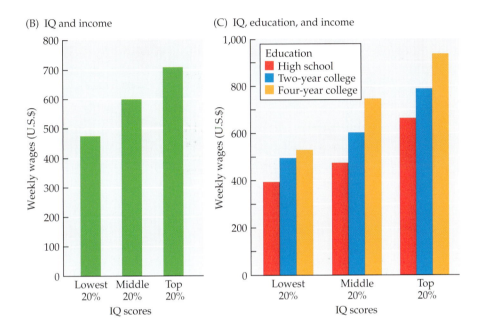

(B) IQ and income

(C) IQ, education, and income

Education
High school
Two-year college
Four-year college

Perhaps the most interesting aspect of IQ tests that you might not already know or suspect is that they demonstrate that humans have been getting progressively smarter, at least in terms of the intelligence measured by IQ tests, since the tests began. Recall I said that the tests are designed so that the average score will always be 100. In fact, since the first IQ tests were made, psychologists have been forced to keep raising the bar, requiring more and more correct answers to more and more difficult questions to qualify for a score of 100. This continuous raising of average IQ in the population has been called the **Flynn effect**. This has been demonstrated in many populations around the world (Flynn, 1987). Because of the Flynn effect, the standardization for each revision of IQ tests has had to be altered so that the test takers must perform

FIGURE 10.25 Occupation and IQ scores correlate overall with occupations However, a wide range of IQs is seen in each. Interestingly, the very widest range of IQs is seen in science and engineering jobs. (After Hauser, 2002.)

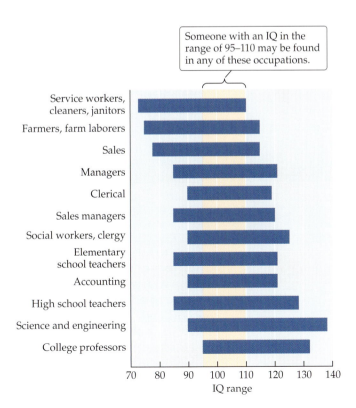

Someone with an IQ in the range of 95–110 may be found in any of these occupations.

Service workers, cleaners, janitors
Farmers, farm laborers
Sales
Managers
Clerical
Sales managers
Social workers, clergy
Elementary school teachers
Accounting
High school teachers
Science and engineering
College professors

70   80   90   100   110   120   130   140
IQ range

FIGURE 10.26 The Flynn effect (A after Flynn, 1998; B after Pietschnig & Voracek, 2015.)

(A)

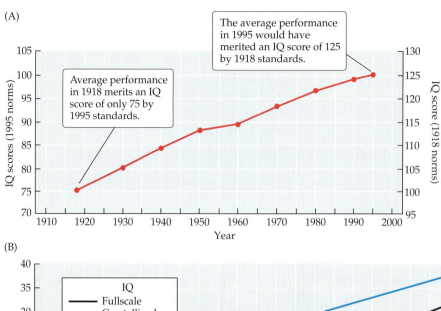

(B)

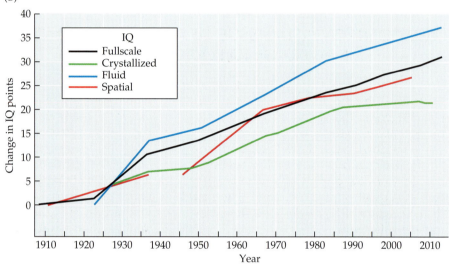

Humans have been getting progressively smarter, at least in terms of the intelligence measured by IQ tests, since the tests began.

By Flynn's reckoning, IQ scores are rising because widespread exposure to abstract concepts and reasoning at early ages is actually raising human intelligence.

better and better just to reach the average score of 100. In other words, performance on an IQ test that would have resulted in a score of 100 in 1932 would only net a score of about 80 in a 1991 version (Neisser, 1997). It is as if the world population has gained an average of 20 IQ points in that time, averaging about 6 IQ points per decade (**FIGURE 10.26A**). The gains are seen in each of the several subdivisions of intelligence tests, not just one component (**FIGURE 10.26B**) (Pietschnig & Voracek, 2015). Obviously this is too fast to represent any change in the gene pool.

In general, the increase in average IQ is happening because there are fewer people receiving very low scores. In other words, while a few more people are getting high scores, a lot fewer people are getting low scores (Colom et al., 2005). This trend may reflect increased overall prosperity, including improved nutrition for a wider range of the population. Most experts, including James Flynn himself, believe the higher scores are due to increasing intellectual stimulation worldwide (Flynn, 2007). For example, modern occupations are more intellectually demanding than the subsistence farming that was once common, even in industrialized societies. Also, the development of widespread, cheap media—first radio, then television, and now smart phones—has opened up a world of information, including abstract discussions of issues and concepts (like "the economy," "intelligent design," or "random sample") that most people were never exposed to in the past. That sort of abstract reasoning is precisely what IQ tests strive to measure. By Flynn's reckoning, IQ scores are rising because widespread exposure to abstract concepts and reasoning at early ages is actually raising human intelligence. Among other things, this finding illustrates that the environment can have a powerful effect on IQ.

How's It Going?

1. What is the average IQ score, and how are scores distributed across the population?
2. What is the Flynn effect, and what are some factors that might be causing it?

■ 10.11 ■

THE CONTROVERSIAL ISSUE OF RACIAL DIFFERENCES IN AVERAGE IQ PERFORMANCE

If you are now convinced that IQ tests measure something that really matters in life, then you should be curious about the developmental roots of this type of intelligence. First, as you surely anticipate from what we learned in the first seven chapters in this book, genes certainly play a role in IQ performance. As has been shown many times, people who are more closely related genetically have more similar IQ scores. The easiest comparison to understand is that between **monozygotic twins**, who are genetically identical, and **dizygotic twins** (also called fraternal twins), who share only 50% of their genes. While the correlation of IQ scores is +0.58 in dizygotic twins, it is even higher in monozygotic twins: +0.85 (**FIGURE 10.27**). Presumably the greater correlation of IQ in monozygotic twins is because they are more closely related to each other than dizygotic twins. If genes were the *only* factor affecting intelligence, we would expect monozygotic twins to always have the same IQ, providing a correlation coefficient of +1.0, and of course that's not what we see. But there's no denying that genes affect IQ, as they must if IQ reflects brain function, because genes play such a major role in early brain development, as we've seen.

On the other hand, it's clear that the environment affects IQ, too, as the Flynn effect shows. For example, one of the most consistent correlations of IQ is with socioeconomic status (SES)—people of higher SES tend to have

monozygotic twins A pair of twins derived from a single fertilized egg, who therefore share all the same genes.

dizygotic twins A pair of twins derived from two fertilized eggs, who are therefore no more genetically related than any other pair of siblings.

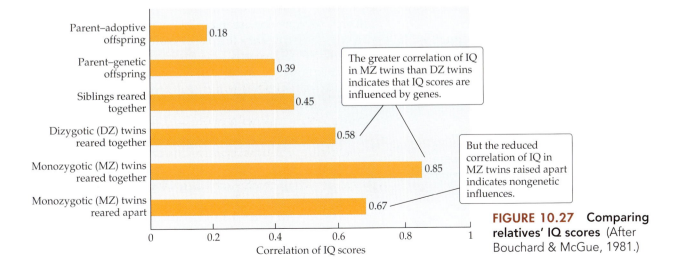

FIGURE 10.27 **Comparing relatives' IQ scores** (After Bouchard & McGue, 1981.)

higher IQs. At least some of this correlation is due to environmental effects on IQ, because adoption studies consistently find that babies from lower-class families, when adopted into middle-class homes, end up with higher IQs than their birth parents. In fact, the adopted children develop an IQ equivalent to that of their adoptive parents, gaining 12–18 IQ points when they're taken out of poverty. The higher the SES of the adoptive family, the more IQ points the adopted children gain (Duyme, Dumaret, & Tomkiewicz, 1999).

> Adoption studies consistently find that babies from lower-class families, when adopted into middle-class homes, end up with higher IQs than their birth parents.

Now we come to the controversial question of whether, since the intelligence measured by IQ is important and since genes affect IQ, the well-publicized average differences in IQ scores between groups of people are due to differences in their genes. Administration of IQ and achievement tests to large numbers of people consistently show that, among Americans for example, Jews score higher than Asians, who score higher than whites, who score higher than Hispanics and African Americans. Again, I remind you that these are *average* differences between *group*s, so you cannot accurately gauge any *individual's* IQ by her race—millions of African Americans have higher IQ scores than the average white American.

But the rank order of these mean scores is consistently found, so we can ask, Are these average differences due to differences in genes or differences in the environment? First, in terms of simple logic, because nongenetic factors (lead poisoning, the Flynn effect) have been demonstrated to affect IQ, knowing that genes affect intelligence does not tell us *anything* about why average IQs differ between any two groups. White Americans of a century ago would only score 80 on today's IQ tests, but their genes are not significantly different from those of white Americans today. Second, environmental differences between these groups may also affect average IQ performance. In other words, we know that many other factors affect IQ, and these may account for the differences in average scores.

For example, the average difference in IQ scores between black Americans and white Americans shrank dramatically in the first half of the twentieth century (Flynn, 2007), as conditions for black Americans improved. The gap between black and white appears to be continuing to shrink (**FIGURE 10.28**), and only time will tell whether it will disappear altogether. Again, this change is too fast to be due to any change in the gene pool for either group. We also know that poverty and malnutrition affect IQ, so it's possible that differences in SES between black and white Americans may at least contribute to differences in average IQ. For example, because African American children are more likely to live in poverty, and children living in impoverished regions (based on postal zip codes) are much more likely to have high blood levels of the heavy metal lead than children in affluent regions (McClure, Niles, & Kaufman, 2016), and because lead is known to reduce IQ scores (**FIGURE 10.29**), the lower scores of African Americans might be due to greater exposure to lead during childhood. Likewise, perhaps other environmental toxins, including those that have not yet been identified, are also more prevalent in impoverished regions of the country, which would also affect IQ scores of African Americans more commonly than white Americans.

Finally, the pervasive stereotype in American culture that African Americans are less intelligent actually affects IQ measurement. **Stereotype threat** is an emotional state experienced when test takers are exposed to a derogatory

stereotype threat The emotional state experienced when exposed to a derogatory stereotype about a group to which you belong.

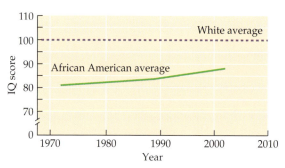

FIGURE 10.28 **Shrinking racial differences in IQ scores at the end of the twentieth century** (After Dickens & Flynn, 2006.)

IQ score
150
140
130
120
110
100
90
80
70
60
50
0
5
10
15
20
25
30
Blood lead concentration (μg/dL)
Note that no child with a lead concentration above 15 μg/dL has an IQ of 100 or more.

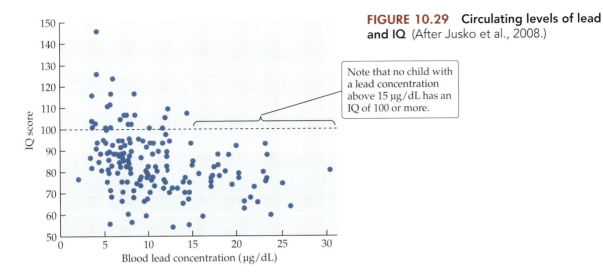

FIGURE 10.29 Circulating levels of lead and IQ (After Jusko et al., 2008.)

stereotype about a particular group. African American and white students were given the same intelligence test under two conditions. If they were told beforehand that African Americans don't tend to do as well as other groups, then only the African American students experienced an increase in blood pressure, and only their performance was lowered. The two groups performed identically when they were not presented with the stereotype threat beforehand (Blascovich, Spencer, Quinn, & Steele, 2001) (**FIGURE 10.30**). In another experiment, if students were simply asked to report their race before taking a test, that was enough to cause African American students (and not white students) to perform more poorly (Steele & Aronson, 1995). In a similar vein, girls told beforehand that there was no sex difference in performance on a math test did better on the test, equaling the performance of the boys. If they were told beforehand that there was a sex difference on test performance, only the girls' performance was impaired (Spencer et al., 1999). In the meantime, the sex difference in math performance in Western societies has shrunk over the years, and today girls score better than boys in some countries (Guiso, Monte, Sapienza, & Zingales, 2008). Coming back to IQ tests, there is no way to gauge how much stereotype threat contributes to the average differences in scores between black and white Americans.

An ideal experiment to test whether genes are responsible for the difference between black and white Americans would be to take black and white babies and randomly assign them to be raised as either white or black people in a society free of racial bias, but of course such a study is neither ethical nor possible. But a "naturalistic" study came close to such an experiment, as we'll see next.

	African Americans	Non-African Americans
No stereotype threat presented	No effect on blood pressure or performance	No effect on blood pressure or performance
Stereotype threat presented ("African Americans don't tend to do well at this test.")	Higher blood pressure; decreased performance	No effect on blood pressure or performance

FIGURE 10.30 Stereotype threat (After Blascovich et al., 2001.)

RESEARCHERS AT WORK

Does Race Affect the IQ of German Offspring of American GIs?

- ■ **QUESTION**: Is the average difference between black and white Americans due to differences in the genes they carry?

- ■ **HYPOTHESIS**: Black and white children raised in a nonracist environment will still have different average IQ scores.

- ■ **TEST**: Many American soldiers were stationed in U.S. bases in Germany after World War II. The racial difference in IQ seen in the general population in America was also present among American soldiers of that era. Some of these black and white soldiers fathered children by German women, who raised the offspring in Germany. Match the mothers for equivalent socioeconomic status (mostly low), then measure the IQs of these children to see if the 98 children inheriting genes from black fathers have lower scores than the 83 fathered by whites.

- ■ **RESULT**: In fact, the average IQ scores did not differ by race. The interracial children had an IQ of 96.5, while the white children had an IQ of 97.2, a difference that is not statistically significant (and is much less than the 15-point IQ difference between white and black American servicemen of that era) (Eyferth, 1961).

- ■ **CONCLUSION**: The results did not conform to the hypothesis. There was no evidence that children inheriting genes from black fathers had lower IQs than those fathered by white men, at least when they were raised in 1950s Germany. The absence of a difference between the two groups of children may be due to rearing in an environment with fewer racist influences (in a backlash against the Holocaust?) and/or an absence of stereotype threat during testing.

By this point in the book, having learned of the many genetic influences on the developing brain and the importance of neural activity, sensory experience, and social stimulation for the finishing touches of neural development, you should not be surprised to learn that human intelligence is affected by both genes and the environment. I hope this means that you are now equipped with the expertise and understanding to think critically about claims of "biological destiny," which you will surely encounter in life. Many people may repeat the claim that the nature-nurture debate is closed, that both factors are important for a person's full human development, and so on. But few of those people are as well informed as you are to understand *why* that debate should have been settled long ago. Unfortunately, so many of your fellow citizens are unaware of this so-called settlement that I guarantee you will encounter people thinking in outmoded ways, that genes entirely determine all behaviors, or the opposite, that genes have absolutely no influence on any behaviors. My deepest hope is that after reading this book, you will never fall into either camp. This won't make your life any easier, but it may help you better understand our world.

> You are now equipped with the expertise and understanding to think critically about claims of "biological destiny."

How's It Going?

1. What is the evidence that genes influence IQ performance?
2. What is stereotype threat, and how does it apply to IQ tests?
3. Describe the results of a study looking for an influence of genes from white Americans versus black Americans on the IQs of offspring raised outside the United States.

SUMMARY

■ *Instinct* and *innate* are vague, rarely defined terms that fail to recognize that all behaviors are influenced by genes and that every behavior can be modified by experience. A more useful criterion to designate more complex behaviors is to ask whether they require social interactions during development to properly unfold. In many species, especially those that require parental care, many behaviors are lost or severely distorted without social experience. **See Figures 10.1 and 10.2**

■ In mammals and birds, parental care is required for survival. Rat maternal care can reduce methylation of the *glucocorticoid receptor* gene in the brain, resulting in offspring that have a less severe stress response as adults. Female rats reared by an attentive dam are more likely to be attentive to their own pups in turn. The altered methylation of the glucocorticoid receptor gene caused by maternal care is sometimes called an *epigenetic* effect, but there are inconsistencies in the use of this term, as with *instinct* and *innate*. **See Figures 10.3 and 10.4, Box 10.1**

■ *Sudden infant death syndrome* (*SIDS*), the unexplained respiratory failure of infants, is less common among babies who are placed on their backs to sleep. Interestingly, those babies are also slightly delayed in reaching milestones in motor development, presumably because they have less experience on all fours. **See Figure 10.5**

■ Young birds *imprint* on the first slowly moving object they see, which is typically their mother. In many species, adults will attempt to mate with animals resembling the birds that reared them as chicks, a result of *sexual imprinting*. Young birds appear to learn the safest migratory route by accompanying older birds. **See Figures 10.6–10.8**

■ Many examples of *observational learning* have been documented in the wild, including monkeys dipping yams in water, birds raiding milk bottles, and birds attacking hibernating bats or surfacing whales. We have no way of knowing what other foraging behaviors of animals originated in such innovations that spread by observational learning. **See Figure 10.9**

■ In many songbird species, males memorize the song they hear as nestlings, whether of their biological father or a foster father, and later produce a variant of that song as their own. The young males first produce *plastic song* and then, guided by auditory feedback and memory of the song they heard as nestlings, fine-tune their own song, which is then said to be *crystallized*. Birds are predisposed to incorporate elements of the song of their own species rather than that of other species. **See Figures 10.10–10.13**

■ The *FOXP2* gene appears to be important for vocal communication, as it is important for birdsong learning

and ultrasonic vocalization in rodents, and mutant alleles of the gene result in specific language deficits in humans. In contrast, people with *Williams syndrome* are verbally fluent but may otherwise be mentally impaired. **See Figure 10.14**

■ Children reach language milestones in a relatively fixed sequence. Across cultures, people speak *parentese* to babies. Newborns appear to be capable of discriminating all the known *phonemes* that occur in human languages, but as they grow, they become better at discriminating the phonemes they hear and may lose the ability to discriminate other phonemes. **See Figures 10.15 and 10.16, Table 10.1**

■ The loss of the ability to discriminate phonemes not heard in infancy makes it more difficult for adults to learn additional languages and may have contributed to the progressive loss of phonemes in languages as humans migrated across the globe. **See Figures 10.17 and 10.18**

■ Monkeys raised without a live, interactive mother, which Harry Harlow regarded as a source of love, display very abnormal behavior as adults; essentially they are unable to socially interact with others. Growing up with peers ameliorates much of the behavioral consequences of being motherless. Children who form a secure *attachment* to a caregiver have healthier, more positive social interactions as adults. **See Figure 10.19**

■ Social isolation during the juvenile period has wide-ranging effects on development of the brain and behavior, including stunted expression of neuregulin, resulting in reduced myelination. **See Figures 10.20–10.23**

■ Scores on *IQ* tests of abstract reasoning positively correlate with health, longevity, and success in school and the marketplace. Nevertheless, people with IQs between 95 and 110 can be found in almost any profession. The *Flynn effect* is the progressive improvement seen across the world in IQ performance since the tests began, which cannot be accounted for by changes in the genome. **See Figures 10.24–10.26**

■ Comparison of relatives demonstrates that genes influence IQ performance. However, environmental factors, including socioeconomic status (SES) and exposure to toxins like lead, also influence IQ. **See Figures 10.27 and 10.29**

■ Differences in average IQ scores across human groups may be due, in part, to the effects of *stereotype threat* for African Americans. The difference in average IQ between white and black Americans shrank dramatically in the first half of the twentieth century, and it appears to be shrinking still. Among children born to women in 1950s Germany and raised there, those fathered by black American servicemen scored as high on IQ tests as those fathered by white Americans. **See Figures 10.28 and 10.30**

Go to the Companion Website
sites.sinauer.com/fond
for animations, flashcards, and other review tools.

IMMANUEL KANT AND THE CRITIQUE OF PURE REASON

Immanuel Kant (1724–1804) was born in the German city of Königsberg in Prussia (today called Kaliningrad, Russia) to a modest family—his father was a saddle maker. As a child, he was educated at a local Protestant school, where he detested the religious teaching but devoured Latin classics. By the time Kant began his college education at the University of Königsberg, he was fascinated by philosophy, mathematics, and physics. Afterward he supported himself as a private tutor until he began teaching at the same college, where for years he was paid not by the college, but by the students directly. That meant he had to teach a lot, lecturing 20 hours per week, and had to be a good teacher to attract students (Rohlf, 2014). Kant was a prolific writer. His most famous work, *Critique of Pure Reason* (1781), mainly concerns *epistemology* (the study of what we know and how we come to know it). The book's title refers to the question of whether we can ever know anything "purely" as the rationalists envisioned, meaning without depending on any experience. Kant tried to reconcile rationalist and empiricist ideas, and his fame rests on the fact that most philosophers then, and today, think he succeeded.

Kant noted that all knowledge can be classified into two categories: those ideas that do not depend on any experience, if there are such, are present **a priori** (Latin for "from what comes before"), while things we learn through experience can be classified as **a posteriori** (Latin for "from what comes after"). The distinction between a priori knowledge and a posteriori knowledge is easiest seen when you think about propositions. You know that $2 + 2 = 4$, or that "if all men are mortal, and Socrates is a man, then Socrates is mortal," without getting out your smart phone to Google it. Those are things you know, and know are true, a priori. On the other hand, if you want to know if it's raining outside, you'll need to use your senses to check up on that somehow (a glance at either your smart phone app or a window may tell you), so that would be a posteriori knowledge.

Immanuel Kant (1724–1804)

a priori   Latin for "from what comes before," referring to things that the mind must know before having any experiences and that are required before the mind can even have an experience.

a posteriori   Latin for "from what comes after," referring to things that we learn through our senses in our experience of the world.

It's one thing to classify everything you presently know as a priori versus a posteriori knowledge, but it's much more difficult to decide whether you *first* learned something with or without relying on experience. For that matter, if you started out knowing nothing, how did you learn anything at all? Using logic, Kant demolished so-called proofs that humans have an immortal soul, which rationalists like Plato and Descartes claimed provide us with a priori knowledge. Then Kant wrestled with the question of how a developing individual who knew absolutely nothing could ever come to know anything, and he concluded that some key concepts shaping our every experience—*time* and *space*—cannot possibly be learned through experience. How could you teach someone about either of these concepts if they had absolutely no notion of them? Try explaining birth, life, and death to someone who does not experience the passing of time, or explaining what a football team is doing to someone who isn't aware of the difference between here and there.

More tellingly, how can you have *any* sensory experience that does not take place at a particular *time* and have a particular duration? Also, every sensory experience you have seems to come from a particular source in *space* (even if it's inside your stomach). Kant concluded that these are indeed a priori concepts—time and space—that are necessary components of any particular experience and so could not have been learned through our senses. He was convinced that these properties of cognition must be already "built into" the very act of experience, a part of our cognitive apparatus (we would call it the brain). Not only are the concepts of time and space there before any experience, but they are *required*—we couldn't have experience without them. For that reason, these aspects of experience are also said to be *transcendent*, meaning they surpass or extend beyond ordinary experience. If these transcendent qualities of *space and time are built into our cognitive apparatus before our minds can take in any experience*, then the rationalists were correct in saying we are born with some aspects of knowledge. But that doesn't mean that *everything* we know, the Pythagorean theorem, what horses and beds look like, which faces are handsome, is already present at birth. Rather, we gain all that other, a posteriori information through our senses, as the empiricists predicted, which could only happen because our cognitive apparatus imposes the a priori concepts of space and time onto each experience.

THE A PRIORI EMBODIMENT OF SPACE AND TIME

The events we described in the first seven chapters of this book beautifully confirmed rationalist ideas that knowledge, or more properly information, is already inside us before birth. All of those processes create a staggeringly complex nervous system even before there is any possibility of sensory experience (because there are not yet any sensory organs). But in the last three chapters, where we saw how neuronal activity, sensory experience, and social experience come to guide neural development, determining which synaptic connections are lost and which are weakened or strengthened, we've laid bare the mechanism by which the empiricist idea is also vindicated. Indeed we do take in a great deal of information through our senses, and that is *critical* for brain development and therefore mental development. Thus the first part of the book was about information that unfolds without experience—a priori knowledge—while the last three chapters have been about information provided by neural activity and experience—a posteriori knowledge.

Chapters 1–7 revealed how information in the genome is enough for an embryo to develop a breathtakingly intricate nervous system before sensory experience is possible. Up to that point, the only contributions of the outside world are nutrients and whatever physical conditions embryos of that particular species need to stay alive. According to Kant, there can be no experience without space and time embedded in that experience, and those notions are built into our cognitive faculties a priori. If we explore the nervous system built solely from genetic specifications before sensory systems arise, we can already see the origins of Kant's a priori, transcendent qualities that form any sensation.

> According to Kant, there can be no experience without space and time embedded in that experience, and those notions are built into our cognitive faculties a priori.

Space

We've seen that several parts of the developing nervous system are laid out in a detailed spatial map, specifically the distribution of touch and pain receptors along the three dimensions of our external skin, as well as the two-dimensional array of photoreceptors in the retinas, which respond to visual images in three-dimensional space (**FIGURE E.1**). The spatial relationship between these various sensory arrays is faithfully reproduced by axons that form topographic maps of innervation in the brain. Given these **topographic projections**, it is impossible for a sensory signal to arrive *without* carrying information about space. Which toe was just pinched, in what part of the retina did that shadow fall? Put simply, the spatial relationship between two photoreceptors on the retina is faithfully reproduced between corresponding neurons in visual cortex. So information about space is *built into* the signal, and we saw that genes are initially responsible for that roughly correct topographic projection, as in the kitten visual system before birth, and that it is reinforced by spontaneous activity. But recall from Chapter 9 that as soon as the kitten's eyes open, visual *experience* sharpens those topographic projections and also guides development of visual circuits to detect, for example, horizontal lines versus vertical lines. Without that crucial visual experience early in life, the cat will be blind,

topographic projections
The property of axonal projections in which the pattern of input represents a spatial array, as in the projection of retinal axons onto the brain.

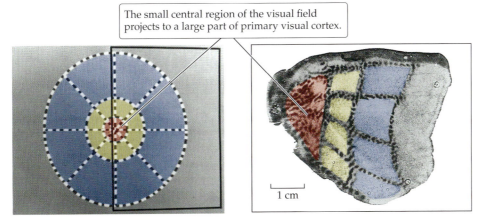

The small central region of the visual field projects to a large part of primary visual cortex.

1 cm

FIGURE E.1 Topographic projection of visual space onto cortex In this experiment, the visual field indicated on the left was demonstrated to project topographically onto the "flattened" primary visual cortex of a monkey on the right. Thus the transcendent concept of space is built into our cognitive apparatus for vision. In other species, the rough outline of this topographic projection is laid down before birth, but we know from people like Mike May (see Chapter 9) that visual experience early in life is required for us to make sense of that information (From Tootell et al., 1988.)

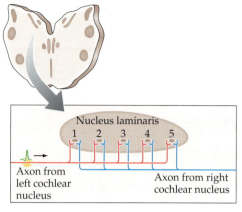

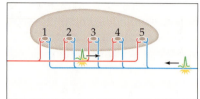

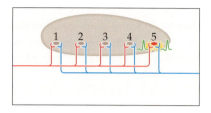

1 A sound occurring to the left of the bird's midline is detected by the left cochlea slightly earlier than the right cochlea.

2 Monaural neurons of the left cochlear nucleus of the brainstem become active, sending action potentials along their axons toward the nucleus laminaris.

3 Slightly later, neurons of the right cochlear nucleus respond to the sound and send action potentials toward the nucleus laminaris…

4 …with the result that the action potential from the left side and the one from the right side arrive simultaneously at neuron 5. Neuron 5 is thus a coincidence detector that signals a particular location to the left of midline. For simplicity, only 5 neurons are shown; in reality, thousands of such neurons make up a detailed map of auditory space.

FIGURE E.2 The classic (Jeffress) model of sound localization in the auditory brainstem of birds

despite all that a priori information provided by the genome alone. To answer Molyneux's question, no, the man born blind and suddenly made able to see will not be able to tell the cube from the sphere by sight alone. A visual system constructed by genes alone, as in kittens that have just opened their eyes, does not work very well and, without visual experience, will become useless. To build a visual system of the quality we enjoy, and which helped our ancestors reproduce, you must have early visual experience, as empiricists claimed.

It's not just touch and vision that have spatial information built in—all the other sensory systems are also assembled to provide information about the spatial source of the stimulus, even before sensory receptors are working. Information from the two nostrils and two ears converges in the brain to convey whether the stimulus is coming from our left or our right.

In fact, a famous model of how we know where a sound came from illustrates how both space and time contribute to sensation. Lloyd Jeffress (1900–1986) proposed a time-delay model in which signals from the two ears in response to a sound could be compared to reveal where the sound came from. In this model, signals from the left and right ear converged on an array of neurons laid out in a row. Input from one ear alone was insufficient to make any target neuron fire—it only reached threshold when it got two inputs, one from each ear, at about the same time. When the sound came from the midline, information from the left and right converged in the middle of the array, and so only the middle neuron fired. If the sound came from the left, then the signal from the left ear entered the brain first and so traveled past the center of the array before converging on the signal from the right ear. In that case, only the neuron on the right side of the array fired (**FIGURE E.2**). In this way, the brain could learn where the sound came from by monitoring which neuron in the array fired. When Jeffress proposed what he called the *coincidence detector* model in 1948 (Jeffress, 1948), it was merely theoretical, but once other scientists applied neurophysiological tools to auditory neural systems, the key assumptions of his model were borne out (Joris,

Other sensory systems are also assembled to provide information about the spatial source of the stimulus.

Smith, & Yin, 1998). This is how our nervous system tells where a sound originated in space. Just as the visual system of kittens is roughly assembled before visual experience, such coincidence detectors in barn owls are present early in life and probably assembled before birth, using genetic information alone. But recall from Chapter 9 that in barn owls these circuits for detecting the location of sound are sharpened by experience by the use of visual information to adjust the circuits.

Time

Note that in the Jeffress model, information about *space* is provided by analysis of *time*, specifically the time it takes for the axons to conduct action potentials. Thus the delay caused by neural conduction unavoidably carries information about time. Multiple sclerosis is so devastating because the drastic alteration of myelination and therefore neural conduction distorts timing information crucial for coordinating sensation and movement. But timing isn't just a crucial aspect of neural conduction along axons. *All neural analysis relies upon the timed convergence of different bits of information onto a given neuron*. In the Jeffress model, no neuron in the array will fire unless two inputs arrive at about the same time. All neuronal analysis relies on such **temporal summation**, the arrival of enough presynaptic input within a particular period of time to get the target neuron above threshold to fire. In this sense, the notion of time is indeed built into not only sensory experience, but also the coordination of motor processes. Just as the nervous system uses visual experience to sharpen our ability to locate objects in space, it uses experience to sharpen our sense of time. We saw in Chapter 6 how synapses in immature nervous systems start out sluggish and slow and become increasingly rapid and brief, and it's entirely possible that experience is required for that sharpening to take place.

Lloyd A. Jeffress (1900–1986) (Courtesy of the Briscoe Center, University of Texas, Austin.)

temporal summation The property of neurons that fire only when sufficient synaptic stimulation occurs within a particular period of time to reach threshold.

WHAT DOES ALL THIS PHILOSOPHY STUFF HAVE TO DO WITH THIS BOOK?

Even in his day, Kant knew that his reconciliation of rationalist and empiricist thinking also reconciled the two camps in the nature versus nurture debate (although it wasn't called that—it was the question of whether the aristocracy are superior to commoners by virtue of their breeding). The rationalist-empiricist and nature-nurture debates are two aspects of the same debate. Kant reconciled the first debate by identifying a priori knowledge that is required, but he pointed out how limited that knowledge was and that the whole point of that a priori knowledge was to permit sensory experience to teach us everything else a posteriori. In this book, we've reconciled the nature-nurture debate by detailing the many wondrous processes at work early in brain development, entirely programmed by the genes we inherited, and then by examining all the machinery that *evolved to exploit experience to shape the final processes*, making a brain far more complex than anything that could have been directed by genes alone. Yes, nature matters as the genetic program assembles the basic framework providing us a priori concepts of space and time, but that is a necessary prerequisite for nurture to pack our brains with a posteriori information gained from experience.

The rationalist-empiricist and nature-nurture debates are two aspects of the same debate.

And if humans are the dominant species on this planet, it is because we evolved to spend a ridiculously long part of our lives learning about the world, bringing ever greater complexity, and more accurate information about the world, into our brains (see Figure 4.16A). To the extent that being born with brains that are still very embryonic in character, as "externalized fetuses," confers an adaptive advantage, it is because that allows the very plastic, fetal-like brain to be even more extensively molded by experience, especially social experience.

But here's the discouraging part. Kant settled this debate in the eighteenth century through use of a sharp intellect and unassailable logic, and then developmental neuroscientists settled it again in the twentieth century through critical experimentation and use of amazing technology that allowed us to watch the brain grow. For anyone paying attention, there's no more room for debate—yes genes are important and yes experience is important, too. In fact, the genes alone would make a worthless brain without experience, and experience isn't even possible without a basic neural architecture provided by the genes. But public discourse, from Kant's time to today, continues to be dominated by people arguing one side or the other of this debate. We still have people publicly arguing that early education can't benefit children growing up in poverty, or that racial differences in average IQ could only be due to differences in genetic endowment, or that genes can't have anything to do with intelligence, or political attitudes, or sexual orientation. If the nature-nurture debate really has been settled, then either these people didn't get the memo or they didn't read it.

> Public discourse, from Kant's time to today, continues to be dominated by people arguing one side or the other of this debate.

If any member of our society *should* understand that both genes and the environment matter, it is those who take a course in developmental neuroscience—you. Perhaps I should have been satisfied telling you about this marvelous, seemingly impossible process by referring only to the stunning collection of scientific experiments probing the process. Certainly that by itself provides a compelling resolution of the nature-nurture debate, ironclad proof that as important as genes are, experience plays a crucial role, too. But I included these philosophical asides because I also wanted you to know that this same conclusion had already been reached, through logic alone, centuries earlier. Of course developmental neuroscientists found that genes and experience both crucially guide brain development. Given the results of philosophical work on epistemology, that had to be the case. Kant might have predicted it a priori.

Molecular Biology
BASIC CONCEPTS AND IMPORTANT TECHNIQUES

In case it has been a while since you had a class involving molecular biology, this appendix is intended as a brief reminder of concepts you should have encountered before.

■ A.1 ■
GENES CARRY INFORMATION THAT ENCODES THE SYNTHESIS OF PROTEINS

The information for making all of our proteins could, in theory, be stored in any sort of format—on sheets of paper, a DVD, an iPod—but organisms on this planet store their genetic information in a chemical called **deoxyribonucleic acid**, or **DNA**. Each molecule of DNA consists of a long strand of chemicals called **nucleotides** strung one after the other. DNA has only four nucleotides: guanine, cytosine, thymine, and adenine (abbreviated G, C, T, and A). The particular sequence of nucleotides (e.g., GCTTACC or TGGTCC or TGA) holds the information that will eventually make a protein. Because many millions of these nucleotides can be joined one after the other, a tremendous amount of information can be stored in very little space—on a single molecule of DNA.

A set of nucleotides that has been strung together can snuggle tightly against another string of nucleotides if it has the proper sequence: T nucleotides preferentially link with A nucleotides, and G nucleotides link with C nucleotides. Thus, T nucleotides are said to be complementary to A nucleotides, and C nucleotides are complementary to G nucleotides. In fact, most of the time our DNA consists not of a single strand of nucleotides, but of two complementary strands of nucleotides wrapped around one another.

The two strands of nucleotides are said to **hybridize** with one another, coiling slightly to form the famous double helix. The double-stranded DNA twists and coils further, becoming visible in microscopes as **chromosomes**, which resemble twisted lengths of yarn. Humans and many other organisms

FIGURE A.1 **Duplication of DNA** Before cell division, the genome must be duplicated as illustrated here so that each daughter cell has the full complement of genetic information.

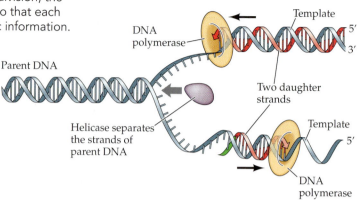

deoxyribonucleic acid (DNA) A nucleic acid that is present in the chromosomes of cells and encodes hereditary information.

nucleotide A portion of a DNA or RNA molecule that is composed of a single base and the adjoining sugar-phosphate unit of the strand.

hybridization The process by which a string of nucleotides becomes linked to a complementary series of nucleotides.

chromosome A complex of condensed strands of DNA and associated protein molecules; found in the nucleus of cells.

eukaryote Any organism whose cells have the genetic material contained within a nuclear envelope.

nucleus Here, the spherical central structure of a cell that contains the chromosomes.

ribonucleic acid (RNA) A nucleic acid that implements information found in DNA.

transcription The process during which mRNA forms bases complementary to a strand of DNA. The resulting message (called a *transcript*) is then used to translate the DNA code into protein molecules.

messenger RNA (mRNA) A strand of RNA that carries the code of a section of a DNA strand to the cytoplasm.

are known as **eukaryotes** because we store our chromosomes in a membranous sphere called a **nucleus** (plural *nuclei*) inside each cell. You may remember that the ability of DNA to exist as two complementary strands of nucleotides is crucial for the duplication of the chromosomes (**FIGURE A.1**). With very few exceptions, every cell in your body has a faithful copy of all the DNA you received from your parents.

The information from DNA is used to assemble another molecule—**ribonucleic acid**, or **RNA**—that serves as a template for later steps in protein synthesis. Like DNA, RNA is made up of a long string of four different nucleotides. For RNA, those nucleotides are G and C (which, you recall, are complementary to each other) and A and U (uracil), which are also complementary to each other. Note that the T nucleotide is found only in DNA, and the U nucleotide is found only in RNA.

When a particular gene becomes active, the double strand of DNA unwinds enough so that one strand becomes free of the other and becomes available to special cellular machinery (including an enzyme called *transcriptase*) that begins **transcription**—the construction of a specific string of RNA nucleotides that are complementary to the exposed strand of DNA (**FIGURE A.2**). This length of RNA goes by several names: **messenger RNA** (**mRNA**), **transcript**, or sometimes, *message*. Each DNA nucleotide encodes a specific RNA nucleotide (an RNA G for every DNA C, an RNA C for every DNA G, an RNA U for every DNA A, and an RNA A for every DNA T). Transcription stops when the assembly reaches a trio of DNA nucleotides, called a **stop codon**, that signals the end. This transcript is made in the nucleus where the DNA resides; then the mRNA molecule moves to the cytoplasm, where protein molecules are assembled.

In the cytoplasm are special organelles, called **ribosomes**, that attach themselves to a molecule of mRNA, "read" the sequence of RNA nucleotides, and, using that information, begin linking together amino acids to form a protein molecule. The structure and function of a protein molecule depend on which particular amino acids are put together and in what order. The decoding of an RNA transcript to manufacture a particular protein is called **translation** (see Figure A.2), as distinct from *transcription*, the construction of the mRNA molecule.

Each trio of RNA nucleotides, or **codon**, encodes one of 20 or so different amino acids. Special molecules associated with the ribosome recognize the codon and bring a molecule of the appropriate amino acid so that the ribosome can fuse that amino acid to the previous one. If the resulting string of amino acids is short (say, 50 amino acids or so), it is called a **peptide**; if it is long, it is called a *protein*. Thus the ribosome assembles a very particular sequence of amino acids at the behest of a very particular sequence of RNA nucleotides,

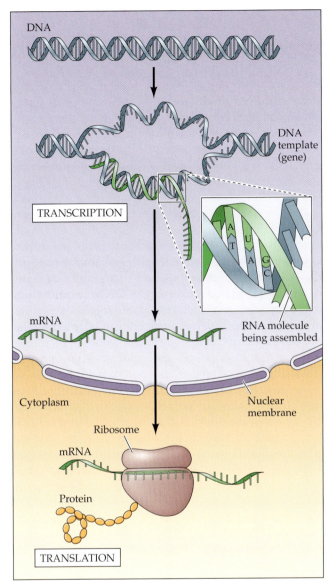

FIGURE A.2 DNA makes RNA, and RNA makes protein

which were themselves encoded in the DNA inherited from our parents. In short, the secret of life is that DNA makes RNA, and RNA makes protein.

There are fascinating amendments to this short story. Often the information from separate stretches of DNA is spliced together to make a single transcript; so-called alternative splicing can create different transcripts from the same gene. Sometimes a protein is modified extensively after translation ends; special chemical processes can cleave long proteins to create one or several active peptides.

Within a population, different individuals will inherit slightly different versions of any given gene, including versions that result in different amino acids at any particular position in the protein. These different versions of a given gene are known as **alleles**. Different alleles may produce proteins that vary in structure and function, including protein products that are completely dysfunctional, or even harmful to the cell and therefore the individual.

Keep in mind that each cell has the complete library of genetic information (collectively known as the **genotype**) but expresses only a fraction of all the proteins encoded in that DNA.

transcript   The mRNA strand that is produced when a stretch of DNA is "read."

stop codon   A trio of nucleotides in DNA to mark the end of transcription.

ribosomes   Structures in the cell body where genetic information is translated to produce proteins.

translation   The process by which amino acids are linked together (directed by an mRNA molecule) to form protein molecules.

codon   A set of three nucleotides that uniquely encodes one particular amino acid.

peptide   A short string of amino acids. Longer strings of amino acids are called *proteins*.

allele   A particular version of a given gene. Different alleles may differ in the functionality of the protein produced.

genotype   The total genetic makeup of an individual, typically determined at conception.

■ A.2 ■
MOLECULAR BIOLOGISTS HAVE CRAFTILY ENSLAVED MICROORGANISMS AND ENZYMES

clone   Here, the reproduction of a gene so that it can be sequenced and/or manipulated.

DNA sequencing   The process by which the order of nucleotides in a gene, or amino acids in a protein, is determined.

polymerase chain reaction (PCR)   Also called *gene amplification*. A method for reproducing a particular RNA or DNA sequence manyfold, allowing amplification for sequencing or manipulating the sequence.

transgenic   Referring to an organism in which foreign DNA has been deliberately inserted.

Molecular biologists have found ways to incorporate DNA from other species into the DNA of microorganisms such as bacteria and viruses. After the foreign DNA is incorporated, the microorganisms are allowed to reproduce rapidly, producing more and more copies of the (foreign) gene of interest. At this point the gene is said to be **cloned** because the researcher can make as many copies as she likes. To ensure that the right gene is being cloned, the researcher generally clones many, many different genes—each into different bacteria—and then "screens" the bacteria rapidly to find the rare one that has incorporated the gene of interest.

When enough copies of the DNA have been made, the microorganisms are ground up and the DNA extracted. If sufficient DNA has been generated, chemical steps can then determine the exact sequence of nucleotides found in that stretch of DNA—a process known as **DNA sequencing**. Once the sequence of nucleotides has been determined, the sequence of complementary nucleotides in the messenger RNA for that gene can be inferred. The sequence of mRNA nucleotides tells the investigator the sequence of amino acids that will be made from that transcript because biologists know which amino acid is encoded by each trio of DNA nucleotides. For example, scientists discovered the amino acid sequence of neurotransmitter receptors by this process.

The business of obtaining many copies of DNA has been boosted by a technique called the **polymerase chain reaction**, or **PCR**. This technique exploits a special type of polymerase enzyme that, like other such enzymes, induces the formation of a DNA molecule that is complementary to an existing single strand of DNA (see Figure A.1). Because this particular polymerase enzyme (called *Taq polymerase*) evolved in bacteria that inhabit geothermal hot springs, it can function in a broad range of temperatures. By heating double-stranded DNA, we can cause the two strands to separate, making each strand available to polymerase enzymes that, when the temperature is cooled enough, construct a new "mate" for each strand so that they are double-stranded again. The first PCR yields only double the original number of DNA molecules; repeating the process results in four times as many molecules as at first. Repeatedly heating and cooling the DNA of interest in the presence of this heat-resistant polymerase enzyme soon yields millions of copies of the original DNA molecule, which is why this process is also referred to as *gene amplification*. In practice, PCR usually requires the investigator to provide primers: short nucleotide sequences synthesized to hybridize on either side of the gene of interest to amplify that particular gene more than others.

With PCR, sufficient quantities of DNA are produced for chemical analysis or other manipulations, such as introducing DNA into cells. For example, we might inject some of the DNA encoding a protein of interest into a fertilized mouse egg (a zygote) and then return the zygote to a pregnant mouse to grow. Occasionally the injected DNA becomes incorporated into the zygote's genome, resulting in a **transgenic** mouse that carries and expresses the foreign gene.

Suppose we want to know whether a particular individual or a particular species carries a certain gene. Because all cells contain a complete copy of the genome, we can gather DNA from just about any kind of cell population: blood, skin, or muscle, for example. After the cells are ground up, a chemical extraction procedure isolates the DNA (discarding the RNA and protein). Finding a particular gene in that DNA just boils down to finding a particular sequence of DNA nucleotides. To do that, we can exploit the tendency of nucleic acids to hybridize with one another.

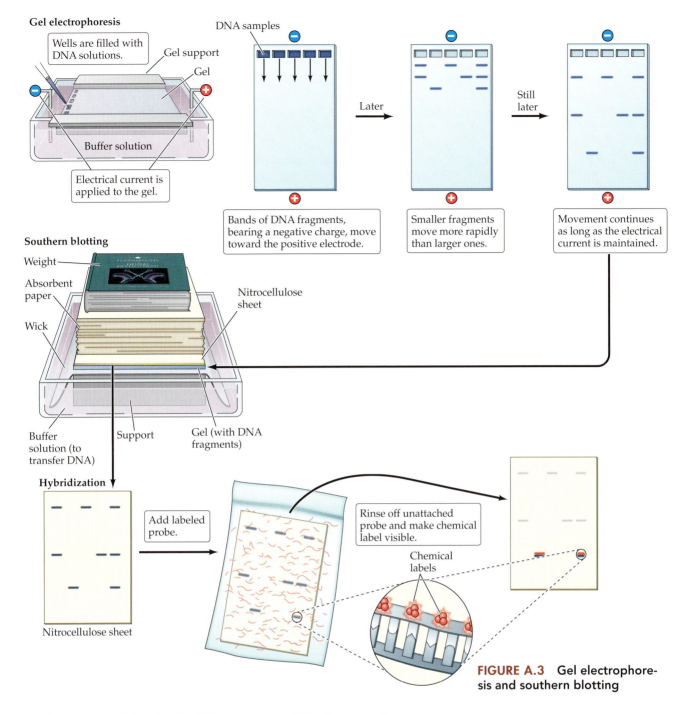

Gel electrophoresis

Wells are filled with DNA solutions.

Gel support

Gel

Electrical current is applied to the gel.

Buffer solution

DNA samples

Later

Bands of DNA fragments, bearing a negative charge, move toward the positive electrode.

Still later

Smaller fragments move more rapidly than larger ones.

Movement continues as long as the electrical current is maintained.

Southern blotting

Weight

Absorbent paper

Wick

Buffer solution (to transfer DNA)

Support

Gel (with DNA fragments)

Nitrocellulose sheet

Hybridization

Add labeled probe.

Rinse off unattached probe and make chemical label visible.

Chemical labels

Nitrocellulose sheet

FIGURE A.3 Gel electrophoresis and southern blotting

If we were looking for the DNA sequence GCT, for example, we could manufacture the sequence CGA (there are machines to do that), which would then stick to (hybridize with) any DNA sequence of GCT. The manufactured sequence CGA is called a **probe** because it is made to include a label (a colorful or radioactive molecule) that lets us track its location. Of course, such a short length of nucleotides will be found in many genes. In order for a probe to recognize one particular gene, it has to be about 15 nucleotides long.

When we extract DNA from an individual, it's convenient to let enzymes cut up the very long stretches of DNA into more manageable pieces of 1000 to 20,000 nucleotides each. A process called **gel electrophoresis** uses electrical current to separate these millions of pieces more or less by size (**FIGURE A.3**). Large pieces move slowly through a tube of gelatin-like material, and small pieces move rapidly. The gel is then placed on top of a sheet of paperlike material called *nitrocellulose*. When fluid is allowed to flow through the gel and nitrocellulose, DNA molecules are pulled out of the gel and deposited on

probe A manufactured sequence of DNA or RNA that is made to include a label (a colorful or radioactive molecule) that lets us track its location.

gel electrophoresis A method of separating molecules of differing size or electrical charge by forcing them to flow through a gel.

blotting Transferring DNA, RNA, or protein fragments to nitrocellulose following separation via gel electrophoresis. The blotted substance can then be labeled.

Southern blot A method of detecting a particular DNA sequence in the genome of an organism, by separating DNA with gel electrophoresis, blotting the separated DNAs onto nitrocellulose, and then using a nucleotide probe to hybridize with, and highlight, the gene of interest.

Northern blot A method of detecting a particular RNA transcript in a tissue or organ, by separating RNA from that source with gel electrophoresis, blotting the separated RNAs onto nitrocellulose, and then using a nucleotide probe to hybridize with, and highlight, the transcript of interest.

in situ hybridization A method for detecting particular RNA transcripts in tissue sections by providing a nucleotide probe that is complementary to, and will therefore hybridize with, the transcript of interest.

the waiting nitrocellulose. This process of making a "sandwich" of gel and nitrocellulose and using fluid to move molecules from the former to the latter is called **blotting** (see Figure A.3).

If the gene that we're looking for is among those millions of DNA fragments sitting on the nitrocellulose, our labeled probe should recognize and hybridize to the sequence. The nitrocellulose sheet is soaked in a solution containing our labeled probe; we wait for the probe to find and hybridize with the gene of interest (if it is present), and we rinse the sheet to remove probe molecules that did not find the gene. Then we *visualize* the probe, either by causing the label to show its color or, if radioactive, by letting the probe expose photographic film to identify the locations where the probe has accumulated. In either case, if the probe has found the gene, a labeled band will be evident, corresponding to the size of DNA fragment that contained the gene (see Figure A.3).

This process of looking for a particular sequence of DNA is called a **Southern blot**, named after the man who developed the technique, Edward Southern. Southern blots are useful for determining whether related individuals share a particular gene or for assessing the evolutionary relatedness of different species. The developed blots, with their lanes of labeled bands (see Figure A.3), are often seen in popular-media accounts of "DNA fingerprinting" of individuals.

A method more relevant for our discussions is the **Northern blot** (whimsically named as the opposite of the Southern blot). A Northern blot can identify which tissues are making a particular RNA transcript. If liver cells are making a particular protein, for example, then some transcripts for the gene that encodes that protein should be present. So we can take the liver, grind it up, and use chemical processes to isolate most of the RNA (discarding the DNA and protein). The resulting mixture consists of RNA molecules of many different sizes: long, medium, and short transcripts. Gel electrophoresis will separate the transcripts by size, and we can blot the size-sorted mRNAs onto nitrocellulose sheets; the process is very similar to the Southern blot procedure.

To see whether the particular transcript that we're looking for is among the mRNAs, we construct a labeled probe (of either DNA nucleotides or RNA nucleotides) that is complementary to the mRNA transcript of interest and long enough that it will hybridize only to that particular transcript. We incubate the nitrocellulose in the probe, allow time for the probe to hybridize with the targeted transcript (if present), rinse off any unused probe molecules, and then visualize the probe as before. If the transcript of interest is present, we should see a band on the film (see Figure A.3). The presence of several bands indicates that the probe has hybridized to more than one transcript and we may need to make a more specific probe or alter chemical conditions to make the probe less likely to bind similar transcripts.

Because different gene transcripts are of different lengths, the transcript of interest should have reached a particular point in the electrophoresis gel: small transcripts should have moved far; large transcripts should have moved only a little. If our probe has found the right transcript, the single band of labeling should be at the point that is appropriate for a transcript of that length.

Northern blots can tell us whether a particular *organ* has transcripts for a particular gene product. For example, Northern blot analyses have indicated that thousands of genes are transcribed only in the brain. Presumably the proteins encoded by these genes are used exclusively in the brain. But such results alone are not very informative, because the brain consists of so many different kinds of glial and neuronal cells. We can refine Northern blot analyses somewhat, by dissecting out a particular part of the brain—say, the hippocampus—to isolate mRNAs. Sometimes, though, it is important to know *exactly which cells* are making the transcript. In that case we use **in situ hybridization**.

FIGURE A.4   In situ hybridization

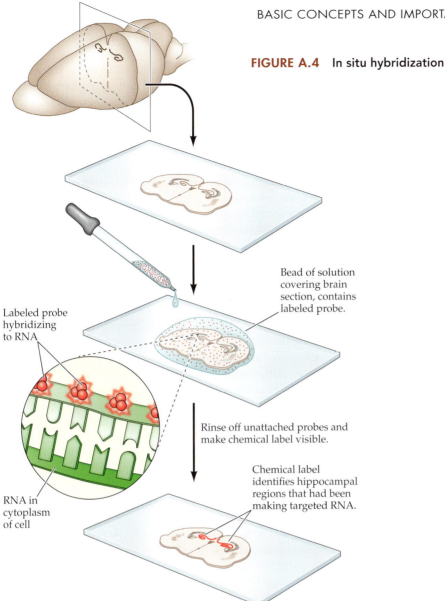

Bead of solution covering brain section, contains labeled probe.

Labeled probe hybridizing to RNA

Rinse off unattached probes and make chemical label visible.

Chemical label identifies hippocampal regions that had been making targeted RNA.

RNA in cytoplasm of cell

With in situ hybridization we use the same sort of labeled probe, constructed of nucleotides that are complementary to (and will therefore hybridize with) the targeted transcript, as in Northern blots. Instead of using the probe to find and hybridize with the transcript on a sheet of nitrocellulose, however, we use the probe to find the transcripts "in place" (*in situ* in Latin)—that is, on a section of tissue. After rinsing off the probe molecules that didn't find a match, we visualize the probe right in the tissue section. Any cells in the section that were transcribing the gene of interest will have transcripts in the cytoplasm that should have hybridized with our labeled probe (**FIGURE A.4**). In situ hybridization therefore can tell us exactly which cells are expressing a particular gene.

Sometimes we wish to study a particular protein rather than its transcript. In such cases we can use antibodies. **Antibodies** are large, complicated proteins that our immune system adds to the bloodstream to identify and fight invading microbes, thereby arresting and preventing disease. But if we inject a rabbit or mouse with a sample of a protein of interest, we can induce the animal to create antibodies that recognize and attach to that particular protein, just as if it were an invader.

Once these antibodies have been purified and chemically labeled, we can use them to search for the target protein. We grind up an organ, isolate the proteins

antibodies   Also called *immunoglobulins*. Large, Y-shaped proteins produced by the immune system that recognize and bind to particular shapes in molecules.

Western blot A method of detecting a particular protein molecule in a tissue or organ, by separating proteins from that source with gel electrophoresis, blotting the separated proteins onto nitrocellulose, and then using an antibody that binds, and highlights, the protein of interest.

immunocytochemistry (ICC) A method for detecting a particular protein in tissues in which an antibody recognizes and binds to the protein and then chemical methods are used to leave a visible reaction product around each antibody.

CRISPR **c**lustered **r**egularly **i**nterspaced **s**hort **p**alindromic **r**epeats. A system of gene manipulation that evolved in single-celled organisms and is exploited by scientists for gene editing.

guide RNA (gRNA) A strand of RNA designed to hybridize with a targeted nucleotide sequence in DNA in order to guide the Cas9 enzyme to break the DNA at that site.

Cas9 **C**RISPR-**as**sociated enzyme **9**. A bacterial enzyme that induces a break in double-stranded DNA as part of the CRISPR system.

(discarding the DNA and RNA), and separate them by means of gel electrophoresis. Then we blot these proteins out of the gel and onto nitrocellulose. Next we use the antibodies to tell us whether the targeted protein is among those made by that organ. If the antibodies identify only the protein we care about, there should be a single band of labeling (if there are two or more, then the antibodies recognize more than one protein). Because proteins come in different sizes, the single band of label should be at the position corresponding to the size of the protein that we're studying. Such blots are called **Western blots**.

To review, Southern blots identify particular DNA pieces (genes), Northern blots identify particular RNA pieces (transcripts), and Western blots identify particular proteins (sometimes called *products*).

If we need to know which particular cells within an organ such as the brain are making a particular protein, we can use the same sorts of antibodies that we use in Western blots, but in this case directed at that protein in tissue sections. We slice up the brain, expose the sections to the antibodies, allow time for them to find and attach to the protein, rinse off unattached antibodies, and use chemical treatments to visualize the antibodies. Cells that were making the protein will be labeled from the chemical treatments.

Because antibodies from the *immune* system are used to identify *cells* with the aid of *chemical* treatment, this method is called **immunocytochemistry**, or **ICC**. This technique can even tell us where, within the cell, the protein is found. Such information can provide important clues about the function of the protein. For example, if the protein is found in axon terminals, it may be a neurotransmitter. The technique may also be called immunohistochemistry to emphasize study of *tissues* ("histo").

■ A.3 ■
GENE EDITING ENABLES THE CREATION OF MODEL ORGANISMS

In Chapter 2 we discuss the use of "knockout" organisms, where a particular gene has been disabled, and "transgenic" organisms, where a new gene has been introduced. These animal models are powerful means of investigating the influence of genes on the nervous system and behavior. A recent breakthrough in molecular biology has made it much easier to produce such models in organisms as diverse as worms and monkeys.

Like many other biotech innovations, such as optogenetics, this one exploits a mechanism that evolved in simple organisms. Unicellular organisms like bacteria become infected with viruses that insert their invasive genes into the host organism's DNA. In response, bacteria have evolved mechanisms to identify and remove, or at least disable, the invasive genes. One mechanism is a stretch of DNA known as **CRISPR**, for **c**lustered **r**egularly **i**nterspaced **s**hort **p**alindromic **r**epeats. The RNA transcribed from a CRISPR site will serve as a guide to target the invasive DNA and so is referred to as a **guide RNA (gRNA)**. The palindromic sequences in the front of the gRNA cause it to fold up in a distinctive shape that binds to an enzyme, an endonuclease called **Cas9** (**C**RISPR-**as**sociated enzyme **9**), while the "tail" of the gRNA is complementary to the invasive sequence and so will hybridize with it. Then the associated Cas9 enzyme breaks the host's double-stranded DNA at that point. This type of break in double-stranded DNA is relatively rare, but all organisms have machinery in place to repair it. The thing is, this emergency repair often introduces an error, either adding or subtracting a few nucleotides. That in turn will introduce a frameshift mutation, with the result that the wrong

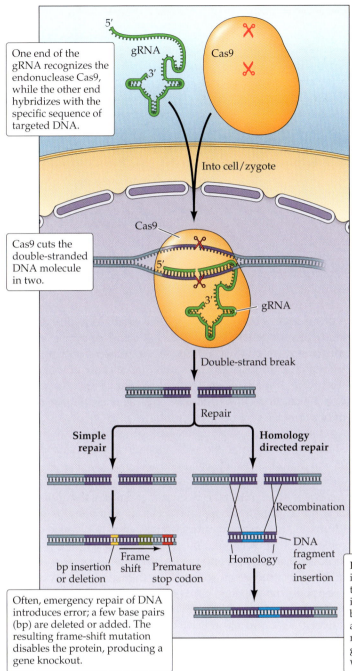

FIGURE A.5   CRISPR gene editing

One end of the gRNA recognizes the endonuclease Cas9, while the other end hybridizes with the specific sequence of targeted DNA.

Into cell/zygote

Cas9 cuts the double-stranded DNA molecule in two.

Double-strand break

Repair

Simple repair

Homology directed repair

Recombination

bp insertion or deletion   Frame shift   Premature stop codon

Homology   DNA fragment for insertion

Often, emergency repair of DNA introduces error; a few base pairs (bp) are deleted or added. The resulting frame-shift mutation disables the protein, producing a gene knockout.

If synthesized DNA is also introduced, it can hybridize with the two ends of the break and introduce a new sequence in between. In this way, a disease allele can be introduced into a model animal, or a dysfunctional gene in a human can be replaced with a functional version.

sequence of amino acids will be assembled and eventually a premature stop codon (the trio of nucleotides that normally signal the end of transcription) will end the transcription entirely (**FIGURE A.5**). The combination of improper amino acids and curtailed transcription means the invasive viral protein won't work, and the infection will be undone.

Eukaryotes like worms and mice lack the CRISPR mechanism and so normally do not produce Cas9. But we can introduce the gene for Cas9 into a zygote (of a worm, a fly, a mouse, or a monkey), and we can also introduce DNA custom made to produce a gRNA directed at whatever gene we choose. In that case, the gRNA will direct Cas9 to break the DNA within the targeted gene. Then the imperfect repair of that DNA will effectively knock out the gene in the individual that results from this zygote. Or the same system can be used to knock out the gene in particular cells in a mature individual. An enormous advantage of the

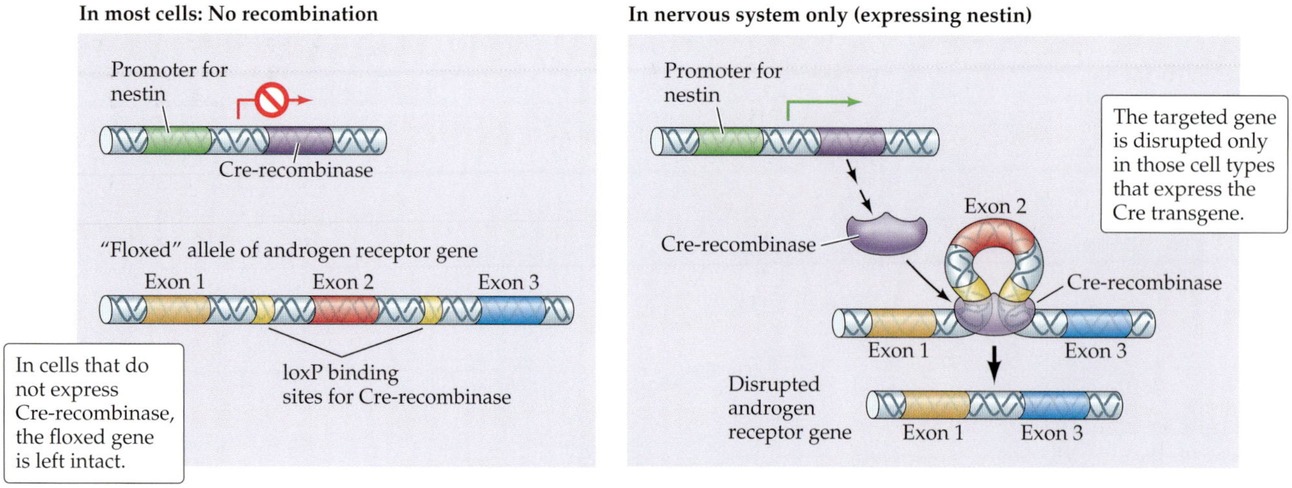

In most cells: No recombination

Promoter for nestin

Cre-recombinase

In cells that do not express Cre-recombinase, the floxed gene is left intact.

"Floxed" allele of androgen receptor gene

Exon 1   Exon 2   Exon 3

loxP binding sites for Cre-recombinase

In nervous system only (expressing nestin)

Promoter for nestin

The targeted gene is disrupted only in those cell types that express the Cre transgene.

Cre-recombinase

Exon 2

Cre-recombinase

Exon 1   Exon 3

Disrupted androgen receptor gene

Exon 1   Exon 3

FIGURE A.6   Cre-Lox systems for knocking out genes in specific cells

loxP   A specific sequence of nucleotides recognized by the enzyme Cre-recombinase. If the enzyme encounters a pair of loxP sites in a gene, it will remove the DNA between the two sites and recombine the gene, usually rendering the gene product dysfunctional.

Cre-recombinase   A bacterial enzyme that recognizes loxP, which is a specific sequence of nucleotides, and recombines the DNA at loxP sites.

CRISPR system is that one can introduce several different gRNAs simultaneously to knock out several different genes at once (Wang et al., 2013).

Scientists have also exploited the CRISPR system for gene editing. In this variant, the scientist again introduces DNA to cause the cell or zygote to express Cas9 and a specific gRNA, but also provides synthesized DNA fragments that are homologous to the cut ends of the host's DNA. Then when the break is repaired, the synthesized DNA fragment recombines into the host's genome, a case of *homology-directed repair*. Now, instead of knocking out the targeted gene, the scientist has replaced that section of the gene, effectively producing a new allele. In this way, a mutated allele can be introduced into the fly, mouse, zebrafish, or monkey, producing a transgenic animal (more specifically, a "knockin" animal, where a new allele has been put in place of the endogenous allele). There is also great hope that this method may someday be used for gene therapy in humans, to replace dysfunctional genes that cause disease, such as the *dystrophin* gene in Duchenne muscular dystrophy.

Another bacterial mechanism to get rid of viral infection is to insert a specific series of nucleotides, called **loxP** sites, to mark the invasive DNA. A bacterial enzyme, called **Cre-recombinase**, then binds to two such loxP sites (as long as they are not too far apart), cuts out the DNA between them, and joins the cut ends of DNA back together, a process called *recombination*. This severe editing renders the viral gene product worthless, thwarting the infection. Multicellular organisms normally do not have this Cre-lox system, but scientists can insert them into the genome of flies/zebrafish/mice to manipulate genes. For example, we can insert lox sites into the DNA of a targeted gene (we might use the CRISPR homology-directed repair to do this) in such a way that the gene still functions normally despite the lox sites. Now we can inject viruses into a particular brain site to infect cells there to express Cre-recombinase. That means the loxP-flanked, or "floxed" gene will be knocked out only in that brain region. Or for another approach, we might cross animals carrying the floxed allele with animals carrying a transgene for Cre-recombinase. This transgene might include a promoter that causes the enzyme to be expressed only in neurons, or only in astrocytes, or only in motor neurons. That means the floxed gene will be disrupted only in those particular classes of cells, not the whole organism (**FIGURE A.6**). Using these methods, scientists are able to test more and more specific hypotheses about how gene expression alters brain development.

GLOSSARY

Numbers or letters in brackets refer to the chapter(s) where the term is introduced.

A

a posteriori   Latin for "from what comes after," referring to things that we learn through our senses in our experience of the world. **[E]**

a priori   Latin for "from what comes before," referring to things that the mind must know before having any experiences and that are required before the mind can even have an experience. **[E]**

acetylcholine (ACh)   The neurotransmitter released by parasympathetic fibers as well as vertebrate motor neurons. **[4, 6]**

acetylcholine receptors (AChRs)   The integral membrane proteins, consisting of several subunits, that respond to the neurotransmitter acetylcholine. **[6]**

acetylcholinesterase   An enzyme that breaks down acetylcholine to halt neurotransmission. **[6]**

achaete-scute complex (AS-C)   The complex of proneural gene products that bind together and serve as transcription factors to direct early differentiation into neurons. **[1]**

actin   A protein that mediates contraction of cell parts, such as filopodia of growth cones. **[5]**

active zone   The portion of the presynaptic terminal that actively releases neurotransmitter. **[6]**

agenesis of the corpus callosum (AgCC)   The condition of being born with the corpus callosum either absent or severely reduced. **[5]**

agrin   A large proteoglycan that serves as a signal and inductive factor at neuromuscular junctions. **[6]**

allele   A particular version of a given gene. Different alleles may differ in the functionality of the protein produced. **[A]**

amblyopia   Reduced visual acuity of one eye, that is not caused by optical or retinal impairments. **[9]**

AMPA   α-**a**mino-3-hydroxy-5-**m**ethyl-4-isoxazole**p**ropionic **a**cid, a synthetic compound that binds one class of ionotropic glutamate receptors with high affinity. **[8]**

amyotrophic lateral sclerosis (ALS)   A neurodegenerative disease characterized by muscle wasting and death of motor neurons. **[7]**

androgen receptor (AR)   A member of the steroid receptor superfamily with a high affinity for androgens such as testosterone. **[7]**

androgens   A class of steroid hormones, including testosterone, found in high concentrations in male vertebrates. **[7]**

anosmia   The condition of being unable to detect odors. **[3, 9]**

Antennapedia   A Hox gene in which mutations result in the formation of a leg where an antenna normally forms. **[2]**

anterior commissure   A relatively small collection of axons, found in the ventral portion of the brain, that communicate between the two cerebral hemispheres. **[5]**

anterograde transport   The transport of materials within axons in the direction of the axonal terminals. **[5]**

anteroventral periventricular nucleus (AVPV)   A hypothalamic region that has a greater volume in females than in males in rats and mice. **[7]**

antibodies   Also called *immunoglobulins*. Large, Y-shaped proteins produced by the immune system that recognize and bind to particular shapes in molecules. **[3, A]**

apoptosis   Cell death that occurs as a natural process during normal development. **[7]**

aromatization   Here, the single-step reaction, catalyzed by the enzyme aromatase, that converts androgens such as testosterone into estrogens such as estradiol. **[7]**

astrotactin   A gene that is expressed by cerebellar granule cells to help them to grip glial fibers for migration. **[3]**

ataxia   Lack of motor coordination. **[6]**

attachment   Here, the strong emotional bond between an infant and one or more caregivers. **[10]**

attractive   Here, referring to materials to which growth cones readily attach. **[5]**

autism spectrum disorder (ASD)   A disorder of social interaction that may be accompanied by problems in communication and severely limited behavior and/or interests. **[5]**

autonomic ganglia   Collections of neurons outside the central nervous system that provide autonomic innervation of body organs. **[3]**

autonomic nervous system   The neural system that regulates activity of many organ systems and consists of the sympathetic nervous system and the parasympathetic nervous system. It is largely outside of our conscious control. **[8]**

autoradiogram   A preparation where tissue "takes its own picture" by exposing chemical film to radioactively labeled markers. **[8]**

autoradiography   A process in which a tissue "takes its own picture" when photographic film is exposed to radioactively labeled markers. **[3]**

axonal transport   The process by which materials are moved within axons, in both directions, along microtubules. **[5]**

B

basal lamina   A particularly thick layer of extracellular matrix that surrounds mature muscle fibers and many organs. **[3, 5, 6]**

Bcl-2   An anti-apoptotic protein in mammals. Its homologue in worms is CED-9. **[7]**

Bergmann glia   Long, slender glial cells in cerebellar cortex that guide neurons migrating from the external granule cell layer to the internal granule cell layer. **[3]**

β-catenin   A transcription factor that also regulates cell adhesion and plays a role in several stages of neural development. **[1, 6]**

beta-galactosidase (β-Gal)   A bacterial enzyme, encoded by the *lacZ* gene, that often serves as a reporter gene or marker in studies of neural development. **[3]**

bicoid   A transcription factor that is concentrated in the presumptive anterior end of the embryo and directs formation of the head. **[2]**

binocular deprivation   A procedure of depriving both eyes of light. **[9]**

binocular   Referring to two eyes. **[8]**

birthdate   Here, the time during development when a given cell underwent its final mitosis before differentiating into a neuron or glial cell. **[3]**

bithorax   A Hox gene in which mutations can result in the doubling of the thorax in *Drosophila*. **[2]**

blastocoel   The hollow, fluid-filled cavity inside a blastula. **[1]**

blastomeres   The individual cells that make up a blastula. **[1]**

blastopore   A small dimple on the surface of a blastula that will invaginate to start forming the primitive gut. **[1]**

blastula   The earliest stage of an embryo, typically a spherical clump of cells. **[1]**

blotting   Transferring DNA, RNA, or protein fragments to nitrocellulose following separation via gel electrophoresis. The blotted substance can then be labeled. **[A]**

bone morphogenetic protein (BMP)   A class of growth factors that act to encourage ectodermal cells to take on an epidermal, rather than a neural, fate. It is part of the transforming growth factor beta (TGFβ) family. **[1]**

brain-derived neurotrophic factor (BDNF)   A factor, originally isolated from pig brains, that supports the developmental survival of DRG cells, but not sympathetic cells. **[7]**

***bride of sevenless* (*boss*)**   A gene in *Drosophila* that encodes the membrane-bound protein boss, which is expressed in photoreceptor 8 and binds sevenless in a nearby cell, inducing it to become photoreceptor 7. **[4]**

bungarotoxin   A toxin found in the venom of certain snakes that irreversibly binds to and blocks nicotinic acetylcholine receptors. **[8]**

C

Caenorhabditis elegans   A microscopic roundworm that offers a valuable model of cell differentiation. **[1]**

calcium-dependent cell adhesion molecules (cadherins)   A class of transmembrane proteins, the adhesive properties of which are sensitive to local levels of calcium. **[5, 6]**

calmodulin   A protein that binds Ca^{2+}, regulating intracellular concentrations of the ion. **[6]**

Cas9   CRISPR **as**sociated enzyme **9**. A bacterial enzyme that induces a break in double-stranded DNA as part of the CRISPR system. **[A]**

caspase-3   One of the last in a cascade of caspases activated during apoptosis. It seems to act as the "executioner" that dooms the cell. **[7]**

caspase-9   One of the earliest in a cascade of caspases leading to apoptosis. **[7]**

caspases   Cysteine-dependent aspartate-directed proteases, including CED-3, which cleave many different proteins. **[7]**

castration   Surgical removal of the gonads, typically testes. **[7]**

caudal   Referring to the tail end. **[2]**

***ced* (cell death abnormal)**   A collection of genes, originally isolated in *C. elegans*, that regulates apoptosis. **[7]**

ced-3 A gene that encodes a protease, CED-3, to dismantle cell machinery as part of the process of apoptosis. **[7]**

cell adhesion molecules (CAMs) A class of molecules, found in extracellular regions, that adhere to some cells and not others. **[3, 5]**

cell-autonomous response The condition when a particular influence acts directly upon a cell to affect it, as opposed to acting first on another cell, which then affects the cell of interest. **[7]**

cell-cell interactions Here, the process by which developing cells communicate with one another and direct each other's fate. **[1]**

cells The basic building blocks of life. **[1]**

cellular differentiation The process by which individual cells in an organism become progressively more specialized and different from one another. **[1]**

cellularization The process in which cell walls are formed between the nuclei in a syncytium. **[2]**

central nervous system (CNS) The brain and spinal cord in vertebrate species, derived from the neural tube. **[1]**

cerebellum A brain region attached to the pons that plays an important role in coordination of movement. **[2]**

chemoaffinity hypothesis Roger Sperry's proposal that axonal growth cones seek out a particular target cell based on chemical signals that mark both. It explains how regenerating retinal cells manage to reestablish their original synaptic contacts with tectal cells. **[5, 9]**

chemotaxis The process of displaying chemotropism. **[5]**

chemotropism The tendency to follow along a particular chemical trail. **[5]**

chimera Here, an individual made up of cells displaying more than one genotype, formed from the combination of cells from two separate zygotes. Typically, such an individual has received genes from more than two parents. **[1, 4]**

chordin A gene that encodes the protein chordin, which exerts an organizer-like effect on ectoderm, shifting it from an epidermal to a neural fate. **[1]**

chromosome A complex of condensed strands of DNA and associated protein molecules; found in the nucleus of cells. **[A]**

ciliary neurotrophic factor (CNTF) A trophic factor that prevents developmental death of neurons in the ciliary ganglia. **[7]**

clone Here, the reproduction of a gene so that it can be sequenced and/or manipulated. **[A]**

codon A set of three nucleotides that uniquely encodes one particular amino acid. **[A]**

colinearity The property of Hox genes in which their order on the chromosome matches the order in which they are expressed along the anterior-posterior axis of the body. **[2]**

collagen A long-chained filamentous protein that contributes to the formation of extracellular matrix and is especially prominent in connective tissue. **[3, 5, 6]**

commissural neuron A neuron that extends its axon across the body midline. **[5]**

commitment Here, the tendency of a cell to take on a particular fate even in the face of particular challenges. Its meaning is restricted to only those challenges that have been tested. **[1]**

concentration gradient The condition in which a particular substance, such as a CAM, is more concentrated at one end of a structure than at another. **[5]**

conditional specification of cell fate The developmental strategy in which each cell's fate depends on environmental conditions, primarily the fate of neighboring cells. **[1]**

conspecific An individual of the same species. **[10]**

construct Here, a stretch of DNA that has been manipulated before being inserted into an organism's genome to create a transgenic animal. **[4]**

contact guidance The process by which growth cones are guided upon direct contact with the membranes of other cells. **[5]**

contralateral Referring to the opposite side of the body or brain. **[8]**

corpus callosum (CC) The large bundle of axons that communicate between the two cerebral hemispheres in placental mammals. **[5]**

cortical plate In developing cortex, the expanding layer of postmitotic cells that settle beneath the marginal zone and above the intermediate zone. It will form layers II–VI. **[3]**

cortical preplate The region between the ventricular zone and marginal zone in developing vertebral cerebral cortex, which develops into the gray matter of the neocortex. **[3]**

Cre-recombinase A bacterial enzyme that recognizes loxP, which is a specific sequence of nucleotides, and recombines the DNA at loxP sites. **[A]**

CRISPR **c**lustered **r**egularly **i**nterspaced **s**hort **p**alindromic **r**epeats. A system of gene manipulation that evolved in single-celled organisms and is exploited by scientists for gene editing. **[1, A]**

crystallization Here, the process by which a young male bird makes the transition from highly variable plastic song to an adult song, which tends to be rather stereotyped. **[10]**

curare A toxin found in various South American plants that blocks nicotinic acetylcholine receptors. **[8]**

D

dam Here, a mother of a domesticated animal. **[3]**

death genes A term for those genes that are activated in cells during apoptosis, including proteases to dismantle proteins and nucleases to fragment DNA. [7]

decapentaplegic (dpp)   The gene that encodes dpp, a bone morphogenetic protein homologue in insects that halts mitosis in the optic disk and promotes differentiation of photoreceptors. It also induces ectodermal cells to an epidermal, rather than a neural, fate. [4]

delamination   Here, the process by which one cell in a cluster of proneural gene–expressing cells detaches from the sheet of neighboring cells to enter the insect body interior. The cell then differentiates into a neuroblast. [1]

Delta   A membrane-bound protein that binds to Notch found on the surface of adjoining cells. Delta directs that target cell away from a neural fate and toward an epidermal fate. [1]

dentate gyrus   A portion of the hippocampal formation. It is known for continued neurogenesis in adulthood. [8]

deoxyribonucleic acid (DNA)   A nucleic acid that is present in the chromosomes of cells and encodes hereditary information. [A]

Diablo   Also known as *Smac*. A protein that promotes apoptosis, in part by blocking the action of IAPs. [7]

diencephalon   The portion of the vertebrate brain that consists of the thalamus and hypothalamus. [2]

dimer   A complex of two proteins that bind together to form a functional unit. [1]

divergence   Here, the gradual change in the structure and function of the different versions of a gene that had been duplicated. [6]

dizygotic twins   A pair of twins derived from two fertilized eggs, who are therefore no more genetically related than any other pair of siblings. [10]

DNA sequencing   The process by which the order of nucleotides in a gene, or amino acids in a protein, is determined. [A]

dorsal lip of the blastopore   An embryonic region of the blastopore that can induce the development of a second nervous system, and therefore a second individual from an embryo. [1]

dorsal root ganglion (DRG)   A collection of neuron cell bodies embedded in vertebrate dorsal roots of the spinal cord that provides neurites to gather sensory information from the periphery and transmit it to the dorsal horn of the spinal cord. The bifurcated axons of the DRG extend one process to the periphery while the other conducts that information to the dorsal horn. [3, 7]

Drosophila melanogaster   The common fruit fly, a valuable model organism for studies of genetics and cellular differentiation. [1]

duplication   Here, replication of a gene in the course of evolution. [6]

E

ectoderm   The outermost of the three germ layers. [1]

embryo   The earliest stage of development of a new individual, typically a spherical collection of cells. [1]

embryonic stem cells   Cells found in embryos that display totipotency. [1]

empiricists   Philosophers who believed that the only way to gain knowledge of the world is through information provided by the senses. [I]

Emx2   A homeobox gene highly expressed in the posterior portion of the developing vertebrate cortex. [2]

endoderm   The innermost of the three germ layers. [1]

Enhancer of split, or E(spl)   A protein that dimerizes to a Notch fragment. Together they suppress proneural gene expression. [1]

enhancer region   The regulatory region that may be upstream or downstream of the structural gene, which plays a role in controlling transcription of that gene. [1]

Enlightenment   The period, roughly 1650–1750, when European thinkers began rebelling against the traditional hierarchies of monarchs and religion in favor of reason and science. [I]

environmental enrichment   The stimulation provided to domesticated animals intended to boost brain development, including social stimulation and access to physical objects. [10]

ephrin receptors   A family of receptor tyrosine kinase (RTK) molecules that bind to ephrins. [5]

ephrins   A family of membrane-bound signaling molecules that bind to ephrin receptors, which are part of the receptor tyrosine kinase (RTK) superfamily. [2, 3, 5, 6]

epidermis   Skin tissue, derived from ectodermal cells that do not become neural tissue. [1]

epigenesis   Here, the process in which successively more, and successively more complex, body structures appear in development. [1]

epigenetic   A variably defined term that often refers to a change in the genome, other than the sequence of nucleotides, that has a lasting effect on expression of a gene (see Box 10.1). [10]

epistemology   The study of knowledge and how we gain it. [P]

epitope   The particular shape of a molecule that a given antibody recognizes and binds. [3]

ErbB   A family of membrane-bound receptors that respond to neuregulins. [6]

estrogen   A class of steroid hormones, including estradiol, secreted by the ovaries. [7]

estrogen receptors   Two members of the steroid receptor superfamily that have a high affinity for estrogens such as estradiol. [7]

eukaryote   Any organism whose cells have the genetic material contained within a nuclear envelope. [A]

external granule cell layer   A layer of granule neurons that migrate to the top of the developing cerebellum before migrating ventrally to form the internal granule cell layer. [3]

extracellular matrix (ECM)   The collection of various long-chain molecules that loosely bind one another to form a layer outside many cell membranes. [5, 6]

F

fasciculation   Here, the tendency of axonal growth cones to adhere to and follow along a preexisting bundle of axons. [5]

fate   Here, the particular structure and function that a given cell takes on in the course of cellular differentiation. [1]

fate map   A representation of which parts of an embryo will give rise to various parts of the adult body. [2]

fetus   A stage of development reached once major organs and body parts are in place. [1]

fibroblast growth factors (FGFs)   A family of proteins that act upon a family of receptor tyrosine kinase (RTK) receptors. They are concentrated in the posterior portions of the developing vertebrate nervous system and demarcate the midbrain/hindbrain junction. [2, 4]

fibronectin   A long-chained glycoprotein that contributes to the formation of the extracellular matrix. [3, 5]

filopodia   The slender, rodlike extensions of membrane produced by growth cones. [5]

5-bromo-2′-deoxyuridine (BrdU)   A synthetic nucleotide that can serve as a substitute for thymidine in the synthesis of DNA but can be readily distinguished from thymidine by the use of antibodies. [3]

floor plate   The ventral portion of the vertebrate neural tube, the developing spinal cord. [2, 5]

Flynn effect   The increase in average human performance on IQ tests that has been observed worldwide over the past century. [10]

follistatin   A gene that encodes the protein follistatin, which exerts an organizer-like effect on ectoderm, shifting it from an epidermal to a neural fate. [1]

forkhead box P2 (FOXP2)   A gene encoding the transcription factor FOXP2, which is implicated in language in humans, song in birds, and ultrasonic vocalizations in mice. [10]

Form   Also called *Platonic Form*. An eternal, perfect "blueprint" of an object or concept that we recognize in its imperfect, highly variable real-life examples. [P]

fragile X syndrome (FXS)   A disorder caused by extended repeats of CGG trinucleotides in the gene named *fragile X mental retardation 1 (FMR1)*. [6]

fraternal birth order effect   The well-established finding that the more brothers a boy's mother carried before him, the more likely he is to grow up to be gay. [7]

Frazzled   A family of membrane-bound receptors that respond to netrins. [5]

Frizzled   A cell surface receptor protein that responds to Wnt. [2]

fruitless (fru)   A gene in *Drosophila* that produces functional transcripts only in males and induces the development of male-specific neural circuits that control male courtship and mating. [7]

G

G proteins   A family of proteins, named for their binding to GDP and GTP, that transmit intracellular signals. [6]

ganglionic eminences (GEs)   Transient bumps along the lateral ventricles of the developing telencephalon, from which neurons migrate tangentially across radial glia. [3]

gap genes   Genes that encode a class of transcription factors, the expression of which is regulated by maternal polarity genes. [2]

gastrula   An embryo that has formed a primitive gut, a tube that passes through the embryo. [1]

gastrulation   The process by which a blastula becomes a gastrula; the formation of a primitive gut. [1]

gel electrophoresis   A method of separating molecules of differing size or electrical charge by forcing them to flow through a gel. [A]

gene   A stretch of DNA that represents a functional unit of inheritance, specifying the structure of one or more proteins. [1]

gene duplication and divergence   The evolutionary process by which a gene duplication is followed by successive divergence in the sequence and function of the two copies. [2]

genotype   The total genetic makeup of an individual, typically determined at conception. [1,A]

germ layers   Here, the three layers of cells formed in the course of gastrulation. [1]

glial fibrillary acidic protein (GFAP)   A structural protein normally expressed in radial glia and astrocytes but not neurons. [3]

gliogenesis   The mitosis of cells that will give rise to glia. [3]

glucocorticoid   A class of steroid hormones, primarily released from the adrenal cortex, named for their role in regulating glucose metabolism. [10]

glucocorticoid receptor   A member of the steroid receptor superfamily that normally binds and responds to glucocorticoids. [10]

gonadotropin-releasing hormone (GnRH)   A protein released by neurons in the hypothalamus into the median eminence, which directs anterior pituitary cells to secrete the gonadotropins: follicle-stimulating hormone (FSH) and luteinizing hormone (LH). [3]

gray matter   The outer portion of the vertebrate cortex predominated by neuronal and glial cell bodies rather than myelin, hence dark in color in postmortem preparations. It is organized in six layers in mammals. [3]

growth cones   The extensions of dendrites and axons that grow away from the cell body to make synaptic contacts. [5]

guide RNA (gRNA)   A strand of RNA designed to hybridize with a targeted nucleotide sequence in DNA in order to guide the Cas9 enzyme to break the DNA at that site. [A]

guidepost cells   Certain cells that seem to serve as targets for axonal growth cones establishing a particular pathway. [5]

Guillain-Barré syndrome   An autoimmune disorder attacking Schwann cells, demyelinating axons and causing varying degrees of weakness and tingling. [6]

gynandromorph   A rare individual that displays both male and female features in a species in which most individuals are either uniformly male or female. [7]

H

halteres   Paired structures that serve as counterweights to maintain balance in the flight of some flies. [2]

Hansen's disease   Formerly called leprosy. A chronic bacterial disease that causes nerve damage and disfiguring sores on the limbs and face. [4]

Hebbian synapse   A synapse that grows stronger when it repeatedly succeeds in driving the postsynaptic cell to fire and grows weaker when it repeatedly fails to drive the postsynaptic cell. [8]

hedgehog (hh)   A gene in *Drosophila* that is a homologue of *Sonic hedgehog (Shh)*. It encodes the secreted protein hedgehog, which influences cell differentiation. [4]

hermaphrodite   An individual capable of reproducing as either a male (producing sperm) or female (producing ova). [1]

heterophilic binding   The property of two different materials that readily bind to one another. [5]

hippocampus   A portion of the limbic system that is known to be important for learning and memory [8]

homeobox   A nucleotide sequence that produces a DNA-binding domain in many transcription factor proteins. It is found in Hox genes and many other transcription factors. [2]

homeotic selector genes or Hox genes   Also called simply *homeotic genes*. A class of genes in which mutations tend to result in swapping out one body part for another, as when mutation of *Antennapedia* results in the formation of legs where antennae normally form. [2]

homophilic binding   The property of a material that readily binds to itself. [5]

hormones   Chemical signals released into circulation by one group of cells that affect cells elsewhere in the body. [7]

hunchback   A transcription factor that is one of the class of gap genes that is inhibited by the nanos protein. [2]

hybridization   The process by which a string of nucleotides becomes linked to a complementary series of nucleotides. [A]

I

immunocytochemistry (ICC)   A method for detecting a particular protein in tissues in which an antibody recognizes and binds to the protein and then chemical methods are used to leave a visible reaction product around each antibody. Sometimes called immunohistochemistry (IHC) to emphasize study of tissues. [A,3]

imprinting   Here, the tendency of young animals, primarily birds, to visually note the characteristics of a parent and then follow that parent. [10]

in situ hybridization   A method for detecting particular RNA transcripts in tissue sections by providing a nucleotide probe that is complementary to, and will therefore hybridize with, the transcript of interest. [A]

induction   Here, the process by which one group of cells directs the differentiation of other, nearby cells. [1]

inhibitors of apoptosis proteins (IAPs)   A family of proteins that block caspases and in other ways avert apoptosis. [7]

innate   A vaguely defined term, roughly equivalent to *instinct*, said to typify some behaviors of some nonhuman animals. [10]

inner cell mass   In mammals, the clump of cells found inside the blastula, which will give rise to the individual's body. The remainder of the blastula will contribute to the placenta and related tissues. [1]

instinct   A vaguely defined term said to typify some behaviors of some nonhuman animals. [10]

integrins   A family of adhesive molecules that are both membrane bound and secreted into the extracellular matrix. [5]

intellectual disability   A lifelong impairment in intellectual function and adaptive behavior, formerly known as mental retardation. [6]

intelligence quotient (IQ)   A measure of intelligence based on performance on certain tests of abstract reasoning. **[10]**

intermediate zone   The layer between the ventricular zone and marginal zone of the developing vertebrate brain. **[3]**

internal granule layer   A layer of small neurons ventral to the Purkinje cell layer in the vertebrate cerebellar cortex. **[3]**

ion channels   Membrane-bound proteins that, when activated, open up to allow select ions to cross the membrane. **[6]**

ionotropic   Referring to a class of neurotransmitter receptor that includes an ion channel to affect the electrical potential across the postsynaptic cell's membrane. **[8]**

ipsilateral   Referring to the same side of the body or brain. **[8]**

Isl-1, Isl-2   Homeotic genes that encode the transcription factors Isl-1 and Isl-2, which are expressed in developing motor neurons. **[4]**

J

jimpy   Mice with an X-linked recessive mutation of the myelin proteolipid protein that results in severe lack of CNS myelin, tremors, and convulsions. **[6]**

K

kainate   A synthetic compound that binds a class of ionotropic glutamate receptors with high affinity. **[8]**

Kallmann syndrome   A condition in which individuals fail to reach puberty and are unable to detect odors. **[3]**

kinases   Enzymes that promote the addition of phosphates to particular sites on proteins, a process called phosphorylation. **[6]**

knockin   Referring to an animal in which an endogenous gene has been deliberately replaced with another allele, often an allele associated with a human disorder. **[4]**

knockout   Here, an animal in which a particular gene has been deliberately removed or disabled. **[1, 4]**

L

L1 cell adhesion molecule (L1)   Also called *L1CAM*. An important cell adhesion molecule. **[5]**

lacZ   A bacterial gene that encodes the enzyme beta-galactosidase. **[3]**

lamellipodia   The sheetlike extensions of membrane produced by growth cones. **[5]**

laminin   A long-chained glycoprotein that readily binds other molecules and is a major component of the extracellular matrix and connective tissue. **[3, 5, 6]**

lateral geniculate nucleus (LGN)   A nucleus of the thalamus that receives projections from the eyes. **[8]**

lateral inhibition   The process by which neighboring cells in a tissue layer inhibit one another, as in the competition between ectodermal cells for a neural fate that is mediated by the Delta-Notch system. **[1]**

layer IV   The fourth layer down from the pial surface in vertebrate neocortex. It is predominated by input from the thalamus and other cortical regions. **[8]**

ligand-gated ion channels   Membrane-bound proteins that open up an ion channel in response to a ligand such as a neurotransmitter. **[6]**

Lim domains   Domains found in various transcription factors that form zinc fingers to bind DNA in the process of gene regulation. **[4]**

long-term potentiation (LTP)   Long-lasting strengthening of synaptic strength seen after the induction of a tetanus in presynaptic afferents. It is observable in several neuronal pathways of the hippocampal formation. **[8]**

lordosis   Literally, the arching of the back that elevates the shoulders and hips; here, a posture displayed by receptive female rodents to permit copulation. **[7]**

loxP   A specific sequence of nucleotides recognized by the enzyme Cre-recombinase. If the enzyme encounters a pair of loxP sites in a gene, it will remove the DNA between the two sites and recombine the gene, usually rendering the gene product dysfunctional. **[A]**

LRP4   A receptor for agrin that forms a complex with MuSK. **[6]**

M

marginal zone   The outermost layer of the developing vertebrate brain. By adulthood it will form the molecular layer of the cerebral cortex. **[3]**

maternal effect   Influences the mother has on ontogeny of offspring apart from the particular genes she contributed. **[1]**

maternal polarity genes   A class of genes provided by the mother such that their products are unevenly distributed in the zygote and thereby specify the anterior-posterior axis. **[2]**

Mauthner cells   Large identifiable neurons found in the brainstem in many fishes and amphibians. **[5]**

mec-3   A gene in *C. elegans* that encodes the protein mec-3, which binds to the protein unc-86. The two proteins together regulate expression of genes important for differentiation into a touch receptor neuron. **[1]**

melanocytes   Neural crest-derived pigment cells that provide color to the skin. **[3]**

meninges   The three layers of tissue protecting the vertebrate central nervous system. **[3]**

mesencephalon   Also called *midbrain*. The middle segment of the embryonic vertebrate brain. It will develop into the adult midbrain. **[2]**

mesoderm   The germ layer that forms between the ectoderm and endoderm. **[1]**

messenger RNA (mRNA)   A strand of RNA that carries the code of a section of a DNA strand to the cytoplasm. **[A]**

metabotropic   Referring to a class of neurotransmitter receptor that uses a second messenger system to biochemically alter the postsynaptic target. **[8]**

metalloproteases   Enzymes that break down proteins and require atoms of a metal such as zinc to function. **[3]**

metazoan   A multicellular animal. **[1]**

metencephalon   A subdivision of the hindbrain that includes the cerebellum and the pons. **[2]**

microglia   The class of glial cells that clean up debris and residue in the nervous system. **[7]**

microtubules   The long filamentous materials formed from tubulin that provide internal structure for cells. **[5]**

mitogen   A substance that promotes mitosis. **[2]**

mitosis   The process of cell division in which both resulting cells receive the full complement of genetic material. **[3]**

mitotic figures   The tangled threads of duplicated chromosomes being pulled apart that are seen in cells undergoing mitosis. **[3]**

mitotic lineage   The sequence of mitosis during ontogeny that gives rise to a particular cell in an individual. In *C. elegans* there is an invariant relationship between mitotic lineage and cell fate. **[1]**

modern synthesis   The fusion of Darwin's theory of evolution by natural selection with Mendel's notion that genes represent discrete units of inheritance that are passed on either whole or not at all. **[1]**

molecular layer   The outermost layer of the vertebrate cerebral cortex, consisting primarily of dendrites and axons with relatively few cell bodies. **[3]**

Molyneux's problem   A famous thought experiment that asks whether an adult enabled to see for the first time would be able to make sense of his or her vision. **[I, 9]**

monocular   Referring to one eye. **[8]**

monocular deprivation   A procedure of depriving one eye of light. **[9]**

monozygotic twins   A pair of twins derived from a single fertilized egg, who therefore share all the same genes. **[10]**

morphogenetic furrow   The prominent indentation of the insect optic disc that moves anteriorly, marking the differentiation of photoreceptors. **[4]**

mosaic specification of fate   A strategy for cellular differentiation in which each cell follows a particular fate no matter what neighboring cells might be doing, as in *C. elegans*. **[1]**

motor neurons   Neurons that send axons out to the periphery to innervate and control muscles. **[2]**

motor unit   A motor neuron and all the muscle fibers that it innervates. **[8]**

multiple sclerosis (MS)   An autoimmune disease attacking CNS myelin, causing highly variable degrees of impairment of motor and sensory processes. **[6]**

muscle-specific kinase (MuSK)   A membrane-bound kinase found exclusively in muscle fibers that serves as a receptor for agrin. **[6]**

myelencephalon   Also called *medulla*. The caudal-most portion of the vertebrate brainstem, which blends into the rostral spinal cord. **[2]**

myelin   The fatty, whitish sheet of membrane wrapped around some axons, provided by a glial cell. **[6]**

myelin basic protein (MBP)   An important component of myelin for both oligodendrocytes and Schwann cells. **[6]**

myelin oligodendrocyte glycoprotein (MOG)   An important membrane-bound glycoprotein component of myelin in the CNS. **[6]**

myoblasts   An embryonic muscle cell of mesodermal origin. **[6]**

myofibers   Also called *muscle fibers*. Multinucleate cells formed from the fusion of several myoblasts that have assembled the molecular machinery for contraction. **[6]**

myotubes   Multinucleate cells formed from the fusion of several myoblasts. **[6]**

N

nanos   A transcription factor that is concentrated in the presumptive posterior end of the embryo. **[2]**

nasal   Referring to the nose. For example, the nasal retina is the medial portion of the retina in vertebrates. **[8]**

neocortex   The six-layered outer region of the mammalian cerebral cortex. **[3]**

neoteny   An instance when a descendant species halts development at what was a juvenile stage of its ancestral species and becomes sexually mature at that stage. **[4]**

nerve growth factor (NGF)   The first isolated and identified neurotrophic factor, critical for the developmental survival of DRG and sympathetic cells. **[7]**

nerve net   A system of interconnected neurons found in jellyfish and related animals. **[1]**

netrins   A family of secreted, diffusible proteins that attract some growth cones while repelling others. **[5]**

neural cadherin (N-cadherin)   Also called *NCad*. A cadherin that is important in neural development. **[3, 5]**

neuronal cell adhesion molecule (NCAM)   Also called *neural-CAM*. An adhesive molecule expressed by many neurons. **[3, 5, 6]**

neural crest   A collection of ectodermal cells that break away from the developing neural tube to lie sandwiched between the tube and overlying ectoderm. **[1]**

neural plate   The earliest stage in the development of the vertebrate nervous system from the ectoderm. **[1]**

neural tube   The early, tube-shaped stage of vertebrate nervous system development. **[1]**

neuregulins   A family of signaling proteins, part of a much larger family of epidermal growth factors, that are membrane bound but can be cleaved to produce a diffusible signal. **[6]**

neurexins   A family of membrane-bound proteins found in presynaptic terminals that bind to neuroligins. **[6]**

neuroblast   A cell that will divide to produce neural cells. **[1, 3]**

neurogenesis   The mitosis of cells that will give rise to neurons. **[3]**

neuroligins   A family of membrane-bound proteins found in postsynaptic sites that bind neurexins. **[6]**

neuromuscular junction (NMJ)   The large chemical synapse found between the terminal of a motor neuron and its muscle target. **[6]**

neuron doctrine   The early twentieth-century proposal that neurons are structurally distinct from one another and communicate across gaps called synapses. **[5]**

neuropilins   A family of membrane receptor proteins that govern growth cone guidance. They bind and respond to semaphorins. **[5, 9]**

neurotransmitter receptors   Proteins embedded in the postsynaptic membrane that bind to neurotransmitter in the cleft and trigger a response in the target cell. **[6]**

neurotrophic   Referring specifically to the capacity of a factor to prevent the death of neural cells. **[7]**

neurotrophin-3 (NT-3)   A member of the neurotrophin family, discovered after NGF and BDNF. **[7]**

neurotrophin-4 (NT-4)   The fourth-discovered member of the neurotrophin family. **[7]**

neurotrophins   The family of neurotrophic factors that includes NGF, BDNF, and the other neurotrophins, which are designated by number. **[7]**

neurotropic   Referring to materials that attract neuronal growth cones. **[5]**

neurula   An embryo that has begun forming a nervous system, typically after the completion of gastrulation. **[1]**

neurulation   The process in which an embryo transitions from gastrula to neurula. **[1]**

NMDA   *N*-methyl-D-aspartate, a synthetic compound that binds with high affinity a class of ionotropic glutamate receptors that are known for imparting a Hebbian-like plasticity. **[8]**

node of Ranvier   The gap between successive segments of myelin in myelinated axons. **[6]**

node   Here, the anterior-most portion of the primitive streak, which will give rise to the brain. **[1]**

noggin   A gene that encodes the protein noggin, which exerts an organizer-like effect on ectoderm, shifting it from an epidermal to a neural fate. **[1]**

Nogo   A membrane-bound protein expressed by oligodendrocytes that inhibits neurite outgrowth. It is the product of a member of the *reticulon* gene family. **[6]**

norepinephrine (NE)   A catecholaminergic neurotransmitter released by most sympathetic fibers to activate the "fight or flight" response in various organs. **[4]**

Northern blot   A method of detecting a particular RNA transcript in a tissue or organ, by separating RNA from that source with gel electrophoresis, blotting the separated RNAs onto nitrocellulose, and then using a nucleotide probe to hybridize with, and highlight, the transcript of interest. **[A]**

Notch   A membrane-bound protein that binds to Delta found on the surface of adjoining cells. **[1]**

notochord   An embryonic rod-shaped structure that is derived from mesodermal tissue in all vertebrates and that induces formation of the ventral neural tube above. **[1, 2]**

nucleotide   A portion of a DNA or RNA molecule that is composed of a single base and the adjoining sugar-phosphate unit of the strand. **[A]**

nucleus   Here, the spherical central structure of a cell that contains the chromosomes. **[A]**

O

observational learning   The process by which an individual first learns how to accomplish some task by watching another individual do it. **[10]**

ocular dominance band   A region of neural tissue that is predominated by input from one eye or the other, such as in layer IV of primary visual cortex. **[8]**

ocular dominance histograms   Plots depicting the distribution of visual cortical neurons in terms of the extent to which they respond to stimuli in both eyes. **[9]**

olfactory bulb   An anterior projection of the brain that terminates in the upper nasal passages and, through small openings in the skull, receives axons from olfactory receptor neurons. **[3, 9]**

olfactory ensheathing glia (OEGs)   Glia found in the olfactory epithelium that guide newly generated olfactory receptor neurons into place. **[3]**

olfactory glomeruli   Spherical clusters of cells in the vertebrate olfactory bulb that process information from a particular class of olfactory sensory neurons and therefore a particular odor. [9]

olfactory placode   A plate-shaped collection of cells outside the developing brain, from which cells migrate through the olfactory bulb and to the rest of the brain. [3]

olfactory receptor protein   A protein embedded in the membrane of olfactory sensory neurons that binds odorants on the extracellular surface and triggers second messenger systems within the neurons. [9]

olfactory sensory neurons (OSNs)   The primary sensory neurons of the olfactory system, which contact and recognize particular odorants. [3, 9]

oligodendrocyte   A glial cell that provides myelination for axons in the central nervous system. [6]

ommatidium   A cluster of photoreceptors and supporting cells that form a single visual unit in the compound eye of an insect. Many ommatidia make up the eye. [4]

ontogeny   The process of individual development; growing up and growing old. [1]

optic nerve   The bundle of axons of retinal ganglion cells that exit the eye and project to the brain. [8]

organizational hypothesis   The proposal that the same testicular steroids that act early in life to permanently masculinize the vertebrate body also permanently masculinize the brain and therefore behavior. [7]

organizer   Here, a hypothesized signal from the dorsal lip of the blastopore that induces formation of a nervous system. [1]

Otx2   A homeobox gene required for development of the vertebrate midbrain and forebrain. [2]

P

p75   A low-affinity receptor for the neurotrophins. [7]

pair-rule genes   Genes that encode a class of transcription factors, the expression of which is regulated by gap genes. [2]

parable of the cave   Plato's thought experiment about the concept of reality that people would have if they grew up seeing only the shadows of objects in the world. [P]

parallel fibers   The long axons from granule neurons of the cerebellum that innervate Purkinje neuron dendrites. [3]

parasympathetic ganglia   The ganglia scattered throughout the body that receive input from either the brainstem or sacral spinal cord. The neurons within the parasympathetic ganglia send their cholinergic axons to innervate various organs. [8]

parasympathetic nervous system   The portion of the autonomic nervous system that generally facilitates relaxation and recuperation of the body. [8]

parentese   The lilting, singsong style of speaking often directed at infants across cultures. [10]

Pax6   A homeobox gene highly expressed in the anterior portion of the developing vertebrate cortex. [2]

peptide   A short string of amino acids. Longer strings of amino acids are called *proteins*. [A]

peripheral myelin protein-22 (PMP22)   An important structural component of myelin in the peripheral nervous system. [6]

peripheral nervous system (PNS)   The entire nervous system other than the central nervous system; it includes the enteric and autonomic nervous system. It is derived from neural crest cells. [1,3]

phagocytes   Immune-related cells that attack and dismantle debris and invading microbes. [7]

pharynx   The tube lining the mouth that connects to the rest of the digestive system. [1]

phenotype   The sum total of physical characteristics that an individual displays at a particular time. [1]

philosophy   The study of the fundamental nature of knowledge, existence, and reality itself. [P]

phoneme   A particular speech sound used in language. Different human languages may use different subsets of phonemes. [10]

pia mater   Also called simply *pia*. The innermost layer of the vertebrate meninges, found along the outer surface of the brain. [3]

pioneer neurons   Neurons that are the first in a region to send out axonal growth cones to establish a path that many later-arising axons will follow. [5]

plastic song   An early stage in birdsong learning in which the song is highly variable. [10]

pleiotropy   The phenomenon in which a single gene plays a role in several, seemingly unrelated, traits. [3]

plexins   A family of receptors that bind and respond to semaphorins. [5]

polylysines   Long-chain molecules consisting of many lysines strung together. [5]

polymerase chain reaction (PCR)   Also called *gene amplification*. A method for reproducing a particular RNA or DNA sequence manyfold, allowing amplification for sequencing or manipulating the sequence. [A]

polyneuronal innervation   The condition of having more than one neuron innervate a target, such as a muscle fiber in newborn rats. [8]

pons   The portion of the brainstem caudal to the midbrain, to which the cerebellum is attached. [2]

postsynaptic densities (PSDs)   Dense regions of a postsynaptic site that are specialized to detect and respond to neurotransmitter from the presynaptic terminal. [6]

postsynaptic density (PSD) proteins   Also called *scaffolding proteins*. A large family of proteins that bind many other proteins to anchor the components for postsynaptic responses. [6]

postsynaptic region   The portion of a neuronal target specialized to respond to a presynaptic terminal. **[6]**

potassium-chloride cotransporter   A specialized membrane-bound protein that pushes both potassium and chloride ions out of cells. **[6]**

preformationism   The notion that all the structures of an individual are present in microscopic form at conception, so development consists of simple growth of structures already present. **[1]**

prefrontal cortex   The portion of the vertebrate cortex found at the extreme anterior pole. **[8]**

presynaptic terminal   The region of the axon terminal at a synapse. **[6]**

primary visual cortex (V1)   Also called *striate cortex*. The region of the occipital lobe where most visual information first arrives in the cortex. **[8]**

primitive streak   The beginnings of the nervous system in the vertebrate embryo, marking the midline of the developing individual. **[1]**

probe   A manufactured sequence of DNA or RNA that is made to include a label (a colorful or radioactive molecule) that lets us track its location. **[A]**

Probst bundles   Malformations caused by growing axons that fail to produce a corpus callosum. **[5]**

promoter region   The regulatory region that tends to be upstream from the structural gene. **[1]**

proneural genes   A collection of genes that tend to be expressed in cells that will go on to differentiate into neurons and glia. **[1]**

prosencephalon   Also called *forebrain*. The most anterior aspect of the embryonic vertebrate brain. It will develop into the telencephalon and diencephalon. **[2]**

prototype   A model or concept that best typifies members of a category; a "best example." **[I]**

protozoan   A single-celled microscopic animal. **[1]**

Purkinje cells   The large, multipolar neurons that form a single layer in the vertebrate cerebellar cortex. **[3]**

pyknotic   A characteristic appearance of cells undergoing apoptosis. A pyknotic cell looks as if the nucleus is collapsing upon itself. **[7]**

R

radial glial cells   Also called simply *radial glia*. Long, slender glial cells that stretch from the ventricular surface to the pial surface in the vertebrate cerebral cortex. **[3]**

rapsyn   A cytoplasmic protein that binds the intracellular components of several acetylcholine receptors, effectively anchoring them together. **[6]**

rationalist philosophers   Philosophers who believed that the only way to understand the true nature of the world is by use of the intellect and reason, rather than the senses. **[P]**

realizator genes   A class of genes, the expression of which is controlled by Hox genes, that direct the actual construction of particular body parts. **[2]**

receptivity   Here, the willingness of a female to allow a male to mate with her. **[7]**

receptor tyrosine kinase (RTK)   A family of membrane-bound signaling proteins that, when activated by a ligand, phosphorylate tyrosine sites on particular intracellular proteins. **[4]**

redundancy   Here, the phenomenon in which several genes play a role in a process such that loss of one may have a relatively minor effect. **[3]**

reelin   A gene that encodes the protein reelin and when dysfunctional results in very a disorganized cerebellum and cerebrum in mice. **[3]**

regulatory portion of the gene   The stretches of DNA adjacent to the structural gene, which play a role in regulating transcription of that gene. **[1]**

repulsive   Here, referring to materials to which growth cones will not attach. Repulsive signals often cause the collapse of filopodia that contact them. **[5]**

retinal ganglion cells (RGCs)   The class of retinal neurons that send their axons out the optic nerve to transmit visual information to the brain. **[8]**

retinoic acid (RA)   A steroid-like molecule, concentrated in the posterior end of vertebrate embryos, that promotes development of posterior structures. It is a powerful teratogen. **[2]**

retinoic acid receptor (RAR)   A member of the steroid receptor superfamily that serves as a receptor for retinoic acid. **[2]**

retinoic acid response element (RARE)   A specific sequence of DNA nucleotides that is bound by the retinoic acid–retinoic acid receptor complex, thereby regulating expression of the associated gene. **[2]**

retrograde messenger   Here, a chemical signal emanating from the postsynaptic cell that affects the presynaptic cell. **[8]**

retrograde transport   The transport of materials within axons in the direction of the cell body. **[5]**

rhombencephalon   Also called *hindbrain*. The caudal-most segment of the embryonic vertebrate brain. It will develop into the metencephalon (pons and cerebellum) and myelencephalon (medulla). **[2]**

rhombomeres   A group of prominently segmented portions of the embryonic rhombencephalon. **[2]**

ribonucleic acid (RNA)   A nucleic acid that implements information found in DNA. **[A]**

ribosomes   Structures in the cell body where genetic information is translated to produce proteins. **[A]**

roof plate   The dorsal portion of the vertebrate neural tube. **[2]**

rostral   Referring to the head end. **[2]**

rostral migratory stream (RMS)   A collection of cells that migrate from the anterior horn of the lateral ventricles to the olfactory bulbs in adult mammals. **[3]**

roundabout (Robo)   A family of membrane-bound receptors that respond to diffusible Slit proteins. [5]

S

savant   A person with extraordinary talent in a specific endeavor, such as calculations, music, or memory. [5]

scaffolding proteins   A family of proteins that anchor intracellular portions of transmembrane proteins, such as neurotransmitter receptors in post-synaptic sites. [6]

Schwann cells   Glial cells that provide myelin sheaths for axons in the peripheral nervous system and cap neuromuscular junctions. [4, 6]

segment polarity genes   Genes that encode a class of signaling factors, the expression of which is regulated by pair-rule genes. [2]

self-regulation   Also known simply as *regulation*. Here, the process by which embryos manage to compensate for missing or damaged cells and nevertheless produce an entire individual. [1]

semaphorins   A family of secreted and membrane-bound molecules that often serve to repulse growth cones, including those that express neuropilin or plexins, and so direct them away from a boundary. [5, 9]

sensitive period   Here, the period during ontogeny when a particular manipulation must be made to affect neural development. [9]

sevenless   A gene that encodes a receptor tyrosine kinase (RTK) and that when stimulated triggers differentiation of photoreceptor 7 in *Drosophila*. [4]

sex-determining region of the Y (Sry)   The gene on the Y chromosome of mammals that promotes the indifferent gonad to develop as a testis. [7]

sexual differentiation   The process by which females and males diverge from each other in structure during ontogeny. [7]

sexual dimorphism   A structural difference between males and females of a species. [7]

sexual imprinting   The tendency of many birds and mammals to prefer mating with individuals that resemble their parents. [10]

sexually dimorphic nucleus of the preoptic area (SDN-POA)   A nucleus in the hypothalamus of many mammals that has a greater volume in males than females. [7]

signal peptide   A particular sequence of N-terminal amino acids that directs the full protein to the cell's secretory pathway. [2]

SKN-1   A transcription factor in *C. elegans* that is more concentrated at one end of an egg or zygote, which will give rise to the posterior half of the individual. [1]

Slit   A family of diffusible proteins that mark the midline in the developing nervous system. [5]

SNARE complex   A combination of SNAREs that anchor a vesicle into position for fusion to the presynaptic membrane and release of neurotransmitter. [6]

SNAREs   **S**oluble-**N**-ethylmaleimide-sensitive **a**ttachment **re**ceptors, a family of long-chain proteins that regulate membrane fusion. [6]

sog   A gene that encodes the protein sog, an insect homologue of chordin, which blocks BMP signaling to direct ectodermal cells to differentiate into a neural, rather than epidermal, fate. [1]

somites   Paired blocks of mesoderm found on either side of the neural tube. [3]

Sonic hedgehog (Shh)   A gene that encodes the signaling protein that is secreted by the notochord and induces formation of the floor plate and the differentiation of motor neurons in the vertebrate neural tube. [2]

Southern blot   A method of detecting a particular DNA sequence in the genome of an organism, by separating DNA with gel electrophoresis, blotting the separated DNAs onto nitrocellulose, and then using a nucleotide probe to hybridize with, and highlight, the gene of interest. [A]

spinal nucleus of the bulbocavernosus (SNB)   A sexually dimorphic collection of motor neurons in the spinal cord that innervate muscles attached to the penis in rats and mice. [7]

spontaneous abortion   The accidental loss of an embryo, sometimes called a miscarriage. [2]

stereotype threat   The emotional state experienced when exposed to a derogatory stereotype about a group to which you belong. [10]

stop codon   A trio of nucleotides in DNA to mark the end of transcription. [A]

strabismus   A deviation of one or both eyes, such that they do not converge on the same region of visual space. [9]

structural portion of the gene   The portion of DNA in a gene that encodes for a particular sequence of amino acids and therefore a particular protein. [1]

subgranular zone   The portion of the dentate gyrus where cells divide in adulthood to contribute new neurons to the overlying granular layer. [3]

subsynaptic nuclei   The group of nuclei within a muscle fiber that lie closest to the neuromuscular junction. [6]

subventricular zone (SVZ)   The region just next to the ventricular zone, where many cells divide to provide neurons and glia to the developing vertebrate brain and, in at least some brain regions, new neurons in adulthood. [3]

sudden infant death syndrome (SIDS)   Also called *crib death*. The unexplained failure to breathe that sometimes kills infants in the first year of life. [10]

superoxide dismutase (SOD)   A protein that normally neutralizes free radicals. Mutations of this gene can cause ALS in humans. [7]

sweat glands   Specialized structures that release sweat onto the skin surface to reduce body temperature. They are activated by the release of acetylcholine from sympathetic fibers. **[4]**

sympathetic ganglia   The two chains of interconnected ganglia alongside the vertebral column that receive input from the sympathetic preganglionic neurons in the spinal cord. The neurons within the sympathetic ganglia send their axons out to innervate various organs, where they usually release norepinephrine. **[7, 8]**

sympathetic nervous system   The portion of the autonomic nervous system that generally prepares the body for action. **[8]**

synapse rearrangement   The process in which some synapses are withdrawn while new synapses form. **[8]**

synaptic cleft   The gap between the presynaptic terminal and postsynaptic region of a synapse. **[6]**

synaptic vesicles   Roughly spherical containers of neurotransmitter found in presynaptic terminals. **[6]**

synaptotagmin   An enzyme that binds to vesicle membranes and serves as a calcium sensor to trigger fusion of the membrane and release of neurotransmitter. **[6]**

syncytium   A single cell containing several nuclei. **[2]**

T

tabula rasa   Latin for "blank slate," the idea that we enter the world with minds that are empty of innate ideas, and so we must gain information through experience. **[I]**

telencephalon   The anterior-most portion of the vertebrate brain, consisting of the cerebral cortex and related subcortical structures such as the basal ganglia and hippocampus. **[2]**

temporal   Here, referring to the side of the head. For example, the temporal retina is the lateral portion of the retina in vertebrates. **[8]**

temporal summation   The property of neurons that fire only when sufficient synaptic stimulation occurs within a particular period of time to reach threshold. **[E]**

teratogen   A substance that causes malformations in development. **[2]**

terminal Schwann cell   Specialized Schwann cells that surround and cap neuromuscular junctions, effectively isolating them from other influences. **[6]**

testosterone   The principle androgen secreted from the vertebrate testes, found in higher concentrations in males than females. **[7]**

tetanus   A sustained burst of rapid neuronal firing. **[8]**

tetrodotoxin (TTX)   A toxin found in the ovaries of pufferfish that blocks voltage-dependent sodium channels and thus prevents conduction of action potentials. **[8]**

TGFβ receptors   A class of receptors that bind transforming growth factors, including BMP. **[1]**

thymidine   A nucleotide used in the synthesis of DNA. Because thymidine is not used in RNA, it can serve as a DNA-specific marker. **[3]**

topographic projections   The property of axonal projections in which the pattern of input represents a spatial array, as in the projection of retinal axons onto the brain. **[E]**

totipotency   Total potency; the ability of early embryonic cells to differentiate into any type of cell in the individual. **[1]**

transcript   The mRNA strand that is produced when a stretch of DNA is "read." **[A]**

transcription   The process during which mRNA forms bases complementary to a strand of DNA. The resulting message (called a *transcript*) is then used to translate the DNA code into protein molecules. **[A]**

transcription factors   Proteins that bind to DNA and regulate the extent to which genes are expressed. **[1]**

transgene   A gene that has been artificially introduced into a model organism. **[1]**

transgenic   Referring to an organism in which foreign DNA has been deliberately inserted. **[1, 4, A]**

translation   The process by which amino acids are linked together (directed by an mRNA molecule) to form protein molecules. **[A]**

transmembrane proteins   Proteins that, due to having several hydrophobic domains, are embedded in a cell membrane. They include ion channels and neurotransmitter receptors. **[6]**

transsynaptic transport   The transfer of a chemical marker across a synapse. **[8]**

trembler   Mice with an autosomal dominant mutation of peripheral myelin protein-22 that results in demyelination and impaired locomotion. **[6]**

TrkA   The first-identified neurotrophin receptor, which has a high affinity for NGF. **[7]**

TrkB   The second-identified neurotrophin receptor, which has a high affinity for BDNF and the other neurotrophins except NGF. **[7]**

TrkC   The third-identified neurotrophin receptor, which has a high affinity for neurotrophin-3. **[7]**

trophic   Here, referring to the capacity of a factor to prevent the death of a cell, as if "feeding" it. **[7]**

tropic   Here, referring to the capacity of a factor to attract a growing neuronal process. **[7]**

tropomyosin receptor kinase (Trk)   A subfamily of the receptor tyrosine kinase (RTK) family that serve as receptors for neurotrophins. **[7]**

tubulin   The specialized protein that assembles to form microtubules providing structure to the cytoplasm. **[5]**

TUNEL   TdT-dUTP nick end labeling, a method to enzymatically label the various "nicks" in DNA that has been fragmented in preparation for apoptosis. **[7]**

tyrosine kinases Enzymes that phosphorylate particular tyrosine molecules found in particular proteins. **[6]**

U

Ultrabithorax A Hox gene complex that affects the fate of cells in the thorax of *Drosophila*. **[2]**

unc-5 A family of membrane-bound receptors that respond to netrins. **[5]**

unc-86 A gene in *C. elegans* that encodes the protein unc-86, which directs certain late-dividing cells to become touch receptor neurons. **[1]**

V

ventricular zone The regions adjacent to the ventricles of the brain and central canal of the spinal cord, where cell division continues throughout life. **[3]**

vital dyes Relatively nontoxic synthetic dyes that can be used to label living cells. **[2]**

voltage-gated ion channels Membrane-bound proteins that open up an ion channel in response to particular membrane potentials. **[6]**

W

Western blot A method of detecting a particular protein molecule in a tissue or organ, by separating proteins from that source with gel electrophoresis, blotting the separated proteins onto nitrocellulose, and then using an antibody that binds, and highlights, the protein of interest. **[A]**

whisker barrel A column of somatosensory cortex that receives sensory information from a single whisker. **[9]**

white matter The inner portion of the vertebrate brain, consisting primarily of myelinated axons coursing to or from the cerebral cortex, hence light in color in postmortem preparations. **[3, 8]**

Williams syndrome A disorder characterized by fluent linguistic function but poor performance on standard IQ tests and great difficulty with spatial processing. **[10]**

Wnt A gene that encodes the secreted protein Wnt, which is concentrated in the posterior end of vertebrate embryos. **[2]**

X

Xenopus laevis The African clawed frog, a valuable vertebrate model species. **[1]**

Z

zinc fingers Relatively small stretches of amino acids found in various transcription factors that bind to DNA in the process of gene regulation. **[4]**

zygote A fertilized egg; the single cell that will divide and grow to form a new individual. **[1]**

REFERENCES

A

Abmayr, S. M., & Pavlath, G. K. (2012). Myoblast fusion: Lessons from flies and mice. *Development, 139*(4), 641–656. doi:10.1242/dev.068353

Adameyko, I., Lallemend, F., Aquino, J. B., Pereira, J. A., Topilko, P., Muller, T., … Ernfors, P. (2009). Schwann cell precursors from nerve innervation are a cellular origin of melanocytes in skin. *Cell, 139*(2), 366–379. doi:10.1016/j.cell.2009.07.049

Adams, J. M. & Cory, S. (1998). The Bcl-2 protein family: Arbiters of cell survival. *Science, 281*, 1322–1326.

Adams, M. D., Celniker, S. E., Holt, R. A., Evans, C. A., Gocayne, J. D., Amanatides, P. G., … Venter, J. C. (2000). The genome sequence of *Drosophila melanogaster. Science, 287*(5461), 2185–2195.

Adams, N. C., Tomoda, T., Cooper, M., Dietz, G., & Hatten, M. E. (2002). Mice that lack astrotactin have slowed neuronal migration. *Development, 129*(4), 965–972.

Aimone, J. B., Li, Y., Lee, S. W., Clemenson, G. D., Deng, W., & Gage, F. H. (2014). Regulation and function of adult neurogenesis: From genes to cognition. *Physiological Reviews, 94*(4), 991–1026. doi:10.1152/physrev.00004.2014

Ainsworth, M. D. S., Blehar, M. C., Waters, E., & Wall, S. (1978). *Patterns of attachment: A psychological study of the strange situation.* Hillsdale, NJ: Erlbaum.

Akitaya, T., & Bronner-Fraser, M. (1992). Expression of cell adhesion molecules during initiation and cessation of neural crest cell migration. *Developmental Dynamics, 194*(1), 12–20. doi:10.1002/aja.1001940103

Albelda, S. M., & Buck, C. A. (1990). Integrins and other cell adhesion molecules. *FASEB Journal, 4*(11), 2868–2880.

Albert, D. J. (2011). What's on the mind of a jellyfish? A review of behavioural observations on *Aurelia* sp. jellyfish. *Neuroscience and Biobehavioral Review, 35*(3), 474–482. doi:10.1016/j.neubiorev.2010.06.001

Alberts, B., Johnson, A., Lewis, J., Raff, M., Roberts, K., & Walter, P. (2007). *Molecular biology of the cell* (5th ed.). New York: Garland Science.

Albuquerque, E. X., Pereira, E. F., Alkondon, M., & Rogers, S. W. (2009). Mammalian nicotinic acetylcholine receptors: From structure to function. *Physiological Reviews, 89*(1), 73–120. doi:10.1152/physrev.00015.2008

Alexander, G. M., & Hines, M. (2002). Sex differences in response to children's toys in nonhuman primates (*Cercopithecus aethiops sabaeus*). *Evolution and Human Behavior, 23*, 467–479.

Alford, J. R., Funk, C. L., & Hibbing, J. R. (2005). Are political orientations genetically transmitted? *American Political Science Review, 99*(2), 153–167.

Allard, S., Adin, P., Gouedard, L., di Clemente, N., Josso, N., Orgebin-Crist, M. C., … Xavier, F. (2000). Molecular mechanisms of hormone-mediated Mullerian duct regression: Involvement of beta-catenin. *Development, 127*(15), 3349–3360.

Alonso, M. C., & Cabrera, C. V. (1988). The *achaete-scute* gene complex of *Drosophila melanogaster* comprises four homologous genes. *EMBO Journal, 7*(8), 2585–2591.

Altman, J. (1963). Autoradiographic investigation of cell proliferation in the brains of rats and cats. *The Anatomical Record, 145*, 573–591.

Altman, J. (1969). Autoradiographic and histological studies of postnatal neurogenesis. IV. Cell proliferation and migration in the anterior forebrain, with special reference to persisting neurogenesis in the olfactory bulb. *Journal of Comparative Neurology, 137*(4), 433–457. doi:10.1002/cne.901370404

Altman, J., & Das, G. D. (1965). Autoradiographic and histological evidence of postnatal hippocampal neurogenesis in rats. *Journal of Comparative Neurology, 124*(3), 319–335.

Alvarado-Mallart, R. M. (2005). The chick/quail transplantation model: Discovery of the isthmic organizer center. *Brain Research. Brain Research Reviews, 49*(2), 109–113. doi:10.1016/j.brainresrev.2005.03.001

Alvarez-Buylla, A., & Garcia-Verdugo, J. M. (2002). Neurogenesis in adult subventricular zone. *Journal of Neuroscience, 22*(3), 629–634.

American Speech-Language-Hearing Association. (n.d.) How does your child hear and talk? Retrieved from www.asha.org/public/speech/development/chart.htm.

Anchan, R. M., Drake, D. P., Haines, C. F., Gerwe, E. A., & LaMantia, A.-S. (1997). Disruption of local retinoid-mediated gene expression accompanies abnormal development in the mammalian olfactory pathway. *Journal of Comparative Neurology, 379*, 171–184.

Andjus, P. R., Zhu, L., Cesa, R., Carulli, D., & Strata, P. (2003). A change in the pattern of activity affects the developmental regression of the Purkinje cell polyinnervation by climbing fibers in the rat cerebellum. *Neuroscience, 121*(3), 563–572.

Angevine, J. B., Jr., & Sidman, R. L. (1961). Autoradiographic study of cell migration during histogenesis of cerebral cortex in the mouse. *Nature, 192*, 766–768.

Ansberry, C. (2010, November 20). Erasing a hurtful label from the books. *The Wall Street Journal*. Retrieved from www.wsj.com.

Aristotle. (350 BCE). *On the generation of animals*.

Arnold, A. P. (1980). Sexual differences in the brain. *American Scientist, 68*, 165–173.

Asmus, S. E., Parsons, S., & Landis, S. C. (2000). Developmental changes in the transmitter properties of sympathetic neurons that innervate the periosteum. *Journal of Neuroscience, 20*(4), 1495–1504.

Atkinson, Q. D. (2011). Phonemic diversity supports a serial founder effect model of language expansion from Africa. *Science, 332*(6027), 346–349. doi:10.1126/science.1199295

Avery, L., & Horvitz, H. R. (1987). A cell that dies during wild-type *C. elegans* development can function as a neuron in a *ced-3* mutant. *Cell, 51*(6), 1071–1078.

Axel, R., & Carniol, K. (2014). A conversation with Richard Axel. *Cold Spring Harbor Symposia on Quantitative Biology, 79*, 258–259. doi:10.1101/sqb.2014.79.03

B

Baccaglini, P. I., & Spitzer, N. C. (1977). Developmental changes in the inward current of the action potential of Rohon-Beard neurones. *Journal of Physiology, 271*(1), 93–117.

Bachiller, D., Klingensmith, J., Kemp, C., Belo, J. A., Anderson, R. M., May, S. R., … De Robertis, E. M. (2000). The organizer factors Chordin and Noggin are required for mouse forebrain development. *Nature, 403*(6770), 658–661. doi:10.1038/35001072

Bagemihl, B. (2000). *Biological exuberance: Animal homosexuality and natural diversity*. New York: Stonewall Inn Editions.

Bailey, J. M., & Bell, A. P. (1993). Familiality of female and male homosexuality. *Behavior Genetics, 23*(4), 313–322.

Baker, N. E., & Zitron, A. E. (1995). *Drosophila* eye development: *Notch* and *Delta* amplify a neurogenic pattern conferred on the morphogenetic furrow by *scabrous*. *Mechanisms of Development, 49*(3), 173–189.

Balice-Gordon, R. J., & Lichtman, J. W. (1994). Long-term synapse loss induced by focal blockade of postsynaptic receptors. *Nature, 372*(6506), 519–524.

Ball, E. E., Hayward, D. C., Saint, R., & Miller, D. J. (2004). A simple plan: Cnidarians and the origins of developmental mechanisms. *Nature Reviews Genetics, 5*, 567–577.

Balthazart, J., & Ball, G. F. (2014). Endogenous versus exogenous markers of adult neurogenesis in canaries and other birds: Advantages and disadvantages. *Journal of Comparative Neurology, 522*(18), 4100–4120. doi:10.1002/cne.23661

Banks, M. S., Aslin, R. N., & Letson, R. D. (1975). Sensitive period for the development of human binocular vision. *Science, 190*(4215), 675–677.

Bansal, A., Singer, J. H., Hwang, B. J., Xu, W., Beaudet, A., & Feller, M. B. (2000). Mice lacking specific nicotinic acetylcholine receptor subunits exhibit dramatically altered spontaneous activity patterns and reveal a limited role for retinal waves in forming ON and OFF circuits in the inner retina. *Journal of Neuroscience, 20*(20), 7672–7681.

Barde, Y. A., Edgar, D., & Thoenen, H. (1982). Purification of a new neurotrophic factor from mammalian brain. *EMBO Journal, 1*(5), 549–553.

Barinaga, M. (1995). Receptors find work as guides. *Science, 269*, 1668–1670.

Barr, M. S., & Corballis, M. C. (2002). The role of the anterior commissure in callosal agenesis. *Neuropsychology, 16*(4), 459–471.

Basham, M. E., Nordeen, E. J., & Nordeen, K. W. (1996). Blockade of NMDA receptors in the anterior forebrain impairs sensory acquisition in the zebra finch (*Poephila guttata*). *Neurobiology of Learning and Memory, 66*(3), 295–304. doi:10.1006/nlme.1996.0071

Basu, R., Taylor, M. R., & Williams, M. E. (2015). The classic cadherins in synaptic specificity. *Cell Adhesion & Migration, 9*(3), 193–201. doi:10.1080/19336918.2014.1000072

Baumann, N., & Pham-Dinh, D. (2001). Biology of oligodendrocyte and myelin in the mammalian central nervous system. *Physiological Reviews, 81*(2), 871–927.

Beach, F. A. (1955). The descent of instinct. *Psychological Review, 62*(6), 401–410.

Bear, M. F. (1996). Progress in understanding NMDA-receptor-dependent synaptic plasticity in the visual cortex. *Journal of Physiology (Paris), 90*(3–4), 223–227.

Becalska, A. N., & Gavis, E. R. (2009). Lighting up mRNA localization in *Drosophila* oogenesis. *Development, 136*(15), 2493–2503. doi:10.1242/dev.032391

Bedard, A., & Parent, A. (2004). Evidence of newly generated neurons in the human olfactory bulb. *Brain Research. Developmental Brain Research, 151*(1–2), 159–168. doi:10.1016/j.devbrainres.2004.03.021

Belluscio, L., Gold, G. H., Nemes, A., & Axel, R. (1998). Mice deficient in G(olf) are anosmic. *Neuron, 20*(1), 69–81.

Ben-Ari, Y., Gaiarsa, J. L., Tyzio, R., & Khazipov, R. (2007). GABA: A pioneer transmitter that excites immature neurons and generates primitive oscillations. *Physiological Reviews, 87*(4), 1215–1284. doi:10.1152/physrev.00017.2006

Bennett, E. L., Diamond, M. C., Krech, D., & Rosenzweig, M. R. (1964). Chemical and anatomical plasticity of brain. *Science, 146*(3644), 610–619.

Bentley, D., & Caudy, M. (1983). Pioneer axons lose directed growth after selective killing of guidepost cells. *Nature, 304*(5921), 62–65.

Bentley, D., & Keshishian, H. (1982). Pathfinding by peripheral pioneer neurons in grasshoppers. *Science, 218*(4577), 1082–1088. doi:10.1126/science.218.4577.1082

Benzer, S. (1973). Genetic dissection of behavior. *Scientific American, 229*(6), 24–37.

Bergan, J. F., & Knudsen, E. I. (2009). Visual modulation of auditory responses in the owl inferior colliculus. *Journal of Neurophysiology, 101*(6), 2924–2933. doi:10.1152/jn.91313.2008

Berger, T., Rubner, P., Schautzer, F., Egg, R., Ulmer, H., Mayringer, I., … Reindl, M. (2003). Antimyelin antibodies as a predictor of clinically definite multiple sclerosis after a first demyelinating event. *New England Journal of Medicine, 349*(2), 139–145. doi:10.1056/NEJMoa022328

Berkemeier, L. R., Winslow, J. W., Kaplan, D. R., Nikolics, K., Goeddel, D. V., & Rosenthal, A. (1991). Neurotrophin-5: A novel neurotrophic factor that activates trk and trkB. *Neuron, 7*(5), 857–866.

Berry, M., & Rogers, A. W. (1965). The migration of neuroblasts in the developing cerebral cortex. *Journal of Anatomy, 99*(Pt. 4), 691–709.

Bian, W. J., Miao, W. Y., He, S. J., Qiu, Z., & Yu, X. (2015). Coordinated spine pruning and maturation mediated by inter-spine competition for cadherin/catenin complexes. *Cell, 162*(4), 808–822. doi:10.1016/j.cell.2015.07.018

Bibel, M., & Barde , Y.-A. (2000). Neurotrophins: Key regulators of cell fate and cell shape in the vertebrate nervous system. *Genes & Development, 14,* 2919–2937.

Billeter, J.-C., Rideout, E. J., Dornan, A. J., & Goodwin, S. F. (2006). Control of male sexual behavior in *Drosophila* by the sex determination pathway. *Current Biology, 16,* R766–R776.

Blanchard, R. (2001). Fraternal birth order and the maternal immune hypothesis of male homosexuality. *Hormones and Behavior, 40*(2), 105–114. doi:10.1006/hbeh.2001.1681

Blanchard, R., Cantor, J. M., Bogaert, A. F., Breedlove, S. M., & Ellis, L. (2006). Interaction of fraternal birth order and handedness in the development of male homosexuality. *Hormones and Behavior, 49*(3), 405–414. doi:10.1016/j.yhbeh.2005.09.002

Blascovich, J., Spencer, S. J., Quinn, D., & Steele, C. (2001). African Americans and high blood pressure: The role of stereotype threat. *Psychological Science, 12*(3), 225–229.

Bliss, T. V., & Gardner-Medwin, A. R. (1973). Long-lasting potentiation of synaptic transmission in the dentate area of the unanaesthetized rabbit following stimulation of the perforant path. *Journal of Physiology, 232*(2), 357–374.

Bliss, T. V., & Lomo, T. (1973). Long-lasting potentiation of synaptic transmission in the dentate area of the anaesthetized rabbit following stimulation of the perforant path. *Journal of Physiology, 232*(2), 331–356.

Blum, D. (2002). *Love at Goon Park.* Philadelphia: Perseus.

Bobek, M., Hampl, R., Peske, L., Pojer, F., Simek, J., & Bures, S. (2008). African Odyssey project satellite tracking of black storks *Ciconia nigra* breeding at a migratory divide. *Journal of Avian Biology, 39,* 500–506.

Boer, M. C., Joosten, S. A., & Ottenhoff, T. H. (2015). Regulatory T-cells at the interface between human host and pathogens in infectious diseases and vaccination. *Frontiers in Immunology, 6,* 217. doi:10.3389/fimmu.2015.00217

Bogaert, A. F. (2006). Biological versus nonbiological older brothers and men's sexual orientation. *Proceedings of the National Academy of Sciences USA, 103*(28), 10771–10774. doi:10.1073/pnas.0511152103

Bogin, B. (1997). Evolutionary hypotheses for human childhood. *Yearbook of Physical Anthropology, 40,* 63–89.

Bookstein, F. L., Sampson, P. D., Connor, P. D., & Streissguth, A. P. (2002). Midline corpus callosum is a neuroanatomical focus of fetal alcohol damage. *Anatomical Record, 269*(3), 162–174. doi:10.1002/ar.10110

Borgstein, J., & Grootendorst, C. (2002). Clinical picture: Half a brain. *Lancet, 359*(9305), 473.

Bottjer, S. W., Glaessner, S. L., & Arnold, A. P. (1985). Ontogeny of brain nuclei controlling song learning and behavior in zebra finches. *Journal of Neuroscience, 5*(6), 1556–1562.

Bottjer, S. W., Miesner, E. A., & Arnold, A. P. (1984). Forebrain lesions disrupt development but not maintenance of song in passerine birds. *Science, 224*(4651), 901–903.

Bouchard, T. J., Jr., & McGue, M. (1981). Familial studies of intelligence: A review. *Science, 212,* 1055–1059.

Bowerman, B., Draper, B. W., Mello, C. C., & Priess, J. R. (1993). The maternal gene *skn-1* encodes a protein that is distributed unequally in early *C. elegans* embryos. *Cell, 74*(3), 443–452.

Boylan, K. (2015). Familial amyotrophic lateral sclerosis. *Neurologic Clinics, 33*(4), 807–830. doi:10.1016/j.ncl.2015.07.001

Boysson-Bardies, B. (2001). *How language comes to children.* Cambridge, MA: MIT Press.

Bradford, D., Cole, S. J., & Cooper, H. M. (2009). Netrin-1: Diversity in development. *International Journal of Biochemistry & Cell Biology, 41*(3), 487–493. doi:10.1016/j.biocel.2008.03.014

Brann, J. H., & Firestein, S. J. (2014). A lifetime of neurogenesis in the olfactory system. *Frontiers in Neuroscience, 8,* 182. doi:10.3389/fnins.2014.00182

Breedlove, S. M., & Arnold, A. P. (1980). Hormone accumulation in a sexually dimorphic motor nucleus of the rat spinal cord. *Science, 210*(4469), 564–566.

Breedlove, S. M., Jordan, C. L., & Arnold, A. P. (1983). Neurogenesis of motor neurons in the sexually dimorphic spinal nucleus of the bulbocavernosus in rats. *Brain Research, 285*(1), 39–43.

Brenner, S. (1963). Letter to Perutz. In W.B. Wood and the Community of *C. elegans* Researchers, 1998 (Eds.), *The nematode Caenorhabditis elegans* (pp. X–XI). Cold Spring Harbor, NY: Cold Spring Harbor Laboratory Press.

Brenowitz, E. A., Arnold, A. P., & Levin, R. N. (1985). Neural correlates of female song in tropical duetting birds. *Brain Research, 343*(1), 104–112.

Bresler, T., Shapira, M., Boeckers, T., Dresbach, T., Futter, M., Garner, C. C., … Ziv, N. E. (2004). Postsynaptic density assembly is fundamentally different from presynaptic active zone assembly. *Journal of Neuroscience, 24*(6), 1507–1520. doi:10.1523/JNEUROSCI.3819-03.2004

Brett, B. (2004). *Uproar's your only music.* Toronto: Exile Editions.

Brickley, S. G., Dawes, E. A., Keating, M. J., & Grant, S. (1998). Synchronizing retinal activity in both eyes disrupts binocular map development in the optic tectum. *Journal of Neuroscience, 18*(4), 1491–1504.

Briggs, R., & King, T. J. (1952). Transplantation of living nuclei from blastula cells into enucleated frogs' eggs. *Proceedings of the National Academy of Sciences USA, 38*(5), 455–463.

Brinkmann, B. G., Agarwal, A., Sereda, M. W., Garratt, A. N., Muller, T., Wende, H., … Nave, K. A. (2008). Neuregulin-1/ErbB signaling serves distinct functions in myelination of the peripheral and central nervous system. *Neuron, 59*(4), 581–595. doi:10.1016/j.neuron.2008.06.028

Britton, W. J. (1998). The management of leprosy reversal reactions. *Leprosy Review, 69*(3), 225–234.

Britton, W. J., & Lockwood, D. N. (2004). Leprosy. *Lancet, 363*(9416), 1209–1219. doi:10.1016/S0140-6736(04)15952-7

Brockes, J. P., & Gates, P. B. (2014). Mechanisms underlying vertebrate limb regeneration: Lessons from the salamander. *Biochemical Society Transactions, 42*(3), 625–630. doi:10.1042/BST20140002

Brown, K. E., Arends, J. J., Wasserstrom, S. P., Zantua, J. B., Jacquin, M. F., & Woolsey, T. A. (1995). Developmental transformation of dendritic arbors in mouse whisker thalamus. *Brain Research. Developmental Brain Research, 86*(1–2), 335–339.

Brown, M. C., Jansen, J. K., & Van Essen, D. (1976). Polyneuronal innervation of skeletal muscle in newborn rats and its elimination during maturation. *Journal of Physiology, 261*(2), 387–422.

Brownell, A. G., & Slavkin, H. C. (1980). Role of basal lamina in tissue interactions. *Renal Physiology, 3*(1–6), 193–204.

Brunelli, G., Spano, P., Barlati, S., Guarneri, B., Barbon, A., Bresciani, R., & Pizzi, M. (2005). Glutamatergic reinnervation through peripheral nerve graft dictates assembly of glutamatergic synapses at rat skeletal muscle. *Proceedings of the National Academy of Sciences USA, 102*(24), 8752–8757. doi:10.1073/pnas.0500530102

Brunet, L. J., Gold, G. H., & Ngai, J. (1996). General anosmia caused by a targeted disruption of the mouse olfactory cyclic nucleotide-gated cation channel. *Neuron, 17*(4), 681–693.

Brunso-Bechtold, J. K., & Hamburger, V. (1979). Retrograde transport of nerve growth factor in chicken embryo. *Proceedings of the National Academy of Sciences USA, 76*(3), 1494–1496.

Büchner, L. (1880/2013). *Mind in animals.* London: Forgotten Books.

Bueker, E. D. (1948). Implantation of tumors in the hind limb field of the embryonic chick and the developmental response of the lumbosacral nervous system. *The Anatomical Record, 102*(3), 369–389.

Bueker, E. D., & Hilderman, H. L. (1953). Growth-stimulating effects of mouse sarcomas I, 37, and 180 on spinal and sympathetic ganglia of chick embryos as contrasted with effects of other tumors. *Cancer, 6*(2), 397–415.

Bunge, R. P., Bunge, M. B., & Eldridge, C. F. (1986). Linkage between axonal ensheathment and basal lamina production by Schwann cells. *Annual Review of Neuroscience, 9*, 305–328. doi:10.1146/annurev.ne.09.030186.001513

Burden, S. J., Sargent, P. B., & McMahan, U. J. (1979). Acetylcholine receptors in regenerating muscle accumulate at original synaptic sites in the absence of the nerve. *Journal of Cell Biology, 82*(2), 412–425.

Burgard, E. C., & Hablitz, J. J. (1993). Developmental changes in NMDA and non-NMDA receptor-mediated synaptic potentials in rat neocortex. *Journal of Neurophysiology, 69*(1), 230–240.

Burr, C. (1993). Homosexuality and biology. *The Atlantic, 272*(3), 47–65.

Busetto, G., Buffelli, M., Cangiano, L., & Cangiano, A. (2003). Effects of evoked and spontaneous motoneuronal firing on synapse competition and elimination in skeletal muscle. *Journal of Neurocytology, 32*(5–8), 795–802.

Busetto, G., Buffelli, M., Tognana, E., Bellico, F., & Cangiano, A. (2000). Hebbian mechanisms revealed by electrical stimulation at developing rat neuromuscular junctions. *Journal of Neuroscience, 20*(2), 685–695.

Bush, J. O., & Soriano, P. (2009). Ephrin-B1 regulates axon guidance by reverse signaling through a PDZ-dependent mechanism. *Genes & Development, 23*(13), 1586–1599. doi:10.1101/gad.1807209

Buss, R. R., Sun, W., & Oppenheim, R. W. (2006). Adaptive roles of programmed cell death during nervous system development. *Annual Review of Neuroscience, 29*, 1–35. doi:10.1146/annurev.neuro.29.051605.112800

C

Calabrese, B., Wilson, M. S., & Halpain, S. (2006). Development and regulation of dendritic spine synapses. *Physiology (Bethesda), 21*, 38–47. doi:10.1152/physiol.00042.2005

Calaora, V., Rogister, B., Bismuth, K., Murray, K., Brandt, H., Leprince, P., … Dubois-Dalcq, M. (2001). Neuregulin signaling regulates neural precursor growth and the generation of oligodendrocytes in vitro. *Journal of Neuroscience, 21*(13), 4740–4751.

Cameron, H. A., & Glover, L. R. (2015). Adult neurogenesis: Beyond learning and memory. *Annual Review of Psychology, 66*, 53–81. doi:10.1146/annurev-psych-010814-015006

Campenot, R. B. (1977). Local control of neurite development by nerve growth factor. *Proceedings of the National Academy of Sciences USA, 74*(10), 4516–4519.

Campenot, R. B. (1981). Regeneration of neurites on long-term cultures of sympathetic neurons deprived of nerve growth factor. *Science, 214*, 579–581.

Cantor, J. M., Blanchard, R., Paterson, A. D., & Bogaert, A. F. (2002). How many gay men owe their sexual orientation to fraternal birth order? *Archives of Sexual Behavior, 31*(1), 63–71.

Cao, X., Cui, Z., Feng, R., Tang, Y. P., Qin, Z., Mei, B., & Tsien, J. Z. (2007). Maintenance of superior learning and memory function in NR2B transgenic mice during ageing. *European Journal of Neuroscience, 25*(6), 1815–1822.

Casarosa, S., Fode, C., & Guillemot, F. (1999). *Mash1* regulates neurogenesis in the ventral telencephalon. *Development, 126*(3), 525–534.

Castillo-Melendez, M., Yawno, T., Jenkin, G., & Miller, S. L. (2013). Stem cell therapy to protect and repair the developing brain: A review of mechanisms of action of cord blood and amnion epithelial derived cells. *Frontiers in Neuroscience, 7*, 194. doi:10.3389/fnins.2013.00194

Ceci, S. J., & Williams, W. M. (1997). Schooling, intelligence, and income. *American Psychologist, 52*, 1051–1058.

Chapman, B., Jacobson, M. D., Reiter, H. O., & Stryker, M. P. (1986). Ocular dominance shift in kitten visual cortex caused by imbalance in retinal electrical activity. *Nature, 324*(6093), 154–156. doi:10.1038/324154a0

Chapman, E. R. (2002). Synaptotagmin: A Ca^{2+} sensor that triggers exocytosis? *Nature Reviews Molecular Cell Biology, 3*, 498–508.

Chen, D., Zhao, M., & Mundy, G. R. (2004). Bone morpho-genetic proteins. *Growth Factors, 22*(4), 233–241. doi:10.1080/08977190412331279890

Chen, M. S., Huber, A. B., van der Haar, M. E., Frank, M., Schnell, L., Spillmann, A. A., … Schwab, M. E. (2000). Nogo-A is a myelin-associated neurite outgrowth inhibitor and an antigen for monoclonal antibody IN-1. *Nature, 403*(6768), 434–439. doi:10.1038/35000219

Chenn, A., & Walsh, C. A. (2002). Regulation of cerebral cortical size by control of cell cycle exit in neural precursors. *Science, 297*, 365–369.

Cheselden, W. (1728). An account of some observations made by a young gentleman, who was born blind, or lost his sight so early, that he had no remembrance of ever having seen, and was couched between 13 and 14 years of age. *Philosophical Transactions, 35*(402), 447–450.

Chiang, C., Litingtung, Y., Lee, E., Young, K. E., Corden, J. L., Westphal, H., & Beachy, P. A. (1996). Cyclopia and defective axial patterning in mice lacking *Sonic hedgehog* gene function. *Nature, 383*(6599), 407–413. doi:10.1038/383407a0

Chin, J. L. (2004). *The psychology of prejudice and discrimination: Disability, religion, physique and other traits.* Westport, CT: Greenwood Publishing Group.

Cholfin, J. A., & Rubenstein, J. L. (2007). Patterning of frontal cortex subdivions by Fgf17. *Proceedings of the National Academy of Sciences USA, 104*(18), 7652–7657. doi:10.1073/pnas.0702225104

Chomsky, N. (1957). *Syntactic structures.* The Hague/Paris: Mouton & Co.

Christian, L., Bahudhanapati, H., & Wei, S. (2013). Extracellular metalloproteinases in neural crest development and craniofacial morphogenesis. *Critical Reviews in Biochemistry and Molecular Biology, 48*(6), 544–560. doi:10.3109/10409238.2013.838203

Chun, L. L., & Patterson, P. H. (1977). Role of nerve growth factor in the development of rat sympathetic neurons in vitro. III. Effect on acetylcholine production. *Journal of Cell Biology, 75*, 712–718.

Claude, P., Hawrot, E., Dunis, D. A., & Campenot, R. B. (1982). Binding, internalization, and retrograde transport of 125I-nerve growth factor in cultured rat sympathetic neurons. *Journal of Neuroscience, 2*(4), 431–442.

Clement, A. M., Nguyen, M. D., Roberts, E. A., Garcia, M. L., Boillee, S., Rule, M., … Cleveland, D. W. (2003). Wild-type nonneuronal cells extend survival of *SOD1* mutant motor neurons in ALS mice. *Science, 302*(5642), 113–117. doi:10.1126/science.1086071

Cline, H. T., Debski, E. A., & Constantine-Paton, M. (1987). N-methyl-D-aspartate receptor antagonist desegregates eye-specific stripes. *Proceedings of the National Academy of Sciences USA, 84*(12), 4342–4345.

Cobb, M. (2000). Reading and writing The Book of Nature: Jan Swammerdam (1637–1680). *Endeavour, 24*(3), 122–128.

Cohen, S., Levi-Montalcini, R., & Hamburger, V. (1954). A nerve growth-stimulating factor isolated from sarcomas 37 and 180. *Proceedings of the National Academy of Sciences USA, 40*(10), 1014–1018.

Colom, R., Lluis-Font, J. M., & Andres-Pueyo, A. (2005). The generational intelligence gains are caused by decreasing variance in the lower half of the distribution: Supporting evidence for the nutrition hypothesis. *Intelligence, 33*, 83–91.

Conel, J. L. (1939). *The postnatal development of the human cerebral cortex: Vol. 1. The cortex of the newborn.* Cambridge, MA: Harvard University Press.

Conel, J. L. (1947). *The postnatal development of the human cerebral cortex: Vol. 3. The cortex of the three-month infant.* Cambridge, MA: Harvard University Press.

Conel, J. L. (1959). *The postnatal development of the human cerebral cortex: Vol. 6. The cortex of the twenty-four-month infant.* Cambridge, MA: Harvard University Press.

Constantine-Paton, M. (1981). Induced ocular-dominance zones in tectal cortex. In F. O. Schmitt, F. G. Worden, G. Adelman, & S. G. Dennis (Eds.), *The organization of the cerebral cortex: Proceedings of a Neurosciences Research Program Colloquium* (pp. 47–67). Cambridge, Massachusetts: MIT Press.

Constantine-Paton, M., & Capranica, R. R. (1975). Central projection of optic tract from translocated eyes in the leopard frog (*Rana pipiens*). *Science, 189*(4201), 480–482.

Constantine-Paton, M., & Law, M. I. (1978). Eye-specific termination bands in tecta of three-eyed frogs. *Science, 202*(4368), 639–641.

Constantine-Paton, M., & Law, M. I. (1982). The development of maps and stripes in the brain. *Scientific American, 7*(6), 62–70.

Conwit, R. A., Stashuk, D., Tracy, B., McHugh, M., Brown, W. F., & Metter, E. J. (1999). The relationship of motor unit size, firing rate and force. *Clinical Neurophysiology, 110*(7), 1270–1275.

Cooke, B. M., Chowanadisai, W., & Breedlove, S. M. (2000). Post-weaning social isolation of male rats reduces the volume of the medial amygdala and leads to deficits in adult sexual behavior. *Behavioural Brain Research, 117*(1–2), 107–113.

Cooke, B. M., Hegstrom, C. D., & Breedlove, S. M. (2002). Photoperiod-dependent response to androgen in the medial amygdala of the Siberian hamster, *Phodopus sungorus. Journal of Biological Rhythms, 17*(2), 147–154.

Cooke, J. E., & Moens, C. B. (2002). Boundary formation in the hindbrain: Eph only it were simple. *Trends in Neuroscience, 25*(5), 260–267.

Cooke, J. E., Kemp, H. A., & Moens, C. B. (2005). EphA4 is required for cell adhesion and rhombomere-boundary formation in the zebrafish. *Current Biology, 15*(6), 536–542. doi:10.1016/j.cub.2005.02.019

Cooper, R. M., & Zubek, J. P. (1958). Effects of enriched and restricted early environments on the learning ability of bright and dull rats. *Canadian Journal of Psychology, 12*(3), 159–164.

Corbin, J. G., Nery, S., & Fishell, G. (2001). Telencephalic cells take a tangent: Non-radial migration in the mammalian forebrain. *Nature Neuroscience, 4 Suppl.*, 1177–1182. doi:10.1038/nn749

Corsi, A. K. (2006). A biochemist's guide to *Caenorhabditis elegans. Analytical Biochemistry, 359*(1), 1–17. doi:10.1016/j.ab.2006.07.033

Cowan, W. M. (1979). The development of the brain. *Scientific American, 241*(3), 112–133.

Cowan, W. M. (2001). Viktor Hamburger and Rita Levi-Montalcini: The path to the discovery of nerve growth factor. *Annual Review of Neuroscience, 24*, 551–600. doi:10.1146/annurev.neuro.24.1.551

Cowan, W. M., & Wenger, E. (1967). Cell loss in the trochlear nucleus of the chick during normal development and after radical extirpation of the optic vesicle. *Journal of Experimental Zoology, 164*, 267–280.

Cragg, B. G. (1975). The development of synapses in the visual system of the cat. *Journal of Comparative Neurology, 160*, 147–166.

Craig, A. M., & Kang, Y. (2007). Neurexin-neuroligin signaling in synapse development. *Current Opinions in Neurobiology, 17*(1), 43–52. doi:10.1016/j.conb.2007.01.011

Crasto, C., Singer, M. S., & Shepherd, G. M. (2001). The olfactory receptor family album. *Genome Biology, 2*(10), reviews1027.1–reviews1027.4.

Creanza, N., Ruhlen, M., Pemberton, T. J., Rosenberg, N. A., Feldman, M. W., & Ramachandran, S. (2015). A comparison of worldwide phonemic and genetic variation in human populations. *Proceedings of the National Academy of Sciences USA, 112*(5), 1265–1272. doi:10.1073/pnas.1424033112

Crossley, P. H., Martinez, S., & Martin, G. R. 1996. Midbrain development induced by FGF8 in the chick embryo. *Nature, 380*, 66–68.

Crowley, J. C., & Katz, L. C. (2002). Ocular dominance development revisited. *Current Opinions in Neurobiology, 12*(1), 104–109.

Culican, S. M., Nelson, C. C., & Lichtman, J. W. (1998). Axon withdrawal during synapse elimination at the neuromuscular junction is accompanied by disassembly of the postsynaptic specialization and withdrawal of Schwann cell processes. *Journal of Neuroscience, 18*(13), 4953–4965.

Currie, D. A., & Bate, M. (1995). Innervation is essential for the development and differentiation of a sex-specific adult muscle in *Drosophila melanogaster*. *Development, 121*(8), 2549–2557.

Curtiss, S. (1977). *Genie: A psycholinguistic study of a modern-day wild child.* San Diego: Academic Press.

Curtis, M. A., Kam, M., Nannmark, U., Anderson, M. F., Axell, M. Z., Wikkelso, C., … Eriksson, P. S. (2007). Human neuroblasts migrate to the olfactory bulb via a lateral ventricular extension. *Science, 315*(5816), 1243–1249. doi:10.1126/science.1136281

Cvekl, A., & Piatigorsky, J. (1996). Lens development and crystallin gene expression: Many roles for Pax-6. *BioEssays, 18*, 621–630.

Cyert, M. S. (2001). Genetic analysis of calmodulin and its targets in *Saccharomyces cerevisiae*. *Annual Review of Genetics, 35*, 647–672. doi:10.1146/annurev.genet.35.102401.091302

D

D'Alessandro, J. S., Yetz-Aldape, J., & Wang, E. A. (1994). Bone morphogenetic proteins induce differentiation in astrocyte lineage cells. *Growth Factors, 11*(1), 53–69. doi:10.3109/08977199409015051

D'Arcangelo, G., Miao, G. G., Chen, S. C., Soares, H. D., Morgan, J. I., & Curran, T. (1995). A protein related to extracellular matrix proteins deleted in the mouse mutant reeler. *Nature, 374*(6524), 719–723. doi:10.1038/374719a0

d'Huy, J. (2013). Polyphemus (*Aa. Th.* 1137) a phylogenetic reconstruction of a prehistoric tale. *New Comparative Mythology, 1*(1). Retrieved January 4, 2016, from www.academia.edu/2433320/2013._Polyphemus_Aa._Th._1137_._A_phylogenetic_reconstruction_of_a_prehistoric_tale._-_Nouvelle_Mythologie_Compar%C3%A9e_New_Comparative_Mythology_1_3-18.

Dalva, M. B., McClelland, A. C., & Kayser, M. S. (2007). Cell adhesion molecules: Signalling functions at the synapse. *Nature Reviews Neuroscience, 8*(3), 206–220. doi:10.1038/nrn2075

Darabid, H., Perez-Gonzalez, A. P., & Robitaille, R. (2014). Neuromuscular synaptogenesis: Coordinating partners with multiple functions. *Nature Reviews Neuroscience, 15*(11), 703–718.

Darwin, C. (1859). *On the origin of species by means of natural selection.* London: John Murray.

Dave, K. A., & Bordey, A. (2009). GABA increases Ca^{2+} in cerebellar granule cell precursors via depolarization: Implications for proliferation. *IUBMB Life, 61*(5), 496–503. doi:10.1002/iub.185

Davidson, J. M., Camargo, C. A., & Smith, E. R. (1979). Effects of androgen on sexual behavior in hypogonadal men. *Journal of Clinical Endocrinology & Metabolism, 48*(6), 955–958. doi:10.1210/jcem-48-6-955

Davis, A. P., Witte, D. P., Hsieh-Li, H. M., Potter, S. M., & Capecchi, M. R. (1995). Absence of radius and ulna in mice lacking *hoxa-11* and *hoxd-11*. *Nature, 375*, 791–795.

Davis, B. E., Moon, R. Y., Sachs, H. C., & Ottolini, M. C. (1998). Effects of sleep position on infant motor development. *Pediatrics, 102*(5), 1135–1140.

Davis, E. C., Popper, P., & Gorski, R. A. (1996). The role of apoptosis in sexual differentiation of the rat sexually dimorphic nucleus of the preoptic area. *Brain Research, 734*(1–2), 10–18.

Daw, N. W., Gordon, B., Fox, K. D., Flavin, H. J., Kirsch, J. D., Beaver, C. J., … Czepita, D. (1999). Injection of MK-801 affects ocular dominance shifts more than visual activity. *Journal of Neurophysiology, 81*(1), 204–215.

de Lange, C. (2012, January 7). The man who saves Stephen Hawking's voice. *New Scientist,* issue 2846. Retrieved from www.newscientist.com.

de Luis, O., Valero, M. C., & Jurado, L. A. (2000). *WBSCR14*, a putative transcription factor gene deleted in Williams-Beuren syndrome: Complete characterisation of the human gene and the mouse ortholog. *European Journal of Human Genetics, 8*(3), 215–222. doi:10.1038/sj.ejhg.5200435

De Robertis, E. M. (2006). Spemann's organizer and self-regulation in amphibian embryos. *Nature Reviews Molecular Cell Biology, 7*, 296–302.

Dean, C., & Dresbach, T. (2006). Neuroligins and neurexins: Linking cell adhesion, synapse formation, and cognitive function. *Trends in Neuroscience, 29*, 21–29.

Dean, C., Scholl, F. G., Choih, J., DeMaria, S., Berger, J., Isacoff, E., & Scheiffele, P. (2003). Neurexin mediates the assembly of presynaptic terminals. *Nature Neuroscience, 6*(7), 708–716. doi:10.1038/nn1074

DeCasper, A. J., & Fifer, W. P. (1980). Of human bonding: Newborns prefer their mothers' voices. *Science, 208*(4448), 1174–1176.

Demir, E., & Dickson, B. J. (2005). *fruitless* splicing specifies male courtship behavior in *Drosophila. Cell, 121*(5), 785–794. doi:10.1016/j.cell.2005.04.027

Deng, W., Lee, J., Wang, H., Miller, J., Reik, A., Gregory, P. D., … Blobel, G. A. (2012). Controlling long range economic interactions at a native locus by targeted tethering of a looping factor. *Cell, 149*, 1233–1244.

Dennis, M. J., Ziskind-Conhaim, L., & Harris, A. J. (1981). Development of neuromuscular junctions in rat embryos. *Developmental Biology, 81*(2), 266–279.

Denniston, L. (2012, March 12). Argument preview: Youthful crimes, life sentences. [Web log post]. Retrieved from www.scotusblog.com/2012/03/argument-preview-youthful-crimes-life-sentences

Dent, E. W., & Gertler, F. B. (2003). Cytoskeletal dynamics and transport in growth cone motility and axon guidance. *Neuron, 40*(2), 209–227.

Descartes, R. (1637). *Discourse on the method of rightly conducting one's reason and of seeking truth in the sciences.*

Dessain, S., Gross, C. T., Kuziora, M. A., & McGinnis, W. (1992). *Antp*-type homeodomains have distinct DNA-binding specificities that correlate with their different regulatory functions in embryos. *EMBO Journal, 11*, 991–1002.

Deuchar, E. (1976). Regeneration of amputated limb-buds in early rat embryos. *Journal of Embryology and Experimental Morphology, 35*(2), 345–354.

Diamond, M. C., Krech, D., & Rosenzweig, M. R. (1964). The effects of an enriched environment on the histology of the rat cerebral cortex. *Journal of Comparative Neurology, 123*, 111–120.

Dickens, W. T., & Flynn, J. R. (2006). Black Americans reduce the racial IQ gap: Evidence from standardization samples. *Psychological Science, 17*, 913–920.

Dickson, B. J. (2001). Rho GTPases in growth cone guidance. *Current Opinions in Neurobiology, 11*(1), 103–110.

Dickson, B. J., & Gilestro, G. F. (2006). Regulation of commissural axon pathfinding by Slit and its Robo receptors. *Annual Review of Cell and Developmental Biology, 22*, 651–675. doi:10.1146/annurev.cellbio.21.090704.151234

Dimachkie, M. M., & Barohn, R. J. (2013). Guillain-Barre syndrome and variants. *Neurologic Clinics, 31*(2), 491–510. doi:10.1016/j.ncl.2013.01.005

Doe, C. Q. (1992). Molecular markers for identified neuroblasts and ganglion mother cells in the *Drosophila* central nervous system. *Development, 116*(4), 855–863.

Doe, C. Q., & Goodman, C. S. (1985). Early events in insect neurogenesis. I. Development and segmental differences in the pattern of neuronal precursor cells. *Developmental Biology, 111*(1), 193–205.

Dolen, G., Osterweil, E., Rao, B. S., Smith, G. B., Auerbach, B. D., Chattarji, S., & Bear, M. F. (2007). Correction of fragile X syndrome in mice. *Neuron, 56*(6), 955–962. doi:10.1016/j.neuron.2007.12.001

Douglas, D. S., & Popko, B. (2009). Mouse forward genetics in the study of the peripheral nervous system and human peripheral neuropathy.

Neurochemical Research, 34(1), 124–137. doi:10.1007/s11064-008-9719-4

Dreger, A. D. (2005). *One of us: Conjoined twins and the future of normal.* Cambridge, MA: Harvard University Press.

Driever, W., & Nusslein-Volhard, C. (1988). The bicoid protein determines position in the *Drosophila* embryo in a concentration-dependent manner. *Cell, 54*(1), 95–104.

Driever, W., Siegel, V., & Nusslein-Volhard, C. (1990). Autonomous determination of anterior structures in the early *Drosophila* embryo by the bicoid morphogen. *Development, 109*(4), 811–820.

Du, J. L., & Poo, M. M. (2004). Rapid BDNF-induced retrograde synaptic modification in a developing retinotectal system. *Nature, 429*(6994), 878–883. doi:10.1038/nature02618

Dufour, S., Duband, J. L., Humphries, M. J., Obara, M., Yamada, K. M., & Thiery, J. P. (1988). Attachment, spreading and locomotion of avian neural crest cells are mediated by multiple adhesion sites on fibronectin molecules. *EMBO Journal, 7*(9), 2661–2671.

Duggan, A., Ma, C., & Chalfie, M. (1998). Regulation of touch receptor differentiation by the *Caenorhabditis elegans mec-3* and *unc-86* genes. *Development, 125*(20), 4107–4119.

Duong, T. D., & Erickson, C. A. (2004). MMP-2 plays an essential role in producing epithelial-mesenchymal transformations in the avian embryo. *Developmental Dynamics, 229*(1), 42–53. doi:10.1002/dvdy.10465

Durham, D., & Woolsey, T. A. (1984). Effects of neonatal whisker lesions on mouse central trigeminal pathways. *Journal of Comparative Neurology, 223*(3), 424–447. doi:10.1002/cne.902230308

Duyme, M., Dumaret, A. C., & Tomkiewicz, S. (1999). How can we boost IQs of "dull children"? A late adoption study. *Proceedings of the National Academy of Sciences USA, 96*(15), 8790–8794.

E

Eccles, J. C., Llinás, R., & Sasaki, K. (1966). The excitatory synaptic action of climbing fibres on the Purkinje cells of the cerebellum. *Journal of Physiology (London), 182*, 268–296.

Eimas, P. D., Siqueland, E. R., Jusczyk, P., & Vigorito, J. (1971). Speech perception in infants. *Science, 171*(3968), 303–306.

Elias, G. M., Funke, L., Stein, V., Grant, S. G., Bredt, D. S., & Nicoll, R. A. (2006). Synapse-specific and developmentally regulated targeting of AMPA receptors by a family of MAGUK scaffolding proteins. *Neuron, 52*(2), 307–320. doi:10.1016/j.neuron.2006.09.012

Emes, R. D., & Grant, S. G. (2012). Evolution of synapse complexity and diversity. *Annual Reviews in Neuroscience, 35*, 111–131. doi:10.1146/annurev-neuro-062111-150433

Eriksson, P. S., Perfilieva, E., Bjork-Eriksson, T., Alborn, A. M., Nordborg, C., Peterson, D. A., & Gage, F. H. (1998). Neurogenesis in the adult human hippocampus. *Nature Medicine, 4*(11), 1313–1317. doi:10.1038/3305

Estok, P., Zsebok, S., & Siemers, B. M. (2010). Great tits search for, capture, kill and eat hibernating bats. *Biology Letters, 6*(1), 59–62. doi:10.1098/rsbl.2009.0611

Evans, M. (2011). Discovering pluripotency: 30 years of mouse embryonic stem cells. *Nature Reviews Molecular Cell Biology, 12*(10), 680–686. doi:10.1038/nrm3190

Ewert, P. H. (1930). A study of the effect of inverted retinal stimulation upon spatially coordinated behavior. *Genetic Psychology Monographs, 7*, 177–363.

Eyferth, K. (1961). Performance of different groups of occupation children on the Hamburg-Wechsler intelligence test for children (HAWIK). *Archiv fur die gesamte Psychologie, 113*, 222–241.

F

Fairbanks, D. J., & Rytting, B. (2001). Mendelian controversies: A botanical and historical review. *American Journal of Botany, 88*(5), 737–752.

Falconer, D. S. (1951). Two new mutants, 'trembler' and 'reeler', with neurological actions in the house mouse (*Mus musculus* L.). *Journal of Genetics, 50*(2), 192–201.

Falconer, D. S. (1951). Two new mutants, 'trembler' and 'reeler', with neurological actions in the house mouse (*Mus musculus* L.). *Journal of Genetics, 50*(2), 192–201.

Falk, D. (2004). Prelinguistic evolution in early hominins: Whence motherese? *Journal of Behavioral and Brain Science, 27*(4), 491–503; discussion 503–483.

Falls, D. L. (2003). Neuregulins and the neuromuscular system: 10 years of answers and questions. *Journal of Neurocytology, 32*(5–8), 619–647. doi:10.1023/B:NEUR.0000020614.83883.be

Fawcett, S. L., Wang, Y. Z., & Birch, E. E. (2005). The critical period for susceptibility of human stereopsis. *Investigative Ophthalmology & Visual Science, 46*(2), 521–525. doi:10.1167/iovs.04-0175

Fehon, R. G., Kooh, P. J., Rebay, I., Regan, C. L., Xu, T., Muskavitch, M. A., & Artavanis-Tsakonas, S. (1990). Molecular interactions between the protein products of the neurogenic loci *Notch* and *Delta*, two EGF-homologous genes in *Drosophila*. *Cell, 61*(3), 523–534.

Feller, M. B., Butts, D. A., Aaron, H. L., Rokhsar, D. S., & Shatz, C. J. (1997). Dynamic processes shape spatiotemporal properties of retinal waves. *Neuron, 19*(2), 293–306.

Feltri, M. L., Poitelon, Y., & Previtali, S. C. (2015). How Schwann cells sort axons: New concepts. *Neuroscientist, 22*(3), 252–265. doi:10.1177/1073858415572361

Fields, R. D. (2015). A new mechanism of nervous system plasticity: Activity-dependent myelination. *Nature Reviews Neuroscience, 16*(12), 756–767.

Fine, I., Wade, A. R., Brewer, A. A., May, M. G., Goodman, D. F., Boynton, G. M., … MacLeod, D. I. (2003). Long-term deprivation affects visual perception and cortex. *Nature Neuroscience, 6*(9), 915–916. doi:10.1038/nn1102

Finlay, B. L., Darlington, R. B., & Nicastro, N. (2001). Developmental structure in brain evolution. *Behavioral and Brain Sciences, 24*(2), 263–278; discussion 278–308.

Fishell, G., & Hatten, M. E. (1991). Astrotactin provides a receptor system for CNS neuronal migration. *Development, 113*(3), 755–765.

Fisher, S. E., & Marcus, G. F. (2006). The eloquent ape: Genes, brains and the evolution of language. *Nature Reviews Genetics, 7*(1), 9–20. doi:10.1038/nrg1747

Fishman, R. B., & Breedlove, S. M. (1985). The androgenic induction of spinal sexual dimorphism is independent of supraspinal afferents. *Brain Research, 355*(2), 255–258.

Fishman, R. B., & Breedlove, S. M. (1992). Local perineal implants of anti-androgen block masculinization of the spinal nucleus of the bulbocavernosus. *Brain Research. Developmental Brain Research, 70*(2), 283–286.

Flanagan, J. G., & Vanderhaeghen, P. (1998). The ephrins and Eph receptors in neural development. *Annual Review of Neuroscience, 21*, 309–345. doi:10.1146/annurev.neuro.21.1.309

Flynn, J. R. (1987). Massive IQ gains in 14 nations: What IQ tests really measure. *Psychological Bulletin, 101*, 171–191.

Flynn, J. R. (1998). IQ gains over time: Toward finding the causes. In U. Neisser (Ed.), *The rising curve: Long-term gains in IQ and related measures* (p. 37). Washington, DC: American Psychological Association.

Flynn, J. R. (2007). *What is intelligence? Beyond the Flynn effect.* Cambridge, UK: Cambridge University Press.

Fode, C., Ma, Q., Casarosa, S., Ang, S. L., Anderson, D. J., & Guillemot, F. (2000). A role for neural determination genes in specifying the dorsoventral identity of telencephalic neurons. *Genes & Development, 14*(1), 67–80.

Forger, N. G. (2006). Cell death and sexual differentiation of the nervous system. *Neuroscience, 138*(3), 929–938. doi:10.1016/j.neuroscience.2005.07.006

Forger, N. G., & Breedlove, S. M. (1986). Sexual dimorphism in human and canine spinal cord: Role of early androgen. *Proceedings of the National Academy of Sciences USA, 83*(19), 7527–7531.

Forger, N. G., & Breedlove, S. M. (1987a). Motoneuronal death during human fetal development. *Journal of Comparative Neurology, 264*(1), 118–122. doi:10.1002/cne.902640109

Forger, N. G., & Breedlove, S. M. (1987b). Seasonal variation in mammalian striated muscle mass and motoneuron morphology. *Journal of Neurobiology, 18*(2), 155–165. doi:10.1002/neu.480180204

Forger, N. G., Rosen, G. J., Waters, E. M., Jacob, D., Simerly, R. B., & de Vries, G. J. (2004). Deletion of Bax eliminates sex differences in the mouse forebrain. *Proceedings of the National Academy of Sciences USA, 101*(37), 13666–13671. doi:10.1073/pnas.0404644101

Forni, P. E., Taylor-Burds, C., Melvin, V. S., Williams, T., & Wray, S. (2011). Neural crest and ectodermal cells intermix in the nasal placode to give rise to GnRH-1 neurons, sensory neurons, and olfactory ensheathing cells. *Journal of Neuroscience, 31*(18), 6915–6927. doi:10.1523/JNEUROSCI.6087-10.2011

Fothergill, T., Donahoo, A. L., Douglass, A., Zalucki, O., Yuan, J., Shu, T., … Richards, L. J. (2014). Netrin-DCC signaling regulates corpus callosum formation through attraction of pioneering axons and by modulating Slit2-mediated repulsion. *Cerebral Cortex, 24*(5), 1138–1151. doi:10.1093/cercor/bhs395

Francis, D. D., Szegda, K., Campbell, G., Martin, W. D., & Insel, T. R. (2003). Epigenetic sources of behavioral differences in mice. *Nature Neuroscience, 6*, 445–446.

Frank, L. G., Glickman, S. E., & Licht, P. (1991). Fatal sibling aggression, precocial development, and androgens in neonatal spotted hyenas. *Science, 252*(5006), 702–704.

Franklin, R. J., & French-Constant, C. (2008). Remyelination in the CNS: From biology to therapy. *Nature Reviews Neuroscience, 9*(11), 839–855. doi:10.1038/nrn2480

Fransen, E., Van Camp, G., Vits, L., & Willems, P. J. (1997). L1-associated diseases: Clinical geneticists divide, molecular geneticists unite. *Human Molecular Genetics, 6*(10), 1625–1632.

Freeman, L. M., Watson, N. V., & Breedlove, S. M. (1996). Androgen spares androgen-insensitive motoneurons from apoptosis in the spinal nucleus of the bulbocavernosus in rats. *Hormones and Behavior, 30*(4), 424–433. doi:10.1006/hbeh.1996.0047

French, C. A., & Fisher, S. E. (2014). What can mice tell us about *Foxp2* function? *Current Opinions in Neurobiology, 28*, 72–79. doi:10.1016/j.conb.2014.07.003

Frey, S. H., Bogdanov, S., Smith, J. C., Watrous, S., & Breidenbach, W. C. (2008). Chronically deafferented sensory cortex recovers a grossly typical organization after allogenic hand transplantation. *Current Biology, 18*(19), 1530–1534. doi:10.1016/j.cub.2008.08.051

Frisby, J. P. (1980). *Seeing: Illusion, brain and mind.* Oxford, UK: Oxford University Press.

G

Gailey, D. A., & Hall, J. C. (1989). Behavior and cytogenetics of *fruitless* in *Drosophila melanogaster*: Different courtship defects caused by separate, closely linked lesions. *Genetics, 121*(4), 773–785.

Gao, W. Q., Liu, X. L., & Hatten, M. E. (1992). The *weaver* gene encodes a nonautonomous signal for CNS neuronal differentiation. *Cell, 68*(5), 841–854.

Garcia-Bellido, A. (1975). Genetic control of wing disc development in *Drosophila. Ciba Foundation Symposium, 0*(29), 161–182.

Gasser, U. E., & Hatten, M. E. (1990). Neuron-glia interactions of rat hippocampal cells in vitro: Glial-guided neuronal migration and neuronal regulation of glial differentiation. *Journal of Neuroscience, 10*(4), 1276–1285.

Gautam, M., Noakes, P. G., Mudd, J., Nichol, M., Chu, G. C., Sanes, J. R., & Merlie, J. P. (1995). Failure of postsynaptic specialization to develop at neuromuscular junctions of rapsyn-deficient mice. *Nature, 377*(6546), 232–236. doi:10.1038/377232a0

Gavrieli, Y., Sherman, Y., & Ben-Sasson, S. A. (1992). Identification of programmed cell death in situ via specific labeling of nuclear DNA fragmentation. *Journal of Cell Biology, 119*(3), 493–501.

Gaze, R. M., Keating, M. J., Ostberg, A., & Chung, S. H. (1979). The relationship between retinal and tectal growth in larval *Xenopus*: Implications for the development of the retino-tectal projection. *Journal of Embryology and Experimental Morphology, 53*, 103–143.

Gazzaniga, M. S. (2000). Cerebral specialization and interhemispheric communication: Does the corpus callosum enable the human condition? *Brain, 123*(Pt. 7), 1293–1326.

Ge, W. P., Zhou, W., Luo, Q., Jan, L. Y., & Jan, Y. N. (2009). Dividing glial cells maintain differentiated properties including complex morphology and functional synapses. *Proceedings of the National Academy of Sciences USA, 106*(1), 328–333. doi:10.1073/pnas.0811353106

Geschwind, D. H., & Flint, J. (2015). Genetics and genomics of psychiatric disease. *Science, 349*(6255), 1489–1494. doi:10.1126/science.aaa8954

Gilbert, S. F., & Barresi, M. J. F. (2016.) *Developmental biology* (11th ed.) Sunderland, MA: Sinauer Associates.

Glucksmann, A. (1951). Cell deaths in normal vertebrate ontogeny. *Biological Reviews of the Cambridge Philosophical Society, 26*(1), 59–86.

Glusman, G., Yanai, I., Rubin, I., & Lancet, D. (2001). The complete human olfactory subgenome. *Genome Research, 11*(5), 685–702. doi:10.1101/gr.171001

Godement, P., Salaun, J., & Imbert, M. (1984). Prenatal and postnatal development of retinogeniculate and retinocollicular projections in the mouse. *Journal of Comparative Neurology, 230*(4), 552–575.

Godfrey, E. W., Nitkin, R. M., Wallace, B. G., Rubin, L. L., & McMahan, U. J. (1984). Components of Torpedo electric organ and muscle that cause aggregation of acetylcholine receptors on cultured muscle cells. *Journal of Cell Biology, 99*(2), 615–627.

Goebel, T. (2007). The missing years for modern humans. *Science, 315*, 194–196.

Gogtay, N., Giedd, J. N., Lusk, L., Hayashi, K. M., Greenstein, D., Vaituzis, A. Z., … Thompson, P. M. (2004). Dynamic mapping of human cortical development during childhood through early adulthood. *Proceedings of the National Academy of Sciences USA, 101*, 8174–8179.

Goldman, D., Brenner, H. R., & Heinemann, S. (1988). Acetylcholine receptor alpha-, beta-, gamma-, and delta-subunit mRNA levels are regulated by muscle activity. *Neuron, 1*(4), 329–333.

Goldman, S. A., & Nottebohm, F. (1983). Neuronal production, migration, and differentiation in a vocal control nucleus of the adult female canary brain. *Proceedings of the National Academy of Sciences USA, 80*(8), 2390–2394.

Gomez, T. M., & Zheng, J. Q. (2006). The molecular basis for calcium-dependent axon pathfinding. *Nature Reviews Neuroscience, 7*(2), 115–125. doi:10.1038/nrn1844

Gorin, P. D., & Johnson, E. M. (1979). Experimental autoimmune model of nerve growth factor deprivation: Effects on developing peripheral sympathetic and sensory neurons. *Proceedings of the National Academy of Sciences USA, 76*(10), 5382–5386.

Gorski, R. A., Gordon, J. H., Shryne, J. E., & Southam, A. M. (1978). Evidence for a morphological sex difference within the medial preoptic area of the rat brain. *Brain Research, 148*(2), 333–346.

Gottlieb, G. (1981). Development of species identification in ducklings: VIII. Embryonic versus postnatal critical period for the maintenance of species-typical perception. *Journal of Comparative Physiological Psychology, 95*(4), 540–547.

Gottlieb, G. (2007). Probabilistic epigenesis. *Developmental Science, 10*(1), 1–11. doi:10.1111/j.1467-7687.2007.00556.x

Gottlieb, G., & Kuo, Z. Y. (1965). Development of behavior in the duck embryo. *Journal of Comparative Physiological Psychology, 59*, 183–188.

Gotz, R., Koster, R., Winkler, C., Raulf, F., Lottspeich, F., Schartl, M., & Thoenen, H. (1994). Neurotrophin-6 is a new member of the nerve growth factor family. *Nature, 372*(6503), 266–269. doi:10.1038/372266a0

Gould, E. (2007). How widespread is adult neurogenesis in mammals? *Nature Reviews Neuroscience, 8*(6), 481–488. doi:10.1038/nrn2147

Gould, E., Cameron, H. A., Daniels, D. C., Woolley, C. S., & McEwen, B. S. (1992). Adrenal hormones suppress cell division in the adult rat dentate gyrus. *Journal of Neuroscience, 12*(9), 3642–3650.

Gould, E., McEwen, B. S., Tanapat, P., Galea, L. A., & Fuchs, E. (1997). Neurogenesis in the dentate gyrus of the adult tree shrew is regulated by psychosocial stress and NMDA receptor activation. *Journal of Neuroscience, 17*(7), 2492–2498.

Gould, E., Reeves, A. J., Graziano, M. S., & Gross, C. G. (1999). Neurogenesis in the neocortex of adult primates. *Science, 286*(5439), 548–552.

Gould, E., Tanapat, P., McEwen, B. S., Flugge, G., & Fuchs, E. (1998). Proliferation of granule cell precursors in the dentate gyrus of adult monkeys is diminished by stress. *Proceedings of the National Academy of Sciences USA, 95*(6), 3168–3171.

Gould, S. J. (1977a). *Ever since Darwin.* New York: W. W. Norton.

Gould, S. J. (1977b). *Ontogeny and phylogeny.* New York: W.W. Norton.

Gould, T. W., & Oppenheim, R. W. (2011). Motor neuron trophic factors: Therapeutic use in ALS? *Brain Research. Brain Research Reviews, 67*(1–2), 1–39. doi:10.1016/j.brainresrev.2010.10.003

Gould, T. W., Buss, R. R., Vinsant, S., Prevette, D., Sun, W., Knudson, C. M., … Oppenheim, R. W. (2006). Complete dissociation of motor neuron death from motor dysfunction by *Bax* deletion in a mouse model of ALS. *Journal of Neuroscience, 26*(34), 8774–8786. doi:10.1523/JNEUROSCI.2315-06.2006

Goulding, M. D., Lumsden, A., & Gruss, P. (1993). Signals from the notochord and floor plate regulate the region-specific expression of two Pax genes in the developing spinal cord. *Development, 117*(3), 1001–1016.

Graf, E. R., Zhang, X., Jin, S. X., Linhoff, M. W., & Craig, A. M. (2004). Neurexins induce differentiation of GABA and glutamate postsynaptic specializations via neuroligins. *Cell, 119*(7), 1013–1026. doi:10.1016/j.cell.2004.11.035

Grainger, R. M. (1992). Embryonic lens induction: Shedding light on vertebrate tissue determination. *Trends in Genetics, 8*(10), 349–355.

Granger, N., Franklin, R. J., & Jeffery, N. D. (2014). Cell therapy for spinal cord injuries: What is really going on? *Neuroscientist, 20*(6), 623–638. doi:10.1177/1073858413514635

Graziadei, P. P., & Graziadei, G. A. (1979). Neurogenesis and neuron regeneration in the olfactory system of mammals. I. Morphological aspects of differentiation and structural organization of the olfactory sensory neurons. *Journal of Neurocytology, 8*(1), 1–18.

Greenough, W. T., (1976). Enduring brain effects of differential experience and training. In M. R. Rosenzweig and E. L. Bennett (Eds.), *Neural mechanisms of learning and memory* (pp. 255–278). Cambridge, MA: MIT Press.

Greenough, W. T., Volkmar, F. R., & Juraska, J. M. (1973). Effects of rearing complexity on dendritic branch-ing in frontolateral and temporal cortex of the rat. *Experimental Neurology, 41*(2), 371–378.

Gregory, R. L., & Wallace, J. G. (1963). Recovery from early blindness: A case study. *Experimental Psychology Society Monograph, 2*, 1–44.

Grimbos, T., Dawood, K., Burriss, R. P., Zucker, K. J., & Puts, D. A. (2010). Sexual orientation and the second to fourth finger length ratio: A meta-analysis in men and women. *Behavioral Neuroscience, 124*(2), 278–287. doi:10.1037/a0018764

Gruneberg, H. (1947). *Animal genetics and medicine.* New York: Paul B. Hoeber.

Grunt, J. A., & Young, W. C. (1953). Consistency of sexual behavior patterns in individual male guinea pigs following castration and androgen therapy. *Journal of Comparative and Physiological Psychology, 46*(2), 138–144.

Guidry, G., & Landis, S. C. (1998). Target-dependent development of the vesicular acetylcholine transporter in rodent sweat gland innervation. *Developmental Biology, 199*(2), 175–184. doi:10.1006/dbio.1998.8929

Guillemot, F., & Zimmer, C. (2011). From cradle to grave: The multiple roles of fibroblast growth factors in neural development. *Neuron, 71*(4), 574–588. doi:10.1016/j.neuron.2011.08.002

Guiso, L., Monte, F., Sapienza, P., & Zingales, L. (2008). Diversity. Culture, gender, and math. *Science, 320*(5880), 1164–1165. doi:10.1126/science.1154094

Gundersen, R. W., & Barrett, J. N. (1979). Neuronal chemotaxis: Chick dorsal-root axons turn toward high concentrations of nerve growth factor. *Science, 206*(4422), 1079–1080.

Gurdon, J. B., Laskey, R. A., & Reeves, O. R. (1975). The developmental capacity of nuclei transplanted from keratinized skin cells of adult frogs. *Journal of Embryology and Experimental Morphology, 34*(1), 93–112.

Gurney, M. E., Pu, H., Chiu, A. Y., Dal Canto, M. C., Polchow, C. Y., Alexander, D. D., … et al. (1994). Motor neuron degeneration in mice that express a human Cu,Zn superoxide dismutase mutation. *Science, 264*(5166), 1772–1775.

Gutmann, L., & Shy, M. (2015). Update on Charcot-Marie-Tooth disease. *Current Opinion in Neurology, 28*(5), 462–467. doi:10.1097/WCO.0000000000000237

Gwinner, E. (1990). *Bird migration: Physiology and ecophysiology.* Heidelberg: Springer-Verlag.

H

Hackman, D. A., Farah, M. J., & Meaney, M. J. (2010). Socioeconomic status and the brain: Mechanistic insights from human and animal research. *Nature Reviews Neuroscience, 11*(9), 651–659. doi:10.1038/nrn2897

Haesler, S., Rochefort, C., Georgi, B., Licznerski, P., Osten, P., & Scharff, C. (2007). Incomplete and inaccurate vocal imitation after knockdown of *FoxP2* in songbird basal ganglia nucleus Area X. *PLOS Biology, 5*(12), e321. doi:10.1371/journal.pbio.0050321

Hafen, E., Basler, K., Edstroem, J. E., & Rubin, G. M. (1987). *Sevenless*, a cell-specific homeotic gene of *Drosophila*, encodes a putative transmembrane receptor with a tyrosine kinase domain. *Science, 236*(4797), 55–63.

Hagihara, K. M., Murakami, T., Yoshida, T., Tagawa, Y., & Ohki, K. (2015). Neuronal activity is not required for the initial formation and maturation of visual selectivity. *Nature Neuroscience, 18*(12), 1780–1788. doi:10.1038/nn.4155

Haidt, J., & Joseph, C. (2007). The moral mind: How five sets of innate intuitions guide the development of many culture-specific virtues, and perhaps even modules. In P. Carruthers, S. Laurence, & S. Stich (Eds.), *The innate mind* (Vol. 3) (pp. 367–392). New York: Oxford University Press.

Hailman, J. P. (1969). How an instinct is learned. *Scientific American, 221*(6), 98–108.

Halbleib, J. M., & Nelson, W. J. (2006). Cadherins in development: Cell adhesion, sorting, and tissue morphogenesis. *Genes & Development, 20*(23), 3199–3214. doi:10.1101/gad.1486806

Hall, Z. W. (1995). Laminin b2 (S-laminin): A new player at the synapse. *Science, 269*, 362–363.

Hall, Z. W., & Sanes, J. R. (1993). Synaptic structure and development: The neuromuscular junction. *Cell, 72* Suppl., 99–121.

Hallbook, F., Ibanez, C. F., & Persson, H. (1991). Evolutionary studies of the nerve growth factor family reveal a novel member abundantly expressed in *Xenopus* ovary. *Neuron, 6*(5), 845–858.

Hamburger, V. (1934). The effects of wing-bud extirpation on the development of the central nervous system in chick embryos. *Journal of Experimental Zoology, 68*, 449–494.

Hamburger, V. (1939). Motor and sensory hyperplasia following limb-bud transplantation in chick embryos. *Physiological Zoology, 12*(3), 268–284.

Hamburger, V. (1948). The mitotic patterns in the spinal cord of the chick embryo and their relation to histogenetic processes. *Journal of Comparative Neurology, 88*(2), 221–283.

Hamburger, V. (1975). Cell death in the development of the lateral motor column of the chick embryo. *Journal of Comparative Neurology, 160*, 535–546.

Hamburger, V. (1981). Historical landmarks in neurogenesis. *Trends in Neurosciences, 4*, 151–155.

Hamburger, V. (1984). Hilde Mangold, co-discoverer of the organizer. *Journal of the History of Biology, 17*(1), 1–11.

Hamburger, V. (1988). *The heritage of experimental embryology: Hans Spemann and the organizer.* Oxford: Oxford University Press.

Hamburger, V., & Levi-Montalcini, R. (1949). Proliferation, differentiation and degeneration in the spinal ganglia of the chick embryo under normal and experimental conditions. *Journal of Experimental Zoology, 111*(3), 457–501.

Hansen, M. J., Dallal, G. E., & Flanagan, J. G. (2004). Retinal axon response to ephrin-As shows a graded, concentration-dependent transition from growth promotion to inhibition. *Neuron, 42*, 717–730.

Harlow, H. F., Dodsworth, R. O., & Harlow, M. K. (1965). Total social isolation in monkeys. *Proceedings of the National Academy of Sciences USA, 54*(1), 90–97.

Harris, W. A., Stark, W. S., & Walker, J. A. (1976). Genetic dissection of the photoreceptor system in the compound eye of *Drosophila melanogaster*. *Journal of Physiology, 256*(2), 415–439.

Harrison, R. G. (1910). The outgrowth of the nerve fiber as a mode of protoplasmic movement. *Journal of Experimental Zoology, 9*(4), 787–846.

Hashimoto, K., Ichikawa, R., Kitamura, K., Watanabe, M., & Kano, M. (2009). Translocation of a "winner" climbing fiber to the Purkinje cell dendrite and subsequent elimination of "losers" from the soma in developing cerebellum. *Neuron, 63*, 106–118.

Hastings, L. (1990). Sensory neurotoxicology: Use of the olfactory system in the assessment of toxicity. *Neurotoxicology and Teratology, 12*(5), 455–459.

Hatten, M. E. (1990). Riding the glial monorail: A common mechanism for glial-guided neuronal migration in different regions of the developing mammalian brain. *Trends in Neuroscience, 13*(5), 179–184.

Hatten, M. E., & Mason, C. A. (1990). Mechanisms of glial-guided neuronal migration in vitro and in vivo. *Experientia, 46*(9), 907–916.

Hatten, M. E., Liem, R. K., & Mason, C. A. (1986). Weaver mouse cerebellar granule neurons fail to migrate on wild-type astroglial processes in vitro. *Journal of Neuroscience, 6*(9), 2676–2683.

Hauser, C. A., Joyner, A. L., Klein, R. D., Learned, T. K., Martin, G. R., & Tjian, R. (1985). Expression of homologous homeo-box-containing genes in differentiated human teratocarcinoma cells and mouse embryos. *Cell, 43*(1), 19–28.

Hauser, R. M. (2002, August). *Meritocracy, cognitive ability, and the sources of occupational success.* Paper presented at the meeting of the American Sociological Association, Chicago, IL. Retrieved from www.ssc.wisc.edu/cde/cdewp/98-07.pdf

Hebb, D. O. (1949). *The organization of behavior: A neuropsychological theory.* New York: John Wiley & Sons.

Hedgecock, E. M., Sulston, J. E., & Thomson, J. N. (1983). Mutations affecting programmed cell deaths in the nematode *Caenorhabditis elegans*. *Science, 220*(4603), 1277–1279.

Hedges, S. B., Dudley, J., & Kumar, S. (2006). TimeTree: A public knowledge-base of divergence times among organisms. *Bioinformatics, 22*(23), 2971–2972. doi:10.1093/bioinformatics/btl505

Hegarty, S. V., O'Keeffe, G. W., & Sullivan, A. M. (2013). BMP-Smad 1/5/8 signalling in the development of the nervous system. *Progress in Neurobiology, 109*, 28–41. doi:10.1016/j.pneurobio.2013.07.002

Heimer, L. (1994.) *The human brain and spinal cord: functional neuroanatomy and dissection guide.* (2nd ed.) New York: Springer.

Held, R., Ostrovsky, Y., de Gelder, B., Gandhi, T., Ganesh, S., Mathur, U., & Sinha, P. (2011). The newly sighted fail to match seen with felt. *Nature Neuroscience, 14*(5), 551–553. doi:10.1038/nn.2795

Helm, B., & Gwinner, E. (2006). Migratory restlessness in an equatorial nonmigratory bird. *PLOS Biology, 4*(4), e110. doi:10.1371/journal.pbio.0040110

Hemmati-Brivanlou, A., Kelly, O. G., & Melton, D. A. (1994). Follistatin, an antagonist of activin, is expressed in the Spemann organizer and displays direct neuralizing activity. *Cell, 77*(2), 283–295.

Henneman, E., Somjen, G., & Carpenter, D. O. (1965). Functional significance of cell size in spinal motoneurons. *Journal of Neurophysiology, 28*, 560–580.

Henriques, A., Pitzer, C., & Schneider, A. (2010). Neurotrophic growth factors for the treatment of amyotrophic lateral sclerosis: Where do we stand? *Frontiers in Neuroscience, 4*, 32. doi:10.3389/fnins.2010.00032

Herculano-Houzel, S. (2012). The remarkable, yet not extraordinary, human brain as a scaled-up primate brain and its associated cost. *Proceedings of the National Academy of Sciences USA, 109 Suppl. 1*, 10661–10668. doi:10.1073/pnas.1201895109

Hess, E. J. (1996). Identification of the weaver mouse mutation: The end of the beginning. *Neuron, 16*(6), 1073–1076.

Hetts, S. W., Sherr, E. H., Chao, S., Gobuty, S., & Barkovich, A. J. (2006). Anomalies of the corpus callosum: An MR analysis of the phenotypic spectrum of associated malformations. *American Journal of Roentgenology, 187*(5), 1343–1348. doi:10.2214/AJR.05.0146

Hibbard, E. (1965). Orientation and directed growth of Mauthners cell axons form duplicated vestibular nerve roots. *Experimental Neurology, 13*(3), 289–301.

Hidalgo-Sanchez, M., Millet, S., Bloch-Gallego, E., & Alvarado-Mallart, R. M. (2005). Specification of the meso-isthmo-cerebellar region: The Otx2/Gbx2 boundary. *Brain Research. Brain Research Reviews, 49*(2), 134–149. doi:10.1016/j.brainresrev.2005.01.010

Hill, R. S., & Walsh, C. A. (2005). Molecular insights into human brain evolution. *Nature, 437*, 64–67.

His, W. (1887). Die Entwicklung der ersten Nervenbahnen beim menschlichen Embryo. Übersichtliche Darstellung. *Arch. Anat. Physiol. Anat. Abth., 92*, 368–378.

Hobert, O. (2011). Regulation of terminal differentiation programs in the nervous system. *Annual Review of Cell and Developmental Biology, 27*, 681–696. doi:10.1146/annurev-cellbio-092910-154226

Hoch, W., Campanelli, J. T., Harrison, S., & Scheller, R. H. (1994). Structural domains of agrin required for clustering of nicotinic acetylcholine receptors. *EMBO Journal, 13*(12), 2814–2821.

Hodgkin, J. (1998). Seven types of pleiotropy. *International Journal of Developmental Biology, 42*(3), 501–505.

Hofer, S., & Frahm, J. (2006). Topography of the human corpus callosum revisited: Comprehensive fiber tractography using diffusion tensor magnetic resonance imaging. *Neuroimage, 3*(3), 989–994.

Hohn, A., Leibrock, J., Bailey, K., & Barde, Y. A. (1990). Identification and characterization of a novel member of the nerve growth factor/brain-derived neurotrophic factor family. *Nature, 344*(6264), 339–341. doi:10.1038/344339a0

Holland, P. W. (2013). Evolution of homeobox genes. *Wiley Interdisciplinary Reviews: Developmental Biology, 2*(1), 31–45. doi:10.1002/wdev.78

Hollyday, M., & Hamburger, V. (1976). Reduction of the naturally occurring motor neuron loss by enlargement of the periphery. *Journal of Comparative Neurology, 170*(3), 311–320. doi:10.1002/cne.901700304

Holtfreter, J. (1944). Neural differentiation of ectoderm through exposure to saline solution. *Journal of Experimental Zoology, 95*(3), 307–343.

Hooke, R. (1665). *Micrographia: or some physiological descriptions of minute bodies made by magnifying glasses with observations and inquiries thereupon.* London: Martyn and Allestry.

Horton, S., Meredith, A., Richardson, J. A., & Johnson, J. E. (1999). Correct coordination of neuronal differentiation events in ventral forebrain requires the bHLH factor MASH1. *Molecular and Cellular Neuroscience, 14*(4–5), 355–369. doi:10.1006/mcne.1999.0791

Hotta, Y., & Benzer, S. (1970). Genetic dissection of the *Drosophila* nervous system by means of mosaics. *Proceedings of the National Academy of Sciences USA, 67*(3), 1156–1163.

Hubel, D. H., & Wiesel, T. N. (1965). Binocular interaction in striate cortex of kittens reared with artificial squint. *Journal of Neurophysiology, 28*(6), 1041–1059.

Huber, A. B., Kolodkin, A. L., Ginty, D. D., & Cloutier, J. F. (2003). Signaling at the growth cone: Ligand-receptor complexes and the control of axon growth and guidance. *Annual Review of Neuroscience, 26*, 509–563. doi:10.1146/annurev.neuro.26.010302.081139

Huber, E., Webster, J. M., Brewer, A. A., MacLeod, D. I., Wandell, B. A., Boynton, G. M., … Fine, I. (2015). A lack of experience-dependent plasticity after more than a decade of recovered sight. *Psychological Science, 26*(4), 393–401. doi:10.1177/0956797614563957

Huberman, A. D., Feller, M. B., & Chapman, B. (2008). Mechanisms underlying development of visual maps and receptive fields. *Annual Review of Neuroscience, 31*, 479–509.

Huberman, A. D., Speer, C. M., & Chapman, B. (2006). Spontaneous retinal activity mediates development of ocular dominance columns and binocular receptive fields in v1. *Neuron, 52*(2), 247–254.

Huettl, R.-E., Soellner, H., Bianchi, E., Novitch, B. G., & Huber, A. B. (2011). Npn-1 contributes to axon-axon interactions that differentially control sensory and motor innervation of the limb. *PLOS Biology, 9*(2), e1001020.

Hughes, A. (1961). Cell degeneration in the larval ventral horn of *Xenopus laevis* (Daudin). *Journal of Embryology and Experimental Morphology, 9*, 269–284.

Hume, R. I., Role, L. W., & Fischbach, G. D. (1983). Acetylcholine release from growth cones detected with patches of acetylcholine receptor-rich membranes. *Nature, 305*(5935), 632–634.

Hunter, A. G. W. (2005). Agenesis of the corpus callosum. In R. E. Stevenson & J. G. Hall (Eds.), *Human malformations and related anomalies* (pp. 581–604). New York: Oxford University Press.

Huttenlocher, P. R., deCourten, C., Garey, L. J., & Van der Loos, H. (1982). Synaptogenesis in human visual cortex: Evidence for synapse elimination during normal development. *Neuroscience Letters, 33*, 247–252.

Hynes, R. O. (1992). Integrins: Versatility, modulation, and signaling in cell adhesion. *Cell, 69*(1), 11–25.

Hynes, R. O. (2002). Integrins: Bidirectional, allosteric signaling machines. *Cell, 110*(6), 673–687.

I

Ibanez, C. F. (1996). Neurotrophin-4: The odd one out in the neurotrophin family. *Neurochemical Research, 21*(7), 787–793.

Imai, T., Sakano, H., & Vosshall, L. B. (2010). Topographic mapping—The olfactory system. *Cold Spring Harbor Perspectives in Biology, 2*(8), a001776. doi:10.1101/cshperspect.a001776

Imai, T., Yamazaki, T., Kobayakawa, R., Kobayakawa, K., Abe, T., Suzuki, M., & Sakano, H. (2009). Pre-target axon sorting establishes the neural map topography. *Science, 325*(5940), 585–590. doi:10.1126/science.1173596

Immelmann, K. (1972). Sexual and other long-term aspects of imprinting in birds and other species. *Advances in the Study of Behavior, 4*, 147–174.

Immelmann, K. (1975). Ecological significance of imprinting and early learning. *Annual Review of Ecology and Systematics, 6*, 15–37.

Ingham, P. (1988). The molecular genetics of embryonic pattern formation in *Drosophila. Nature, 355*, 25–34.

Innocenti, G. M., & Price, D. J. (2005). Exuberance in the development of cortical networks. *Nature Reviews Neuroscience, 6*(12), 955–965. doi:10.1038/nrn1790

Innocenti, G. M., Manger, P. R., Masiello, I., Colin, I., & Tettoni, L. (2002). Architecture and callosal connections of visual areas 17, 18, 19 and 21 in the ferret (*Mustela putorius*). *Cerebral Cortex, 12*(4), 411–422.

Ip, N. Y., Ibanez, C. F., Nye, S. H., McClain, J., Jones, P. F., Gies, D. R., … Squinto, S. P. (1992). Mammalian neurotrophin-4: Structure, chromosomal localization, tissue distribution, and receptor specificity. *Proceedings of the National Academy of Sciences USA, 89*(7), 3060–3064.

Ipulan, L. A., Suzuki, K., Sakamoto, Y., Murashima, A., Imai, Y., Omori, A., … Yamada, G. (2014). Nonmyocytic androgen receptor regulates the sexually dimorphic development of the embryonic bulbocavernosus muscle. *Endocrinology, 155*(7), 2467–2479. doi:10.1210/en.2014-1008

Irgens, L. M. (1984). The discovery of *Mycobacterium leprae*: A medical achievement in the light of evolving scientific methods. *American Journal of Dermatopathology, 6*(4), 337–343.

J

Jackson, C. A., Peduzzi, J. D., & Hickey, T. L. (1989). Visual cortex development in the ferret. I. Genesis and migration of visual cortical neurons. *Journal of Neuroscience, 9*(4), 1242–1253.

Jacobson, M. (1991). *Developmental neurobiology* (2nd ed.). New York: Plenum.

Jeeves, M. A., & Temple, C. M. (1987). A further study of language function in callosal agenesis. *Brain and Language, 32*(2), 325–335.

Jeffress, L. A. (1948). A place theory of sound localization. *Journal of Comparative Physiological Psychology, 41*(1), 35–39. Retrieved from www.ncbi.nlm.nih.gov/pubmed/18904764.

Jennings, B. H. (2011). *Drosophila*—A versatile model in biology and medicine. *Materials Today, 14*(5), 190–195.

Johns, P. R., & Easter, S. S., Jr. (1977). Growth of the adult goldfish eye. II. Increase in retinal cell number. *Journal of Comparative Neurology, 176*(3), 331–341. doi:10.1002/cne.901760303

Johnson, J. S., & Newport, E. L. (1989). Critical period effects in second language learning: The influence of maturational state on the acquisition of English as a second language. *Cognitive Psychology, 21*, 60–99.

Johnson, R. L., & Scott, M. P. (1998). New players and puzzles in the hedgehog signaling pathway. *Current Opinion in Genetics and Development, 8*, 450–456.

Johnson, T., & Barton, N. (2005). Theoretical models of selection and mutation on quantitative traits. *Philosophical Transactions of the Royal Society B: Biological Sciences, 360*(1459), 1411–1425. doi:10.1098/rstb.2005.1667

Johnstone, O., & Lasko, P. (2001). Translational regulation and RNA localization in *Drosophila* oocytes and embryos. *Annual Review of Genetics, 35*, 365–406. doi:10.1146/annurev.genet.35.102401.090756

Joris, P. X., Smith, P. H., & Yin, T. C. (1998). Coincidence detection in the auditory system: 50 years after Jeffress. *Neuron, 21*(6), 1235–1238. Retrieved from www.ncbi.nlm.nih.gov/pubmed/9883717.

Junghans, D., Haas, I. G., & Kemler, R. (2005). Mammalian cadherins and protocadherins: About cell death, synapses and processing. *Current Opinions in Cell Biology, 17*(5), 446–452. doi:10.1016/j.ceb.2005.08.008

Jusko, T. A., Henderson, C. R., Jr., Lanphear, B. P., Cory-Slechta, D. A., Parsons, P. J., & Canfield, R. L. (2008). Blood lead concentrations <10 µg/dL and child intelligence at 6 years old. *Environmental Health Perspectives, 116*, 243–248.

K

Kaila, K., Price, T. J., Payne, J. A., Puskarjov, M., & Voipio, J. (2014). Cation-chloride cotransporters in neuronal development, plasticity and disease. *Nature Reviews Neuroscience, 15*(10), 637–654. doi:10.1038/nrn3819

Kamiguchi, H., & Lemmon, V. (1997). Neural cell adhesion molecule L1: Signaling pathways and growth cone motility. *Journal of Neuroscience Research, 49*(1), 1–8.

Kane, M. J., & Engle, R. W. (2002). The role of prefrontal cortex in working-memory capacity, executive attention, and general fluid intelligence: An individual-differences perspective. *Psychonomic Bulletin & Review, 9*(4), 637–671.

Kaneko, N., Marin, O., Koike, M., Hirota, Y., Uchiyama, Y., Wu, J. Y., … Sawamoto, K. (2010). New neurons clear the path of astrocytic processes for their rapid migration in the adult brain. *Neuron, 67*(2), 213–223. doi:10.1016/j.neuron.2010.06.018

Kaplan, M. S. (1981). Neurogenesis in the 3-month-old rat visual cortex. *Journal of Comparative Neurology, 195*(2), 323–338. doi:10.1002/cne.901950211

Kaplan, M. S., & Hinds, J. W. (1977). Neurogenesis in the adult rat: Electron microscopic analysis of light radioautographs. *Science, 197*(4308), 1092–1094.

Kaslin, J., Ganz, J., & Brand, M. (2008). Proliferation, neurogenesis and regeneration in the non-mammalian vertebrate brain. *Philosophical Transactions of the Royal Society of London B: Biological Sciences, 363*(1489), 101–122. doi:10.1098/rstb.2006.2015

Kaufman, T. C., Seeger, M. A., & Olsen, G. (1990). Molecular and genetic organization of the *Antennapedia* gene complex of *Drosophila melanogaster. Advances in Genetics, 27*, 309–362.

Kawai, M. (1965). Newly acquired pre-cultural behavior of the natural troop of Japanese monkeys on Koshima islet. *Primates, 6*, 1–30.

Kazdoba, T. M., Leach, P. T., Silverman, J. L., & Crawley, J. N. (2014). Modeling fragile X syndrome in the *Fmr1* knockout mouse. *Intractable & Rare Diseases Research, 3*(4), 118–133. doi:10.5582/irdr.2014.01024

Keating, M. J., Dawes, E. A., & Grant, S. (1992). Plasticity of binocular visual connections in the frog, *Xenopus laevis*: Reversibility of effects of early visual deprivation. *Experimental Brain Research, 90*(1), 121–128.

Keino-Masu, K., Masu, M., Hinck, L., Leonardo, E. D., Chan, S. S., Culotti, J. G., & Tessier-Lavigne, M. (1996). Deleted in Colorectal Cancer (DCC) encodes a netrin receptor. *Cell, 87*(2), 175–185.

Kempermann, G., Kuhn, H. G., & Gage, F. H. (1997). More hippocampal neurons in adult mice living in an enriched environment. *Nature, 386*(6624), 493–495. doi:10.1038/386493a0

Kennedy, T. E., Serafini, T., de la Torre, J. R., & Tessier-Lavigne, M. (1994). Netrins are diffusible chemotropic factors for commissural axons in the embryonic spinal cord. *Cell, 78*(3), 425–435.

Kenneson, A., Zhang, F., Hagedorn, C. H., & Warren, S. T. (2001). Reduced FMRP and increased *FMR1* transcription is proportionally associated with CGG repeat number in intermediate-length and premutation carriers. *Human Molecular Genetics, 10*(14), 1449–1454.

Kerr, J. F., Wyllie, A. H., & Currie, A. R. (1972). Apoptosis: A basic biological phenomenon with wide-ranging implications in tissue kinetics. *British Journal of Cancer, 26*(4), 239–257.

Kessel, M. (1993). Reversal of axonal pathways from rhombomere 3 correlates with extra Hox expression domains. *Neuron, 10*(3), 379–393.

Khazipov, R., Ragozzino, D., & Bregestovski, P. (1995). Kinetics and Mg^{2+} block of *N*-methyl-D-aspartate receptor channels during postnatal development of hippocampal CA3 pyramidal neurons. *Neuroscience, 69*(4), 1057–1065.

Kidd, T., Bland, K. S., & Goodman, C. S. (1999). Slit is the midline repellent for the Robo receptor in *Drosophila*. *Cell, 96*, 785–794.

Kim, E., & Jung, H. (2015). Local protein synthesis in neuronal axons: Why and how we study. *Biochemistry and Molecular Biology Reports, 48*(3), 139–146.

Kim, J. E., Li, S., GrandPre, T., Qiu, D., & Strittmatter, S. M. (2003). Axon regeneration in young adult mice lacking Nogo-A/B. *Neuron, 38*(2), 187–199.

Kim, J. H., & Juraska, J. M. (1997). Sex differences in the development of axon number in the splenium of the rat corpus callosum from postnatal day 15 through 60. *Brain Research. Developmental Brain Research, 102*(1), 77–85.

Kimble, J. (1981). Alterations in cell lineage following laser ablation of cells in the somatic gonad of *Caenorhabditis elegans*. *Developmental Biology, 87*(2), 286–300.

Kimble, J., & Hirsh, D. (1979). The postembryonic cell lineages of the hermaphrodite and male gonads in *Caenorhabditis elegans*. *Developmental Biology, 70*(2), 396–417.

Kimura, K., Ote, M., Tazawa, T., & Yamamoto, D. (2005). *Fruitless* specifies sexually dimorphic neural circuitry in the *Drosophila* brain. *Nature, 438*(7065), 229–233. doi:10.1038/nature04229

King, T. J. (1966). Nuclear transplantation in amphibia. *Methods in Cell Physiology, 2*, 1–36.

Kleinschmidt, A., Bear, M. F., & Singer, W. (1987). Blockade of "NMDA" receptors disrupts experience-dependent plasticity of kitten striate cortex. *Science, 238*(4825), 355–358.

Klose, M., & Bentley, D. (1989). Transient pioneer neurons are essential for formation of an embryonic peripheral nerve. *Science, 245*(4921), 982–984.

Knöll, B., Weinl, C., Nordheim, A., & Bonhoeffer, F. (2007). Stripe assay to examine axonal guidance and cell migration. *Nature Protocols, 2*, 1216–1224.

Knudsen, E. I. (1985). Experience alters the spatial tuning of auditory units in the optic tectum during a sensitive period in the barn owl. *Journal of Neuroscience, 5*(11), 3094–3109.

Knudsen, E. I. (1999). Mechanisms of experience-dependent plasticity in the auditory localization pathway of the barn owl. *Journal of Comparative Physiology A, 185*(4), 305–321.

Kolodkin, A. L., Matthes, D. J., & Goodman, C. S. (1993). The semaphorin genes encode a family of transmembrane and secreted growth cone guidance molecules. *Cell, 75*(7), 1389–1399.

Kornack, D. R., & Rakic, P. (1999). Continuation of neurogenesis in the hippocampus of the adult macaque monkey. *Proceedings of the National Academy of Sciences USA, 96*(10), 5768–5773.

Kornack, D. R., & Rakic, P. (2001). Cell proliferation without neurogenesis in adult primate neocortex. *Science, 294*(5549), 2127–2130. doi:10.1126/science.1065467

Koropouli, E., & Kolodkin, A. L. (2014). Semaphorins and the dynamic regulation of synapse assembly, refinement, and function. *Current Opinion in Neurobiology, 27*, 1–7. doi:10.1016/j.conb.2014.02.005

Krech, D., Rosenzweig, M. R., & Bennett, E. L. (1962). Relations between chemistry and problem-solving among rats raised in enriched and impoverished environments. *Journal of Comparative and Physiological Psychology, 55*, 801–807.

Krull, C. E., Lansford, R., Gale, N. W., Collazo, A., Marcelle, C., Yancopoulos, G. D., … Bronner-Fraser, M. (1997). Interactions between Eph-related receptors and ligands confer rostrocaudal pattern to trunk neural crest migration. *Current Biology, 7*, 571–580.

Kuhl, P. K., Stevens, E., Hayashi, A., Deguchi, T., Kiritani, S., & Iverson, P. (2006). Infants show a facilitation effect for native language phonetic perception between 6 and 12 months. *Developmental Science, 9*(2), F13–F21. doi:10.1111/j.1467-7687.2006.00468.x

Kuida, K., Haydar, T. F., Kuan, C. Y., Gu, Y., Taya, C., Karasuyama, H., … Flavell, R. A. (1998). Reduced apoptosis and cytochrome c–mediated caspase activation in mice lacking caspase 9. *Cell, 94*(3), 325–337.

Kuida, K., Zheng, T. S., Na, S., Kuan, C., Yang, D., Karasuyama, H., … Flavell, R. A. (1996). Decreased apoptosis in the brain and premature lethality in CPP32-deficient mice. *Nature, 384*(6607), 368–372. doi:10.1038/384368a0

Kumada, T., & Komuro, H. (2004). Completion of neuronal migration regulated by loss of Ca(2+) transients. *Proceedings of the National Academy of Sciences USA, 101*(22), 8479–8484. doi:10.1073/pnas.0401000101

Kummer, T. T., Misgeld, T., & Sanes, J. R. (2006). Assembly of the postsynaptic membrane at the neuromuscular junction: Paradigm lost. *Current Opinions in Neurobiology, 16*(1), 74–82. doi:10.1016/j.conb.2005.12.003

Kuo, Z. Y. (1921). Giving up instincts in psychology. *Journal of Philosophy, 18*, 645–664.

Kuo, Z. Y. (1922). How are instincts acquired? *Psychological Review, 29*(5).

L

Laflamme, M. (2010). Palaeontology: Wringing out the oldest sponges. *Nature Geoscience, 3*, 597–598.

Lai, C. S., Fisher, S. E., Hurst, J. A., Vargha-Khadem, F., & Monaco, A. P. (2001). A forkhead-domain gene is mutated in a severe speech and language disorder. *Nature, 413*(6855), 519–523. doi:10.1038/35097076

Lammer, E. J., Chen, D. T., Hoar, R. M., Agnish, N. D., Benke, P. J., Braun, J. T., … Sun, S. S. (1985). Retinoic acid embryopathy. *New England Journal of Medicine, 313*(14), 837–841. doi:10.1056/NEJM198510033131401

Lance-Jones, C., & Landmesser, L. (1980). Motoneurone projection patterns in the chick hind limb following early partial reversals of the spinal cord. *Journal of Physiology, 302*(1), 581–602.

Landis, S. C. (1983). Neuronal growth cones. *Annual Review of Physiology, 45*, 567–580. doi:10.1146/annurev.ph.45.030183.003031

Landis, S. C. (1996). The development of cholinergic sympathetic neurons: A role for neuropoietic cytokines? *Perspectives on Developmental Neurobiology, 4*(1), 53–63.

Landmesser, L., & Pilar, G. (1974). Synaptic transmission and cell death during normal ganglionic development. *Journal of Physiology* (London), *241*, 737–749.

Larsen, P. D., & Osborn, A. G. (1982). Computed tomographic evaluation of corpus callosum agenesis and associated malformations. *Journal of Computed Tomography, 6*(3), 225–230.

Larsen, W. J. (1993). *Human embryology.* New York: Churchill Livingston.

Lassonde, M. C., Sauerwein, H. C., & Lepore, F. (2003). Agenesis of the corpus callosum. In E. Zaidel & M. Iacoboni (Eds.), *The parallel brain: The cognitive neuroscience of the corpus callosum* (pp. 357–369). Cambridge, MA: MIT Press.

Lassonde, M., Sauerwein, H., Chicoine, A. J., & Geoffroy, G. (1991). Absence of disconnexion syndrome in callosal agenesis and early callosotomy: Brain reorganization or lack of structural specificity during ontogeny? *Neuropsychologia, 29*(6), 481–495.

Lauder, J. M., & Oppenheim, R. (2001). Obituary. Viktor Hamburger (1900–2001). *Nature, 412*(6846), 496. doi:10.1038/35087701

Law, M. I., & Constantine-Paton, M. (1981). Anatomy and physiology of experimentally produced striped tecta. *Journal of Neuroscience, 1*(7), 741–759.

Lawlor, D. A., Clark, H., & Leon, D. A. (2007). Associations between childhood intelligence and hospital admissions for unintentional injuries in adulthood: The Aberdeen Children of the 1950s cohort study. *American Journal of Public Health, 97*(2), 291–297. doi:10.2105/AJPH.2005.080168

Lawrence, P. A., & Johnston, P. (1986). The muscle pattern of a segment of *Drosophila* may be determined by neurons and not by contributing myoblasts. *Cell, 45*(4), 505–513.

Le Douarin, N. M. (1973). A Feulgen-positive nucleolus. *Experimental Cell Research, 77*(1), 459–468.

Le Douarin, N. M. (1980). The ontogeny of the neural crest in avian embryo chimaeras. *Nature, 286*(5774), 663–669.

Le Douarin, N. M. (1982). *The neural crest.* New York: Cambridge University Press.

Le Douarin, N. M. (2004). The avian embryo as a model to study the development of the neural crest: A long and still ongoing study. *Mechanisms of Development, 121*, 1089–1102.

Le Douarin, N. M., Renaud, D., Teillet, M. A., & Le Douarin, G. H. (1975). Cholinergic differentiation of presumptive adrenergic neuroblasts in interspecific chimeras after heterotopic transplantations. *Proceedings of the National Academy of Sciences USA, 72*(2), 728–732.

Leber, S. M., Breedlove, S. M., & Sanes, J. R. (1990). Lineage, arrangement, and death of clonally related motor neurons in chick spinal cord. *Journal of Neuroscience, 10*(7), 2451–2462.

Ledford, H. (2008). Language: Disputed definitions. *Nature, 455*(7216), 1023–1028. doi:10.1038/4551023a

Lee, J. K., Kim, J. E., Sivula, M., & Strittmatter, S. M. (2004). Nogo receptor antagonism promotes stroke recovery by enhancing axonal plasticity. *Journal of Neuroscience, 24*(27), 6209–6217. doi:10.1523/JNEUROSCI.1643-04.2004

Lee, J. Y., & Petratos, S. (2013). Multiple sclerosis: Does Nogo play a role? *Neuroscientist, 19*(4), 394–408. doi:10.1177/1073858413477207

Lee, K. J., & Jessell, T. M. (1999). The specification of dorsal cell fates in the vertebrate central nervous system. *Annual Review Neuroscience, 22*, 261–294. doi:10.1146/annurev.neuro.22.1.261

Le Grand, R., Mondloch, C. J., Maurer, D., & Brent, H. P. (2001). Early visual experience and face processing. *Nature, 410*, 890.

Lehrman, D. S. (1953). A critique of Konrad Lorenz's theory of instinctive behavior. *Quarterly Review of Biology, 28*(4), 337–363.

Lemmon, M. A., & Schlessinger, J. (2010). Cell signaling by receptor tyrosine kinases. *Cell, 141*(7), 1117–1134.

Lemmon, V., Burden, S. M., Payne, H. R., Elmslie, G. J., & Hlavin, M. L. (1992). Neurite growth on different substrates: Permissive versus instructive influences and the role of adhesive strength. *Journal of Neuroscience, 12*(3), 818–826.

Lennington, J. B., Yang, Z., & Conover, J. C. (2003). Neural stem cells and the regulation of adult neurogenesis. *Reproductive Biology and Endocrinology*, 1:99, doi:10.1186/1477-7827-1-99

Leuner, B., Gould, E., & Shors, T. J. (2006). Is there a link between adult neurogenesis and learning? *Hippocampus, 16*(3), 216–224. doi:10.1002/hipo.20153

Leuner, B., Kozorovitskiy, Y., Gross, C. G., & Gould, E. (2007). Diminished adult neurogenesis in the marmoset brain precedes old age. *Proceedings of the National Academy of Sciences USA, 104*(43), 17169–17173. doi:10.1073/pnas.0708228104

LeVay, S., Stryker, M. P., & Shatz, C. J. (1978). Ocular dominance columns and their development in layer IV of the cat's visual cortex: A quantitative study. *Journal of Comparative Neurology, 179*, 223–244.

LeVay, S., Wiesel, T. N., & Hubel, D. H. (1980). The development of ocular dominance columns in normal and visually deprived monkeys. *Journal of Comparative Neurology, 191*, 1–51.

Levi-Montalcini, R. (1963). Growth and differentiation in the nervous system. In J. Allen (Ed.), *The nature of biological diversity* (pp. 261–296). New York: McGraw-Hill.

Levi-Montalcini, R. (1972). The morphological effects of immunosympathectomy. In G. Steiner & E. Schönbaum (Eds.), *Immunosympathectomy* (pp. 55–78). Amsterdam: Elsevier.

Levi-Montalcini, R. (1988). *In praise of imperfection: My life and work.* New York: Basic Books.

Levi-Montalcini, R., & Cohen, S. (1956). In vitro and in vivo effects of a nerve growth-stimulating agent isolated from snake venom. *Proceedings of the National Academy of Sciences USA, 42*(9), 695–699.

Levi-Montalcini, R., & Hamburger, V. (1951). Selective growth stimulating effects of mouse sarcoma on the sensory and sympathetic nervous system of the chick embryo. *Journal of Experimental Zoology, 116*(2), 321–361.

Levi-Montalcini, R., & Levi, G. (1944). Correlazioni nello sviluppo tra varie parti del sistema nervosa. *Commentarii of the Pontifical Academy of Sciences, 8*, 526–569.

Levitt, P., Cooper, M. L., & Rakic, P. (1983). Early divergence and changing proportions of neuronal and glial precursor cells in the primate cerebral ventricular zone. *Developmental Biology, 96*(2), 472–484.

Lewis, E. B. (1978). A gene complex controlling segmentation in *Drosophila. Nature, 276*(5688), 565–570.

Li, Y., Field, P. M., & Raisman, G. (1997). Repair of adult rat corticospinal tract by transplants of olfactory ensheathing cells. *Science, 277*(5334), 2000–2002.

Lichtman, J. W. (1977). The reorganization of synaptic connexions in the rat submandibular ganglion during post-natal development. *Journal of Physiology, 273*(1), 155–177.

Lichtman, J. W., & Colman, H. (2000). Synapse elimination and indelible memory. *Neuron, 25*(2), 269–278.

Liedtke, W. B., McKinley, M. J., Walker, L. L., Zhang, H., Pfenning, A. R., Drago, J., … Denton, D. A. (2011). Relation of addiction genes to hypothalamic gene changes subserving genesis and gratification of a classic instinct, sodium appetite. *Proceedings of the National Academy of Sciences USA, 108*(30), 12509–12514. doi:10.1073/pnas.1109199108

Liem, K. F., Jr., Tremml, G., Roelink, H., & Jessell, T. M. (1995). Dorsal differentiation of neural plate cells induced by BMP-mediated signals from epidermal ectoderm. *Cell, 82*(6), 969–979.

Lin, A. C., & Holt, C. E. (2008). Function and regulation of local axonal translation. *Current Opinions in Neurobiology, 18*(1), 60–68. doi:10.1016/j.conb.2008.05.004

Lin, L. F., Doherty, D. H., Lile, J. D., Bektesh, S., & Collins, F. (1993). GDNF: A glial cell line-derived neurotrophic factor for midbrain dopaminergic neurons. *Science, 260*(5111), 1130–1132.

Lin, S., Landmann, L., Ruegg, M. A., & Brenner, H. R. (2008). The role of nerve- versus muscle-derived factors in mammalian neuromuscular junction formation. *Journal of Neuroscience, 28*(13), 3333–3340. doi:10.1523/JNEUROSCI.5590-07.2008

Lin, W., Burgess, R. W., Dominguez, B., Pfaff, S. L., Sanes, J. R., & Lee, K. F. (2001). Distinct roles of nerve and muscle in postsynaptic differentiation of the neuromuscular synapse. *Nature, 410*(6832), 1057–1064. doi:10.1038/35074025

Lindsay, R. M., Thoenen, H., & Barde, Y. A. (1985). Placode and neural crest-derived sensory neurons are responsive at early developmental stages to brain-derived neurotrophic factor. *Developmental Biology, 112*(2), 319–328.

Linney, E., & LaMantia, A.-S. (1994). Retinoid signaling in mouse embryos. *Advances in Developmental Biology, 3*, 73–114.

Lino, M. M., Schneider, C., & Caroni, P. (2002). Accumulation of *SOD1* mutants in postnatal motoneurons does not cause motoneuron pathology or motoneuron disease. *Journal of Neuroscience, 22*(12), 4825–4832.

Linquist, S., Machery, E., Griffiths, P. E., & Stotz, K. (2011). Exploring the folkbiological conception of human nature. *Philosophical Transactions of the Royal Society B: Biological Sciences, 366*(1563), 444–453. doi:10.1098/rstb.2010.0224

Lise, M. F., & El-Husseini, A. (2006). The neuroligin and neurexin families: From structure to function at the synapse. *Cellular and Molecular Life Sciences, 63*(16), 1833–1849. doi:10.1007/s00018-006-6061-3

Liu, D. W., & Westerfield, M. (1992). Clustering of muscle acetylcholine receptors requires motoneurons in live embryos, but not in cell culture. *Journal of Neuroscience, 12*(5), 1859–1866.

Liu, D., Diorio, J., Tannenbaum, B., Caldji, C., Francis, D., Freedman, A., … Meaney, M. J. (1997). Maternal care, hippocampal glucocorticoid receptors, and hypothalamic-pituitary-adrenal responses to stress. *Science, 277*(5332), 1659–1662.

Liu, J., Dietz, K., DeLoyht, J. M., Pedre, X., Kelkar, D., Kaur, J., … Casaccia, P. (2012). Impaired adult myelination in the prefrontal cortex of socially isolated mice. *Nature Neuroscience, 15*(12), 1621–1623. doi:10.1038/nn.3263

Livesey, F. J., & Cepko, C. L. (2001). Vertebrate neural cell-fate determination: Lessons from the retina. *Nature Reviews Neuroscience, 2*(2), 109–118. doi:10.1038/35053522

Lloyd, T. E., & Taylor, J. P. (2010). Flightless flies: *Drosophila* models of neuromuscular disease. *Annals of the New York Academy of Sciences, 1184*, e1–20.

Locke, J. L. (1993). *The child's path to spoken language.* Cambridge, MA: Harvard University Press.

Lovegrove, B., Simoes, S., Rivas, M. L., Sotillos, S., Johnson, K., Knust, E., … Hombria, J. C. (2006). Coordinated control of cell adhesion, polarity, and cytoskeleton underlies Hox-induced organogenesis in *Drosophila. Current Biology, 16*(22), 2206–2216. doi:10.1016/j.cub.2006.09.029

Lowery, L. A., & Van Vactor, D. (2009). The trip of the tip: Understanding the growth cone machinery. *Nature*

Reviews Molecular Cell Biology, 10(5), 332–343. doi:10.1038/nrm2679

Lozano, R., Rosero, C. A., & Hagerman, R. J. (2014). Fragile X spectrum disorders. *Intractable & Rare Disease Research, 3*(4), 134–146. doi:10.5582/irdr.2014.01022

Luo, Y., Raible, D., & Raper, J. A. (1993). Collapsin: A protein in brain that induces the collapse and paralysis of neuronal growth cones. *Cell, 75*(2), 217–227.

Luskin, M. B., & Shatz, C. J. (1985). Neurogenesis of the cat's primary visual cortex. *Journal of Comparative Neurology, 242*(4), 611–631. doi:10.1002/cne.902420409

Lynch, G., & Baudry, M. (1991). Reevaluating the constraints on hypotheses regarding LTP expression. *Hippocampus, 1*(1), 9–14.

Lynch, J. W. (2004). Molecular structure and function of the glycine receptor chloride channel. *Physiological Reviews, 84*(4), 1051–1095. doi:10.1152/physrev.00042.2003

Lynch, J. W. (2009). Native glycine receptor subtypes and their physiological roles. *Neuropharmacology, 56*(1), 303–309. doi:10.1016/j.neuropharm.2008.07.034

Lyons, J. I., Kerr, G. R., & Mueller, P. W. (2015). Fragile X syndrome: Scientific background and screening technologies. *Journal of Molecular Diagnostics, 17*(5), 463–471. doi:10.1016/j.jmoldx.2015.04.006

M

Mabie, P. C., Mehler, M. F., Marmur, R., Papavasiliou, A., Song, Q., & Kessler, J. A. (1997). Bone morpho-genetic proteins induce astroglial differentiation of oligodendroglial-astroglial progenitor cells. *Journal of Neuroscience, 17*(11), 4112–4120.

MacDermott, A. B., & Westbrook, G. L. (1986). Early development of voltage-dependent sodium currents in cultured mouse spinal cord neurons. *Developmental Biology, 113*(2), 317–326.

MacDonald, B. T., Tamai, K., & He, X. (2009). Wnt/b-catenin signaling: Components, mechanisms, and disease. *Developmental Cell, 17*, 9–26.

Macmillan, M. (2000). *An odd kind of fame.* Cambridge, MA: MIT Press.

Maden, M. (2006). Retinoids and spinal cord development. *Journal of Neurobiology, 66*(7), 726–738.

Maeda, R. K., & Karch, F. (2010). *Cis*-regulation in the *Drosophila bithorax* complex. *Advances in Experimental Medicine and Biology, 689*, 17–40.

Magrassi, L., & Graziadei, P. P. (1995). Cell death in the olfactory epithelium. *Anatomy and Embryology, 192*(1), 77–87.

Maier, E. C., Saxena, A., Alsina, B., Bronner, M. E., & Whitfield, T. T. (2014). Sensational placodes: Neurogenesis in the otic and olfactory systems. *Developmental Biology, 389*(1), 50–67. doi:10.1016/j.ydbio.2014.01.023

Maisonpierre, P. C., Belluscio, L., Squinto, S., Ip, N. Y., Furthe, M. E., Lindsay, R. M., & Yancopoulos, G. D. (1990). Neurotrophin-3: A neurotrophic factor related to NGF and BDNF. *Science, 247*, 1446–1451.

Makinodan, M., Rosen, K. M., Ito, S., & Corfas, G. (2012). A critical period for social experience-dependent oligodendrocyte maturation and myelination. *Science, 337*(6100), 1357–1360. doi:10.1126/science.1220845

Malenka, R. C., & Bear, M. F. (2004). LTP and LTD: An embarrassment of riches. *Neuron, 44*(1), 5–21.

Mallo, M., Wellik, D. M., & Deschamps, J. (2010). Hox genes and regional patterning of the vertebrate body plan. *Developmental Biology, 344*(1), 7–15. doi:10.1016/j.ydbio.2010.04.024

Mameli, M., & Bateson, P. (2006). Innateness and the sciences. *Biology and Philosophy, 21*(2), 155–188.

Manthorpe, M., Barbin, G., & Varon, S. (1982). Isoelectric focusing of the chick eye ciliary neuronotrophic factor. *Journal of Neuroscience Research, 8*(2–3), 233–239. doi:10.1002/jnr.490080213

Mardinly, A. R., Spiegel, I., Patrizi, A., Centofante, E., Bazinet, J. E., Tzeng, C. P., … Greenberg, M. E. (2016). Sensory experience regulates cortical inhibition by inducing IGF1 in VIP neurons. *Nature, 531*(7594), 371–375. doi:10.1038/nature17187

Mariani, J., & Changeux, J.-P. (1981). Ontogenesis of olivocerebellar relationships. I. Studies by intracellular recordings of the multiple innervation of Purkinje cells by climbing fibers in the developing rat cerebellum. *Journal of Neuroscience, 1*(/), 696–702.

Mariette, M. M., & Buchanan, K. L. (2016). Prenatal acoustic communication programs offspring for high posthatching temperatures in a songbird. *Science, 353*(6301), 812–814. doi:10.1126/science.aaf7049

Marin-Teva, J. L., Dusart, I., Colin, C., Gervais, A., van Rooijen, N., & Mallat, M. (2004). Microglia promote the death of developing Purkinje cells. *Neuron, 41*(4), 535–547.

Marler, P. (1997). Three models of song learning: Evidence from behavior. *Journal of Neurobiology, 33*(5), 501–516.

Marler, P., & Peters, S. (1977). Selective vocal learning in a sparrow. *Science, 198*(4316), 519–521. doi:10.1126/science.198.4316.519

Marler, P., & Peters, S. (1981). Sparrows learn adult song and more from memory. *Science, 213*(4509), 780–782. doi:10.1126/science.213.4509.780

Marler, P., & Sherman, V. (1983). Song structure with-out auditory feedback: Emendations of the auditory template hypothesis. *Journal of Neuroscience, 3*, 517–531.

Marler, P., & Sherman, V. (1985). Innate differences in sing-ing behaviour of sparrows reared in isolation from adult conspecific song. *Animal Behavior, 33*, 57–71.

Maron, C. F., Beltramino, L., Di Martino, M., Chirife, A., Seger, J., Uhart, M., … Rowntree, V. J. (2015). Increased wounding of southern right whale (*Eubalaena australis*) calves by kelp gulls (*Larus dominicanus*) at Peninsula Valdes, Argentina. *PLOS One, 10*(10), e0139291. doi:10.1371/journal.pone.0139291

Martin, D. P., Schmidt, R. E., DiStefano, P. S., Lowry, O. H., Carter, J. G., & Johnson, E. M., Jr. (1988). Inhibitors of protein synthesis and RNA synthesis prevent neuro-nal death caused by nerve growth factor deprivation. *Journal of Cell Biology, 106*(3), 829–844.

Martin, P. T., Ettinger, A. J., & Sanes, J. R. (1995). A synap-tic localization domain in the synaptic cleft protein laminin beta 2 (s-laminin). *Science, 269*(5222), 413–416.

Martin-Zanca, D., Hughes, S. H., & Barbacid, M. (1986). A human oncogene formed by the fusion of truncated tropomyosin and protein tyrosine kinase sequences. *Nature, 319*(6056), 743–748. doi:10.1038/319743a0

Martinez, S., Crossley, P. H., Cobos, I., Rubenstein, J. L., & Martin, G. R. (1999). FGF8 induces formation of an ectopic isthmic organizer and isthmocerebellar development via a repressive effect on *Otx2* expression. *Development, 126*(6), 1189–1200.

Masaki, T., Qu, J., Cholewa-Waclaw, J., Burr, K., Raaum, R., & Rambukkana, A. (2013). Reprogramming adult Schwann cells to stem cell–like cells by leprosy bacilli promotes dissemination of infection. *Cell, 152*(1–2), 51–67. doi:10.1016/j.cell.2012.12.014

Maselli, R. A., Arredondo, J., Ferns, M. J., & Wollmann, R. L. (2012). Synaptic basal lamina-associated congenital myasthenic syndromes. *Annals of the New York Academy of Sciences, 1275*, 36–48. doi:10.1111/j.1749-6632.2012.06807.x

Matsuo, I., Kuratani, S., Kimura, C., Takeda, N., & Aizawa, S. (1995). Mouse Otx2 functions in the formation and patterning of rostral head. *Genes and Development, 9*(21), 2646–2658.

Maurel, P., & Salzer, J. L. (2000). Axonal regulation of Schwann cell proliferation and survival and the initial events of myelination requires PI 3-kinase activity. *Journal of Neuroscience, 20*(12), 4635–4645.

McAllister, A. K., Katz, L. C., & Lo, D. C. (1997). Opposing roles for endogenous BDNF and NT-3 in regulating cortical dendritic growth. *Neuron, 18*(5), 767–778.

McClure, L. F., Niles, J. K., & Kaufman, H. W. (2016). Blood lead levels in young children: US, 2009–2015. *Journal of Pediatrics, 175*, 173–181. doi:10.1016/j.jpeds.2016.05.005

McConnell, S. K., & Kaznowski, C. E. (1991). Cell cycle dependence of laminar determination in developing neocortex. *Science, 254*(5029), 282–285.

McCoy, T. (2015, February 24). How Stephen Hawking, diagnosed with ALS decades ago, is still alive. *The Washington Post*. Retrieved from www.washingtonpost.com.

McFadden, D., & Pasanen, E. G. (1998). Comparison of the auditory systems of heterosexuals and homosexuals: Click-evoked otoacoustic emissions. *Proceeding of the National Academy of Sciences USA, 95*(5), 2709–2713.

McGee, A. W., Yang, Y., Fischer, Q. S., Daw, N. W., & Strittmatter, S. M. (2005). Experience-driven plasticity of visual cortex limited by myelin and Nogo receptor. *Science, 309*(5744), 2222–2226. doi:10.1126/science.1114362

McGinnis, W., Levine, M. S., Hafen, E., Kuroiwa, A., & Gehring, W. J. (1984). A conserved DNA sequence in homoeotic genes of the *Drosophila Antennapedia* and *bithorax* complexes. *Nature, 308*(5958), 428–433.

McGowan, P. O., Sasaki, A., D'Alessio, A. C., Dymov, S., Labonte, B., Szyf, M., … Meaney, M. J. (2009). Epigenetic regulation of the glucocorticoid receptor in human brain associates with childhood abuse. *Nature Neuroscience, 12*(3), 342–348. doi:10.1038/nn.2270

McKeown, S. J., Wallace, A. S., & Anderson, R. B. (2013). Expression and function of cell adhesion molecules during neural crest migration. *Developmental Biology, 373*(2), 244–257. doi:10.1016/j.ydbio.2012.10.028

Meaney, M. J., Stewart, J., Poulin, P., & McEwen, B. S. (1983). Sexual differentiation of social play in rat pups is mediated by the neonatal androgen-receptor system. *Neuroendocrinology, 37*(2), 85–90.

Mei, L., & Nave, K. A. (2014). Neuregulin-ERBB signaling in the nervous system and neuropsychiatric diseases. *Neuron, 83*(1), 27–49. doi:10.1016/j.neuron.2014.06.007

Menezes, J. R., Smith, C. M., Nelson, K. C., & Luskin, M. B. (1995). The division of neuronal progenitor cells during migration in the neonatal mammalian forebrain. *Molecular and Cellular Neuroscience, 6*(6), 496–508. doi:10.1006/mcne.1995.0002

Mensch, S., Baraban, M., Almeida, R., Czopka, T., Ausborn, J., El Manira, A., & Lyons, D. A. (2015). Synaptic vesicle release regulates myelin sheath number of individual oligodendrocytes in vivo. *Nature Neuroscience, 18*(5), 628–630.

Merino, R., Macias, D., Ganan, Y., Economides, A. N., Wang, X., Wu, Q., … Hurle, J. M. (1999). Expression and function of Gdf-5 during digit skeletogenesis in the embryonic chick leg bud. *Developmental Biology, 206*(1), 33–45. doi:10.1006/dbio.1998.9129

Merlie, J. P., & Sanes, J. R. (1985). Concentration of acetylcholine receptor mRNA in synaptic regions of adult muscle fibres. *Nature, 317*(6032), 66–68.

Merzenich, M. M., & Jenkins, W. M. (1993). Reorganization of cortical representations of the hand following alterations of skin inputs induced by nerve injury, skin island transfers, and experience. *Journal of Hand Therapy, 6*, 89–104.

Merzenich, M. M., Nelson, R. J., Stryker, M. P., Cynader, M. S., Schoppmann, A., & Zook, J. M. (1984). Somatosensory cortical map changes following digit amputation in adult monkeys. *Journal of Comparative Neurology, 224*(4), 591–605. doi:10.1002/cne.902240408

Mets, M. B., Beauchamp, C., & Haldi, B. A. (2003). Binocularity following surgical correction of strabismus in adults. *Transactions of the American Ophthalmological Society, 101*, 201–205; discussion 205–207.

Mettke-Hofmann, C., & Gwinner, E. (2003). Long-term memory for a life on the move. *Proceedings of the National Academy of Sciences USA, 100*(10), 5863–5866. doi:10.1073/pnas.1037505100

Meyer, B. U., Roricht, S., & Niehaus, L. (1998). Morphology of acallosal brains as assessed by MRI in six patients leading a normal daily life. *Journal of Neurology, 245*(2), 106–110.

Meyer, R. L., & Wolcott, L. L. (1987). Compression and expansion without impulse activity in the retinotectal projection of goldfish. *Journal of Neurobiology, 18*(6), 549–567. doi:10.1002/neu.480180606

Michel, J. B., Shen, Y. K., Aiden, A. P., Veres, A., Gray, M. K., Google Books Team, … Aiden, E. L. (2011). Quantitative analysis of culture using millions of digitized books. *Science, 331*(6014), 176–182. doi:10.1126/science.1199644

Miller v. Alabama, 567 U.S. (2012).

Miller, B. R., & Hen, R. (2015). The current state of the neurogenic theory of depression and anxiety. *Current Opinion in Neurobiology, 30*, 51–58. doi:10.1016/j.conb.2014.08.012

Miller, M. W. (1995). Relationship of the time of origin and death of neurons in rat somatosensory cortex: Barrel versus septal cortex and projection versus local circuit neurons. *Journal of Comparative Neurology, 355*(1), 6–14. doi:10.1002/cne.903550104

Miller, R. H., & Mi, S. (2007). Dissecting demyelination. *Nature Neuroscience, 10*(11), 1351–1354. doi:10.1038/nn1995

Miller, T. M., Kim, S. H., Yamanaka, K., Hester, M., Umapathi, P., Arnson, H., … Kaspar, B. K. (2006). Gene transfer demonstrates that muscle is not a primary target for non-cell-autonomous toxicity in familial amyotrophic lateral sclerosis. *Proceedings of the National Academy of Sciences USA, 103*(51), 19546–19551. doi:10.1073/pnas.0609411103

Millet, S., Bloch-Gallego, E., Simeone, A., & Alvarado-Mallart, R. M. (1996). The caudal limit of *Otx2* gene expression as a marker of the midbrain/hindbrain boundary: A study using in situ hybridisation and chick/quail homotopic grafts. *Development, 122*(12), 3785–3797.

Millet, S., Campbell, K., Epstein, D. J., Losos, K., Harris, E., & Joyner, A. L. (1999). A role for *Gbx2* in repression of *Otx2* and positioning the mid/hindbrain organizer. *Nature, 401*(6749), 161–164. doi:10.1038/43664

Mills, K. L., Lalonde, F., Clasen, L. S., Giedd, J. N., & Blakemore, S. J. (2014). Developmental changes in the structure of the social brain in late childhood and adolescence. *Social Cognitive and Affective Neuroscience, 9*(1), 123–131. doi:10.1093/scan/nss113

Ming, G. L., & Song, H. (2011). Adult neurogenesis in the mammalian brain: Significant answers and significant questions. *Neuron, 70*(4), 687–702. doi:10.1016/j.neuron.2011.05.001

Ming, G. L., Song, H. J., Berninger, B., Holt, C. E., Tessier-Lavigne, M., & Poo, M. M. (1997). cAMP-dependent growth cone guidance by netrin-1. *Neuron, 19*(6), 1225–1235.

Miri, A., Azim, E., & Jessell, T. M. (2013). Edging toward entelechy in motor control. *Neuron, 80*(3), 827–834. doi:10.1016/j.neuron.2013.10.049

Missias, A. C., Chu, G. C., Klocke, B. J., Sanes, J. R., & Merlie, J. P. (1996). Maturation of the acetylcholine receptor in skeletal muscle: Regulation of the AChR gamma-to-epsilon switch. *Developmental Biology, 179*(1), 223–238. doi:10.1006/dbio.1996.0253

Mizutani, C. M., & Bier, E. (2008). EvoD/Vo: The origins of BMP signalling in the neuroectoderm. *Nature Reviews Genetics, 9*(9), 663–677. doi:10.1038/nrg2417

Montagu, A. (1989). *Growing young* (2nd ed.). South Hadley, Massachusetts: Bergin and Garvey.

Moore, C. L., Dou, H., & Juraska J. M. (1992). Maternal stimulation affects the number of motor neurons in a sexually dimorphic nucleus of the lumbar spinal cord. *Brain Research. 572*(1–2), 52–56.

Moore, K. L., & Persaud, T. V. N. (1993). *Before we are born: Essentials of embryology and birth defects.* Philadelphia: W. B. Saunders.

Morgan, T. (1895). Half embryos and whole embryos from one of the first two blastomeres. *Anatomischer Anzeiger, 10*, 623–638.

Morgan, T. H. (1911). Chromosomes and associative inheritance. *Science, 34*(880), 636–638. doi:10.1126/science.34.880.636

Morgan, T. H., & Bridges, C. B. (1919). *The origin of gynandromorphs* (No. 278) (pp. 1–122). Washington, DC: Carnegie Institute of Washington.

Morishita, H., & Yagi, T. (2007). Protocadherin family: Diversity, structure, and function. *Current Opinions in Cell Biology, 19*(5), 584–592. doi:10.1016/j.ceb.2007.09.006

Morris, R. G. (1989). Synaptic plasticity and learning: Selective impairment of learning rats and blockade of long-term potentiation in vivo by the *N*-methyl-D-aspartate receptor antagonist AP5. *Journal of Neuroscience, 9*(9), 3040–3057.

Mosconi, T., Woolsey, T. A., & Jacquin, M. F. (2010). Passive vs. active touch-induced activity in the developing whisker pathway. *European Journal of Neuroscience, 32*(8), 1354–1363. doi:10.1111/j.1460-9568.2010.07396.x

Muir-Robinson, G., Hwang, B. J., & Feller, M. B. (2002). Retinogeniculate axons undergo eye-specific segregation in the absence of eye-specific layers. *Journal of Neuroscience, 22*(13), 5259–5264.

Muller, H. J. (1928). The measurement of gene mutation rate in *Drosophila*, its high variability, and its dependence upon temperature. *Genetics, 13*(4), 279–357.

Munnamalai, V., & Suter, D. M. 2009. Reactive oxygen species regulate F-actin dynamics in neuronal growth cones and neurite outgrowth. *Journal of Neurochemistry, 108*(3), 644–661.

Munte, T. F., Altenmuller, E., & Jancke, L. (2002). The musician's brain as a model of neuroplasticity. *Nature Reviews Neuroscience, 3*(6), 473–478. doi:10.1038/nrn843

Murray, A. J., Tucker, S. J., & Shewan, D. A. (2009). cAMP-dependent axon guidance is distinctly regulated by Epac and protein kinase A. *Journal of Neuroscience, 29*(49), 15434–15444. doi:10.1523/JNEUROSCI.3071-09.2009

Muzio, L., & Mallamaci, A. (2003). *Emx1, emx2* and *pax6* in specification, regionalization and arealization of the cerebral cortex. *Cerebral Cortex, 13*(6), 641–647.

N

Naef, A. (1926.) Uber die Urformen der Anthropomorphen und die Stammesgaschichte des Mendenschädels. *Naturwissenschaften, 14*, 445–452.

National Institutes of Health. (2014, February). *Speech and language developmental milestones* (NIH Pub. No. 13-4781). Bethesda, MD. Retrieved from http://www.nidcd.nih.gov/health/voice/Pages/speechandlanguage.aspx.

Nave, K. A., & Salzer, J. L. (2006). Axonal regulation of myelination by neuregulin 1. *Current Opinions in Neurobiology, 16*(5), 492–500. doi:10.1016/j.conb.2006.08.008

Neisser, U. (1997). Rising scores on intelligence tests. *American Scientist, 85*, 440–447.

Nelson, D. A., & Marler, P. (1994). Selection-based learning in bird song development. *Proceedings of the National Academy of Sciences USA, 91*(22), 10498–10501.

Nikolic, M., Gardner, H. A., & Tucker, K. L. (2013). Postnatal neuronal apoptosis in the cerebral cortex: Physiological and pathophysiological mechanisms. *Neuroscience, 254*, 369–378. doi:10.1016/j.neuroscience.2013.09.035

Nilsson, A. S., Fainzilber, M., Falck, P., & Ibanez, C. F. (1998). Neurotrophin-7: A novel member of the neurotrophin family from the zebrafish. *FEBS Letters, 424*(3), 285–290.

Nobel Media. (2015). *The Nobel Foundation—Statutes* (par.10). Retrieved from www.nobelprize.org/nobel_organizations/nobelfoundation/statutes.html.

Noctor, S. C., Martinez-Cerdeno, V., & Kriegstein, A. R. (2008). Distinct behaviors of neural stem and progenitor cells underlie cortical neurogenesis. *Journal of Comparative Neurology, 508*(1), 28–44. doi:10.1002/cne.21669

Noctor, S. C., Martinez-Cerdeno, V., Ivic, L., & Kriegstein, A. R. (2004). Cortical neurons arise in symmetric and asymmetric division zones and migrate through specific phases. *Nature Neuroscience, 7*(2), 136–144. doi:10.1038/nn1172

Nordeen, E. J., Nordeen, K. W., Sengelaub, D. R., & Arnold, A. P. (1985). Androgens prevent normally occurring cell death in a sexually dimorphic spinal nucleus. *Science, 229*(4714), 671–673.

Nottebohm, F., & Arnold, A. P. (1976). Sexual dimorphism in vocal control areas of the songbird brain. *Science, 194*(4261), 211–213.

Nottebohm, F., Nottebohm, M. E., & Crane, L. (1986). Developmental and seasonal changes in canary song and their relation to changes in the anatomy of song-control nuclei. *Behavioral and Neural Biology, 46*(3), 445–471.

Nudel, R., & Newbury, D. F. (2013). FOXP2. *Wiley Interdisciplinary Reviews: Cognitive Science, 4*(5), 547–560. doi:10.1002/wcs.1247

Nusse, R., & Varmus, H. (2012). Three decades of Wnts: A personal perspective on how a scientific field developed. *EMBO Journal, 31*(12), 2670–2684. doi:10.1038/emboj.2012.146

Nusslein-Volhard, C. (1991). Determination of the embryonic axes of *Drosophila. Development Supplement, 1,* 1–10.

Nusslein-Volhard, C., & Wieschaus, E. (1980). Mutations affecting segment number and polarity in *Drosophila. Nature, 287*(5785), 795–801.

O

O'Dowd, D. K., Ribera, A. B., & Spitzer, N. C. (1988). Development of voltage-dependent calcium, sodium, and potassium currents in *Xenopus* spinal neurons. *Journal of Neuroscience, 8*(3), 792–805.

O'Leary, D. D. M., & Wilkinson, D. G. (1999). Eph receptors and ephrins in neural development. *Current Opinion in Neurobiology, 9,* 65–73.

Oelgeschlager, M., Kuroda, H., Reversade, B., & De Robertis, E. M. (2003). Chordin is required for the Spemann organizer transplantation phenomenon in *Xenopus* embryos. *Developmental Cell, 4*(2), 219–230.

Okabe, S., & Hirokawa, N. (1990). Turnover of fluorescently labelled tubulin and actin in the axon. *Nature, 343*(6257), 479–482. doi:10.1038/343479a0

Okado, N., & Oppenheim, R. W. (1984). Cell death of motoneurons in the chick embryo spinal cord. IX. The loss of motoneurons following removal of afferent inputs. *Journal of Neuroscience, 4*(6), 1639–1652.

Olavarria, J. F., & Van Sluyters, R. C. (1995). Overall pattern of callosal connections in visual cortex of normal and enucleated cats. *Journal of Comparative Neurology, 363*(2), 161–176. doi:10.1002/cne.903630202

Olson, M. D., & Meyer, R. L. (1994). Normal activity-dependent refinement in a compressed retinotectal projection in goldfish. *Journal of Comparative Neurology, 347*(4), 481–494. doi:10.1002/cne.903470402

Olson, P. L. (1991). Driver perception response time. *Accident Reconstruction Journal, 3,* 16–21, 29.

Oomen, C. A., Girardi, C. E. N., Cahyadi, R., Verbeek, E. C., Krugers, H., Marian Joëls, M., & Lucassen, P. J. 2009. Opposite effects of early maternal deprivation on neurogenesis in male versus female rats.*PLoS ONE, 4*(1),e3675.

Opendak, M., & Gould, E. (2015). Adult neurogenesis: A substrate for experience-dependent change. *Trends in Cognitive Science, 19*(3), 151–161. doi:10.1016/j.tics.2015.01.001

Oppenheim, R. W. (2001). Viktor Hamburger (1900–2001): Journey of a neuroembryologist to the end of the millennium and beyond. *Neuron, 31*(2), 179–190.

Oppenheim, R. W., Bursztajn, S., & Prevette, D. (1989). Cell death of motoneurons in the chick embryo spinal cord. XI. Acetylcholine receptors and synaptogenesis in skeletal muscle following the reduction of motoneuron death by neuromuscular blockade. *Development, 107*(2), 331–341.

Ostrovsky, Y., Andalman, A., & Sinha, P. (2006). Vision following extended congenital blindness. *Psychological Science, 17*(12), 1009–1014. doi:10.1111/j.1467-9280.2006.01827.x

Ostrovsky, Y., Meyers, E., Ganesh, S., Mathur, U., & Sinha, P. (2009). Visual parsing after recovery from blindness. *Psychological Science, 20*(12), 1484–1491. doi:10.1111/j.1467-9280.2009.02471.x

Otteson, D. C., & Hitchcock, P. F. (2003). Stem cells in the teleost retina: Persistent neurogenesis and injury-induced regeneration. *Vision Research, 43*(8), 927–936.

P

Paaby, A. B., & Rockman, M. V. (2013). The many faces of pleiotropy. *Trends in Genetics, 29*(2), 66–73. doi:10.1016/j.tig.2012.10.010

Palacios, I. M. (2007). How does an mRNA find its way? Intracellular localisation of transcripts. *Seminars in Cell and Developmental Biology 18*(2), 163–170.

Paredes, M. F., James, D., Gil-Perotin, S., Kim, H., Cotter, J. A., Ng, C.,…Alvarez-Buylla, A. (2016). Extensive migration of young neurons into the infant human frontal lobe. *Science, 354*(6308): aaf7073-1–7.

Park, H., & Poo, M. M. (2013). Neurotrophin regulation of neural circuit development and function. *Nature Reviews Neuroscience, 14*(1), 7–23. doi:10.1038/nrn3379

Parnavelas, J. G. (2000). The origin and migration of cortical neurones: New vistas. *Trends in Neuroscience, 23*(3), 126–131.

Pasinelli, P., & Brown, R. H. (2006). Molecular biology of amyotrophic lateral sclerosis: Insights from genetics. *Nature Reviews Neuroscience, 7*(9), 710–723. doi:10.1038/nrn1971

Pasterski, V. L., Geffner, M. E., Brain, C., Hindmarsh, P., Brook, C., & Hines, M. (2005). Prenatal hormones and postnatal socialization by parents as determinants of male-typical toy play in girls with congenital adrenal hyperplasia. *Child Development, 76*(1), 264–278. doi:10.1111/j.1467-8624.2005.00843.x

Paton, J. A., & Nottebohm, F. N. (1984). Neurons generated in the adult brain are recruited into functional circuits. *Science, 225*(4666), 1046–1048.

Paul, D. (2000). A double-edged sword. *Nature, 405*(6786), 515. doi:10.1038/35014676

Paul, L. K., Brown, W. S., Adolphs, R., Tyszka, J. M., Richards, L. J., Mukherjee, P., & Sherr, E. H. (2007). Agenesis of the corpus callosum: Genetic, developmental and functional aspects of connectivity. *Nature Reviews Neuroscience, 8*(4), 287–299. doi:10.1038/nrn2107

Pearlman, A. L., & Sheppard, A. M. (1996). Extracellular matrix in early cortical development. *Progress in Brain Research, 108*, 117–134.

Pencea, V., Bingaman, K. D., Freedman, L. J., & Luskin, M. B. (2001). Neurogenesis in the subventricular zone and rostral migratory stream of the neonatal and adult primate forebrain. *Experimental Neurology, 172*(1), 1–16. doi:10.1006/exnr.2001.7768

Penn, A. A., Riquelme, P. A., Feller, M. B., & Shatz, C. J. (1998). Competition in retinogeniculate patterning driven by spontaneous activity. *Science, 279*(5359), 2108–2112.

Perez, J. D., Rubinstein, N. D., Fernandez, D. E., Santoro, S. W., Needleman, L. A., Ho-Shing, O., … Dulac, C. (2015). Quantitative and functional interrogation of parent-of-origin allelic expression biases in the brain. *eLife, 4*, e07860. doi:10.7554/eLife.07860

Peschansky, V. J., Pastori, C., Zeier, Z., Motti, D., Wentzel, K., Velmeshev, D., … Wahlestedt, C. (2015). Changes in expression of the long non-coding RNA FMR4 associate with altered gene expression during differentiation of human neural precursor cells. *Frontiers in Genetics, 6*, 263. doi:10.3389/fgene.2015.00263

Petanjek, Z., Judaš, M., Šimic, G., Rasin, M. R., Uylings, H. B. M., Rakic, P., & Kostovic, I. (2011). Extraordinary neoteny of synaptic spines in the human prefrontal cortex. *Proceedings of the National Academy of Sciences USA, 108*(32), 13281–13286.

Peters, A., Palay, S. L., & Webster, H. deF. (1991). *The fine structure of the nervous system: Neurons and their supporting cells* (3rd ed.). New York: Oxford University Press.

Philippidou, P., & Dasen, J. S. (2013). Hox genes: Choreographers in neural development, architects of circuit organization. *Neuron, 80*(1), 12–34. doi:10.1016/j.neuron.2013.09.020

Phoenix, C. H., Goy, R. W., Gerall, A. A., & Young, W. C. (1959). Organizing action of prenatally administered testosterone propionate on the tissues mediating mating behavior in the female guinea pig. *Endocrinology, 65*(3), 369–382. doi:10.1210/endo-65-3-369

Pietschnig, J., & Voracek, M. (2015). One century of global IQ gains: A formal meta-analysis of the Flynn effect (1909–2013). *Perspectives on Psychological Science, 10*(3), 282–306. doi:10.1177/1745691615577701

Pin, T., Eldridge, B., & Galea, M. P. (2007). A review of the effects of sleep position, play position, and equipment use on motor development in infants. *Developmental Medicine and Child Neurology, 49*(11), 858–867. doi:10.1111/j.1469-8749.2007.00858.x

Pines, M. (ed.). (1992.) *From egg to adult: What worms, flies, and other creatures can teach us about the switches that control human development* (pp. 30–38). Bethesda, MD: Howard Hughes Medical Institute.

Pittman, R., & Oppenheim, R. W. (1979). Cell death of motoneurons in the chick embryo spinal cord. IV. Evidence that a functional neuromuscular interaction is involved in the regulation of naturally occurring cell death and the stabilization of synapses. *Journal of Comparative Neurology, 187*(2), 425–446.

Placzek, M., Jessell, T. M., & Dodd, J. (1993). Induction of floor plate differentiation by contact-dependent, homeogenetic signals. *Development, 117*(1), 205–218.

Placzek, M., Tessier-Lavigne, M., Yamada, T., Jessell, T., & Dodd, J. (1990). Mesodermal control of neural cell identity: Floor plate induction by the notochord. *Science, 250*, 985–988.

Plomin, R., Corley, R., DeFries, J. C., & Fulker, D. W. (1990). Individual differences in television viewing in early childhood: Nature as well as nurture. *Psychological Science, 1*(6), 371–377.

Poiani, A. (2010). *Animal homosexuality: A biosocial perspective.* Cambridge: Cambridge University Press.

Porman, P. E., & Savage-Smith, E. (2007). *Medieval Islamic medicine.* Edinburgh: Edinburgh University Press.

Pothos, E. M., & Wills, A. J. (2011). *Formal approaches in categorization.* Cambridge: Cambridge University Press.

Poulson, D. F. (1937). Chromosomal deficiencies and the embryonic development of *Drosophila melanogaster. Proceedings of the National Academy of Sciences USA, 23*(3), 133–137.

Pradel, J., & White, R. A. (1998). From selectors to realizators. *International Journal of Developmental Biology, 42*(3), 417–421.

Pramatarova, A., Laganiere, J., Roussel, J., Brisebois, K., & Rouleau, G. A. (2001). Neuron-specific expression of mutant superoxide dismutase 1 in transgenic mice does not lead to motor impairment. *Journal of Neuroscience, 21*(10), 3369–3374.

Priess, J. R., & Thomson, J. N. (1987). Cellular interactions in early *C. elegans* embryos. *Cell, 48*(2), 241–250.

PRO-ED Inc. (1999). *Speech and language milestone chart.* Retrieved from http://www.ldonline.org/article/6313.

Probst, M. (1901). Über den Bau des vollständing balkenlosen Grosshirns sowie uber mikrogyrie und heterotopie der grauen Substanz. *Archiv für Psychiatrie und Nervenkrankheiten, 34*(3), 709–786.

Pultz, M. A., Diederich, R. J., Cribbs, D. L., & Kaufman, T. C. (1988). The *proboscipedia* locus of the *Antennapedia* complex: A molecular and genetic analysis. *Genes and Development, 2*(7), 901–920.

Purves, D. (1975). Functional and structural changes in mammalian sympathetic neurones following interruption of their axons. *Journal of Physiology, 252*(2), 429–463.

Purves, D. (1983). Modulation of neuronal competition by postsynaptic geometry in autonomic ganglia. *Trends in Neurosciences, 6*(1), 10–16.

Purves, D. (1994). *Neural activity and the growth of the brain.* New York: Cambridge University Press.

Purves, D., & Hume, R. I. (1981). The relation of postsynaptic geometry to the number of presynaptic axons that innervate autonomic ganglion cells. *Journal of Neuroscience, 1*(5), 441–452.

Purves, D., & Lichtman, J. W. (1978). Formation and maintenance of synaptic connections in autonomic ganglia. *Physiological Reviews, 58*(4), 821–862.

Purves, D., & Lichtman, J. W. (1985). *Principles of neural development*. Sunderland, MA: Sinauer Associates.

Purves, D., & Sanes, J. R. (1987). The 1986 Nobel Prize in Physiology or Medicine. *Trends in Neurosciences, 10*(6), 231–235.

Puts, D. A., Jordan, C. L., & Breedlove, S. M. (2006). O brother, where art thou? The fraternal birth-order effect on male sexual orientation. *Proceedings of the National Academy of Sciences USA, 103*(28), 10531–10532. doi:10.1073/pnas.0604102103

Q

Quignon, P., Rimbault, M., Robin, S., & Galibert, F. (2012). Genetics of canine olfaction and receptor diversity. *Mammalian Genome, 23*(1–2), 132–143. doi:10.1007/s00335-011-9371-1

Quinonez, S. C., & Innis, J. W. (2014). Human HOX gene disorders. *Molecular Genetics and Metabolism, 111*(1), 4–15. doi:10.1016/j.ymgme.2013.10.012

R

Rakic, P. (1971a). Guidance of neurons migrating to the fetal monkey neocortex. *Brain Research, 33*, 471–476.

Rakic, P. (1971b). Neuron-glia relationship during granule cell migration in developing cerebellar cortex: A Golgi and electron microscopic study in *Macacus rhesus*. *Journal of Comparative Neurology, 141*(3), 283–312. doi:10.1002/cne.901410303

Rakic, P. (1974.) Neurons in rhesus monkey visual cortex: Systematic relation between time of origin and eventual disposition. *Science, 183*, 425–427.

Rakic, P. (1985). Limits of neurogenesis in primates. *Science, 227*(4690), 1054–1056.

Rakic, P. (2009). Evolution of the neocortex: A perspective from developmental biology. *Nature Reviews Neuroscience, 10*(10), 724–735. doi:10.1038/nrn2719

Ramón y Cajal, S. (1890). A quelle epoque apparaissent les expansions des cellule nerveuses de la moelle epinere du poulet? *Anatomischer Anzeiger, 5*(21–22), 609–613; 631–639.

Ramón y Cajal, S. (1904). *Histoiogie du Systeme Nerveux de L 'Homme et des Vertebres* (Vol. 1). Madrid: Consejo Superior de Investigaciones Cientificas.

Ramón y Cajal, S. (1928). *Degeneration and regeneration in the nervous system*. New York: Hafner.

Ramus, F., Hauser, M. D., Miller, C., Morris, D., & Mehler, J. (2000). Language discrimination by human newborns and by cotton-top tamarin monkeys. *Science, 288*(5464), 349–351.

Rana, Z. A., Gundersen, K., & Buonanno, A. (2009). The ups and downs of gene regulation by electrical activity in skeletal muscles. *Journal of Muscle Research and Cell Motility, 30*(7–8), 255–260. doi:10.1007/s10974-010-9200-2

Rand, M. N., & Breedlove, S. M. (1987). Ontogeny of functional innervation of bulbocavernosus muscles in male and female rats. *Brain Research, 430*(1), 150–152.

Ransohoff, R. M. (2012). Animal models of multiple sclerosis: The good, the bad and the bottom line. *Nature Neuroscience, 15*(8), 1074–1077. doi:10.1038/nn.3168

Rash, B. G., & Richards, L. J. (2001). A role for cingulate pioneering axons in the development of the corpus callosum. *Journal of Comparative Neurology, 434*(2), 147–157.

Raymond, P. A., & Easter, S. S., Jr. (1983). Postembryonic growth of the optic tectum in goldfish. I. Location of germinal cells and numbers of neurons produced. *Journal of Neuroscience, 3*(5), 1077–1091.

Ready, D. F., Hanson, T. E., & Benzer, S. (1976). Development of the *Drosophila* retina, a neurocrystalline lattice. *Developmental Biology, 53*(2), 217–240.

Reape, T. J., & McCabe, P. F. (2008). Apoptotic-like programmed cell death in plants. *New Phytologist, 180*(1), 13–26. doi:10.1111/j.1469-8137.2008.02549.x

Reaume, A. G., Elliott, J. L., Hoffman, E. K., Kowall, N. W., Ferrante, R. J., Siwek, D. F., … Snider, W. D. (1996). Motor neurons in Cu/Zn superoxide dismutase-deficient mice develop normally but exhibit enhanced cell death after axonal injury. *Nature Genetics, 13*(1), 43–47. doi:10.1038/ng0596-43

Redfern, P. A. (1970). Neuromuscular transmission in newborn rats. *Journal of Physiology, 209*(3), 701–709.

Redline, R. W., Neish, A., Holmes, L. B., & Collins, T. (1992). Homeobox genes and congenital malformations. *Laboratory Investigation, 66*(6), 659–670.

Reh, T. A., & Constantine-Paton, M. (1985). Eye-specific segregation requires neural activity in three-eyed *Rana pipiens*. *Journal of Neuroscience, 5*(5), 1132–1143.

Reichert, H. (2002). Conserved genetic mechanisms for embryonic brain patterning. *International Journal of Developmental Biology, 46*(1), 81–87.

Reil, J. C. (1812). Die vordere commissur im grossen gehirn. *Archiv für die Physiologie., 11*, 89–100.

Reiner, W. G. (2004). Psychosexual development in genetic males assigned female: The cloacal exstrophy experience. *Child and Adolescent Psychiatric Clinics of North America, 13*(3), 657–674, ix. doi:10.1016/j.chc.2004.02.009

Reiner, W. G., & Gearhart, J. P. (2004). Discordant sexual identity in some genetic males with cloacal exstrophy assigned to female sex at birth. *New England Journal of Medicine, 350*(4), 333–341. doi:10.1056/NEJMoa022236

Reinke, R., & Zipursky, S. L. (1988). Cell-cell interaction in the *Drosophila* retina: The *bride of sevenless* gene is required in photoreceptor cell R8 for R7 cell development. *Cell, 55*(2), 321–330.

Reiter, L. T., Potocki, L., Chien, S., Gribskov, M., & Bier, E. (2001). A systematic analysis of human disease-associated gene sequences in *Drosophila melanogaster*. *Genome Research, 11*(6), 1114–1125. doi:10.1101/gr.169101

Rice, D. S., & Curran, T. (2001). Role of the reelin signaling pathway in central nervous system development. *Annual Review of Neuroscience, 24*, 1005–1039. doi:10.1146/annurev.neuro.24.1.1005

Richards, L. J., Plachez, C., & Ren, T. (2004). Mechanisms regulating the development of the corpus callosum and its agenesis in mouse and human. *Clinical Genetics, 66*(4), 276–289. doi:10.1111/j.1399-0004.2004.00354.x

Richardson, M. K., Hanken, J., Gooneratne, M. L., Pieau, C., Raynaud, A., Selwood, L., & Wright, G. M. (1997). There is no highly conserved embryonic stage in the vertebrates: Implications for current theories of evolution and development. *Anatomy and Embryology (Berlin), 196*(2), 91–106.

Richardus, J. H., & Oskam, L. (2015). Protecting people against leprosy: Chemoprophylaxis and immuno-prophylaxis. *Clinics in Dermatology, 33*(1), 19–25. doi:10.1016/j.clindermatol.2014.07.009

Richter, J. D., Bassell, G. J., & Klann, E. (2015). Dysregulation and restoration of translational homeostasis in fragile X syndrome. *Nature Reviews Neuroscience, 16*, 595–605. doi:10.1038/nrn4001

Riddle, R. D., Johnson, R. L., Laufer, E., & Tabin, C. (1993). *Sonic hedgehog* mediates the polarizing activity of the ZPA. *Cell, 75*(7), 1401–1416.

Rimer, M. (2007). Neuregulins at the neuromuscular synapse: Past, present, and future. *Journal of Neuroscience Research, 85*(9), 1827–1833. doi:10.1002/jnr.21237

Ripolles, P., Rojo, N., Grau-Sanchez, J., Amengual, J. L., Camara, E., Marco-Pallares, J., … Rodriguez-Fornells, A. (2015). Music supported therapy promotes motor plasticity in individuals with chronic stroke. *Brain Imaging and Behavior.* doi:10.1007/s11682-015-9498-x

Robb, L., & Tam, P. P. L. (2004). Gastrula organiser and embryonic patterning in the mouse. *Seminars in Cell and Developmental Biology, 15*, 543–554.

Robberecht, W., & Philips, T. (2013). The changing scene of amyotrophic lateral sclerosis. *Nature Reviews Neuroscience, 14*(4), 248–264. doi:10.1038/nrn3430

Roche, F. K., Marsick, B. M., & Letourneau, P. C. (2009). Protein synthesis in distal axons is not required for growth cone responses to guidance cues. *Journal of Neuroscience, 29*(3), 638–652. doi:10.1523/JNEUROSCI.3845-08.2009

Roebuck, T. M., Mattson, S. N., & Riley, E. P. (1998). A review of the neuroanatomical findings in children with fetal alcohol syndrome or prenatal exposure to alcohol. *Alcoholism, Clinical and Experimental Research, 22*(2), 339–344.

Roelink, H., Augsburger, A., Heemskerk, J., Korzh, V., Norlin, S., Ruiz i Altaba, A., … Jessell, T. M., et al. (1994). Floor plate and motor neuron induction by *vhh-1*, a vertebrate homolog of *hedgehog* expressed by the notochord. *Cell, 76*(4), 761–775.

Rogers, E. M., Brennan, C. A., Mortimer, N. T., Cook, S., Morris, A. R., & Moses, K. (2005). *Pointed* regulates an eye-specific transcriptional enhancer in the *Drosophila hedgehog* gene, which is required for the movement of the morphogenetic furrow. *Development, 132*(21), 4833–4843. doi:10.1242/dev.02061

Rohlf, M. (2014). Immanuel Kant. In *Stanford Encyclopedia of Philosophy.* Retrieved from plato.stanford.edu/archives/sum2014/entries/kant/.

Role, L. W., Matossian, V. R., O'Brien, R. J., & Fischbach, G. D. (1985). On the mechanism of acetylcholine receptor accumulation at newly formed synapses on chick myotubes. *Journal of Neuroscience, 5*(8), 2197–2204.

Rosen, D. R., Siddique, T., Patterson, D., Figlewicz, D. A., Sapp, P., Hentati, A., … Brown, R. H., Jr. (1993). Mutations in Cu/Zn superoxide dismutase gene are associated with familial amyotrophic lateral sclerosis. *Nature, 362*(6415), 59–62. doi:10.1038/362059a0

Rosenzweig, M. R., Krech, D., Bennett, E. L., & Diamond, M. C. (1962). Effects of environmental complexity and training on brain chemistry and anatomy: A replication and extension. *Journal of Comparative and Physiological Psychology, 55*, 429–437.

Roth, K. A., Kuan, C., Haydar, T. F., D'Sa-Eipper, C., Shindler, K. S., Zheng, T. S., … Rakic, P. (2000). Epistatic and independent functions of caspase-3 and Bcl-X(L) in developmental programmed cell death. *Proceedings of the National Academy of Sciences USA, 97*(1), 466–471.

Rubin, G. M. (1989). Development of the *Drosophila* retina: Inductive events studied at single-cell resolution. *Cell, 57*, 519–520.

Rudell, J. B., & Ferns, M. J. (2013). Regulation of muscle acetylcholine receptor turnover by beta subunit tyrosine phosphorylation. *Developmental Neurobiology, 73*(5), 399–410. doi:10.1002/dneu.22070

Ryan, T. J., & Grant, S. G. (2009). The origin and evolution of synapses. *Nature Reviews Neuroscience, 10*(10), 701–712. doi:10.1038/nrn2717

S

Salz, H. K. (2011). Sex determination in insects: A binary decision based on alternative splicing. *Current Opinion in Genetics & Development, 21*(4), 395–400. doi:10.1016/j.gde.2011.03.001

Sanders, A. R., Martin, E. R., Beecham, G. W., Guo, S., Dawood, K., Rieger, G., … Bailey, J. M. (2015). Genome-wide scan demonstrates significant linkage for male sexual orientation. *Psychological Medicine, 45*(7), 1379–1388. doi:10.1017/S0033291714002451

Sanes, J. R. 1983. Roles of extracellular matrix in neural development. *Annual Review of Physiology, 45*, 581–600.

Sanes, J. R. (1989). Extracellular matrix molecules that influence neural development. *Annual Review of Neuroscience, 12*, 491–516.

Sanes, J. R., Johnson, Y. R., Kotzbauer, P. T., Mudd, J., Hanley, T., Martinou, J. C., & Merlie, J. P. (1991). Selective expression of an acetylcholine receptor-lacZ transgene in synaptic nuclei of adult muscle fibers. *Development, 113*(4), 1181–1191.

Sanes, J. R., Marshall, L. M., & McMahan, U. J. (1978). Reinnervation of muscle fiber basal lamina after removal of myofibers. Differentiation of regenerating axons at original synaptic sites. *Journal of Cell Biology, 78*(1), 176–198.

Santarelli, L., Saxe, M., Gross, C., Surget, A., Battaglia, F., Dulawa, S., … Hen, R. (2003). Requirement of hippocampal neurogenesis for the behavioral effects of antidepressants. *Science, 301*(5634), 805–809. doi:10.1126/science.1083328

Sato, M., Suzuki, T., & Nakai, Y. (2013). Waves of differentiation in the fly visual system. *Developmental Biology, 380*(1), 1–11. doi:10.1016/j.ydbio.2013.04.007

Satterlie, R. A. (2011). Do jellyfish have central nervous systems? *Journal of Experimental Biology, 214*(Pt. 8), 1215–1223. doi:10.1242/jeb.043687

Saxe, M. D., Battaglia, F., Wang, J. W., Malleret, G., David, D. J., Monckton, J. E., … Drew, M. R. (2006). Ablation of hippocampal neurogenesis impairs contextual fear conditioning and synaptic plasticity in the dentate gyrus. *Proceedings of the National Academy of Sciences USA 103*(46), 17501–17506. doi:10.1073/pnas.0607207103

Scheiffele, P., Fan, J., Choih, J., Fetter, R., & Serafini, T. (2000). Neuroligin expressed in nonneuronal cells triggers presynaptic development in contacting axons. *Cell, 101*(6), 657–669.

Scheller, R. H., & Axel, R. (1984). How genes control an innate behavior. *Scientific American, 250*(3), 54–62.

Schiaffino, S., Sandri, M., & Murgia, M. (2007). Activity-dependent signaling pathways controlling muscle diversity and plasticity. *Physiology (Bethesda), 22,* 269–278. doi:10.1152/physiol.00009.2007

Schmid, S. M., Kott, S., Sager, C., Huelsken, T., & Hollmann, M. (2009). The glutamate receptor subunit delta2 is capable of gating its intrinsic ion channel as revealed by ligand binding domain transplantation. *Proceedings of the National Academy of Sciences USA, 106*(25), 10320–10325.

Schmidt, J. T. (1978). Retinal fibers alter tectal positional markers during the expansion of the retinal projection in goldfish. *Journal of Comparative Neurology, 177*(2), 279–295. doi:10.1002/cne.901770207

Schmidt, J. T. (1983). Regeneration of the retinotectal projection following compression onto a half tectum in goldfish. *Journal of Embryology and Experimental Morphology, 77,* 39–51.

Schmidt, J. T., & Edwards, D. L. (1983). Activity sharpens the map during the regeneration of the retinotectal projection in goldfish. *Brain Research, 269*(1), 29–39.

Schmidt, J. T., & Eisele, L. E. (1985). Stroboscopic illumination and dark rearing block the sharpening of the regenerated retinotectal map in goldfish. *Neuroscience, 14*(2), 535–546.

Scholze, A. R., & Barres, B. A. (2012). A Nogo signal coordinates the perfect match between myelin and axons. *Proceedings of the National Academy of Sciences USA, 109*(4), 1003–1004. doi:10.1073/pnas.1120301109

Schotzinger, R. J., & Landis, S. C. (1988). Cholinergic phenotype developed by noradrenergic sympathetic neurons after innervation of a novel cholinergic target in vivo. *Nature, 335*(6191), 637–639. doi:10.1038/335637a0

Schwab, M. E., & Thoenen, H. (1985). Dissociated neurons regenerate into sciatic but not optic nerve explants in culture irrespective of neurotrophic factors. *Journal of Neuroscience, 5*(9), 2415–2423.

Schwanzel-Fukuda, M., & Pfaff, D. W. (1989). Origin of luteinizing hormone-releasing hormone neurons. *Nature, 338*(6211), 161–164. doi:10.1038/338161a0

Scott, M. P., & Weiner, A. J. (1984). Structural relationships among genes that control development: Sequence homology between the *Antennapedia, Ultrabithorax,* and *fushi tarazu* loci of *Drosophila. Proceedings of the National Academy of Sciences USA, 81*(13), 4115–4119.

Searle, L. V. (1949). The organization of hereditary maze-brightness and maze-dullness. *Genetic Psychology Monographs, 39*(2), 297–325.

Sebastian, A., Jung, P., Krause-Utz, A., Lieb, K., Schmahl, C., & Tuscher, O. (2014). Frontal dysfunctions of impulse control: A systematic review in borderline personality disorder and attention-deficit/hyperactivity disorder. *Frontiers in Human Neuroscience, 8,* 698.

Seeger, M. A., & Kaufman, T. C. (1990). Molecular analysis of the *bicoid* gene from *Drosophila pseudoobscura*: Identification of conserved domains within coding and noncoding regions of the *bicoid* mRNA. *EMBO Journal, 9*(9), 2977–2987.

Seib, D. R., & Martin-Villalba, A. (2014). Neurogenesis in the normal ageing hippocampus: A mini-review. *Gerontology.* doi:10.1159/000368575

Seidel, F. (1952). Die Entwicklungspotenzen einer isolierten Blastomere des Zweizellenstadiums im Saugetierei. *Naturwissenschaften, 39,* 355–356.

Serafini, G., Hayley, S., Pompili, M., Dwivedi, Y., Brahmachari, G., Girardi, P., & Amore, M. (2014). Hippocampal neurogenesis, neurotrophic factors and depression: Possible therapeutic targets? *CNS & Neurological Disorders—Drug Targets, 13*(10), 1708–1721.

Serafini, T., Kennedy, T. E., Galko, M. J., Mirzayan, C., Jessell, T. M., & Tessier-Lavigne, M. (1994). The netrins define a family of axon outgrowth-promoting proteins homologous to *C. elegans* UNC-6. *Cell, 78*(3), 409–424.

Serizawa, S., Miyamichi, K., Takeuchi, H., Yamagishi, Y., Suzuki, M., & Sakano, H. (2006). A neuronal identity code for the odorant receptor-specific and activity-dependent axon sorting. *Cell, 127*(5), 1057–1069. doi:10.1016/j.cell.2006.10.031

Serrano, P. A., Beniston, D. S., Oxonian, M. G., Rodriguez, W. A., Rosenzweig, M. R., & Bennett, E. L. (1994). Differential effects of protein kinase inhibitors and activators on memory formation in the 2-day-old chick. *Behavioral and Neural Biology, 61*(1), 60–72.

Shalem, O., Sanjana, N. E., & Zhang, F. (2015). High-throughput functional genomics using CRISPR-Cas9. *Nature Reviews Genetics, 16*(5), 299–311.

Shatz, C. J. (1983). The prenatal development of the cat's retinogeniculate pathway. *Journal of Neuroscience, 3*(3), 482–499.

Shatz, C. J. (1992). The developing brain. *Scientific American, 267*(3), 60–67.

Shepherd, I. T., Luo, Y., Lefcort, F., Reichardt, L. F., & Raper, J. A. (1997). A sensory axon repellent secreted from ventral spinal cord explants is neutralized by antibodies raised against collapsin-1. *Development, 124*(7), 1377–1385.

Shimizu, E., Tang, Y. P., Rampon, C., & Tsien, J. Z. (2000). NMDA receptor-dependent synaptic reinforcement as a crucial process for memory consolidation. *Science, 290*(5494), 1170–1174.

Shu, T., Sundaresan, V., McCarthy, M. M., & Richards, L. J. (2003). Slit2 guides both precrossing and postcrossing callosal axons at the midline in vivo. *Journal of Neuroscience, 23*(22), 8176–8184.

Sidman, R. L. (1965). *Catalog of the neurological mutants of the mouse.* Cambridge, MA: Harvard University Press.

Sidman, R. L. (1968). Development of interneuronal connections in brains of mutant mice. In F. D. Carlson (Ed.), *Physiological and biochemical aspects of nervous integration* (pp. 163–193). Englewood, NJ: Prentice-Hall.

Sidman, R. L., Lane, P. W., & Dickie, M. M. (1962). *Staggerer,* a new mutation in the mouse affecting the cerebellum. *Science, 137*(3530), 610–612.

Siffrin, V., Vogt, J., Radbruch, H., Nitsch, R., & Zipp, F. (2010). Multiple sclerosis—candidate mechanisms underlying CNS atrophy. *Trends in Neuroscience, 33*(4), 202–210. doi:10.1016/j.tins.2010.01.002

Sikl, R., Simeccek, M., Porubanova-Norquist, M., Bezdicek, O., Kremlacek, J., Stodulka, P., … Ostrovsky, Y. (2013). Vision after 53 years of blindness. *i-Perception, 4*(8), 498–507. doi:10.1068/i0611

Silver, J., & Miller, J. H. (2004). Regeneration beyond the glial scar. *Nature Reviews Neuroscience, 5*(2), 146–156. doi:10.1038/nrn1326

Silver, J., & Ogawa, M. Y. (1983). Postnatally induced formation of the corpus callosum in acallosal mice on glia-coated cellulose bridges. *Science, 220*(4601), 1067–1069.

Simoes-Costa, M., & Bronner, M. E. (2015). Establishing neural crest identity: A gene regulatory recipe. *Development, 142*(2), 242–257. doi:10.1242/dev.105445

Simpson, J. A., Collins, W. A., Tran, S., & Haydon, K. C. (2007). Attachment and the experience and expression of emotions in romantic relationships: A developmental perspective. *Journal of Personality and Social Psychology, 92*(2), 355–367. doi:10.1037/0022-3514.92.2.355

Singh, N., Morlock, H., & Hanes, S. D. (2011). The *Bin3* RNA methyltransferase is required for repression of caudal translation in the *Drosophila* embryo. *Developmental Biology, 352*(1), 104–115. doi:10.1016/j.ydbio.2011.01.017

Singhal, N., & Martin, P. T. (2011). Role of extracellular matrix proteins and their receptors in the development of the vertebrate neuromuscular junction. *Developmental Neurobiology, 71*(11), 982–1005. doi:10.1002/dneu.20953

Sivanantharajah, L., & Percival-Smith, A. (2015). Differential pleiotropy and HOX functional organization. *Developmental Biology, 398*(1), 1–10. doi:10.1016/j.ydbio.2014.11.001

Skeath, J. B., Panganiban, G., Selegue, J., & Carroll, S. B. (1992.) Gene regulation in two dimensions: the proneural achaete and scute genes are controlled by combinations of axis-patterning genes through a common intergenic control region. *Genes & Development*, 6(12B), 2606–2619.

Smith, A., & Sugar, O. (1975). Development of above normal language and intelligence 21 years after left hemispherectomy. *Neurology, 25*(9), 813–818.

Smith, K. M., Ohkubo, Y., Maragnoli, M. E., Rasin, M. R., Schwartz, M. L., Sestan, N., & Vaccarino, F. M. (2006). Midline radial glia translocation and corpus callosum formation require FGF signaling. *Nature Neuroscience, 9*(6), 787–797. doi:10.1038/nn1705

Smith, N., & Quinton, R. (2012). Kallmann syndrome. *BMJ, 345*, e6971. doi:10.1136/bmj.e6971

Snider, W. D. (1994). Functions of the neurotrophins during nervous system development: What the knockouts are teaching us. *Cell, 77*(5), 627–638.

Southwell, D. G., Paredes, M. F., Galvao, R. P., Jones, D. L., Froemke, R. C., Sebe, J. Y., … Alvarez-Buylla, A. (2012). Intrinsically determined cell death of developing cortical interneurons. *Nature, 491*(7422), 109–113. doi:10.1038/nature11523

Spalding, D. A. (1872). On instinct. *Nature, 6*(154), 485–486.

Spemann, H. (1921). Über die Erzeugung tierischer Chimären durch heteroplastische embryonale Transplantation zwischen *Triton cristatus* und *Triton taeniatus. Archiv für Entwicklungsmechanik der Organismen, 48*, 533–570.

Spemann, H. (1938). *Embryonic development and induction.* New Haven, CT: Yale University Press.

Spemann, H., & Mangold, H. (1924). Ueber die Induktion von Embryonalanlagen durch Implantation artfremder Organisatore [On the induction of embryonic anlagen by implantation of organizers from different species]. *Archiv für Entwicklungsmechanik der Organismen, 100*, 599–638.

Spencer, S. J., Steele, C. M., & Quinn, D. M. (1999). Stereotype threat and women's math performance. *Journal of Experimental Social Psychology: 35*, 4–28.

Sperry, R. W. (1943). Effect of 180 degree rotation of the retinal field on visuomotor coordination. *Journal of Experimental Zoology, 92*(3), 263–279.

Sperry, R. W. (1949). Reimplantation of eyes in fishes (*Bahtygobius soporator*) with recovery of vision. *Proceedings of the Society for Experimental Biology and Medicine, 71*(1), 80.

Sperry, R. W. (1963). Chemoaffinity in the orderly growth of nerve fiber patterns and connections. *Proceedings of the National Academy of Sciences USA, 50*, 703–710.

Spirov, A., Fahmy, K., Schneider, M., Frei, E., Noll, M., & Baumgartner, S. (2009). Formation of the bicoid morphogen gradient: An mRNA gradient dictates the protein gradient. *Development, 136*(4), 605–614. doi:10.1242/dev.031195

Spitzer, N. C. (1979). Ion channels in development. *Annual Review of Neuroscience, 2*, 363–397. doi:10.1146/annurev.ne.02.030179.002051

Spitzer, N. C. (1981). Development of membrane properties in vertebrates. *Trends in Neurosciences, 4*, 169–172.

Spratford, C. M., & Kumar, J. P. (2014). Hedgehog and extramacrochaetae in the *Drosophila* eye: An irresistible force meets an immovable object. *Fly (Austin), 8*(1), 36–42. doi:10.4161/fly.27691

St.-Hilaire, E. G. (1822). Considérations générales sur la vertèbre. *Mémoires du Muséum national d'Histoire Naturelle, 9*, 89–119.

Staubli, U. V. (1995). Parallel properties of long-term potentiation and memory. In J. L. McGaugh, N. M. Wienberger, & G. Lynch (Eds.), *Brain and memory: Modulation and mediation of neuroplasticity* (pp. 303–318). New York: Oxford University Press.

Stearns, F. W. (2010). One hundred years of pleiotropy: A retrospective. *Genetics, 186*(3), 767–773. doi:10.1534/genetics.110.122549

Steele, C. M., & Aronson, J. (1995). Stereotype threat and the intellectual test performance of African Americans. *Journal of Personality and Social Psychology, 69*(5), 797–811.

Steinberg, L. (2013). The influence of neuroscience on US Supreme Court decisions about adolescents' criminal culpability. *Nature Reviews Neuroscience, 14*(7), 513–518.

Stemple, D. L. (2005). Structure and function of the notochord: An essential organ for chordate development. *Development, 132*(11), 2503–2512. doi:10.1242/dev.01812

Stenkamp, D. L., Barthel, L. K., & Raymond, P. A. (1997). Spatiotemporal coordination of rod and cone photo-receptor differentiation in goldfish retina. *Journal of Comparative Neurology, 382*(2), 272–284.

Stepniak, E., Radice, G. L., & Vasioukhin, V. (2009). Adhesive and signaling functions of cadherins and catenins in vertebrate development. *Cold Spring Harbor Perspectives in Biology, 1*(5), a002949. doi:10.1101/cshperspect.a002949

Stewart, T. A., & Mintz, B. (1981). Successful generations of mice produced from an established culture line of euploid teratocarcinoma cells. *Proceedings of the National Academy of Sciences USA, 78*, 6314–6318.

Stratton, G. M. (1896). Some preliminary experiments on vision without inversion on the retinal image. *Psychology Review, 3*, 611–617.

Stratton, G. M. (1897). Vision without inversion of the retinal image [part 1]. *Psychology Review, 4*(4), 341–360.

Straznicky, K., & Gaze, R. M. (1972). The development of the tectum in *Xenopus laevis*: An autoradiographic study. *Journal of Embryology and Experimental Morphology, 28*(1), 87–115.

Stryker, M. P., & Harris, W. A. (1986). Binocular impulse blockade prevents the formation of ocular dominance columns in cat visual cortex. *Journal of Neuroscience, 6*(8), 2117–2133.

Stryker, M. P., Chapman, B., Miller, K. D., & Zahs, K. R. (1990). Experimental and theoretical studies of the organization of afferents to single orientation columns in visual cortex. *Cold Spring Harbor Symposia on Quantitative Biology, 55*, 515–527.

Sulston, J. E., & Horvitz, H. R. (1977). Postembryonic cell lineages of the nematode, *Caenorhabditis elegans*. *Developmental Biology, 56*(1), 110–156.

Sulston, J. E., & White, J. G. (1980). Regulation and cell autonomy during postembryonic development of *Caenorhabditis elegans*. *Developmental Biology, 78*(2), 577–597.

Sulston, J. E., Schierenberg, J., White, J., & Thomson, N. (1983). The embryonic cell lineage of the nematode *Caenorhabditis elegans*. *Developmental Biology, 100*, 64–119.

Suomi, S. J., Delizio, R., & Harlow, H. F. (1976). Social reha-bilitation of separation-induced depressive disorders in monkeys. *American Journal of Psychiatry, 133*(11), 1279–1285. doi:10.1176/ajp.133.11.1279

Suomi, S. J., Harlow, H. F., & Kimball, S. D. (1971). Behavioral effects of prolonged partial social isolation in the rhesus monkey. *Psychological Reports, 29*(3), 1171–1177. doi:10.2466/pr0.1971.29.3f.1171

Suter, U., Welcher, A. A., Ozcelik, T., Snipes, G. J., Kosaras, B., Francke, U., … Shooter, E. M. (1992). Trembler mouse carries a point mutation in a myelin gene. *Nature, 356*(6366), 241–244. doi:10.1038/356241a0

Syed, M. M., Lee, S., Zheng, J., & Zhou, Z. J. (2004). Stage-dependent dynamics and modulation of spontane-ous waves in the developing rabbit retina. *Journal of Physiology, 560*(Pt. 2), 533–549.

Syed, N., Reddy, K., Yang, D. P., Taveggia, C., Salzer, J. L., Maurel, P., & Kim, H. A. (2010). Soluble neuregu-lin-1 has bifunctional, concentration-dependent effects on Schwann cell myelination. *Journal of Neuroscience, 30*(17), 6122–6131. doi:10.1523/JNEUROSCI.1681-09.2010

Syroid, D. E., Maycox, P. R., Burrola, P. G., Liu, N., Wen, D., Lee, K. F., … Kilpatrick, T. J. (1996). Cell death in the Schwann cell lineage and its regulation by neuregulin. *Proceedings of the National Academy of Sciences USA, 93*(17), 9229–9234.

T

Tait, S. W., & Green, D. R. (2010). Mitochondria and cell death: Outer membrane permeabilization and beyond. *Nature Reviews Molecular Cell Biology, 11*(9), 621–632. doi:10.1038/nrm2952

Takahashi, Y., Sipp, D., & Enomoto, H. (2013). Tissue interactions in neural crest cell development and disease. *Science, 341*(6148), 860–863. doi:10.1126/science.1230717

Tanabe, Y., & Jessell, T. M. (1996). Diversity and pattern in the developing spinal cord. *Science, 274*(5290), 1115–1123.

Taneyhill, L. A., & Schiffmacher, A. T. (2013). Cadherin dynamics during neural crest cell ontogeny. *Progress in Molecular Biology and Transitional Science, 116*, 291–315. doi:10.1016/B978-0-12-394311-8.00013-3

Tang, Y. P., Shimizu, E., Dube, G. R., Rampon, C., Kerchner, G. A., Zhuo, M. … Tsien, J. Z. (1999). Genetic enhance-ment of learning and memory in mice. *Nature, 401*(6748), 63–69.

Tantbirojn, P., Taweevisit, M., Sritippayawan, S., & Uerpairojkit, B. (2008). Diabetic fetopathy associated with bilateral adrenal hyperplasia and ambiguous geni-talia: A case report. *Journal of Medical Case Reports, 2*, 251. doi:10.1186/1752-1947-2-251

Tarabeux, J., Kebir, O., Gauthier, J., Hamdan, F. F., Xiong, L., Piton, A. … Krebs, M. O. (2011). Rare mutations in *N*-methyl-D-aspartate glutamate receptors in autism spectrum disorders and schizophrenia. *Translational Psychiatry, 1*, e55.

Task Force on Sudden Infant Death Syndrome & Moon, R. Y. (2011). SIDS and other sleep-related infant deaths: Expansion of recommendations for a safe infant sleeping environment. *Pediatrics, 128*(5), 1030–1039. doi:10.1542/peds.2011-2284

Taylor, J. S., & Raes, J. (2004). Duplication and divergence: The evolution of new genes and old ideas. *Annual Review of Genetics, 38*, 615–643. doi:10.1146/annurev.genet.38.072902.092831

Tayman, J. (2010). *The colony: The harrowing true story of the exiles of Molokai.* New York: Scribner.

Tessier-Lavigne, M., & Goodman, C. S. (1996). The molecu-lar biology of axon guidance. *Science, 274*(5290), 1123–1133.

Thompson, W. (1983). Synapse elimination in neonatal rat muscle is sensitive to pattern of muscle use. *Nature, 302*(5909), 614–616.

Tian, H., & Ma, M. (2008). Activity plays a role in eliminat-ing olfactory sensory neurons expressing multiple odor-ant receptors in the mouse septal organ. *Molecular and Cellular Neuroscience, 38*(4), 484–488. doi:10.1016/j.mcn.2008.04.006

Timmerman, V., Nelis, E., Van Hul, W., Nieuwenhuijsen, B. W., Chen, K. L., Wang, S., … Van Broeckhoven, C. (1992). The peripheral myelin protein gene *PMP-22* is contained within the Charcot-Marie-Tooth disease type 1A duplication. *Nature Genetics, 1*(3), 171–175. doi:10.1038/ng0692-171

Tinbergen, N. (1958). *Curious naturalists*. New York: Basic Books.

Tishkoff, S. A., Reed, F. A., Friedlaender, F. R., Ehret, C., Ranciaro, A., Froment, A., … Williams, S. M. (2009). The genetic structure and history of Africans and African Americans. *Science, 324*(5930), 1035–1044. doi:10.1126/science.1172257

Tomlinson, A. (1988). Cellular interactions in the *Drosophila* eye. *Development, 104*, 183–193.

Tootell, R. B., Switkes, E., Silverman, M. S., & Hamilton, S. L. (1988). Functional anatomy of macaque striate cortex. II. Retinotopic organization. *Journal of Neuroscience, 8*(5), 1531–1568.

Totenberg, N. (2012, March 20). Do Juvenile Killers Deserve Life Behind Bars? [Radio story transcript] Retrieved from www.npr.org/2012/03/20/148538071/do-juvenile-killers-deserve-life-behind-bars

Traynelis, S. F., Wollmuth, L. P., McBain, C. J., Menniti, F. S., Vance, K. M., Ogden, K. … Dingledine, R. (2010). Glutamate receptor ion channels: Structure, regulation, and function. *Pharmacological Review, 62*(3), 405–496.

Treffert, D. A., & Christensen, D. D. (2005). Inside the mind of a savant. *Scientific American, 293*(6), 108–113.

Treisman, J. E. (2013). Retinal differentiation in *Drosophila*. *Wiley Interdisciplinary Reviews: Developmental Biology, 2*(4), 545–557. doi:10.1002/wdev.100

Tryon, R. C. (1940). Genetic differences in maze learning ability in rats. *Yearbook of the National Society for Studies of Education, 39*, 111–119.

Tryon, R. C. (1942). Individual differences. In F. A. Moss (Ed.), *Comparative psychology* (pp. 330–365). New York: Prentice-Hall.

Tsuchida, T., Ensini, M., Morton, S. B., Baldassare, M., Edlund, T., Jessell, T. M., & Pfaff, S. L. (1994). Topographic organization of embryonic motor neurons defined by expression of LIM homeobox genes. *Cell, 79*(6), 957–970.

Turner, D. L., Snyder, E. Y., & Cepko, C. L. (1990). Lineage-independent determination of cell type in the embryonic mouse retina. *Neuron, 4*(6), 833–845.

U

Ungar, D., & Hughson, F. M. (2003). SNARE protein structure and function. *Annual Review of Cell and Developmental Biology, 19*, 493–517. doi:10.1146/annurev.cellbio.19.110701.155609

Urist, M. R., & Strates, B. S. (1971). Bone morphogenetic protein. *Journal of Dental Research, 50*(6), 1392–1406.

Ushkaryov, Y. A., Petrenko, A. G., Geppert, M., & Sudhof, T. C. (1992). Neurexins: Synaptic cell surface proteins related to the alpha-latrotoxin receptor and laminin. *Science, 257*(5066), 50–56.

Usui-Aoki, K., Mikawa, Y., & Yamamoto, D. (2005). Species-specific patterns of sexual dimorphism in the expression of fruitless protein, a neural musculinizing factor in *Drosophila*. *Journal of Neurogenetics, 19*(2), 109–121. doi:10.1080/01677060591007191

V

van Kesteren, R. E., & Spencer, G. E. (2003). The role of neurotransmitters in neurite outgrowth and synapse formation. *Reviews in Neurosciences, 14*(3), 217–231.

van Straaten, H. W., Hekking, J. W., Thors, F., Wiertz-Hoessels, E. L., & Drukker, J. (1985). Induction of an additional floor plate in the neural tube. *Acta Morphologica Neerlando-Scandinavica, 23*(2), 91–97.

van Straaten, H. W., Hekking, J. W., Wiertz-Hoessels, E. J., Thors, F., & Drukker, J. (1988). Effect of the notochord on the differentiation of a floor plate area in the neural tube of the chick embryo. *Anatomy and Embryology (Berlin), 177*(4), 317–324.

Varoqueaux, F., Aramuni, G., Rawson, R. L., Mohrmann, R., Missler, M., Gottmann, K., … Brose, N. (2006). Neuroligins determine synapse maturation and function. *Neuron, 51*(6), 741–754. doi:10.1016/j.neuron.2006.09.003

Varoqueaux, F., Jamain, S., & Brose, N. (2004). Neuroligin 2 is exclusively localized to inhibitory synapses. *European Journal of Cell Biology, 83*(9), 449–456. doi:10.1078/0171-9335-00410

Vassar, R., Ngai, J., & Axel, R. (1993). Spatial segregation of odorant receptor expression in the mammalian olfactory epithelium. *Cell, 74*(2), 309–318.

Veenhoven, R. (2010). Capability and happiness: Conceptual differences and reality links. *Journal of Socio-Economics, 39*, 344–350.

Vela, J. M., Gonzalez, B., & Castellano, B. (1998). Understanding glial abnormalities associated with myelin deficiency in the *jimpy* mutant mouse. *Brain Research Brain Research Reviews, 26*(1), 29–42.

Veraksa, A., & McGinnis, W. (2000). Developmental patterning genes and their conserved functions: From model organisms to humans. *Molecular Genetics and Metabolism, 69*, 85–100.

von Philipsborn, A. C., Jorchel, S., Tirian, L., Demir, E., Morita, T., Stern, D. L., & Dickson, B. J. (2014). Cellular and behavioral functions of fruitless isoforms in *Drosophila* courtship. *Current Biology, 24*(3), 242–251. doi:10.1016/j.cub.2013.12.015

W

Wagner, G. P., & Zhang, J. (2011). The pleiotropic structure of the genotype-phenotype map: The evolvability of complex organisms. *Nature Reviews Genetics, 12*(3), 204–213. doi:10.1038/nrg2949

Waites, C. L., Craig, A. M., & Garner, C. C. (2005). Mechanisms of vertebrates synaptogenesis. *Annual Review of Neuroscience, 28*, 251–274.

Wallace, V. A. (1999). Purkinje-cell-derived Sonic hedgehog regulates granule neuron precursor cell proliferation in the developing mouse cerebellum. *Current Biology, 9*(8), 445–448.

Walsh, F. S., & Doherty, P. (1997). Neural cell adhesion molecules of the immunoglobulin superfamily: Role in axon growth and guidance. *Annual Review of Cell and Developmental Biology, 13*, 425–456. doi:10.1146/annurev.cellbio.13.1.425

Walsh, M. K., & Lichtman, J. W. (2003). In vivo time-lapse imaging of synaptic takeover associated with naturally occurring synapse elimination. *Neuron, 37*, 67–73.

Walter, J., Henke-Fahle, S., & Bonhoeffer, F. (1987). Avoidance of posterior tectal membranes by temporal retinal axons. *Development, 101*, 909–913.

Wang, H. U., & Anderson, D. J. (1997). Eph family transmembrane ligands can mediate repulsive guidance of trunk neural crest migration and motor axon outgrowth. *Neuron, 18*(3), 383–396.

Wang, H., Yang, H., Shivalila, C. S., Dawlaty, M. M., Cheng, A. W., Zhang, F., & Jaenisch, R. (2013). One-step generation of mice carrying mutations in multiple genes by CRISPR/Cas-mediated genome engineering. *Cell, 153*(4), 910–918.

Wang, X., Merzenich, M. M., Sameshima, K., & Jenkins, W. M. (1995). Remodelling of hand representation in adult cortex determined by timing of tactile stimulation. *Nature, 378*(6552), 71–75. doi:10.1038/378071a0

Watson, J. B. (1924). *Behaviorism.* New York: People's Institute Publishing Co., Inc.

Weaver, I. C., Champagne, F. A., Brown, S. E., Dymov, S., Sharma, S., Meaney, M. J., & Szyf, M. (2005). Reversal of maternal programming of stress responses in adult offspring through methyl supplementation: Altering epigenetic marking later in life. *Journal of Neuroscience, 25*(47), 11045–11054. doi:10.1523/JNEUROSCI.3652-05.2005

Werker, J. F., Gilbert, J. H., Humphrey, K., & Tees, R. C. (1981). Developmental aspects of cross-language speech perception. *Child Development, 52*(1), 349–355.

Whalley, L. J., & Deary, I. J. (2001). Longitudinal cohort study of childhood IQ and survival up to age 76. *British Medical Journal, 322*(7290), 819.

Wharton, K. A., Johansen, K. M., Xu, T., & Artavanis-Tsakonas, S. (1985). Nucleotide sequence from the neurogenic locus notch implies a gene product that shares homology with proteins containing EGF-like repeats. *Cell, 43*(3 Pt. 2), 567–581.

Whitaker, D., & McGraw, P. V. (2000). Long-term visual experience recalibrates human orientation perception. *Nature Neuroscience, 3*(1), 13. doi:10.1038/71088

White, J. G., Southgate, E., Thomson, J. N., & Brenner, S. (1986). The structure of the nervous system of the nematode *Caenorhabditis elegans. Philosophical Transactions of the Royal Society B: Biological Sciences, 314*(1165), 1–340.

Whitfield, T. T. (2013). Shedding new light on the origins of olfactory neurons. *eLife, 2*, e00648. doi:10.7554/eLife.00648

Whitlock, J. R., Heynen, A. J., Shuler, M. G., & Bear, M. F. (2006). Learning induces long-term potentiation in the hippocampus. *Science, 313*(5790), 1093–1097.

Wiesel, T. N., & Hubel, D. H. (1965). Comparison of the effects of unilateral and bilateral eye closure on cortical unit responses in kittens. *Journal of Neurophysiology, 28*(6), 1029–1040.

Wilkinson, D. G. (2001). Multiple roles of EPH receptors and ephrins in neural development. *Nature Reviews Neuroscience, 2*, 155–164.

Williams, T. J., Pepitone, M. E., Christensen, S. E., Cooke, B. M., Huberman, A. D., Breedlove, N. J., … Breedlove, S. M. (2000). Finger-length ratios and sexual orientation. *Nature, 404*(6777), 455–456. doi:10.1038/35006555

Wilson, C. (1995). *The invisible world: Early modern philosophy and the invention of the microscope.* Princeton, NJ: Princeton University Press.

Wilson, E. B. (1896). *The cell in development and inheritance.* New York: Macmillan.

Winocur, G., Wojtowicz, J. M., Sekeres, M., Snyder, J. S., & Wang, S. (2006). Inhibition of neurogenesis interferes with hippocampus-dependent memory function. *Hippocampus, 16*(3), 296–304. doi:10.1002/hipo.20163

Witten, I. B., Knudsen, E. I., & Sompolinsky, H. (2008). A Hebbian learning rule mediates asymmetric plasticity in aligning sensory representations. *Journal of Neurophysiology, 100*(2), 1067–1079. doi:10.1152/jn.00013.2008

Wolf, B. B., Schuler, M., Echeverri, F., & Green, D. R. (1999). Caspase-3 is the primary activator of apoptotic DNA fragmentation via DNA fragmentation factor-45/inhibitor of caspase-activated DNase inactivation. *Journal of Biological Chemistry, 274*(43), 30651–30656.

Wolf, C. C. (2013). The mystery of missed connection. *Scientific American Mind, 23*(6), 54–57. doi:10.1038/scientificamericanmind0113-54

Wonders, C., & Anderson, S. A. (2005). Cortical interneurons and their origins. *Neuroscientist, 11*(3), 199–205. doi:10.1177/1073858404270968

Wonders, C. P., & Anderson, S. A. (2006). The origin and specification of cortical interneurons. *Nature Reviews Neuroscience, 7*(9), 687–696. doi:10.1038/nrn1954

Workman, A. D., Charvet, C. J., Clancy, B., Darlington, R. B., & Finlay, B. L. (2013). Modeling transformations of neurodevelopmental sequences across mammalian species. *Journal of Neuroscience, 33*(17), 7368–7383. doi:10.1523/JNEUROSCI.5746-12.2013

Wray, S., Grant, P., & Gainer, H. (1989). Evidence that cells expressing luteinizing hormone-releasing hormone mRNA in the mouse are derived from progenitor cells in the olfactory placode. *Proceedings of the National Academy of Sciences USA, 86*(20), 8132–8136.

Wright, C. V., Cho, K. W., Fritz, A., Burglin, T. R., & De Robertis, E. M. (1987). A *Xenopus laevis* gene encodes both homeobox-containing and homeobox-less transcripts. *EMBO Journal, 6*(13), 4083–4094.

Wu, H., Xiong, W. C., & Mei, L. (2010). To build a synapse: signaling pathways in neuromuscular junction assembly. *Development, 137*(7), 1017–1033. doi:10.1242/dev.038711

X

Xerri, C., Stern, J. M., & Merzenich, M. M. (1994). Alterations of the cortical representation of the rat ventrum induced by nursing behavior. *Journal of Neuroscience, 14*(3, Pt. 2), 1710–1721.

Y

Yamanaka, K., Boillee, S., Roberts, E. A., Garcia, M. L., McAlonis-Downes, M., Mikse, O. R., … Goldstein, L. S. (2008). Mutant *SOD1* in cell types other than motor neurons and oligodendrocytes accelerates onset of disease in ALS mice. *Proceedings of the National Academy of Sciences USA, 105*(21), 7594–7599. doi:10.1073/pnas.0802556105

Yang, T. T., Gallen, C. C., Ramachandran, V. S., Cobb, S., Schwartz, B. J., & Bloom, F. E. (1994). Noninvasive detection of cerebral plasticity in adult human somatosensory cortex. *Neuroreport, 5*(6), 701–704.

Yao, W. D., Rusch, J., Poo, M., & Wu, C. F. (2000). Spontaneous acetylcholine secretion from developing growth cones of *Drosophila* central neurons in culture: Effects of cAMP-pathway mutations. *Journal of Neuroscience, 20*(7), 2626–2637.

Yazdani, U., & Terman, J. R. (2006). The semaphorins. *Genome Biology, 7*(3), 211. doi:10.1186/gb-2006-7-3-211

Yerkes, R. M. (1907). *The dancing mouse: A study in animal behavior.* New York: Macmillan.

Yiu, G., & He, Z. (2006). Glial inhibition of CNS axon regeneration. *Nature Reviews Neurosciences, 7*(8), 617–627. doi:10.1038/nrn1956

York, A. D., Breedlove, S. M., Diamond, M. C., & Greer, E. R. (1989). Housing adult male rats in enriched conditions increases neurogenesis in the dentate gyrus. *Society for Neuroscience Abstracts, 15*, 962.

Yoshida, Y. (2012). Semaphorin signaling in vertebrate neural circuit assembly. *Frontiers in Molecular Neuroscience, 5*, 71. doi:10.3389/fnmol.2012.00071

Young, S. H., & Poo, M. M. (1983). Spontaneous release of transmitter from growth cones of embryonic neurones. *Nature, 305*(5935), 634–637.

Yuan, J., Shaham, S., Ledoux, S., Ellis, H. M., & Horvitz, H. R. (1993). The *C. elegans* cell death gene *ced-3* encodes a protein similar to mammalian interleukin-1 beta-converting enzyme. *Cell, 75*(4), 641–652.

Yuasa, S., Kawamura, K., Ono, K., Yamakuni, T., & Takahashi, Y. (1991). Development and migration of Purkinje cells in the mouse cerebellar primordium. *Anatomy and Embryology, 184*(3), 195–212.

Z

Zachary, I. (2014). Neuropilins: Role in signalling, angiogenesis and disease. *Chemical Immunology and Allergy, 99*, 37–70. doi:10.1159/000354169

Zantua, J. B., Wasserstrom, S. P., Arends, J. J., Jacquin, M. F., & Woolsey, T. A. (1996). Postnatal development of mouse "whisker" thalamus: Ventroposterior medial nucleus (VPM), barreloids, and their thalamocortical relay neurons. *Somatosensory & Motor Research, 13*(3–4), 307–322.

Zhang, J., Lefebvre, J. L., Zhao, S., & Granato, M. (2004). Zebrafish unplugged reveals a role for muscle-specific kinase homologs in axonal pathway choice. *Nature Neuroscience, 7*(12), 1303–1309. doi:10.1038/nn1350

Zhang, Y., Ma, C., Delohery, T., Nasipak, B., Foat, B. C., Bounoutas, A., … Chalfie, M. (2002). Identification of genes expressed in *C. elegans* touch receptor neurons. *Nature, 418*(6895), 331–335. doi:10.1038/nature00891

Zhao, H., & Reed, R. R. (2001). X inactivation of the OCNC1 channel gene reveals a role for activity-dependent competition in the olfactory system. *Cell, 104*(5), 651–660.

Zhou, Q., Chipperfield, H., Melton, D. A., & Wong, W. H. (2007). A gene regulatory network in mouse embryonic stem cells. *Proceedings of the National Academy of Sciences USA, 104*(42), 16438–16443. doi:10.1073/pnas.0701014104

Zhou, Q., Lai, Y., Bacaj, T., Zhao, M., Lyubimov, A. Y., Uervirojnangkoorn, M.… Brunger, A. T. (2015). Architecture of the synaptotagmin-SNARE machinery for neuronal exocytosis. *Nature, 525*, 62–67.

Ziv, N. E., & Garner, C. C. (2004). Cellular and molecular mechanisms of presynaptic assembly. *Nature Reviews Neuroscience, 5*(5), 385–399. doi:10.1038/nrn1370s

Zou, D. J., Feinstein, P., Rivers, A. L., Mathews, G. A., Kim, A., Greer, C. A., … Firestein, S. (2004). Postnatal refinement of peripheral olfactory projections. *Science, 304*(5679), 1976–1979. doi:10.1126/science.1093468

Zup, S. L., Carrier, H., Waters, E. M., Tabor, A., Bengston, L., Rosen, G. J., … Forger, N. G. (2003). Overexpression of Bcl-2 reduces sex differences in neuron number in the brain and spinal cord. *Journal of Neuroscience, 23*(6), 2357–2362.

Zweifel, L. S., Kuruvilla, R. & Ginty, D. D. (2005). Functions and mechanisms of retrograde neurotrophin signalling. *Nature Reviews Neuroscience, 6*, 615–625.

AUTHOR INDEX

SUBJECT INDEX

Page numbers followed by *f* refer to figures; those followed by *t* refer to tables.

sexual orientation in, 263–264
social experience affecting
  development of, 333, 334, 343–
  344, 364–370
spinal motor neuron apoptosis in,
  228f
tactile experience and
  somatosensory cortex in, 331,
  331f
vision reversal experiments in,
  306–307, 308
visual fields in, 313, 314f
Hume, David, 267
hunchback gene and hunchback protein, 61, 62, 62f, 63f, 64
Hybridization, A.1, A.2
  in situ, A.6–A.7
Hydras
  conditional specification of cell fate
    in, 32
  nerve net in, 9f
  self-regulation of embryonic
    development, 30, 31f, 53
Hyenas, spotted, 53
Hypothalamus, 69

I

Immunocytochemistry, 99f, 99–100,
  A.8
Immunoglobulins, 99, A.7–A.8
Immunohistochemistry, 99f, 99–100,
  A.8
Imprinting, 344–346
  sexual, 346, 346f
Impulse control, frontal lobe in,
  302–303
Induction, 34–35, 35f
  definition of, 35
  organizer signals in, 36–43
Inheritance, 12–13
Inhibitors of apoptosis proteins (IAPs),
  243, 243f, 244f
Inhibitory postsynaptic potential
  (IPSP), 212
Innate and instinctive behavior, 334,
  335–337
  definition of, 335
  use of terms in literature, 335, 335f,
    336–337
Inner cell mass, 25–26, 27f
Insectivores, later stages of brain development in, 148, 148f
Insects
  development of body pattern,
    51–52
  Drosophila. See Drosophila
  ectodermal cells in formation of
    neural tissue, 43–48
  maternal factors in body polarity,
    55–59, 56f
  mitotic lineage and cell fate in, 94
  redundancy in, 89
In situ hybridization, A.6–A.7

Instinctive and innate behavior, 334,
  335–337
  definition of, 335
  use of terms in literature, 335, 335f,
    336–337
Integrins, 160, 161f
Intellectual disability
  definition of, 193
  in fragile X syndrome, 185, 193–194
Intelligence, and brain size, 149
Intelligence quotient, 334, 364–370,
  378
  average and standard deviation in,
    364, 365f
  environmental influences on,
    367–368
  Flynn effect in, 364, 365–367, 366f
  income compared to, 364, 365f
  and occupation, 364, 365f
  racial and ethnic differences in,
    368f, 368–370, 369f
  reliability and validity of, 364
  stereotype threat affecting, 368–
    369, 369f, 370
  in twins, 367, 367f
Intergenerational patterns
  in innovative behaviors, 347–348
  in phoneme loss, 355–356
Intermediate zone, 92, 92f, 95f, 95–97,
  96f, 101
Internal granule layer, 95f, 95–96
Interneurons
  GABA-ergic, 103, 105
  olfactory, 107
  periglomerular, 107, 108f
  spinal, differentiation of cells into,
    135
Invertebrates
  ectodermal cells in formation of
    neural tissue in, 43–48
  glutamate as neurotransmitter in,
    202, 207
  insects as. *See Insects*
  long-term potentiation in, 290
  ventral location of nervous system
    in, 43–44, 44f
Ion channels, 209–212
  definition of, 209
  and ionotropic glutamate receptors,
    287–291
  types of, 209, 209f
Ionotropic glutamate receptors,
  287–291
Ipsilateral side, definition of, 292
Isl-1 gene and Isl-1 transcription factor, 135, 136, 136f
Isl-2 gene and Isl-2 transcription factor, 135, 136, 136f
Isolation, social
  amygdala size in, 361, 361f
  compared to enriched environment,
    359–363
  myelination and memory in, 362f,
    362–363, 363f

Isotretinoin, 80

J

Japanese dancing mice, 118
Japanese macaques, 347, 347f
Jeffress, Lloyd A., 376–377, 377f
Jellyfish, 8, 9f
Jimpy mice, 218

K

Kagan, Elena, 303
Kainate receptors, 287
Kallmann syndrome, 123–124
Kant, Immanuel, 373–375, 377, 378
Kinases, 186
knirps gene, 61, 62
Knockin animals, 130, 131, A.10
Knockout animals, 34, 100, 131
  bicoid, 57
  caspase-9, 242f
  β-catenin, 191f, 192
  in CRISPR technology, 34, 131,
    A.8–A.10
  definition of, 34, 130
  ephrin-B1, 181f
  Gbx2, 71f
  neuregulin, 214
  neuroligin genes, 197
  noggin and *noggin/chordin*, 40, 40f
  odor detection blocked in, 325
  Otx2, 70
  Pax6, 72
  Robo, 168f
  Slit, 168f, 169f
  spontaneous retinal activity blocked
    in, 298
  Trk receptor, 238t
Knowledge, 2–5
  Descartes on, 4, 374
  Kant on, 373–375, 377, 378
  Locke on, 268–269
  Plato on, 2–4, 374
  a priori and a posteriori types of,
    373–375, 377, 378
Krox20, 76f
Krüppel gene, 61, 62, 63f, 64f

L

L1 cell adhesion molecule, 156–157,
  157f, 160, 181
lacZ gene, 93–94
Lamellipodia, 152, 154
Laminin, 113, 160
  in axonal growth, 156, 157f
  definition of, 113, 156, 189
Language learning, 352–357, 359
  genes affecting, 352–353
  phonemes in, 354f, 354–356, 355f,
    356f
  of second language, 355, 355f

About the Book

Editor: Sydney Carroll

Production Editor: Kathaleen Emerson

Copy Editor: Lou Doucette

Production Manager: Christopher Small

Book Production: Joanne Delphia

Art: Morales Studio

Photo Researcher: David McIntyre

Book and Cover Design: Joanne Delphia

Book and Cover Manufacturer: LSC Communications